Modern Perspectives of Mobile Robot Systems

Modern Perspectives of Mobile Robot Systems

Edited by **Jared Kroff**

Clanrye International

New Jersey

Published by Clanrye International,
55 Van Reypen Street,
Jersey City, NJ 07306, USA
www.clanryeinternational.com

Modern Perspectives of Mobile Robot Systems
Edited by Jared Kroff

International Standard Book Number: 978-1-63240-366-7 (Hardback)

Printed in China.

Contents

Preface

This book has been an outcome of determined endeavour from a group of educationists in the field. The primary objective was to involve a broad spectrum of professionals from diverse cultural background involved in the field for developing new researches. The book not only targets students but also scholars pursuing higher research for further enhancement of the theoretical and practical applications of the subject.

This book is a compilation of extensive information regarding the field of mobile robots. It elucidates broad topics like the significance of robots for educational purposes, hardware composition, healthcare and medical robots, and localization as well as navigation. It also provides elaborative information regarding independent mobile robot Emmy III and KCLBOT. The book also discusses walking support robots, leg-wheel mechanism as a mobile platform, control system for wheelchairs, I2C bus system, and quadruped robots and more topics like omnidirectional mobile robots, planetary wheeled and ball-wheel mobile robots, etc.

It was an honour to edit such a profound book and also a challenging task to compile and examine all the relevant data for accuracy and originality. I wish to acknowledge the efforts of the contributors for submitting such brilliant and diverse chapters in the field and for endlessly working for the completion of the book. Last, but not the least; I thank my family for being a constant source of support in all my research endeavours.

Editor

Part 1

Robots for Educational Purposes

1

Mobile Robotics in Education and Research

Georgios A. Demetriou
Frederick University
Cyprus

1. Introduction

Mobile robotics is a new field. Mobile robots range from the sophisticated space robots, to the military flying robots, to the lawn mower robots at our back yard. Mobile robotics is based on many engineering and science disciplines, from mechanical, electrical and electronics engineering to computer, cognitive and social sciences (Siegwart & Nourbakhsh, 2004). A mobile robot is an autonomous or remotely operated programmable mobile machine that is capable of moving in a specific environment. Mobile robots use sensors to perceive their environment and make decisions based on the information gained from the sensors.

The autonomous nature of mobile robots is giving them an important part in our society. Mobile robots are everywhere, from military application to domestic applications. The first mobile robots as we know them today were developed during World War II by the Germans and they were the V1 and V2 flying bombs. In the 1950s W. Grey Walter developed Elmer and Elsie, two autonomous robots that were designed to explore their environment. Elmer and Elsie were able to move towards the light using light sensors, thus avoiding obstacles on their way. The evolution of mobile robots continued and in the 1970s Johns Hopkins University develops the "Beast". The Beast used an ultrasound sensor to move around. During the same period the Stanford Cart line follower was developed by Stanford University. It was a mobile robot that was able to follow a white line, using a simple vision system. The processing was done off-board by a large mainframe. The most known mobile robot of the time was developed by the Stanford Research Institute and it was called Shakey. Shakey was the first mobile robot to be controlled by vision. It was able to recognize an object using vision, find its way to the object. Shakey, shown in Figure 1, had a camera, a rangefinder, bump sensors and a radio link.

These robots had limitations due to the lack of processing power and the size of computers, and thus industrial robotics was still dominating the market and research. Industrial manipulators are attached to an off-board computer (controller) for their processing requirements and thus do not require an onboard computer for processing. Unlike industrial robots, mobile robots operate in dynamic and unknown environments and thus require many sensors (i.e. vision, sonar, laser, etc.) and therefore more processing power. Another important requirement of mobile robots is that their processing must be done onboard the moving robot and cannot be done off-board. The computer technology of the time was too bulky and too slow to meet the requirements of mobile robots. Also, sensor technology had to advance further before it could be used reliably on mobile robots.

In the last twenty years we saw a revolution in computer technology. Computers got smaller, a lot faster and less expensive. This met the requirements of mobile robots and as a result we saw an explosion of research and development activities in mobile robotics. Mobile robots are increasingly becoming important in advanced applications for the home, military, industry, space and many others. The mobile robot industry has grown enormously and it is developing mobile robots for all imaginable applications. The vast number of mobile robot applications has forced a natural subdivision of the field based on their working environment: land or surface robots, aquatic/underwater robots, aerial robots and space robots. Land or surface robots are subdivided based on their locomotion: Legged robots, wheeled robots and track robots. Legged robots can be classified as two legged (i.e. humanoids) robots and animal-like robots that can have anywhere from four legs to as many as the application and the imagination of the developer requires.

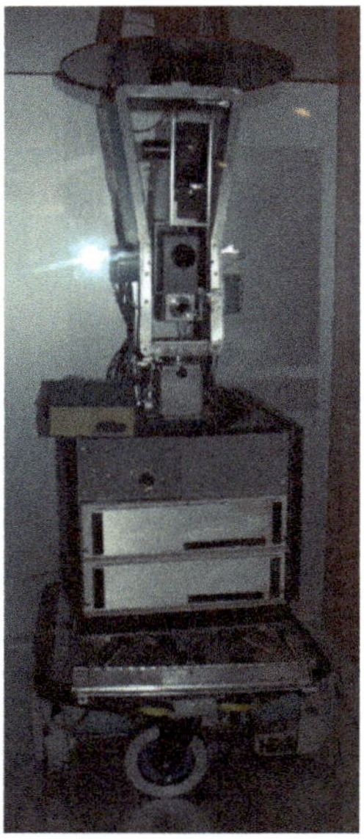

Fig. 1. Shakey the Robot in its display case at the Computer History Museum

The revolution of mobile robotics has increased the need for more mobile robotics engineers for manufacturing, research, development and education. And this in turn has significantly changed the nature of engineering and science education at all levels, from K-12 to graduate school. Most schools and universities have integrated or are integrating robotics courses into their curriculums. Mobile robotics are widely accepted as a multidisciplinary approach to combine and create knowledge in various fields as mechanical engineering, electrical engineering, control, computer science, communications, and even psychology or biology in some cases.

The majority of robotics research is focusing on mobile robotics from surface robots, humanoids, aerial robots, underwater robots, and many more. The development of several less expensive mobile robotic platforms (i.e. VEX Robotics Design System (VEX Robotics Design System, 2011), LEGO Mindstorms (LEGO Education, 2011), Engino Robotics (Engino international website - play to invent, 2011), Fischertechnik (Fischertechnik GmbH, 2011), etc.), the development of countless of robotics software tools and programming languages (i.e. Microsoft Robotics Studio (Microsoft Robotics Developer Studio, 2011), RoboLab (Welcome to the RoboLab, 2011), ROBOTC (ROBOTC.net, 2011)) and the development of many robotic simulators have made robotics more accessible to more educators, students and robot enthusiasts at all levels. This and the fact that using mobile robots in education is an appealing way of promoting research and development in robotics, science and technology, has triggered a revolution of mobile robotics education, research and development. Now educators, researchers, and robot enthusiasts are pursuing innovative robotic, electronic, and advanced mobile programming projects in a variety of different fields.

This chapter provides an overview on mobile robotics education and research. Rather than attempting to cover the wide area of this subject exhaustively, it highlights some key concepts of robotics education at the K-12 and the university levels. It also presents several robotic platforms that can be used in education and research. Various mobile robotic competitions for K-12 and university students will also be presented.

2. Education and research

Since the late 1980s, when robotics was first introduced into the classroom, mobile robotics is used in education at all levels and for various subjects (Malec, 2001). Mobile robotic technology is introduced at the school level as a novelty item and teaching tool. Even though robotics as a field is recognized as a separate educational discipline, it is usually incorporated within the computer science and engineering departments. Robotics courses are becoming core curriculum courses within these departments at most universities. Currently only a small number of universities have pure robotics departments. As the demand for specialized robotics engineers becomes greater, more universities will start to offer robotics degrees and have robotics departments.

It is a known fact that most students, regardless of age and educational background /interests, consider working with robots to be "fun" and "interesting". Mobile robotics is used in education in many ways, but generally there are two approaches on how mobile robots are used (Malec, 2001). The first, and most obvious, approach is using robots to teach courses that are directly related to robotics. These courses are usually introductory robotics courses and teach the basic concepts of mobile robotics. These courses are usually divided into lectures and laboratory sessions. At the lecture sessions students learn concepts such as kinematics, perception, localization, map building and navigation. At the laboratory sessions, students experiment on real or simulated robots. Experiments are done on robotics concepts and most times students are required to use concepts they learned in other courses, such as control and programming. This method of using/teaching robots in education is primarily used at the university level and very few times we see it at the high-school level. The second approach is to use mobile robots as a tool to teach other subjects in engineering, science and even non-related fields such as biology and psychology. Since students enjoy working with robots, learning becomes more interesting. This allows robotics to be incorporated into various disciplines and departments. This method is primarily used at the K-12 levels of education to teach technology related courses. It can easily be used at the first year of university level to teach courses such as programming and control. Mobile robots can also be used outside the science and engineering fields to teach biology students for example about leg locomotion or to create studies for psychology students on how mobile robotics is affecting our personal life.

The majority of the mobile robotics activities are offered at the university level, but over the last few years we have seen many robotics courses, competitions and other robotics activities offered at the K-12 level of education as well. Many K-12 school systems are starting to teach young people using robots. At the high-school level (ages 16 to 18) robotics is often used to teach programming courses. The most commonly used robots at this level are the Lego Mindstorms NXT and the VEX Robotics Design Systems. The nature of mobile robotics allows it to be an effective tool in teaching technology courses to children at the primary and early secondary levels (from the ages of 6 to 15). Companies such as Engino Toy Systems and LEGO have started developing robotics packages that can be used to teach basic physics and technology concepts to young children, even from the age of 6. By introducing robotics at the K-12 level of education, students may be better prepared for the university and more students will be interested in robotics as a field of study.
The results of the success of robotics in the education of young children has triggered many successful local, national and international robotics competitions (i.e. FIRST LEGO League (USFIRST.org - Welcome to FIRST, 2011), RoboCupJunior (RoboCupJunior, 2011)), many robotics workshops and summer camps for students of all levels. These activities have generated more momentum in robotics education and research and already many educational systems are starting to offer robotics education even at the early stages of K-12 education. Companies such as Engino Robotics (Engino international website - play to invent, 2011) and Lego Education WeDo (Lego Education WeDo, 2011) are developing robotic platforms to be used specifically at the primary school level (ages 6-12). This momentum must be carried over to the university level but in most cases it is not. University students must wait until their third or fourth year before they can take their first robotics course.
Most universities only offer an introductory course in robotics and only a few offer more advanced courses. Most advanced robotics courses are offered at the graduate level. Where advanced courses are offered, specific concepts such as vision, advanced navigation, map-building and sensor fusion are targeted. At the undergraduate level, usually students only go beyond the introductory courses only in the form of final year projects or other projects. In many cases, robotics interested students form robotics groups in order to exchange information, to participate in robotics competitions or to just build a mobile robot. By offering robotics courses early at the undergraduate level will create a natural continuation to the K-12 robotics education. By introducing more advanced mobile robotics courses at the undergraduate level, we will have better prepared students for the graduate level, make more research progress in the long run and have more prepared workforce for the robotics industry as well.

3. Teaching robotics at the K-12 level

It has been proven that hands-on education provides better motivation for learning new material. Abstract knowledge is more difficult to comprehend and sometimes not interesting enough for most students. By providing experiments and real-world situations, students become more interested and they learn new topics much easier.
There is a lack of student interest in science, technology, engineering, and math (STEM) topics and increasing attention has been paid to developing innovative tools for improved teaching of STEM. For this reason alone, hands-on education has been imposed on today's K-12 teachers. Mobile robotics has been shown to be a superb tool for hands-on learning, not

only of robotics itself, but of general topics in STEM as well (van Lith, 2007). Mobile robotics is a great way to get kids excited about STEM topics. Students at this level do not need to have a considerable understanding about how robots work. The approach used at this level is to experiment, learn and play. It is also highly effective in developing teamwork and self-confidence. Even children with no interest in technology and sciences still consider robots interesting. Utilizing this interest, robots are used to teach children about robots or using robots as a tool to teach STEM topics. The study of robotics, by its very nature, captures all four legs of STEM very well while creating classroom environments where both knowledge and skill development can flourish without having to compromise one for the other.

But getting students interested in science and other subjects is only one part of the equation, as we must also prepare them for logical thinking and problem solving. At this level it is beneficial to start solving logical problems as the brain is forming in order to develop the required neural plasticity that can be employed over a lifetime of logical thinking and problem solving (Matson et al., 2003). In order to succeed in this, the right tools have to be selected. Choosing the right tools is difficult when competing with high-tech computer games, electronic gadgets and other toys that children are using today. Children today need more stimulation than ever before in human education. Young people are very good at using gadgets and electronic toys, but not many of them are interested in how these devices work or are built (Malec, 2001). We need to stimulate their interest in technology and science in order to make them try to understand the functionality of the devices they use. If they understand, then they will want to develop. Thus it is critical to provide engaging hands-on education to all children as early as possible in order to open their minds to the technology career choices in the STEM areas.

It has been shown that no age is too young for being engaged by robots. Toddlers express deep interest in active machines and toys; robots motivate even children with special social and cognitive needs (Chu et al., 2005). Increasingly, high schools across the world are providing elective robotics courses as well as after-school programs. Gradually, middle schools are starting to get involved, as well. Slowly, even children at the elementary schools are being exposed to robotics (Lego Education WeDo, 2011; Engino international website – play to invent, 2011). There are also many robotics competitions that were designed specifically for these age groups; some of the most visible are FIRST (FIRST, 2011) LEGO Mindstorms (LEGO Education, 2011), VEX Robotics Competition (Competition - VEX Robotics, 2011) and RoboCupJunior (RoboCupJunior, 2011). These competitions increase the interest of students since they add the ingredient of competition.

Generally there is a lack of age appropriate robotics teaching materials for the K-12 level. Normally due to lack of financial resources, schools do not have enough or up-to-date equipment in order to use robotics successfully. Also, because of the broad range of backgrounds of K-12 educators who teach robotics and the lack of appropriate lesson plans, it is critical that ready educational materials be provided in order to teach robotics successfully. There is a lack of available robotics textbooks at this level of education as well. For these reasons, many universities are directly working with school systems to develop robotic material, robotic platforms and appropriate lesson plans in order to aid teachers overcome these problems. Universities must become more involved with K-12 schools and offer activities such as competitions, summer camps, lectures and workshops to students and teachers. K-12 teachers have to be properly educated and trained to use robots in their classrooms.

There are two general categories of mobile robots that are used for education: do-it-yourself (DIY) kits and prebuilt robots. Prebuilt robots are generally more expensive and are only found in university labs, industry and the military. The least expensive prebuild robot cost about 3.000 Euro. Do-it-yourself mobile robots are less expensive and normally cost less than 1.000 Euro. There are many DIY robotic kits that are ideal for K12 education, such as Lego Mindstorms NXT, VEX Robotics Design System and Engino Toy Systems.

Generally students are required to construct and then program the robots. Normally the construction is done by using specific instructions and prepackaged robotic parts, such as LEGO or Engino Toy Systems parts. Most DIY kits are equipped with simple graphical user interfaces (GUI), such as RoboLab (Welcome to the RoboLab, 2011), for students to program the robots. The programming capabilities of these GUIs are limited and are primarily used by students up the middle school (grade 8) level. Students at the high-school level (grades 9 to 12) require more control over their robots, and they often use more advanced programming tools. They use high level languages such as ROBOTC (ROBOTC.net, 2011) or more advanced visual programming languages (VPL) such as Microsoft Robotics Studio (Microsoft Robotics Developer Studio, 2011). Many more high level programming languages and visual programming languages are constantly being developed and this will give students many more options for programming.

Using mobile robotic simulators is another way mobile robotics can be used in education. The goal of simulators at this level is to provide a complete learning experience without the need to have the actual robot hardware. This eliminates the cost of purchasing enough robots to satisfy the needs of all students, since usually one robot is needed per 3-4 students. Simulators are generally less expensive, can be used by all students simultaneously and do not have any hardware costs since they can normally run on existing school computers. In addition to this, the animation of simulation is ideal for today's children who are keen in animation and gaming. Mobile robotic simulators allow students to virtually build mobile robots and program the robot to perform similar functions as the real robot would.

A very good mobile robotics simulator is the RobotC Virtual Worlds by Carnegie Mellon Robotics Academy (Computer Science Social Network, 2011). It allows students to program simulated Lego NXT and VEX robots using the RobotC programing language. The program offers four mobile robotic challenges: Labyrinth Challenge, Maze Challenge, Race Track Challenge and the Gripper Challenge. Students can also venture on to a simulated extraterrestrial planet environment where they can explore different areas such as the Astronaut Camp and the Container Yard.

In order to have success in robotics education, children from the early stages of K-12 education it would be good for them to be exposed to robotics. Elementary school children (grades 1 to 6) may play and program simple robots. Companies such as Engino Toy Systems and Lego Education WeDo are developing robotic platforms that can be used at the elementary school systems. In the early secondary education (grades 7 to 9) children may build simple robots, program them and even participate in local or international robotic competitions. High school students (grades 10 to 12) may design more complex mobile robots, program them using high level languages and compete in competitions to design robots. Generally by introducing robotics at this younger age group, we will have better prepared students for the university and graduate levels.

3.1 Mobile robotic platforms for K-12 education

Most K-12 educational mobile robots come in the form of kits. Most kits contain building blocks, cables and sensors. They normally come equipped with their own programming interface but sometimes we see third party companies developing software for programming these kits. Older students sometimes prefer using high-level languages to program since it allows them to have more control over the robots and to accomplish more complex functions. The complexity of the kids depends on the age group they are targeting. Some of the most commonly used mobile robot platforms are described here. It is impossible to separate these platforms into age-group categories, since most of them are used by all K-12 groups. Some of these kits are so advanced that are even used at the university level. There are many other available educational mobile robots and the list is growing fast, but it is not possible to list and describe all of them here. The list here is only a selected list based on availability, innovation and usage.

3.1.1 Lego Mindstorms NXT

Lego Mindstorms is a line of programmable robotics/construction toys, manufactured by the Lego Group (LEGO Education, 2011). It was first introduced in 1998 and it was called Robotics Invention System (RIS). The next generation was released in 2006 as Lego Mindstorms NXT (Figure 2). The newest version, released on August 5, 2009, is known as Lego Mindstorms NXT 2.0. Lego Mindstorms is primarily used in secondary education but most universities also use the Lego Mindstorms NXT for their introductory courses in mobile robotics or for projects.

Fig. 2. Lego Mindstorms NXT Controller with sensors and a sample mobile robot

Lego Mindstorms NXT is equipped with three servo motors, a light sensor, a sound sensor, an ultrasound sensor and a touch sensor. The NXT 2.0 has two touch sensors, a light, sound and distance sensors, and support for four sensors without using a sensor multiplexor.
The main component in the Lego Mindstorms kit is a brick-shaped computer called the NXT Intelligent Brick. It can take input from up to four sensors and control up to three motors. The brick has a 100x64 pixel monochrome LCD display and four buttons that can be used to navigate a user interface using menus. It also has a speaker that can play sound files. It allows USB and Bluetooth connections to a computer.
Lego Mindstorms NXT comes with the Robolab (Welcome to the RoboLab, 2011) graphical user interface (GUI) programming software, developed at Tufts University using the National Instruments LabVIEW (NI LabVIEW, 2011) as an engine. With Robolab, students

use flowchart like "blocks" to design their program. Students that need to write more advanced programs sometimes prefer to use third party firmware and/or high-level programming languages, including some of the most popular ones used by professionals in the embedded systems industry, like Java, C and C# (i.e. ROBOTC). The programs are downloaded from the computer onto the NXT Brick using the USB port or wirelessly using the Bluetooth connection. Programs can also be run on the computer and wirelessly through Bluetooth can control the NXT brick.

Some of the programming languages that are used to program NXT Brick are:

- NXT-G: Is the programming software that comes bundled with the NXT. This software is suitable for basic programming, such as driving motors, incorporating sensor inputs, doing calculations, and learning simplified programming structures and flow control.
- C# with Microsoft Robotics Developer Studio: Uses the free tools Visual Studio Express and the Robotics Developer Studio and allows programming using the C# language.
- Next Byte Codes (NBC): Is a simple open source language with an assembly language syntax that can be used to program the NXT brick.
- Not eXactly C (NXC): Is a high level open-source language, similar to C. NXC is basically NQC (Not Quite C) for the NXT. It is one of the most widely used third-party programming languages for the NXT.
- ROBOTC: Developed by the Carnegie Mellon Robotic's Academy. ROBOTC runs a very optimized firmware, which allows the NXT to run programs very quickly, and also compresses the files so that you can fit a large amount of programs into your NXT. Like other NXT languages, ROBOTC requires this firmware to be downloaded from the ROBOTC interface in order to run.

Another addition that allows more rigid Lego Mindstorms designs is TETRIX by Pitsco (TETRIX, 2011). The metal building system was designed specifically to work with the LEGO Technic building system through the use of the innovative *Hard Point Connector*. *TETRIX*, combined with custom motor controllers from Hitechnic, enables students to use the power of the Mindstorms technology and incorporate and control powerful DC and servo motors and metal gears. Students can build more versatile and robust robots designed for more sophisticated tasks, all while mastering basic wiring, multi-motor control, and much more.

3.1.2 Fischertechnik ROBO TX training lab

This kit includes the ROBO TX Controller (Figure 3), the construction parts for building mobile robots, the obstacle detector and trail searcher, two encoder motors for exact positioning, one motor XS, one infrared trail sensor and two sensing devices. It also includes the ROBO Pro software for programming. The system can be expanded with the extensive accessories provided by Fischertechnik (Fischertechnik GmbH, 2011).

The ROBO TX Controller is based on a 200Mhz 32-bit processor and is equipped with Bluetooth, eight universal inputs, 8 MB RAM (2 MB flash) and a display. Several ROBO TX Controllers can be coupled together to form more complex systems. The ROBO TX Controller can be purchased separately and can be used in custom designs as well.

This mobile robot kid is suitable for children at the secondary level or university. Everything else is included such as the ROBO interface and ROBO Pro software.

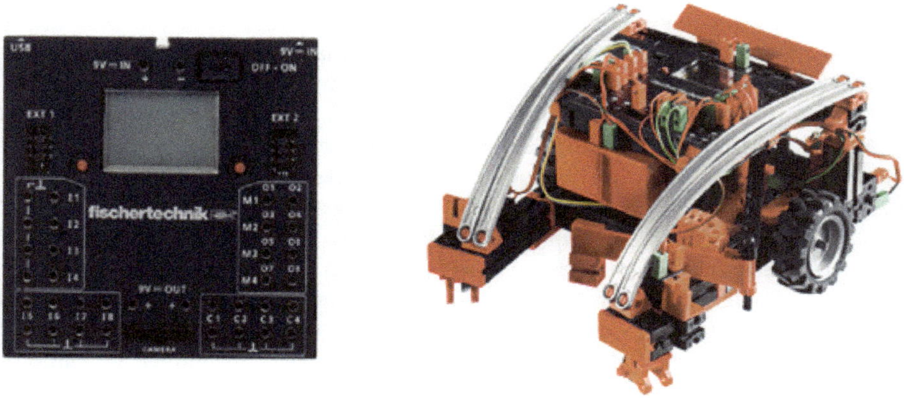

Fig. 3. Fischertechnik ROBO TX Controller and sample mobile robot

3.1.3 VEX robotics design system

The VEX Robotics Design System (VEX Robotics Design System, 2011) is a robotic kit intended to introduce robotics to students. VEX Robotics Design System offers the VEX Classroom Lab Kits that make it easy to bring the VEX Robotics Design System into the classroom. The Classroom Lab Kits include everything you need to design, build, power and operate robots. It comes with four sensors (two bumper sensors and two light switches), four electric motors and a servo motor, and building parts such as wheels and gears. The challenge level for the students can increase by adding expansion kits for advanced sensors, drive systems and pneumatics. Additional sensors (ultrasonic, line tracking, optical shaft encoder, bumper switches, limit switches, and light sensors), wheels (small and large omni-directional wheels, small, medium, and large regulars), tank treads, motors, servos, gears (regular and advanced), chain and sprocket sets, extra transmitters and receivers, programming kit (easyC, ROBOTC, MPLab), extra metal, pneumatics, and rechargeable battery power packs, can all be purchased separately. There are two options on the controllers that can be used with the VEX robotic kits: the basic PIC Microcontroller or the advanced and more powerful CORTEX Microcontroller.

Fig. 4. VEX Robotics Design CORTEX controller and a sample mobile robot

3.1.4 Engino robotics

The engino toy system (Engino international website - play to invent, 2011) was launched in 2007, initially with structural snap fit components and later with gears, pulleys and motors for more complex models (Figure 5). In 2011 the company teamed with Frederic University to develop robotic solutions for primary school students. The robotics module is expected to be officially launched in 2012 and will have 3 modes of operation, making it suitable for both primary and early secondary education (up to grade 8).

The engino controller is based on an ARM 32-bit processor and has four analogue motor outputs, six LED & buzzer outputs, two digital inputs and two analogue inputs. The kit comes equipped with basic building equipment, the controller, motors and basic sensors that include touch, sound, temperature, infrared and light. The controller comes equipped with a USB port that allows it to connect directly to a computer. It can be used to directly control the robots from the computer or to download programs for autonomous functionality.

Fig. 5. Engino Robotics sample mobile robots

The module is designed for three levels of programming, basic, intermediate and advanced. The basic level allows manual programming using the existing buttons on the module and allows recording of steps. Younger students (primary education) can be introduced to robotics by recording their steps and functions and then playing them in sequence. It is restricted to two motor outputs and three LED/buzzer outputs. The second level of programming is done on a computer and can fully program the controller. A specialized GUI allows programming the controller using graphical blocks that represent the blocks on the actual system (i.e. motors, sensors, etc). An innovation of the system is that students can create modules of code reusable code that they can use in other programs. The first two levels of programing are suitable for children at the primary education level. The advanced level programming allows students to use a custom made C like high-level language to create more complex programs. This is suitable for students at the secondary education level.

3.1.5 Lego WeDo

The Lego Education WeDo platform is shown in Figure 6. It powered by the LabVIEW software and is ideal for primary school students. Students can build and program their

own robots using the simple, drag-and-drop software. The WeDo is designed to teach simpler concepts to slightly younger kids than other kits. This kit does not allow mobile robot constructions but it is worth mentioning since it is targeting an age group that not many manufacturers target.
By hooking the robots up to your computer via the included USB hub, the WeDo software allows you to program the robots, controlling its actions, sounds and responses. All the programming is drag-and-drop.

Fig. 6. Lego WeDo parts and a robot

3.2 Mobile robotics competitions for K-12 students

Teaching mobile robotics at K-12 level is never a complete task. Students and teachers strengthen their knowledge by participating in mobile robot competitions. The success of robotics in the education of young children has triggered many successful local, national and international robotics competitions such as the FIRST Lego League (USFIRST.org - Welcome to FIRST, 2011), RoboCup@Home (RoboCup@Home, 2011), RoboCup (RoboCup, 2011) and RoboCupJunior (RoboCupJunior, 2011). Only a few international competitions will be briefly mentioned here since trying to name and describe all will be a never-ending task.

3.2.1 VEX robotics competition

VEX robotics competition (Competition - VEX Robotics, 2011) comes at two levels: VEX Robotics Classroom Competition and VEX Robotics Competition (Figure 7). VEX Robotics Classroom Competition is specifically tailored to bring the magic of robotics competition into the classroom. Robotics is an engaging way to integrate all facets of STEM education into the classroom and head-to-head competition is a natural way to capture students' attention. During the excitement that comes with building and competing with their robots, students will be having too much fun to realize they're learning important STEM concepts and life skills. A single teacher can easily implement all aspects of this program as part of their daily classroom activities.
The VEX Robotics Competition is the largest and fastest growing middle and high school robotics program globally with more than 3,500 teams from 20 countries playing in over 250 tournaments worldwide. Local VEX Robotics Competition events are being held in many

different cities throughout the world. In addition to just having a great time and building amazing robots, through their participation in the VEX Robotics Competition and their work within their team, students will learn many academic and life skills.

Fig. 7. VEX robotics competition

3.2.2 FIRST LEGO league and junior FIRST LEGO league

The FIRST LEGO League (also known by the acronym FLL) is an international competition organized by FIRST for primary and middle school students (ages 9–14 in the USA and Canada, 9–16 elsewhere). It is an annual competition and each year a new challenge is announced that focuses on a different real-world topic related to the sciences. Each challenge within the competition then revolves around that theme. The robotics part of the competition revolves around designing and programming LEGO Robots to complete tasks. Students work out solutions to the various problems they are given and then meet for regional tournaments to share their knowledge, compare ideas, and display their robots. The Junior FIRST LEGO League is a scaled-down robotics program for children of ages 6–9.

3.2.3 RoboCupJunior

RoboCupJunior started in 1998 with a demonstration held at the RoboCup international competition held in Paris, France. RoboCup Junior is closely related to the RoboCup competition. RoboCup is an international robotics competition that aims to develop autonomous football (soccer) robots with the intention of promoting research and education in the field of artificial intelligence. Robot cup is described in section *Mobile Robot Competitions for University Students*.

The programming and engineering-influenced competition introduces the aims and goals of the RoboCupJunior project to the primary and secondary school aged level (typically people under the 18). RoboCupJunior is an educational initiative to promote knowledge and skills in programming, hardware engineering, and the world of 3D through robotics for young minds. It aims to fuse real and virtual robotic technologies towards bridging two prominent areas of the future namely, Interactive Digital Media and Robotics. Those involved create and build robots in a variety of different challenges, and compete against other teams. RobotCupJunior is divided into four challenges: Soccer challenge, Dance challenge, Rescue challenge and CoSpace challenge (Figure 8) (van Lith, 2007).

The Soccer challenge is a competition for youth to design, program and strategize autonomous soccer-playing robots. At the Dance challenge, students create dancing robots which, dressed in costumes, move in creative harmony to music. The Rescue challenge is a platform which involves student programming autonomous robots to rescue "victims" in disaster scenarios. And the CoSpace challenges offer an opportunity for RoboCupJunior participants to explore robotics technology, digital media, and CoSpace concept. It also provides a platform for young minds who are keen in animation and gaming into the world of robotics. It offers an opportunity for junior participants to explore the robotics programming and AI strategies with Simulation based competitions. It comprises of two sub-leagues, namely CoSpace Adventure Challenge and CoSpace Dance Challenge.

Fig. 8. RoboCup Rescue and Soccer challenges

4. Robotics at the university level

Mobile robotics is a compelling subject for engineering and computer science undergraduates, but unfortunately most universities do not offer them at the introductory level. Most computer science and engineering departments do offer an introductory course in classical (industrial) robotics but not mobile robotics. These courses generally concentrate on the introduction to industrial robotics and in most cases they incorporate one chapter on mobile robotics. This chapter normally gives a general introduction to mobile robotics concepts and not much more. Mobile robotics courses are usually left as directed studies or are offered only at the graduate level.

Another issue is the fact that robotic courses are not offered at the first year of university education; they are normally offered at the third or fourth year. This creates a problem because it gives no continuation with the robotics education that students are getting in high school. One argument might be that mobile robotic courses require extensive knowledge of electronics, engineering and control. This is true, but it does not mean that the first year university courses must be so involved. The first year mobile robotics course can be a mobile robotic applications course. Another way to introduce mobile robots in the first year of university education is to use mobile robots with other courses such as programing, control and engineering courses in order to offer experimentation of the materials learned. Students learn more when they apply what they learn in lectures to real world applications such as robotics. More advanced mobile robotic topics can be covered with a second or third year course in mobile robotics.

It is only recently that we see many universities start to offer mobile robotics at the undergraduate level. At K-12 education students are normally concentrating on mobile robotics and not industrial robots since mobile robots are more interesting and more fun for students. Therefore, offering introductory mobile robotics courses at the undergraduate level will be a natural continuation of what students were doing in high-school. At this level robots such as the Lego Mindstorms or the VEX Robotics Design System can be used.
In order to have successful graduate students and develop more research at the undergraduate and graduate levels in mobile robotics, more advanced courses have to be offered. There is a need for a pure mobile robotic course that covers the basic concepts of mobile robotics such as kinematics, perception, localization, map building and navigation. Universities that offer robotics degrees and have robotics departments, more specialized courses can be offered to include topics such as vision, behavior coordination, robot learning, swarm robotics, humanoid robotics etc. at the undergraduate level. Beyond this, faculty may offer directed studies and projects (i.e. final year projects or group projects) to students interested in mobile robotics. This will prepare students for industry work and graduate level work.
Universities normally do not have the same problems as K-12 education. There are normally more resources available either from funding agencies, university funds or research projects. Also there is no need to have many robotic systems since there are fewer students using them compared to K-12 schools. At this level, students can also use simulators such as Microsoft Robotics Developer Studio and others. There are definitely enough teaching materials, textbooks, equipment and simulators that can successfully support mobile robotics courses. There are several excellent textbooks such as AI Robotics by Robin Murphy (Murphy, 2000), Computational Principles of Mobile Robotics by Dudek and M. Jenkin (Dudek & Jenkin, 2011), Introduction to Autonomous Mobile Robots by Roland Siegwart and Illah Nourbakhsh (Siegwart & Nourbakhsh, 2004), etc.
Students must grain most of their mobile robotics education at the university level. At the graduate level, robotics courses must be in the form of research, projects and theses. At this level more research must be done.

4.1 Mobile robotic platforms for the university level

There are many mobile robots available at the university level. A few selected ones will be presented and described in this section. The robots will be described in terms of hardware, capabilities, control and applications. The peripherals that can be used for each robot will also be briefly described, if possible. An effort will be made to describe the possible applications for each of these robots.

4.1.1 Lego Mindstorms NXT and Fischertechnik ROBO TX

The processing power and programming flexibilities that these systems offer make them ideal for University introductory courses. Many universities are using these robots very successfully. These systems were described earlier in the *Mobile robotic platforms for K-12 education* section thus the explanation part will be omitted.
These robotic kits are primarily used at the introductory level to teach the basic concepts and principles of perception, localization, map building and path planning and navigation. We often find these type of mobile robots used for projects and/or to develop prototype ideas. Programming at this level is done using high-level languages such as C# with Microsoft Robotics Developer Studio, NBC, ROBOTC or other high-level languages.

4.1.2 VEX robotics design system

VEX robotics was described earlier in section *Mobile robotic platforms for K-12 education*. Many universities are using VEX robotics because of the rigid design, powerful processor and programming flexibility. This kit, like the Mindstorms kit, is primarily used for introductory courses, projects and prototyping.

4.1.3 iRobot create

The Create robot from iRobot (iRobot: Education & Research, 2011) is intended for educational purposes. This robot is based on the iRobot Roomba vacuum cleaner. In place of the vacuum hardware of the Roomba, the Create includes a cargo bay which houses a 25 pin port that can be used for digital and analog input and output. The Create also possesses a serial port through which sensor data can be read and motor commands can be issued using the iRobot Roomba Open Interface protocol. The platform accepts virtually all accessories designed for iRobot's domestic robots and can also be programmed with the addition of a small "command module" (a microcontroller with a USB connector and four DE-9 expansion ports).

The controller of the iRobot is limited in processing power and thus many choose to utilize an external computer in controlling the Create robot. Since the built-in serial port supports the transmission of sensor data and can receive actuation commands, any embedded computer that supports serial communication can be used as the control computer. Popular choices include the gumstix (Gumstix small open source hardware, 2011) line of computers. In many cases laptop computers are used to control the Create through the serial connection.

The Create is supported both in hardware and in simulation by Microsoft Robotics Developer Studio (RDS). An example of an RDS simulation that contains the Create is shown in Figure 10. The Create is the platform used for Sumo robot competitions. The iRobot Command Module for the Create is not required for RDS and is not used.

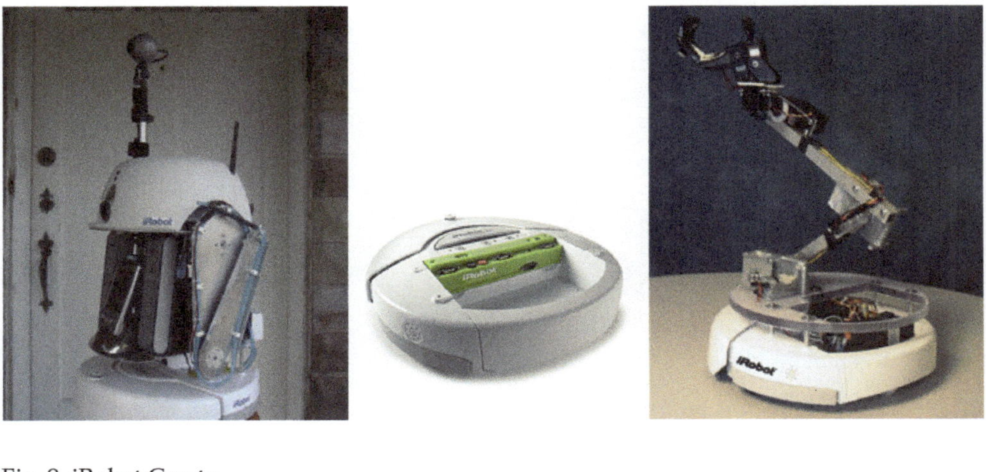

Fig. 9. iRobot Create

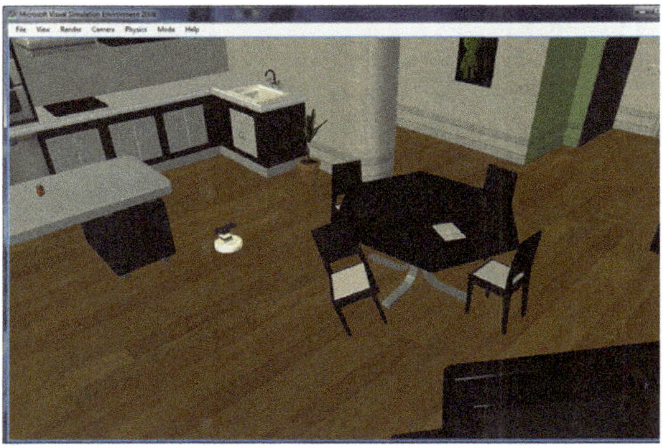

Fig. 10. iRobot Create in Microsoft Robotics Developer Studio Simulation

4.1.4 K-Team SA mobile robots

K-Team (K-Team Corporation, 2011) is a Swiss company that develops, manufactures and markets high quality mobile robots for use in advanced education and research. The Khepera III and Koala II (Figure 10) are found in many university laboratories that specialize in robotics education and research. KoreBot II is the standard controller for the Khepera II and the Koala II. The KoreBot II is a single board controller used for custom robotics developments.

The Khepera III is a miniature mobile robot that has features that can match the performances of much bigger robots. It has upgradable embedded computing power using the KoreBot system, multiple sensor arrays for both long range and short range object detection, swappable battery pack system for optimal autonomy, and exceptional differential drive odometry. Khepera III is able to move on a tabletop but it is also designed to move on rough floor surfaces and carpets.

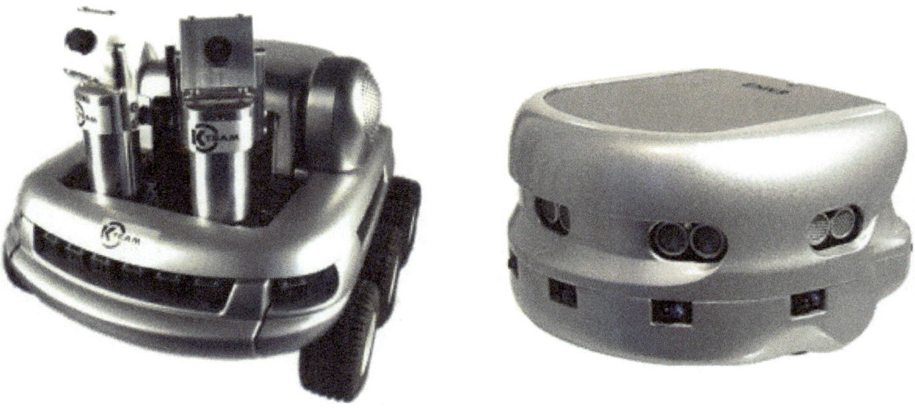

Fig. 11. K-Team SA Koala II and Khepera III

The Khepera III architecture provides exceptional modularity. The robot base can be used with or without a KoreBot II board. Using the KoreBot II, it features a standard embedded Linux Operating System for quick and easy autonomous application development. Without

the KoreBot II, the robot can be remotely operated. It is easily interfaced with any Personal Computer.
The robot includes an array of nine infrared sensors for obstacle detection as well as five ultrasonic sensors for long range object detection. An optional front pair of ground Infrared Sensors is available for line following and table edge detection. Through the KoreBot II, the robot is also able to host standard Compact Flash extension cards, supporting WiFi, Bluetooth, extra storage space, and many others.
Koala is a mid-size robot designed for real-world applications. It is bigger than Khepera, more powerful, and capable of carrying larger accessories. Koala has the functionality necessary for use in practical applications.
In addition to these new features, Koala retains a shape and structure similar to Khepera, such that experiments performed on Khepera can be migrated to the Koala. The BIOS of both robots is compatible, permitting programs written for one robot to be easily adapted and recompiled for the other.
Programming of the robots can be done by standard cross-C language to more sophisticated tools like LabView, MATLAB (MATLAB - The Language OF Technical Computing, 2011) or SysQuake (Calerga - Sysquake, 2011). In general any programming environment capable of communicating over a serial port can also be used to program these robots. KoreBot II GNU (The GNU Operating System, 2011) C/C++ cross-compiler provides a powerful standard tool for complex code compilation and conditional build. It also supports all the standard C library functions and almost all the other Posix libraries.
The K-team robots are commonly used for experiments and research is Navigation, Artificial Intelligence, Multi-Agents System, Control, Collective Behavior and Real-Time Programming, among others.

4.1.5 Adept MobileRobots

Adept MobileRobots (Intelligent Mobile Robotic Platforms, 2011) offers an array of mobile robots; the Seekur Jr, the GuiaBot and the PowerBot. The most frequently used and most popular research mobile robot is the Pioneer 3DX (P3DX). It is an advanced research robot that is controlled by a computer (PC) and has a large range of sensors (including an optional laser range finder), and communicates via WiFi. The Pioneer's versatility, reliability and durability have made it the reference platform for robotics research. Unlike hobby and kit robots, Pioneer is fully programmable.
The base Pioneer 3DX (Figure 13) platform arrives fully assembled with motors with 500-tick encoders, 19cm wheels, tough aluminum body, 8 forward-facing ultrasonic (sonar) sensors, 8 optional real-facing sonar, 1, 2 or 3 hot-swappable batteries, and a complete software development kit. One of the innovations of Pioneer is that it does not have an on board controller. The robot can be controlled either by an optional internal computer (PC) or an external laptop. The base Pioneer 3DX platform can reach speeds of 1.6 meters per second and carry a payload of up to 23 kg. With the optional Laser Mapping & Navigation System and MobileEyes, pioneer can map buildings and constantly update its position within a few cm while traveling within mapped areas. With the appropriate optional accessories, the robot can be remotely viewed, speak and play and hear audio.
Since it is controlled by a computer, it can be programmed in any programming language that the user selects. I addition to this, the P3DX is supported both in hardware and in simulation by Microsoft Robotics Developer Studio. The Pioneer 3 DX is an all-purpose

base, used for research and applications involving mapping, teleoperation, localization, monitoring, reconnaissance, vision, manipulation, autonomous navigation and multi-robot cooperation and other behaviors.

Fig. 12. The Seekur Jr, GuiaBot and PowerBot from Adept MobileRobots

Fig. 13. Adept MobileRobots Pioneer 3DX

4.2 Mobile robot competitions for university students

Robotic competitions attempt to foster several research areas by providing a standard problem where wide range of technologies can be integrated and examined, as well as used for integrated project-oriented education. Additionally, mobile robotic competitions create media interest and this may even generate additional funds from external sources. Normally the work to conceive, build and program the robots is integrated in final graduation projects, extracurricular activities or post-graduation activities. Students are very motivated, because they can integrate most of the knowledge acquired during their courses. Preparing for mobile robot competitions involves interdepartmental co-operations, sometimes not only of the engineering and science departments.

Mobile Robotic competitions are a very motivating way of fostering research, development and education, mainly in Robotics, but also in Science and Technology in general. Mobile robot contests are recognized by the scientific community as a way for the development of research. The practical solutions students find for their projects on their way to a competition will teach them more skills than a course can. There is a wide variety of competitions for robots of various types. The following examples describe a few of the higher profile events.

4.2.1 RoboCup

RoboCup (RoboCup, 2011) is an international robotics competition founded in 1997. RoboCup chose to use soccer as a central topic of research, aiming at innovations to be applied for socially significant problems and industries. The aim is to develop autonomous football (soccer) robots with the intention of promoting research and education in the field of artificial intelligence. The name *RoboCup* is a contraction of the competition's full name, "Robot Soccer World Cup", but there are many other stages of the competition as well. The ultimate goal of the RoboCup project is by 2050, develop a team of fully autonomous humanoid robots that can win against the human world champion team in soccer.

Fig. 14. RoboCup Humanoid and Simulation Competitions

The contest currently has three major competition domains, each with a number of leagues and subleagues: RoboCupSoccer (RoboCupSoccer, 2011), RoboCupRescue (RoboCupRescue, 2011) and RoboCup@Home (RoboCup@Home, 2011). RoboCupSoccer includes a number of sub-competitions: Simulation League, Small Size Robot League, Middle Size Robot League, Standard Platform League and the Humanoid League. Figure 13 shows the humanoid competition and the simulation competition.

The RoboCupRescue Robot League is an international competition for urban search and rescue robots, in which robots compete to find victims in a simulated earthquake environment. RoboCupRescue includes real robot and simulation leagues. RoboCup@Home is a new league inside the RoboCup competitions that focuses on real-world applications and human-machine interaction with autonomous robots. A set of benchmark tests is used to evaluate the robots' abilities and performance in a realistic non-standardized home environment setting. The aim is to foster the development of useful robotic applications that can assist humans in everyday life. The ultimate scenario is the real world itself. To build up the required technologies gradually a basic home environment is provided as a general scenario. In the first years it will consist of a living room and a kitchen but soon it should also involve other areas of daily life, such as a garden/park area, a shop, a street or other public places. Focus lies on the following domains but is not limited to: Human-Robot-Interaction and Cooperation, Navigation and Mapping in dynamic environments, Computer Vision and Object Recognition under natural light conditions, Object Manipulation, Adaptive Behaviors, Behavior Integration, Ambient Intelligence, Standardization and System Integration.

4.2.2 Eurobot

Eurobot (Eurobot, international robotics contest, 2011) is an international amateur robotics contest, created in 1998. It is open to teams of young people, organized either in student projects or in independent clubs. Countries that present more than three teams must organize a national qualification round where only three teams are selected for the final Eurobot competition (Figure 15). These teams could be formed from students as part of their studies or as independent clubs or non-profit organizations. A team must be made up of two or more active participants. Team members may be up to 30 years old, each team may have one supervisor for which this age limit does not apply. The contest aims at interesting the largest public to robotics and at encouraging the group practice of science by youth. The competition includes a conference for the students and the public. Eurobot is an opportunity to unleash technical imagination and exchange ideas, know-how, hints and engineering knowledge around a common challenge.

Fig. 15. Eurobot competitions

4.2.3 European Land-Robot trial (ELROB)

ELROB (Elrob-Website, 2011) is a European event, which demonstrates the abilities of unmanned robots. The ELROB is an annual event and alternates between a military and a civilian focus each year (Figure 16). Only teams from Europe are allowed. But teams of both commercial and academic backgrounds are allowed.

ELROB is designed to assess current technology to solve problems at hand, using whatever strategy to achieve it. The scenarios are designed to simulate real world missions, be it military or civilian ones. There are no artificial constraints set to these scenarios to ease the task for the robots like e.g. very visible road markings. This forces the participating teams and systems to fulfill high requirements set by the real world scenarios.

Fig. 16. ELROB military competition robots

5. Conclusion

Mobile robots are becoming part of our everyday life in many forms. The mobile robot industry, other organizations and universities are developing mobile robots for all imaginable applications. Mobile robot research and development are at their highest. There is a need for specialized mobile robotics engineers and scientists.

The mobile robotics momentum that has been created over the last two decades must not stop. More courses and more activities are needed starting from the early K-12 education. This is now possible because of the development of various types of mobile robots and simulators. The existence of various mobile robotics competitions gives extra motivation to students and educators to do more work in mobile robotics.

Teaching with robots from the early stages of K-12 education will better prepare students for high school mobile robotics education and competitions. At the university level, introductory level mobile robotics courses must be offered. More advanced mobile robotics courses must also be offered that specialize in topics such as vision, localization, navigation, etc. At the graduate level, students must concentrate on research and projects. This will create a continuation on mobile robotics education, from K-12 all the way to graduate school, and thus creating better prepared graduate students, developing more research interest and creating skilled graduates for the many jobs that are opening up in the mobile robotics industry.

6. References

Chu, K.H., Goldman, R. & Sklar, E. (2005). RoboXAP: an agent-based educational robotics simulator, *Agent-based Systems for Human Learning Workshop (ABSHL) at AAMAS-2005,* Utrecht University, The Netherlands, July 2005.

Dudek, G. & Jenkin, M. (2011). *Computational Principles of Mobile Robotics* (Second Edition), Cambridge University Press, ISBN 9780521692120, Cambridge, UK.

Malec, J. (2001). Some Thoughts on Robotics for Education, *2001 AAAI Spring Symposium on Robotics and Education,* Stanford University, USA, March 2001

Matson, E. Pauly, R. & DeLoach, S. (2003). Robotic Simulators to Develop Logic and Critical Thinking Skills in Underserved K-6 Children, *Proceedings of the 38th ASEE Midwest Section Conference,* Rolla, Missouri, USA, September, 2003.

Murphy, R. R. (2000). *Introduction to AI Robotics*, MIT Press, ISBN 0-262-13383-0, Cambridge, MA, USA.

Siegwart, R. & Nourbakhsh, I. (2004). *Introduction to Autonomous Mobile Robots*, MIT Press, ISBN: 0-262-19502-X, Cambridge, MA, USA.

van Lith, P. (2007). Teaching Robotics in Primary and Secondary schools, *Proceedings, ComLab International Conference 2007*, Radovljica, Slovenia, November 30 - December 1, 2007

Calerga - Sysquake. (n.d.). 2011, Available from: http://www.calerga.com/products/Sysquake

Competition - VEX Robotics. (n.d.). 2011, Available from: http://www.vexrobotics.com/competition

Computer Science Social Network. (n.d.). 2011, Available at http://www.cs2n.org/rvw

Elrob-Website: Home/Objectives. (n.d.). 2011, Available from: http://www.elrob.org

Engino international website - play to invent. (n.d.). 2011, Available from: http://www.engino.com

Eurobot, international robotics contest. (n.d.). 2011. Available from :http://www. eurobot.org/eng

Fischertechnik GmbH. (n.d.). 2011, Available from: http://www.fischertechnik.de/en

Gumstix small open source hardware. (n.d.). 2011, Available from: http:// www.gumstix.com

Intelligent Mobile Robotic Platforms for Service Robots, Research and Rapid Prototyping. (n.d.). 2011, Available from: http://www.mobilerobots.com/Mobile_Robots.aspx

iRobot: Education & Research. (n.d.). 2011, Available from: www.irobot.com/create

K-Team Corporation - Mobile Robotics. (n.d.). 2011, Available from: http://www.k-team.com

LEGO Education. (n.d.). 2011, Available from: http://www.legoeducation.com

Lego Education WeDo. (n.d.). 2011, Available from: http://www.legoeducation.us/eng/product/lego_education_wedo_robotics_construction_set/2096

MATLAB - The Language OF Technical Computing. (n.d.). 2011, Available from: http://www.mathworks.com/products/matlab

Microsoft Robotics Developer Studio. (n.d.). 2011, Available from: http:// www.microsoft.com/robotics

NI LabVIEW - Improving the Productivity of Engineers and Scientists. (n.d.). 2011, Available from: http://sine.ni.com/np/app/flex/p/ap/global/lang/en/pg/1/docid/nav-77

RoboCup. (n.d.). 2011, Available from: http://www.robocup.org

RoboCup@Home. (n.d.). 2011, Available from: http://www.ai.rug.nl/robocupathome

RoboCupJunior. (n.d.). 2011, Available from: http:// www.robocup.org/robocup-junior

RoboCupRescue. (n.d.). 2011, Available from: http://www.robocuprescue.org

RoboCupSoccer. (n.d.). 2011, Available from: http://www.robocup.org/robocup-soccer

ROBOTC.net: Home of the Best Programming Language for Educational Robotics. (n.d.). 2011, Available from: http://www.robotc.net

TETRIX Robotics. (n.d.). 2011, Available from: http://www.tetrixrobotics.com

The GNU Operating System. (n.d.). 2011, Available from: http://www.gnu.org

USFIRST.org - Welcome to FIRST. (n.d.). 2011, Available from: http://www.usfirst.org

VEX Robotics Design System. (n.d.). 2011, Available from: http://www.vexrobotics.com

Welcome to the RoboLab. (n.d.). 2011, Available from: http://www.robolabonline.com/home

2

Autonomous Mobile Robot Emmy III

Cláudio Rodrigo Torres, Jair Minoro Abe,
Germano Lambert-Torres and João Inácio da Silva Filho
Universidade Metodista de São Paulo,
University of São Paulo and Paulista University – UNIP,
Federal University of Itajubá, Universidade Santa Cecilia – UNISANTA
Brazil

1. Introduction

In this work we present a description of the Emmy III robot architecture [1], [2], [3], [4] and, also, a summary of the previous projects which have led to the Emmy III robot building [5], [6], [7], [8], [9]. These robots are part of a project of applying the Paraconsistent Annotated Evidential Logic Eτ [12] which allows manipulating concepts like fuzziness, inconsistencies and paracompleteness.

The Emmy III robot is designed to achieve a set point in an environment which is divided into coordinates. The Emmy III robot may be considered as a system which is divided into three other subsystems: the planning subsystem, the sensing subsystem and the mechanical subsystem.

The planning subsystem is responsible for generating the sequence of movements the robot must perform to achieve a set point. The sensing subsystem has the objective of informing the planning subsystem the position of obstacles; the mechanical subsystem is the robot itself, it means, the mobile mechanical platform which carries all devices that come from the other subsystems. This platform must also perform the sequence of movements borne by the planning subsystem.

It is observed that the planning subsystem and the sensing subsystem have already been implemented, but the mechanical subsystem has not been implemented yet.

The sensing subsystem uses the Paraconsistent Artificial Neural Networks - PANN [2], [3]. PANN is a new type of Artificial Neural Networks – ANNs based on Paraconsistent Annotated Evidential Logic Eτ. In the next paragraph we introduce the main basic concepts of the Logic Eτ, as well as some terminologies.

2. Paraconsistent annotated evidential logic Eτ

Generally, Paraconsistent Logics is a new kind of logics that allows contradictions without trivialization. A branch of it, the Paraconsistent Annotated Evidential Logic Eτ, which will be employed in this work, also deals with the concept of fuzziness. Its language consists of propositions in the usual sense p together with annotation constants: (μ, λ) where $\mu, \lambda \in [0, 1]$ (real unitary interval). Thus an atomic formula of the Logic Eτ is of the form $p(\mu, \lambda)$ which can be intuitively read: the favorable evidence expressed by p is μ and the contrary evidence expressed by p is λ. A detailed feature on the subject is found in [12].

The Favorable Evidence Degree (μ) is a value that represents the favorable evidence in which the sentence is true; this value is between 0 and 1.
The Contrary Evidence Degree (λ) is a value that represents the contrary evidence in which the sentence is true; this value is between 0 and 1.
Through the Favorable and Contrary Degrees it is possible to represent the four extreme logic states, as shown in the figure 1.

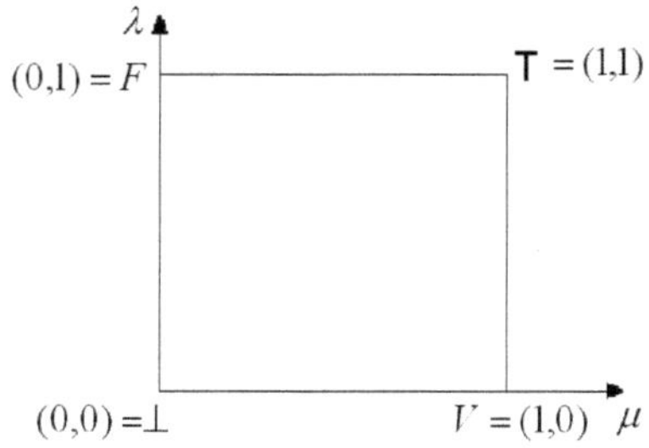

Fig. 1. The extreme logic states

The four extreme logic states are:
- True (V)
- False (F)
- Paracomplete (⊥)
- Inconsistent (T)

In [6] it is proposed the Para-analyzer Algorithm. By this algorithm it is also possible to represent the non-extreme logic state. The figure 2 shows this.

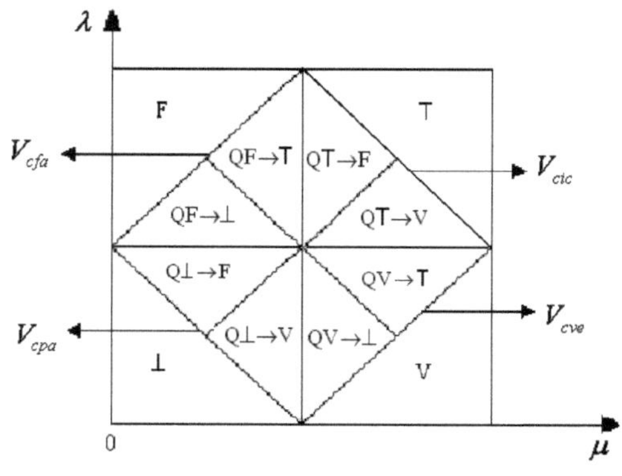

Fig. 2. The non-extreme logic states

The eight non-extreme logic states are:
- Quasi-true tending to Inconsistent - QV→T
- Quasi-true tending to Paracomplete - QV→⊥
- Quasi-false tending to Inconsistent - QF→T
- Quasi-false tending to Paracomplete - QF→⊥
- Quasi-inconsistent tending to True - QT→V
- Quasi-inconsistent tending to False - QT→F

- Quasi-paracomplete tending to True - $Q\perp\rightarrow V$
- Quasi-paracomplete tending to False - $Q\perp\rightarrow F$

It is also defined the Uncertainty Degree: $Gun(\mu, \lambda) = \mu + \lambda - 1$ and the Certainty Degree: $Gce(\mu, \lambda) = \mu - \lambda$ $(0 \leq \mu, \lambda \leq 1)$

Some additional control values are:

- Vcic = maximum value of uncertainty control
- Vcve = maximum value of certainty control
- Vcpa = minimum value of uncertainty control
- Vcfa = minimum value of certainty control

It is described the proposed sensing system in the next section.

3. Paraconsistent artificial neural network

The artificial neural network of the sensing subsystem is composed of two types of cells: Analytic Paraconsistent Artificial Neural Cell - CNAPa, and Passage Paraconsistent Artificial Neural Cell - CNAPpa. The cells are described as it follows:

3.1 Analytic paraconsistent artificial neural cell - CNAPa

The Analytic Paraconsistent Artificial Neural Cell – CNAPa has two inputs (μ_{RA} and μ_{RB}) and two outputs (S1 and S2). There are also two configuration parameter inputs (Ftct and Ftc). The figure 3 shows the graphic representation of this cell.

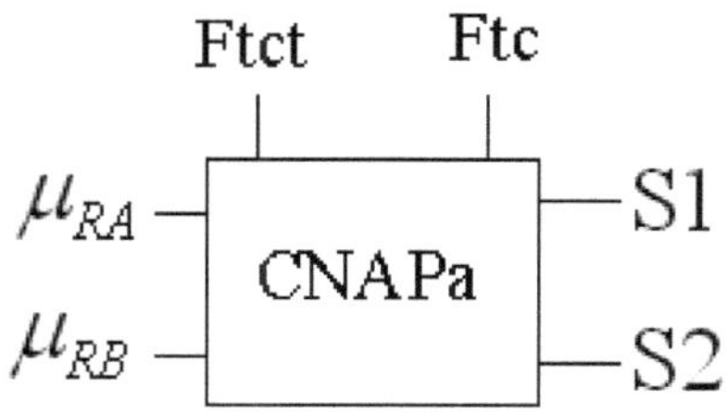

Fig. 3. Graphic representation of the Analytic Paraconsistent Artificial Neural Cell

The input evidence degrees are:

- μ_{RA}, such as: $0 \leq \mu_{RA} \leq 1$
- μ_{RB}, such as: $0 \leq \mu_{RB} \leq 1$

There are also two control values:

- Contradiction Tolerance Factor – Ftct, such as: $0 \leq Ftct \leq 1$
- Certainty Tolerance Factor – Ftc, such as: $0 \leq Ftc \leq 1$

The Analytic Paraconsistent Artificial Neural Cell – CNAPa has two outputs. The out-put 1 (S1) is the Resultant Evidence Degree - μ_E.

- μ_E, such as: $0 \leq \mu_E \leq 1$
- The output 2 (S2) is the Resultant Evidence Interval – φE.
- φE, such as: $0 \leq \varphi E \leq 1$

The Analytic Paraconsistent Artificial Neural Cell calculates the maximum value of certainty - V_{cve}, the minimum value of certainty control - V_{cfa}, the maximum value of uncertainty control - V_{cic}, and the minimum value of uncertainty control - V_{cpa}, by this way:

$$V_{cve} = \frac{1 + Ft_c}{2} \quad (1)$$

$$V_{cfa} = \frac{1 - Ft_c}{2} \quad (2)$$

$$V_{cic} = \frac{1 + Ft_{ct}}{2} \quad (3)$$

$$V_{cpa} = \frac{1 - Ft_{ct}}{2} \quad (4)$$

The Resultant Evidence Degree – μ_E, is determined as:

$$\mu_E = \frac{G_c + 1}{2} \quad (5)$$

As $Gc = \mu - \lambda$, we can say that:

$$\mu_E = \frac{\mu - \lambda + 1}{2} \quad (6)$$

It is called as Certainty Interval (φ) the Certainty Degree interval that can be modified without changing the Uncertainty Degree value. This value is determined as:

$$\varphi = 1 - |Gct| \quad (7)$$

3.2 Passage Paraconsistent Artificial Neural Cell - CNAPpa

The Passage Paraconsistent Artificial Neural Cell – CNAPpa has one input (μ), one out-put (S1) and one parameter control in-put (Ftc). The figure 4 shows the graphic representation of CNAPpa.

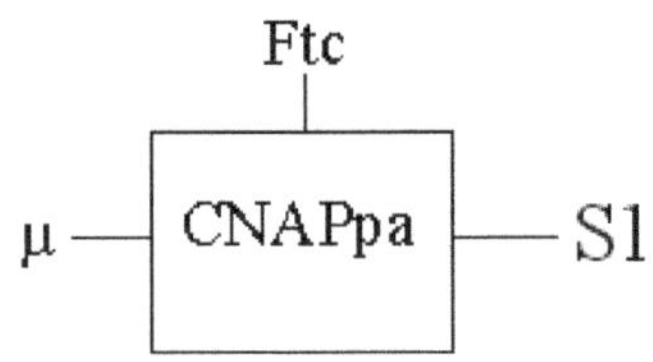

Fig. 4. Graphic representation of the Passage Paraconsistent Artificial Neural Cell

- The in-put is the Favorable Evidence Degree (μ).
- μ, such as: $0 \leq \mu \leq 1$
- The value for the output S1 is the same as for the in-put μ. But, the out-put value may be limited through the parameter control in-put Ftc.

The CNAPpa calculates the maximum value of certainty - V_{cve} and the minimum value of certainty control - V_{cfa} by the equations (1) and (2). And it also determines the Resultant Evidence Degree - μ_E, by the equation (6). For this, λ is considered as the following:

$$\lambda = 1 - \mu \quad (8)$$

The output S1 assumes the same value as in the in-put μ when the following situation is true:

- $[(V_{cve} \leq \mu_E)$ or $(\mu_E \leq V_{cfa})]$
- Otherwise, S1 is 0,5.

4. Autonomous mobile robot Emmy I

The robot Emmy I was the first application of the Paraconsistent Evidential Logics in robotics [8], [9]. The Emmy I robot project finished in 1999 and its results have led to the construction of the Emmy II robot and to the Emmy III project itself.

The Emmy I has two ultra-sonic sensors: one determines the favorable evidence degree and the other determines the contrary evidence degree. The Emmy I controller, named as Paracontrol, allows the Emmy I to act conveniently in "special" situations, as when there is contradictory datum: one sensor may detect an obstacle in front of the robot (for example, a wall) while the other detects the presence of no obstacles (for example, it may be in direction to an opened door). In a situation like that Emmy may stop and turn 45° to the most free direction. Then, if in a new measurement, there is no inconsistency, the robot may take another decision, for example, to go ahead.

The Emmy I robot consists of a circular mobile platform of aluminum with a 30 cm diameter and being 60 cm tall. Its main device is the Paracontrol controller. While moving into a non-structured environment, the Emmy robot gets information about presence/absence of obstacles using a sonar system called Parasonic [17]. The figure 5 shows the autonomous mobile robot Emmy.

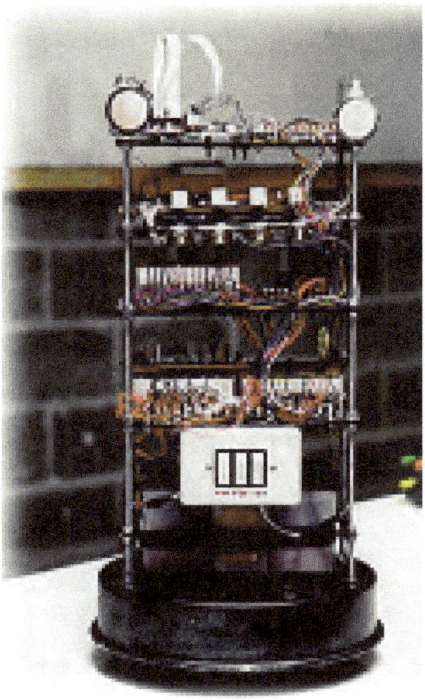

Fig. 5. The autonomous mobile robot Emmy

The Paracontrol [18] is an electronic materialization of the Para-analyzer algorithm [9] [19]. Basically, it is an electronic circuitry which treats logic signals in a context of logic Eτ. Such circuitry compares the logic value entries and determines the logic values output. Favorable evidence and contrary evidence degrees are represented by voltage. Operational amplifiers determine the Certainty and the Uncertainty degrees. The Paracontrol comprises both, analogical and digital systems, and it can be externally adjusted by applying positive and negative voltages. As there are 12 logic states in the Para-analyzer algorithm, the Paracontrol can take 12 different decisions.

Parasonic is an electronic circuitry that the Emmy I robot uses to detect obstacles in its path. Parasonic converts distances from obstacles into electric signals of continuous voltage, ranging from 0 to 5 volts. Parasonic is basically composed of two ultrasonic sensors, type POLAROID 6500, controlled by an 8051 microcontroller. The microcontroller is programmed to carry out the synchronization between the measurements of the two sensors and the change of the distance into electric voltage.

Parasonic generates the favorable evidence degree value (μ) and the contrary evidence degree value (λ). They make a continuous voltage which ranges from 0 to 5 volts. Paracontrol receives these signals from Parasonic. The figure 6 shows the basic structure of the Emmy robot.

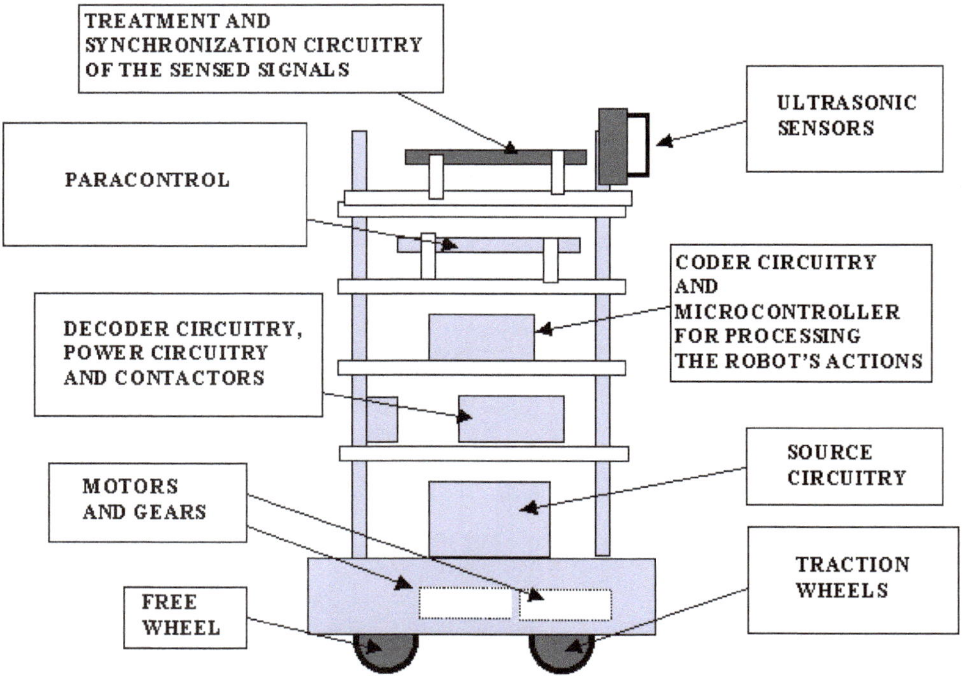

Fig. 6. Basic structure of Emmy robot

In the figure 7 can be seen the main components of the Emmy robot.

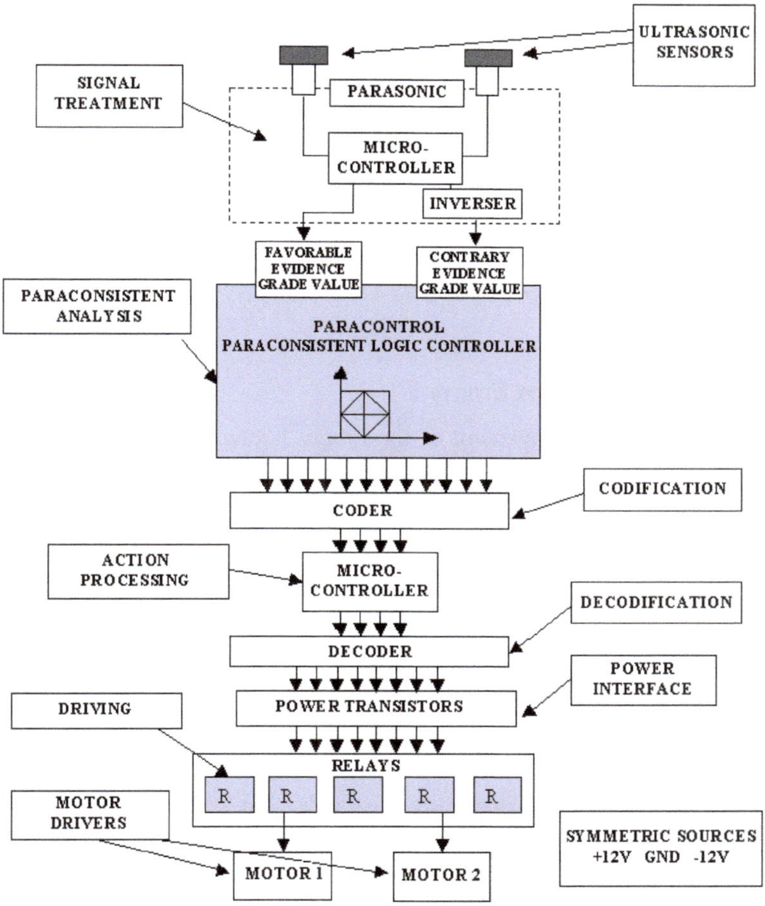

Fig. 7. Main components of the Emmy robot

The description of the Emmy robot components is the following.

- Ultrasonic sensors: two ultrasonic sensors are responsible for emitting ultrasonic waves and for detecting the return of them.
- Signal treatment: in Parasonic there is a microcontroller which sends to the sensors a signal that makes them emit ultrasonic waves. When the ultrasonic waves return, the sensors send the microcontroller a signal. The microcontroller measures the time lasted between the sending and the returning of the ultrasonic waves. So, the microcontroller is able to determine the distance between the sensors and the obstacles in front of them. Paracontrol generates a continuous voltage ranging from 0 to 5 volt, it must be proportional to the distance between the sensor and the obstacle. This signal is considered the favorable evidence degree value on the proposition "The front of the robot is free". In the same way the Paracontrol generates a continuous voltage ranging from 5 to 0 volt, it must be related to the contrary evidence degree.

- Paraconsistent analysis: the Paracontrol makes the logical analysis of the signals according to the logic Eτ.
- Codification: the coding circuitry changes a 12-digit word to a code of 4 digits.
- Action processing: a microcontroller processes the 4 digit code generated by the coding circuitry, determining the sequence of relays which must be actuated for the robot to perform the right movement.
- Decodification: the decoding circuitry changes a 4-digit word into a code of 12 digits.
- Power interface: transistors amplify the signals of the 12-digit-word generated by the decoding circuitry. Then the signals can actuate on the relays.
- Driving: relays are responsible for actuating the DC motors M1 and M2.
- Motor driver: two DC motors are responsible for moving the robot.
- Sources: two batteries composing a ± 12 volt symmetric source feed the Emmy robot electric circuitries.

5. Autonomous mobile robot Emmy II

The Emmy II robot is an improvement of the Emmy I robot. It is an autonomous mobile robot which is able to avoid obstacles while it is moving in any environment.
The platform used to assemble the Emmy II robot is approximately 23cm high and has a diameter of 25cm. The Emmy II robot main components are a microcontroller from the 8051 family, two sonar ranging module (sensors) and two DC motors. The figure 8 shows the Emmy II basic structure.

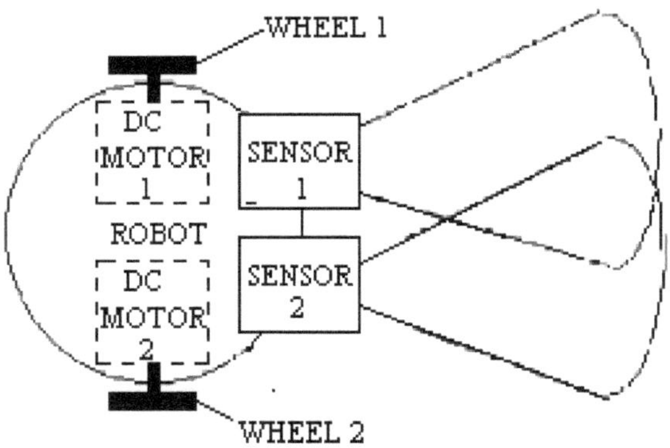

Fig. 8. The Emmy II basic structure

The Emmy II controller system uses six logic states instead of 12 logic states which are used in the Emmy I controller. Moreover, it may present some commands that do not exist in the Emmy I robot:

1. Velocity control: the Emmy II controller allows the robot to brake, turn and accelerate "in a smooth way", what is not possible in the Emmy I robot.
2. The Emmy II controller allows the backward motion. In some situations the robot may move backward or turn around with a fixed wheel having the other spinning around backward. There are not these types of movements in the Emmy I robot.

It can be seen in the figure 9 a simplified block representation of the Emmy II robot.

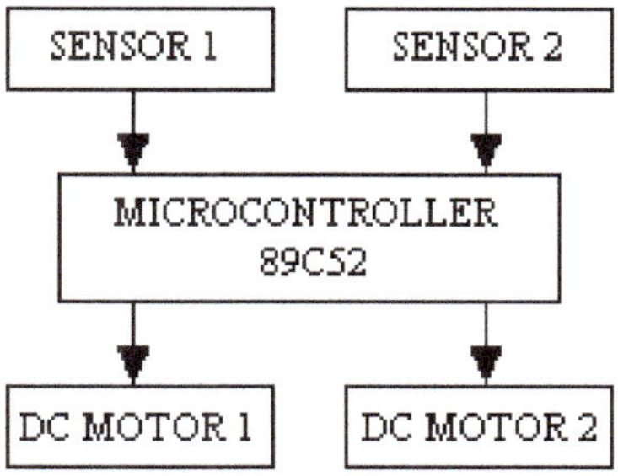

Fig. 9. Emmy II block representation

The figure 10 shows a picture of the Emmy II robot

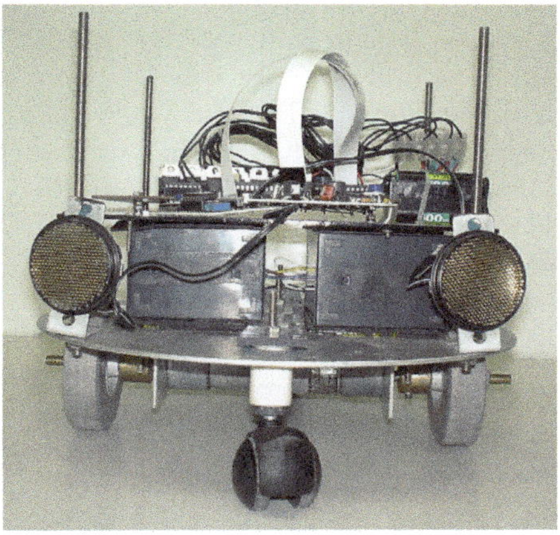

Fig. 10. The front part of the Emmy II robot

It is shown in the figure 11 the lower part of the Emmy II robot.

Fig. 11. The lower part of the Emmy II robot

The sonar ranging modules are responsible for verifing whether there is any obstacle in front of the robot or not. The signals generated by the sonar ranging modules are sent to the microcontroller. These signals are used to determine the favorable evidence degree (μ) and the contrary evidence degree (λ) on the proposition "There is no obstacle in front of the robot". The favorable and contrary evidence degrees are used to determine the robot movements.

The Emmy II possible movements are the following:

- Robot goes ahead. DC motors 1 and 2 are supplied for spinning around forward.
- Robot goes back. DC motors 1 and 2 are supplied for spinning around backward.
- Robot turns right. Just DC motor 1 is supplied for spinning around forward.
- Robot turns left. Just DC motor 2 is supplied for spinning around forward.
- Robot turns right. Just DC motor 2 is supplied for spinning around backward.
- Robot turns left. Just DC motor 1 is supplied for spinning around backward.

The signal generated by the sensor 1 is considered the favorable evidence degree and the signal generated by the sensor 2 is considered the contrary evidence degree for the proposition "There is no obstacle in front of the robot". When there is an obstacle near the sensor 1, the favorable evidence degree is low and when there is an obstacle far from the sensor 1, the favorable evidence degree is high. Otherwise, when there is an obstacle near the sensor 2, the contrary evidence degree is high and when there is an obstacle far from the sensor 2, the contrary evidence degree is low. The Emmy II controller decision of which movement the robot should perform is based on the reticulated showed in the figure 12.

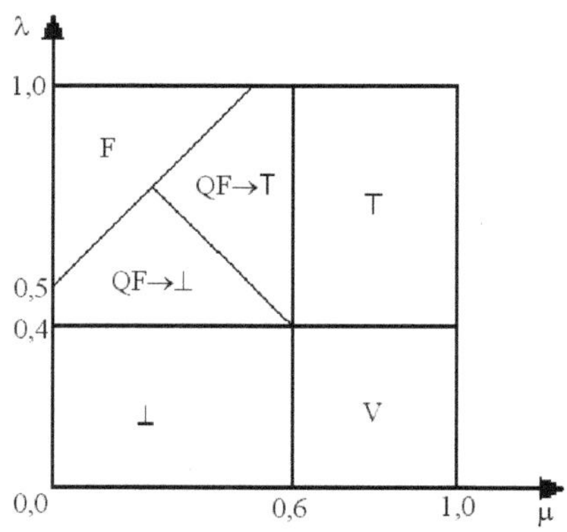

Fig. 12. Lattice of Emmy II controller

The decision for each logic state is the following:

- V state: Robot goes ahead.
- F state: Robot goes back.
- ⊥ state: Robot turns right.
- T state: Robot turns left.
- QF→⊥ state: Robot turns right.
- QF→ T state: Robot turns left.

The justification for each decision is the following:

When the logic state is true (V), it means that the front of the robot is free. So, the robot can go ahead.

In the inconsistency (T), μ and λ are high (i.e., belong to T region). It means that the sensor 1 is far from an obstacle and the sensor 2 is near an obstacle, so the left side is more free than the right side. Then, the behavior should be to turn left by supplying only the DC motor 2 for spinning around forward and keeping the DC motor 1 stopped.
When the Paracompleteness (⊥) is detected, μ and λ are low. It means that the sensor 1 is near an obstacle and the sensor 2 is far from an obstacle, so the right side is more free than the left side. Then, the behavior should be to turn right by supplying only the DC motor 1 for spinning around forward and keeping the DC motor 2 stopped.
In the false state (F) there are obstacles near the front of the robot. Therefore the robot should go back.
In the QF→ T state, the front of the robot is obstructed but the obstacle is not so near as in the false state and the left side is a little bit more free than the right side. So, in this case, the robot should turns left by supplying only the DC motor 1 for spinning around backward and keeping the DC motor 2 stopped.
In the QF→⊥ state, the front of the robot is obstructed but the obstacle is not so near as in the false state and the right side is a little bit freer than the left side. So, in this case, the robot should turns right by supplying only the DC motor 2 for spinning around backward and keeping the DC motor 1 stopped.

5.1 Tests

Aiming to verify Emmy II robot functionally, it has been performed 4 tests. Basically, counting how many collisions there were while the robot moved in an environment as showed in figure 13 composed the tests.

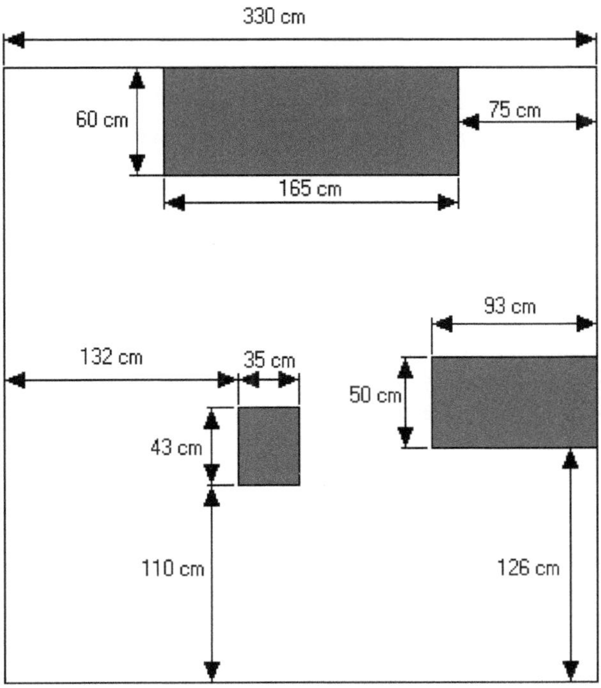

Fig. 13. Environment used to perform the Emmy II tests

The time duration and results for each test have been the following:

Test 1: Duration: 3 minutes and 50 seconds. Result: 13 collisions.

Test 2: Duration: 3 minutes and 10 seconds. Result: 7 collisions.

Test 3: Duration: 3 minutes and 30 seconds. Result: 10 collisions.

Test 4: Duration: 2 minutes and 45 seconds. Result: 10 collisions.

The sonar ranging modules used in the Emmy II robot can't detect obstacles closer than 7,5 cm. The sonar ranging modules transmit sonar pulses and wait for them to return (echo) so that it can determine the distance between the sonar ranging modules and the obstacles; however, sometimes the echo doesn't return, because it reflects to another direction. These are the main causes for the robot collisions:

Test 1: Collisions: 13.

Collisions caused by echo reflection: 4.

Collisions caused by too near obstacles: 9.

Test 2: Collisions: 7.

Collisions caused by echo reflection: 2.

Collisions caused by too near obstacles: 5.

Test 3: Collisions: 10.

Collisions caused by echo reflection: 5.

Collisions caused by too near obstacles: 5.

Test 4: Collisions: 10.

Collisions caused by echo reflection: 4.

Collisions caused by too near obstacles: 6.

There is another robot collision possibility when the robot is going back. As there is no sonar ranging module behind the robot, it may collide.

6. Autonomous mobile robot Emmy III

The aim of the Emmy III autonomous mobile robot is to move from an origin to an end, both predetermined in a non-structured environment. The Emmy III controller considers the environment around the robot divided into cells [15] and a planning subsystem gives the sequence of cells the robot must follow to reach the end cell. These ideas have been applied in [20], [21]. The robot must avoid cells that are supposed to be occupied. A sensing subsystem detects the cells which are occupied. The sensing subsystem uses Paraconsistent Annotated Logic to handle information captured by the sensors. The Emmy III structure is composed of a sensing subsystem, a planning subsystem and a mechanical subsystem as described in the follow.

Sensing subsystem - The environment around the robot is considered as a set of cells. The sensing subsystem has to determine the cells which have obstacles in. But the information captured by the sensors always has an inherent imprecision, which leads to an uncertainty regarding to the actual situation of the cells. In order to manipulate the inconsistent information, the sensing subsystem is based on ParaconsistentAnnotated Evidential Logic Eτ, which captures the information generated by the sensors using the favorable and contrary evidence degrees.

Planning subsystem - The planning subsystem determines a path linking an initial point to an end point in a non-structured environment. For this, the environment around the robot is divided into cells and the planning subsystem gives the sequence of cells that the robot must follow to reach the end cell successfully.

Mechanical subsystem - The Emmy III mechanical part must perform the schedule determined by the planning subsystem. For this, the mechanical subsystem must know the cell occupied by the robot, therefore, a monitoring position makes part of this construction. For each cell that the robot reaches, the possible error of position should be considered.

6.1 Sensing subsystem

The objective of the sensing subsystem is to inform the other robot components about the obstacle position. The proposed sensing subsystem has as its main part Paraconsistent Neural Network [13], [14]. This artificial neural network is based on the Paraconsistent Evidential Logics - Eτ.

The sensing subsystem is a set of electronic components and softwares which are responsible for analyzing the environment around the robot and detecting the obstacle positions. After that, it must inform the other components of the robot the position of the obstacles.

The sensing subsystem may get information from any type of sensor.

In [15] it is presented a method of robot perception and the world's modeling which uses a probabilistic tessellated representation of spatial information called the Occupancy Grid. It is proposed in the chapter a similar method, but instead of using probabilistic representation, it is used Paraconsistent Annoted Evidential Logic Eτ.

The proposed sensing subsystem aims to generate a Favorable Evidence Degree for each environment position. The Favorable Evidence Degree is related to the sentence: there is an obstacle in the analyzed position.

The sensing subsystem is divided into two parts. The first part is responsible for receiving the data from the sensors and sending information to the second part of the system. The second part is Paraconsistent Artificial Neural Network itself. Figure 14 shows this idea.

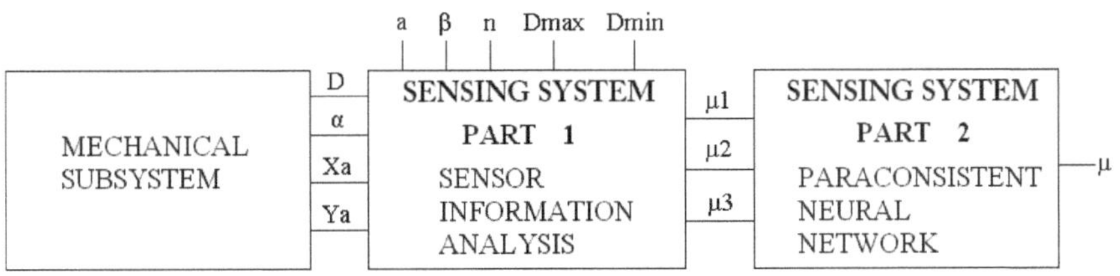

Fig. 14. Representation of the sensing system

The proposed sensing subsystem is prepared to receive data from ultrasonic sensors. The robot sensors are on the mechanical subsystem. So, this subsystem must treat the data generated by the sensors and send information to the first part of the sensing subsystem. The data the mechanical subsystem must send to the first part of the sensing subsystem are: D, α, Xa and Ya.

a. The distance between the sensor and the obstacle (D).
b. The angle between the horizontal axis of the environment and the direction to the front of the sensor (α). Figure 15 shows the angle α.

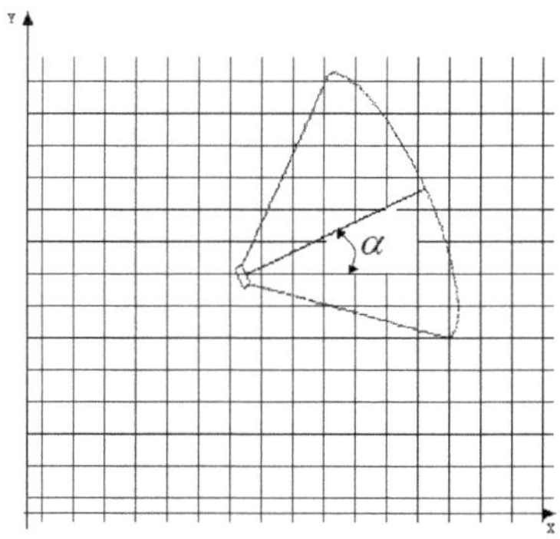

Fig. 15. Angle α

c. The coordinate occupied by the robot (Xa, Ya).

In the first part of the sensing subsystem there are also some configuration parameters, which are:

a. The distance between the environment coordinates (a); it is indicated in the figure 16.

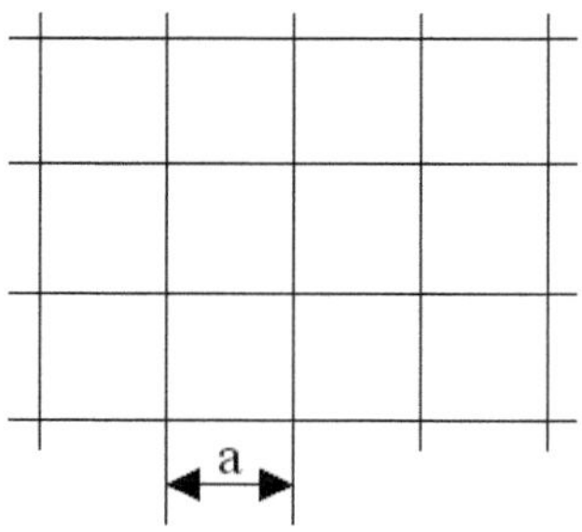

Fig. 16. Distance between coordinates

b. The angle of the ultrasonic sensor conical field of view (β). Figure 17 shows this.

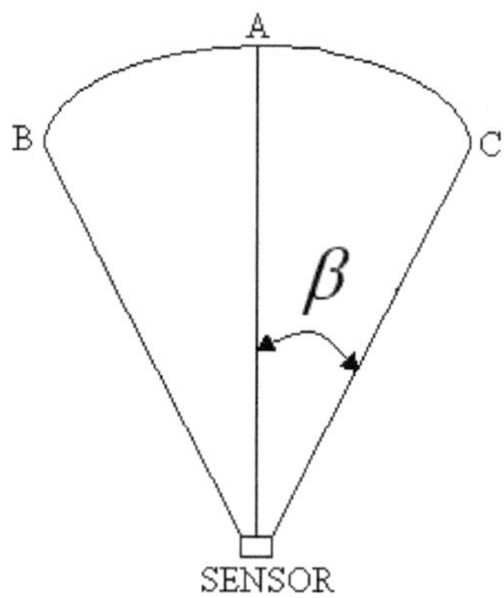

Fig. 17. Ultrasonic sensor conical field of view (β)

c. The number of positions on the arc BC, shown in the figure 17, considered by the system (n).
d. The maximum distance measured by the sensor; the system considers it (Dmax).
e. The minimum distance measured by the sensor; the system considers it (Dmin).

The first part of the sensing system generates three Favorable Evidence Degree, μ1, μ2 and μ3.

The Favorable Evidence Degree μ1 is related to the distance between the sensor and the obstacle. The nearer the obstacle is from the sensor, the bigger μ1 value is.

The Favorable Evidence Degree μ2 is related to the coordinate position on the arc BC shown in the figure 17. As the analyzed coordinate is near from the point A, the μ2 value must be the biggest. And as the analyzed coordinate is near from the points B or C, the μ2 value must be the smallest. The inspiration for this idea comes from [16] which says that the probability for the obstacle be near from the point A is high. And this probability decreases as we analyze the region near from the points B and C.

Eventually, the Favorable Evidence Degree μ3 is the previous value of the coordinate Favorable Evidence Degree.

6.1.1 Paraconsistent artificial neural network architecture

In the figure 18, it is shown Paraconsistent Artificial Neural Network - PANN architecture for the sensing subsystem.

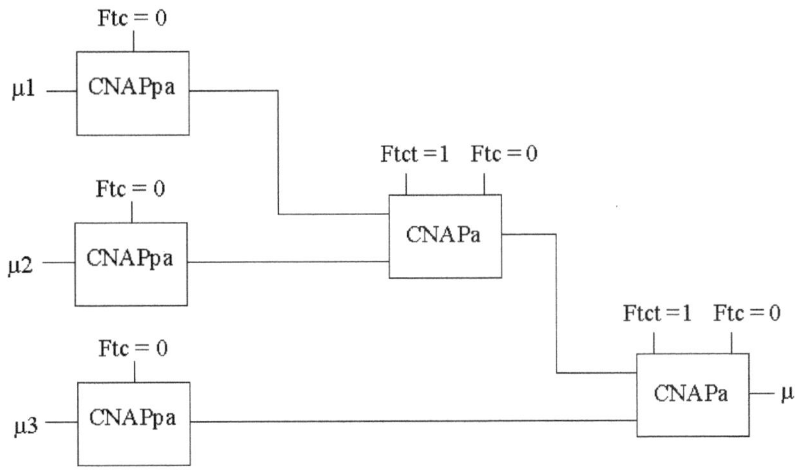

Fig. 18. Chosen Paraconsistent Neural Network Architecture for sensing system.

The PANN output μ is Favorable Evidence Degree for the analyzed position. There is a database which has recorded the μ for each analyzed position. The robot considers each position as a cell.

6.2 Results of the sensing subsystem

The sensing subsystem has been tested by simulating its inputs and analyzing the database generated. The database stores Favorable Evidence Degree in each environment position analyzed. It is shown here the result of three tests. The information from one ultrasonic sensor was considered as the Sensing System inputs.

6.2.1 First test

The configuration parameters of this test have been the following. The distance between the environment coordinates (a): 10. The angle of the ultrasonic sensor conical field of view (β): 30. The number of positions on the arc of the sensor conical field of view considered by the system (n): 10. The maximum distance measured by the sensor; the system considers it (Dmax): 800. The minimum distance measured by the sensor; the system considers it (Dmin): 8.
The mechanical subsystem treats the data from the sensors and generates the sensing subsystem inputs. It has been needed to simulate the sensing subsystem inputs because the mechanical subsystem has not been implemented yet.
Thus, the simulated sensing subsystem data have been the ones described in the follow. The distance between the sensor and the obstacle (D): 200. The angle between the horizontal axis of the environment and the direction to the front of the sensor (α): 30. The coordinate where the robot is (Xa, Ya): (0, 0).
It has been simulated the first measuring of the sensor, then, μ3 has been initially 0.
It is shown in the figure 19 the representation of the coordinates in which sensing system considered to have obstacles in. Summarizing, the figure 10 is a graphical representation of the database generated by sensing subsystem.

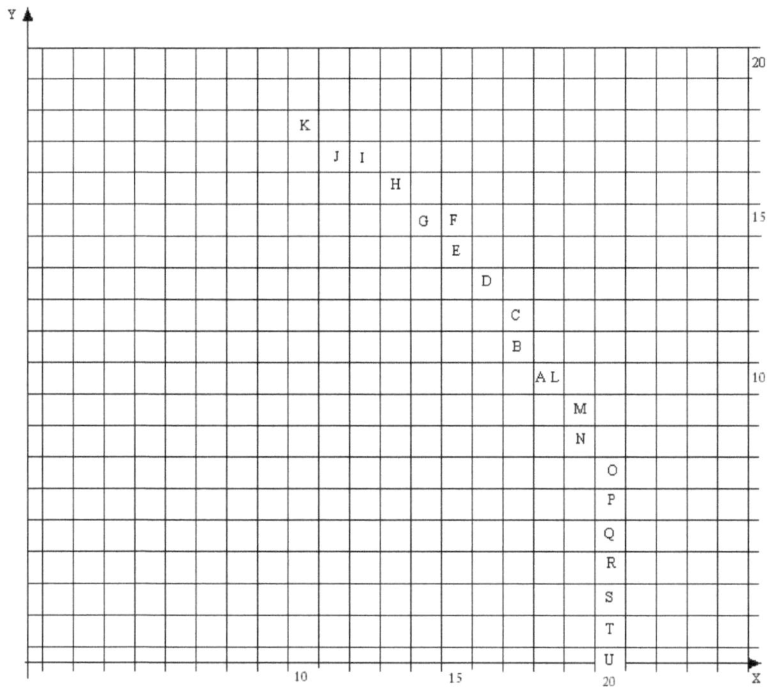

Fig. 19. The graphical representation of the database generated by the first test of sensing subsystem

Coordinate	μ
A (18,10)	0.438
B (17,11)	0.413
C (17,12)	0.388
D (16,13)	0.363
E (15,14)	0.338
F (15,15)	0.313
G (14,15)	0.288
H (13,16)	0.263
I (12,17)	0.238
J (11,17)	0.213
K (10,18)	0.188
L (18,10)	0.413
M (19,9)	0.388
N (19,8)	0.363
O (20,7)	0.338
P (20,6)	0.313
Q (20,5)	0.288
R (20,4)	0.263
S (20,3)	0.238
T (20,2)	0.213
U (20,0)	0.188

Table 1. Results of the first test.

The analyzed coordinates and their Favorable Evidence Degree are shown in the table 1.

6.2.2 Second test

The configuration parameters of this test have been the same as the ones from the first test. The simulated sensing subsystem data have been the ones described in the follow. The distance between the sensor and the obstacle (D): 400. The angle between the horizontal axis of the environment and the direction to the front of the sensor (α): 45. The coordinate where the robot is (Xa, Ya): (0, 0).

It has been simulated the first measuring of the sensor, then, μ3 was initially 0.

It is shown in the figure 20 the graphical representation of the database generated by the sensing subsystem.

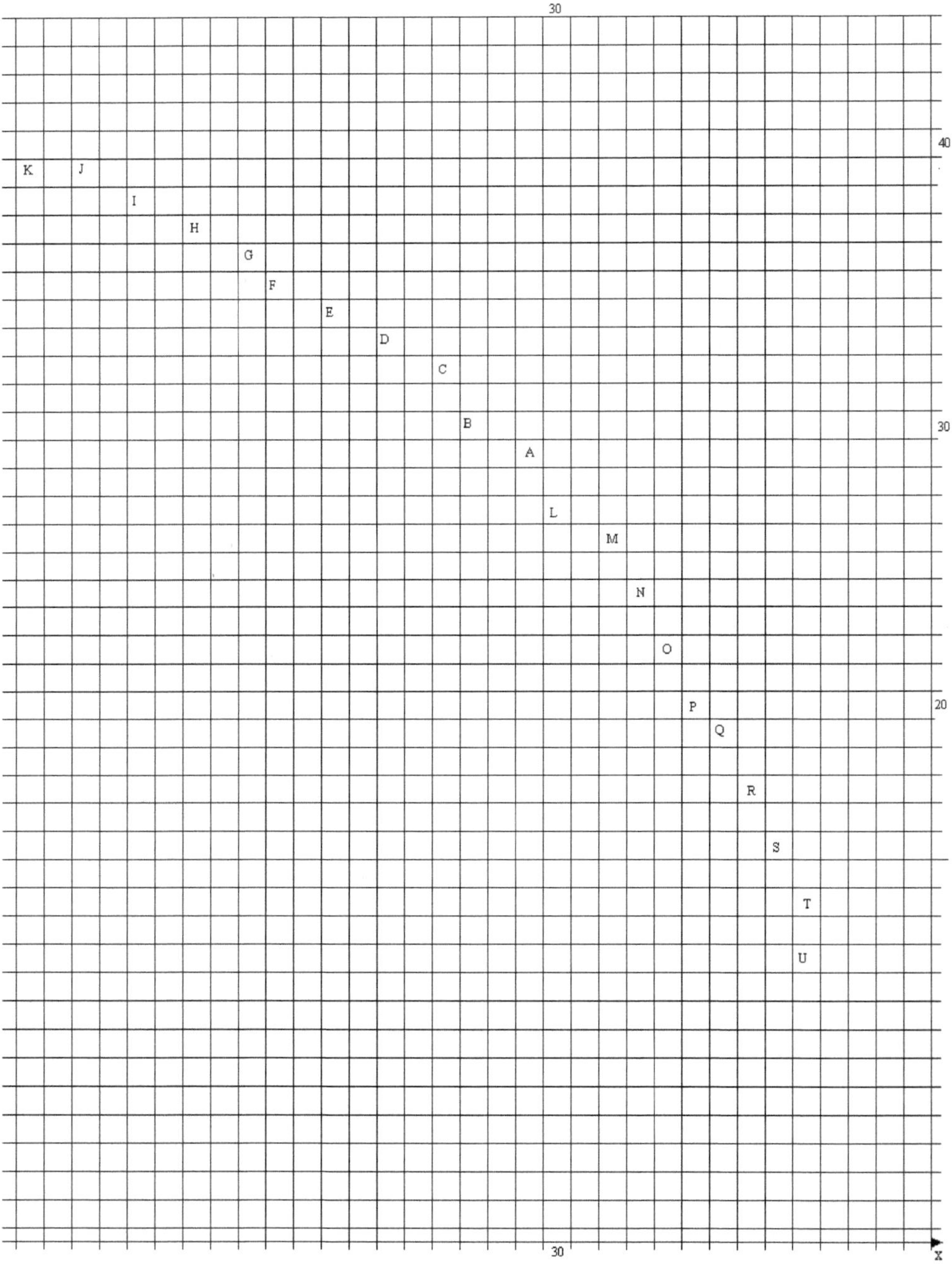

Fig. 20. The graphical representation of the database generated by the second test of the sensing subsystem

The analyzed coordinates and their Favorable Evidence Degree are shown in the table 2.

Coordinate	μ
A (29,29)	0.375
B (27,30)	0.35
C (26,32)	0.325
D (24,33)	0.3
E (22,34)	0.275
F (20,35)	0.25
G (19,36)	0.225
H (17,37)	0.2
I (15,38)	0.175
J (13,39)	0.15
K (11,39)	0.125
L (30,27)	0.35
M (32,26)	0.325
N (33,24)	0.3
O (34,22)	0.275
P (35,20)	0.25
Q (36,19)	0.225
R (37,17)	0.2
S (38,15)	0.175
T (39,13)	0.15
U (39,11)	0.125

Table 2. Results of the second test.

6.2.3 Third test

The configuration parameters and the sensing subsystem data have been the same ones of the second test; then the analyzed coordinates have been the same as the second test. The third test has been done just after the second, therefore, their Favorable Evidence Degree have been different from the one in the second test because $\mu 3$ has been the Favorable Evidence Degree generated by the second test.

The analyzed coordinates and their Favorable Evidence Degree are shown in the table 3.

If it is considered the sequence of positions from K to U as an arc in the three tests; it is perceived that the Favorable Evidence Degree (μ) decreases as the coordinate is farther from the center of the arc. It means that the system is working as desired.

6.3 Planning subsystem

The planning subsystem is responsible for generating the sequence of movements the robot must perform to achieve a set point. The sensing subsystem has the objective of informing the planning subsystem about the position of obstacles; and the mechanical subsystem is the robot itself, it means, the mobile mechanical platform which carries all devices away from the other subsystems. This platform must also perform the sequence of movements which are borne by the planning subsystem.

Coordinate	μ
A (29,29)	0.565
B (27,30)	0.525
C (26,32)	0.49
D (24,33)	0.45
E (22,34)	0.415
F (20,35)	0.375
G (19,36)	0.34
H (17,37)	0.3
I (15,38)	0.265
J (13,39)	0.225
K (11,39)	0.19
L (30,27)	0.525
M (32,26)	0.49
N (33,24)	0.45
O (34,22)	0.415
P (35,20)	0.375
Q (36,19)	0.34
R (37,17)	0.3
S (38,15)	0.265
T (39,13)	0.225
U (39,11)	0.19

Table 3. Results of the third test.

6.4 Mechanical subsystem

The Emmy III mechanical part must perform the schedule which is determined by the planning system. The mechanical subsystem must know the position where it is, therefore, a monitoring position makes part of this construction. In the process, for each cell that the robot reaches, any possible error of position should be considered. Some Emmy III prototypes are described here.

6.4.1 First prototype of the autonomous mobile robot Emmy III

The first prototype is composed of a planning subsystem and a mechanical construction. The planning system considers all cells free.

The planning subsystem asks for the initial point and the aimed point. After that, a sequence of movements is given on a screen. Also a sequence of pulses is sent to the step Motors which are responsible for moving the physical platform of the robot. So, the robot moves from the initial point to the aimed point.

The Figure 21 shows the planning system screen.

The physical construction of the first prototype of the Emmy III robot is basically composed of a circular platform of approximately 286 mm of diameter and two-step motors. The Figure 22 shows the Emmy III first prototype. The planning subsystem is recorded in a notebook. And the communication between the notebook and the physical construction is made through the parallel port. A potency driver is responsible for getting the pulses from the notebook and sending them to the step motors which are responsible for moving the robot.

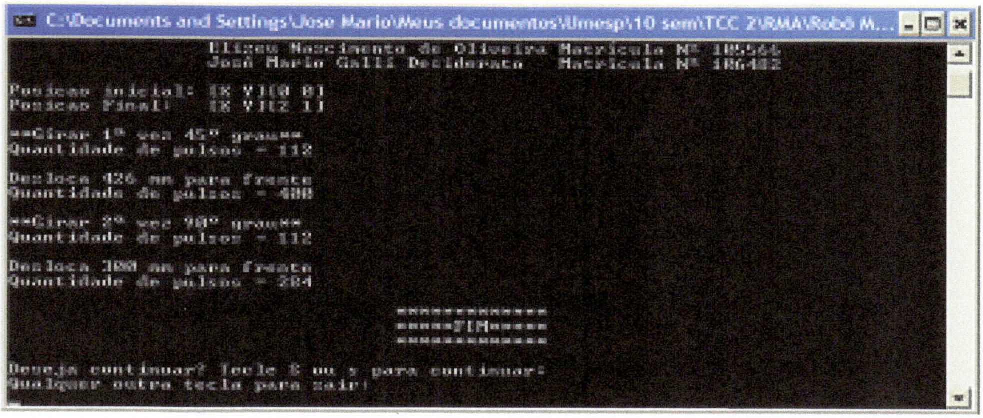

Fig. 21. Planning subsystem screen.

Fig. 22. The first prototype of the Emmy III robot.

6.4.2 Second prototype of the autonomous mobile robot Emmy III

Similarly to the first prototype, the second prototype of the autonomous mobile robot Emmy III is basically composed of a planning subsystem and a mechanical structure. The planning subsystem can be recorded in any personal computer and the communication between the personal computer and the mechanical construction is done through a USB port. The planning system considers the environment around the robot divided into cells. So, it is necessary to inform the planning system about the cell the robot is in, and the aimed cell too. The answer of the planning system is a sequence of cells which the robot must follow to go from the origin cell to the aimed cell.

The planning system considers all cells free. The Figure 23 shows the screen of the planning system.

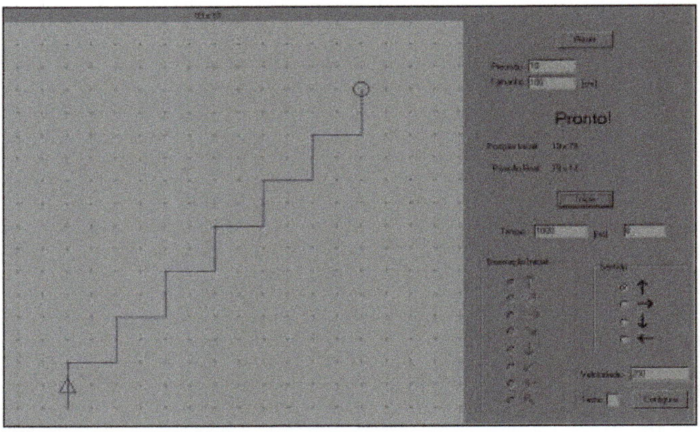

Fig. 23. The output of the planning system - Emmy III

Figure 24 shows the mechanical structure of Emmy III second prototype.

Fig. 24. The mechanical structure of the Emmy III second prototype

The planning system considers all cells free. The mechanical construction is basically composed of a steel structure, two DC motors and three wheels. Each motor has a wheel fixed in its axis and there is a free wheel. There is an electronic circuitry on the steel structure. The main device of the electronic circuitry is the microcontroller PIC18F4550 that is responsible for receiving the schedule from the planning system and activates the DC motors. Also there is a potency driver between the microcontroller and the DC motors.

7. Conclusions

In this work, it is discussed several autonomous mobile robots dubbed Emmy. They are based on a new kind of logic, namely the Paraconsistent Annotated Evidential Logic Eτ. A logical controller - Paracontrol served as basis for control system and in the 3rd prototype it was incorporated the use of Artificial Neural Network, also based on Logic Eτ.
This work presents a proposal of an autonomous mobile robot composed of three modules: sensing subsystem, planning subsystem and mechanical subsystem. The mechanical subsystem has not been implemented yet.

The aim of the sensing subsystem is to inform the planning subsystem the positions in which may have obstacles in. It considers the environment divided into coordinates.
The sensing subsystem is based on the Paraconsistent Artificial Neural Network - PANN. The sensing subsystem neural network is composed of two types of cells: Analytic Paraconsistent Artificial Neural Cell - CNAPa and Passage Paraconsistent Artificial Neural Cell - CNAPpa.
The output of the sensing subsystem is the Favorable Evidence Degree related to the sentence: there is obstacle in the position. In fact, the sensing subsystem generates a database with the Favorable Evidence Degree for each analyzed coordinate.
Some tests were made with the sensing subsystem. The reached results were satisfactory.
The next step is the implementation of the mechanical subsystem and the connection of the three subsystems.

8. References

[1] Torres, Cláudio Rodrigo; Abe, Jair Minoro; Lambert-Torres, Germano; Da Silva Filho, João Inácio & Martins, Helga Gonzaga., J. I. da Silva Filho, H. G. Martins . A sensing system for an autonomous mobile robot based on the paraconsistent artificial neural network. *Lecture Notes in Computer Science*. Berlin/Heidelberg: Springer-Verlag, 2010, v. 6278, p. 154-163.

[2] Torres, Cláudio Rodrigo "Sistema inteligente baseado na lógica paraconsistente anotada Eτ para controle e navegação de robôs móveis autônomos em um ambiente não estruturado", in Portuguese, Ph. D. Thesis, Federal University of Itajubá, Itajubá, MG, Brazil, 2010.

[3] Torres, Cláudio Rodrigo; ABE, J. M. ; Torres, Germano Lambert ; Silva Filho, João Inácio da ; Martins, Helga Gonzaga . Autonomous Mobile Robot Emmy III. In: Nakamatsu, K.; Phillips-Wren, G.; Jain, L.C.; Howlett, R.J.. (Org.). New Advances in Intelligent Decision Technologies. 1 ed. Helderberg: Springer-Verlag, 2009, v. 199, p. 317-327.

[4] Abe, Jair Minoro; Lambert-Torres, Germano; Da Silva Filho, João Inácio; Torres, Cláudio Rodrigo; Martins, Helga Gonzaga. Paraconsistent Autonomous Mobile Robot Emmy III, 6th Congress of Logic Applied to Technology - LAPTEC'2007. Santos, Proceedings of the VI Congress of Logic Applied to Technology. São Paulo - Brazil, 2007.

[5] Abe, Jair Minoro ; Torres, Cláudio Rodrigo ; Lambert-Torres, Germano ; Nakamatsu, K. ; Kondo, M . Intelligent Paraconsistent Logic Controller and Autonomous Mobile Robot Emmy II. Lecture Notes in Computer Science, v. 4252, p. 851-857, 2006.

[6] Abe, Jair Minoro; Torres, Cláudio Rodrigo ; Lambert-Torres, Germano ; Nakamatsu, K. ; Kondo, M . Intelligent Paraconsistent Logic Controller and Autonomous Mobile Robot Emmy II. In: 10th International Conference on Knowledge-Based, Intelligent Information & Engineering Systems, KES2006, 2006, Bournemouth. Proceedings of the 10th International Conference on Knowledge-Based, Intelligent Information & Engineering Systems. Bournemouth - UK : KES Pub., 2006.

[7] Torres, Cláudio Rodrigo ; Lambert-Torres, Germano ; Silva, Luiz Eduardo Borges da ; Abe, Jair Minoro . Intelligent System of Paraconsistent Logic to Control Autonomous Moving Robots. In: 32nd Annual Conference of the IEEE Industrial Electronics Society, IECON'06, 2006, Paris. Proceedings of the 32nd Annual Conference of the IEEE Industrial Electronics Society. Paris : IEEE Press, 2006.

[8] Da Silva Filho, João Inácio & Abe, Jair Minoro. Emmy: A Paraconsistent Autonomous Mobile Robot. In: Laptec' 2001 The 2 Congress of Logic Applied to Technology, 2001, São Paulo -Brazil. Logic, Artificial Inteligence and Robotics. Amsterdam : IOS - Press - Holanda, 2001. v. 1. p. 53-61.

[9] Da Silva Filho, João Inácio, "Métodos de Aplicações da Lógica Paraconsistente Anotada de Anotação com Dois Valores LPA2v com Construção de Algoritmo e Implementação de Circuitos Eletrônicos", in Portuguese, Ph. D. Thesis, University of São Paulo, São Paulo, 1999.

[10] Abe, Jair Minoro Some Aspects of Paraconsistent Systems and Applications. Logique et Analyse, v. 157, p. 83-96, 1997.

[11] Abe, Jair Minoro. A logical system for reasoning with inconsistency. In: 5a Reunião Anual da SBPN'97, 1997, Aguas de Lindoia. Anais da 5a Reunião Anual da SBPN'97Ciência e Cultura na Globalização - Novos Paradigmas. Aguas de Lindoia : SBPN, 1997. p. 196-201.

[12] Abe, Jair Minoro, "Fundamentos da lógica anotada" (Foundations of Annotated Logics), in Portuguese, Ph. D. Thesis, University of São Paulo, São Paulo, 1992.

[13] Da Silva Filho, João Inácio; Lambert-Torres, Germano & Abe, Jair Minoro. Uncertainty Treatment Using Paraconsistent Logic - Introducing Paraconsistent Artificial Neural Networks. 211. ed. Amsterdam: IOS Press, 2010. 328 pp. p.

[14] Da Silva Filho, João Inácio; Abe, Jair Minoro & Lambert-Torres, Germano. "Inteligência artificial com redes de análises paraconsistentes: teoria e aplicação", in Portuguese. Rio de Janeiro: LTC, 2008.

[15] Elfes, A. Using occupancy grids for mobile robot perception and navigation, Comp. Mag., vol. 22, No. 6, pp. 46-57, June 1989.

[16] Boreinstein, J. & Koren, Y. The vector field histogram: fast obstacle avoidance for mobile robots. IEEE Journal of Robotics and Automation. v. 7, p. 278-288, jun. de 1991.

[17] Abe, Jair Minoro & Da Silva Filho, João Inácio. Manipulating Conflicts and Uncertainties in Robotics, *Multiple-Valued Logic and Soft Computing*, V.9, ISSN 1542-3980, 147-169, 2003.

[18] Da Silva Filho, João Inácio & Abe, Jair Minoro. Para-Control: An Analyser Circuit Based On Algorithm For Treatment of Inconsistencies, Proc. of the World Multiconference on Systemics, Cybernetics and Informatics, ISAS, SCI 2001, Vol. XVI, Cybernetics and Informatics: Concepts and Applications (Part I), ISBN 9800775560, 199-203, Orlando, Florida, USA, 2001.

[19] Da Silva Filho, João Inácio & Abe, Jair Minoro., Paraconsistent analyzer module, *International Journal of Computing Anticipatory Systems*, vol. 9, ISSN 1373-5411, ISBN 2-9600262-1-7, 346-352, 2001.

[20] Desiderato, J. M. G. & De Oliveira, E. N., Primeiro Protótipo do Robô Móvel Autônomo Emmy III, in Portuguese, Trabalho de Conclusão de Curso, Universidade Metodista de São Paulo, São Bernardo do Campo - SP, Brazil, 2006.

[21] Maran, L. H. C., Riba, P. A., Collett, R. G. & De Souza, R. R., Mapeamento de um Ambiente Não-Estruturado para Orientação de um Robô Móvel Autônomo Utilizando Redes Neurais Paraconsistente, in Portuguese, Trabalho de Conclusão de Curso, Universidade Metodista de São Paulo, São Bernardo do Campo - SP, Brazil, 2006.

3

The KCLBOT: A Framework of the Nonholonomic Mobile Robot Platform Using Double Compass Self-Localisation

Evangelos Georgiou, Jian Dai and Michael Luck
King's College London
United Kingdom

1. Introduction

The key to effective autonomous mobile robot navigation is accurate self-localization. Without self-localization or with inaccurate self-localization, any non-holonomic autonomous mobile robot is blind in a navigation environment. The KCLBOT [1] is a non-holonomic two wheeled mobile robot that is built around the specifications for 'Micromouse Robot' and the 'RoboCup' competition. These specifications contribute to the mobile robot's form factor and size. The mobile robot holds a complex electronic system to support on-line path planning, self-localization, and even simultaneous localization and mapping (SLAM), which is made possible by an onboard sensor array. The mobile robot is loaded with eight Robot-Electronics SRF05 [2] ultrasonic rangers, and its drive system is supported by Nu-botics WC-132 [3] WheelCommander Motion Controller and two WW-01 [4] WheelWatcher Encoders. The motors for robot are modified continuous rotation servo motors, which are required for the WheelCommander Motion Controller. The rotation of the mobile robot is measured by Robot-Electronics CMPS03 [5] Compass Module. These two modules make the double compass configuration, which supports the self-localization theory presented in this paper. Each individual module provides the bearing of the mobile robot relative to the magnetic field of the earth. The central processing for the mobile robot is managed by a Savage Innovations OOPic-R microcontroller. The OOPic-R has advanced communication modules to enable data exchange between the sensors and motion controller. Communication is managed via a serial bus and an I^2C bus. The electronics are all mounted on two aluminium bases, which make the square structure of the mobile robot. To support the hardware requirements of the novel localisation methodology, the cutting edge technology of a 200 MHz 32-bit ARM 9 processor on a GHI ChipworkX module [6] is employed. The software architecture is based on the Microsoft .NET Micro Framework 4.1 using C# and the Windows Presentation Foundation (WPF).

The combination of hardware electronics and drive mechanics, makes the KCLBOT, as represented in Fig. 1, a suitable platform for autonomous self-localization.

Many different systems have been considered for self-location, from using visual odometry [7] to using a GPS method. While all of these have benefits and detriments, the solution proposed in this paper endeavours to offer significantly more benefits than detriments. Over the years, several solutions to self-localization have been presented. The most common application uses the vehicle's shaft encoders to estimate the distance travelled by the vehicle and deduce the position. Other applications use external reference entities to compute the vehicle's position, like global positioning systems (GPS) or marker beacons. All of these applications come with their respective weaknesses; for example, the shaft encoder assumes no slippage and is subject to accumulative drift; the GPS will not work indoors and is subject to a large margin of accuracy; and the beacons method is subject to the loss of the multi-path component delivery and accuracy is affected by shadowing, fast/slow fading, and the Doppler effect. More accurate applications have been presented using visual odometry but such applications require off-line processing or high computation time for real-time applications. Frederico et al. [8] present an interesting self-localization concept for the Bulldozer IV robot using shaft encoders, an analog compass, and optical position sensor from the PC mouse. This configuration of the vehicle is dependent on a flat surface for visual odometry to be effective; any deviation from the surface will cause inaccuracy. Hofmeister et al. [9] present a good idea for self-localization using visual odometry with a compass to cope with the low resolution visual images. While this is a very good approach to self-localization, the vehicle is dependent on a camera and the computation ability to process images quickly. Haverinen et al. [10] propose an excellent basis for self-localization utilizing ambient magnetic fields for indoor environments, whilst using the Monte Carlo Localization technique.

Fig. 1. The KCLBOT: A Nonholonomic Manoeuvrable Mobile Robot

It is not ideal to have multiple solutions to the position and orientation of the mobile robot and the computational requirement will affect the ability of the solution being available in real-time. Heavy computation also affects the battery life of the mobile robot and this is a critical aspect. In the technique utilizing two compasses, an analytical calculation model is presented over a numerical solution, which allows for minimal computation time over the numerical model. The configuration of the mobile robot employs two quadrature shaft encoders and two digital magnetic compasses to compute the vehicle's position and angular orientation on a two dimensional Cartesian plane. The accuracy of analytical model is benchmarked against visual odometry telemetry. However, this model still suffers from accumulative drift because of the utilization of quadrature shaft encoders. The digital compasses also encounter the same problem with the level of resolution being limited. The ideal solution will not have accumulation of drift error and will not be dependent on the previous configuration values of position and orientation. Such a solution is only available via visual odometry.

2. A manoeuvrable nonholonomic mobile robot

In this section, the experimental vehicle is evaluated by defining its constraints and modelling its kinematic and dynamic behaviour. The constraints are based on holonomic and nonholonomic behaviour of the rotating wheels, and the vehicles pose in a two dimensional Cartesian plane. The equations of motion for the mobile robot are deduced using Lagrangian d'Alembert's principle, with the implementation of Lagrangian multiplier for optimization. The behaviour of the dual motor configuration is deduced using a combination of Newton's law with Kirchhoff's law.

2.1 Configuration constraints and singularities

In the manoeuvrable classification of mobile robots [11], the vehicle is defined as being constrained to move in the vehicle's fixed heading angle. For the vehicle to change manoeuvre configuration, it needs to rotate about itself.

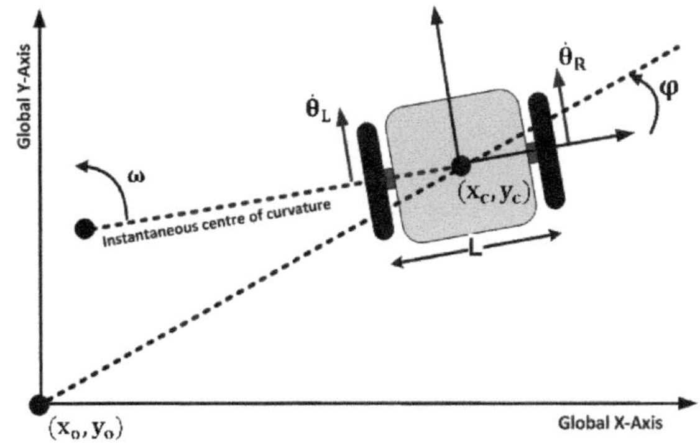

Fig. 2. A typical two wheel mobile robot constrained under the maneuverable classification.

As the vehicle traverses the two dimensional plane, both left and right wheels follow a path that moves around the instantaneous centre of curvature at the same angle, which can be defined as ω, and thus the angular velocity of the left and right wheel rotation can be deduced as follows:

$$\dot{\theta}_L = \omega(icc_r - \frac{L}{2}) \tag{1}$$

$$\dot{\theta}_R = \omega(icc_r + \frac{L}{2}) \tag{2}$$

where L is the distance between the centres of the two rotating wheels, and the parameter icc_r is the distance between the mid-point of the rotating wheels and the instantaneous centre of curvature. Using the velocities equations (1) and (2) of the rotating left and rights wheels, $\dot{\theta}_L$ and $\dot{\theta}_R$ respectively, the instantaneous centre of curvature, icc_r and the curvature angle, ω can be derived as follows:

$$icc_r = \frac{L(\dot{\theta}_R + \dot{\theta}_L)}{2(\dot{\theta}_R - \dot{\theta}_L)} \tag{3}$$

$$\omega = \frac{(\dot{\theta}_R - \dot{\theta}_L)}{L} \tag{4}$$

Using equations (3) and (4), two singularities can be identified. When $\dot{\theta}_R = \dot{\theta}_L$, the radius of instantaneous centre of curvature, $\mathrm{icc_r}$ tends towards infinity and this is the condition when the mobile robot is moving in a straight line. When $\dot{\theta}_R = -\dot{\theta}_L$, the mobile robot is rotating about its own centre and the radius of instantaneous centre of curvature, $\mathrm{icc_r}$, is null. When the wheels on the mobile robot rotate, the quadrature shaft encoder returns a counter tick value; the rotation direction of the rotating wheel is given by positive or negative value returned by the encoder. Using the numbers of tick counts returned, the distance travelled by the rotating left and right wheel can be deduced in the following way:

$$d_L = \frac{L_{ticks}\pi D}{L_{res}} \tag{5}$$

$$d_R = \frac{R_{ticks}\pi D}{R_{res}} \tag{6}$$

where $\mathrm{L_{ticks}}$ and $\mathrm{R_{ticks}}$ depicts the number of encoder pulses counted by left and right wheel encoders, respectively, since the last sampling, and D is defined as the diameter of the wheels. With resolution of the left and right shaft encoders $\mathrm{L_{res}}$ and $\mathrm{R_{res}}$, respectively, it is possible to determine the distance travelled by the left and right rotating wheel, $\mathrm{d_L}$ and $\mathrm{d_R}$. This calculation is shown in equations (5) and (6).

In the field of robotics, holonomicity [12] is demonstrated as the relationship between the controllable and total degrees of freedom of the mobile robot, as presented by the mobile robot configuration in Fig. 3. In this case, if the controllable degrees of freedom are equal to the total degrees of freedom, then the mobile robot is defined as holonomic. Otherwise, if the controllable degrees of freedom are less than the total degrees of freedom, it is nonholonomic. The manoeuvrable mobile robot has three degrees of freedom, which are its position in two axes and its orientation relative to a fixed heading angle. This individual holonomic constraint is based on the mobile robot's translation and rotation in the direction of the axis of symmetry and is represented as follows:

$$\dot{y}_c \cos(\phi) - \dot{x}_c \sin(\phi) - d\dot{\phi} = 0 \tag{7}$$

where, $\mathrm{x_c}$ and $\mathrm{y_c}$ are Cartesian-based coordinates of the mobile robot's centre of mass, which is defined as $\mathrm{P_c}$, and ϕ describes the heading angle of the mobile robot, which is referenced from the global x-axis. To conclude, Equation (7) presents the pose of the mobile robot. The mobile robot has two controllable degrees of freedom, which control the rotational velocity of the left and right wheel and, adversely - with changes in rotation - the heading angle of the mobile robot is affected; these constraints are stated as follows:

$$\dot{y}_c \sin(\phi) + \dot{x}_c \cos(\phi) + L\phi = r\dot{\theta}_r \tag{8}$$

$$\dot{y}_c \sin(\phi) + \dot{x}_c \cos(\phi) - L\phi = r\dot{\theta}_l \tag{9}$$

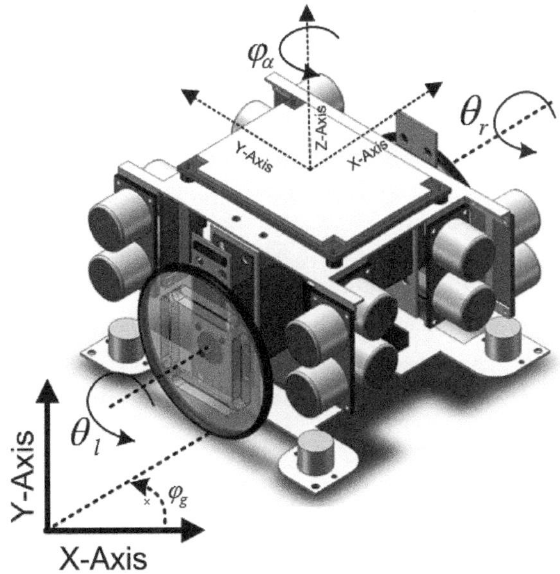

Fig. 3. A Manoeuvrable Nonholonomic mobile robot

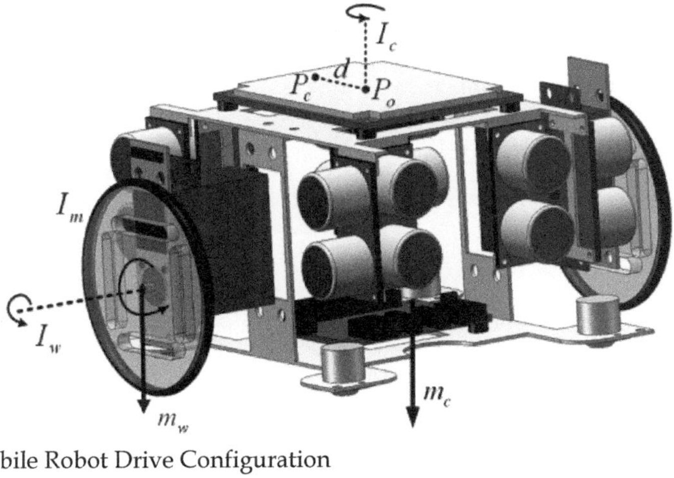

Fig. 4. The Mobile Robot Drive Configuration

where $\dot{\theta}_r$ and $\dot{\theta}_l$ are the angular displacements of the right and left mobile robot wheels, respectively, and where r describes the radius of the mobile robot's driving wheels. As such, the two-wheeled manoeuvrable mobile robot is a nonholonomic system. To conclude, Equation (8) and (9) describe the angular velocity of the mobile robot's left and right wheel.

Symbol	Description of Structured Constant
P_o	The intersection of the axis of symmetry with the mobile robot's driving wheel axis
P_c	The centre of the mass of the mobile robot
d	The distance between P_o and P_c
I_c	The moment of inertia of the mobile robot without the driving wheels and the rotating servo motors about a vertical axis through P_o
I_w	The moment of inertia of each of the wheels and rotating servo motors about the wheel's axis
I_m	The moment of inertia of each of the wheels and rotating servo motors about the diameter of the wheels
m_c	The mass of the mobile robot without the driving wheels and the rotating servo motors
m_w	The mass of each of the mobile robot's wheels and rotating motors

Table 1. The Mobile Robots Constants

Based on the mobile robot drive configuration, presented in Fig. 4., Table 1 describes the structured constants required to provide the physical characteristics of the mobile robots movement.

2.2 Kinematics a dynamics modeling

Using the diagrammatic model expressed by Figures 2 and 3, and the structured constants listed in Table 1, the nonholonomic equations of motion with Lagrangian multiplier are derived using Lagrange - d'Alembert's principle [13] and are specified as follows:

$$M(q)\ddot{q} + V(q,\dot{q}) + G(q) = E(q)u + B^T(q)\lambda_n \tag{10}$$

where $M(q)$ describes an $n \times n$ dimensional inertia matrix, and where $M(q)\ddot{q}$ is represented as follows:

$$\begin{bmatrix} (m_c + 2m_w) & 0 & -m_c d\ddot{\phi}\sin(\phi) & 0 & 0 \\ 0 & (m_c + 2m_w) & m_c d\ddot{\phi}\cos(\phi) & 0 & 0 \\ -m_c d\sin(\phi) & m_c d\cos(\phi) & I_z & 0 & 0 \\ 0 & 0 & 0 & I_w & 0 \\ 0 & 0 & 0 & 0 & I_w \end{bmatrix} \begin{bmatrix} \ddot{x}_c \\ \ddot{y}_c \\ \ddot{\phi}_c \\ \ddot{\theta}_r \\ \ddot{\theta}_l \end{bmatrix} \tag{11}$$

Here, $I_z = (I_c + 2m_w(d^2 + L^2) + 2I_m)$ and $V(q,\dot{q})$ describes an n dimensional velocity dependent force vector and is represented as follows:

$$\begin{bmatrix} -m_c d\dot{\phi}^2 \cos(\phi) \\ -m_c d\dot{\phi}^2 \sin(\phi) \\ 0 \\ 0 \\ 0 \end{bmatrix} \tag{12}$$

$G(q)$ describes the gravitational force vector, which is null and is not taken into consideration, u describes a vector of r dimensions of actuator force/torque, $E(q)$ describes an $n \times r$ dimensional matrix used to map the actuator space into a generalized coordinate space, and $E(q)u$ is specified as follows:

$$\begin{bmatrix} 0 & 0 \\ 0 & 0 \\ 0 & 0 \\ 1 & 0 \\ 0 & 1 \end{bmatrix} \begin{bmatrix} \tau_r \\ \tau_l \end{bmatrix} \tag{13}$$

Where the expression is in terms of coordinates $(q,\dot{q})$ where $q \in \mathbb{R}^n$ is the position vector and $\dot{q} \in \mathbb{R}^n$ is the velocity vector. Where q is defined as $[x_c, y_c, \phi, \theta_r, \theta_l]^T$, the constraints equation can be defined as $A(q)\dot{q} = 0$. Where, $A(q)$, the constraints matrix is expressed as follows:

$$A(q)\dot{q} = \begin{bmatrix} -\sin(\phi) & \cos(\phi) & -d & 0 & 0 \\ -\cos(\phi) & -\sin(\phi) & -b & r & 0 \\ -\cos(\phi) & -\sin(\phi) & b & 0 & r \end{bmatrix} \begin{bmatrix} \dot{x}_c \\ \dot{y}_c \\ \dot{\phi} \\ \dot{\theta}_r \\ \dot{\theta}_l \end{bmatrix} \tag{14}$$

Where finally, $B^T(q) = A^T(q)$ and λ_n describes an m dimensional vector of Lagrangian multipliers and can be described as follows:

$$\begin{bmatrix} -\sin(\phi) & -\cos(\phi) & -\cos(\phi) \\ \cos(\phi) & -\sin(\phi) & -\sin(\phi) \\ -d & -L & L \\ 0 & r & 0 \\ 0 & 0 & r \end{bmatrix} \begin{bmatrix} \lambda_1 \\ \lambda_2 \\ \lambda_3 \end{bmatrix} \tag{15}$$

The purpose of using the Lagrangian multipliers is to optimize the behaviour of the nonholonomic manoeuvrable mobile robot, by providing a strategy for finding the maximum or minimum of the equations' behaviour, subject to the defined constraints.

Equation (10) describes the Lagrangian representation of the KCLBOT, in a state-space model, and Equations (11), (12), (13), (14), and (15) decompose the state-space model.

2.3 The dual-drive configuration

The manoeuvrable mobile robot is configured with two independent direct current (DC) servo motors, set up to be parallel to each other, with the ability for continuous rotation. This configuration allows the mobile robot to behave in a manoeuvrable configuration, is illustrated in Figure 5.

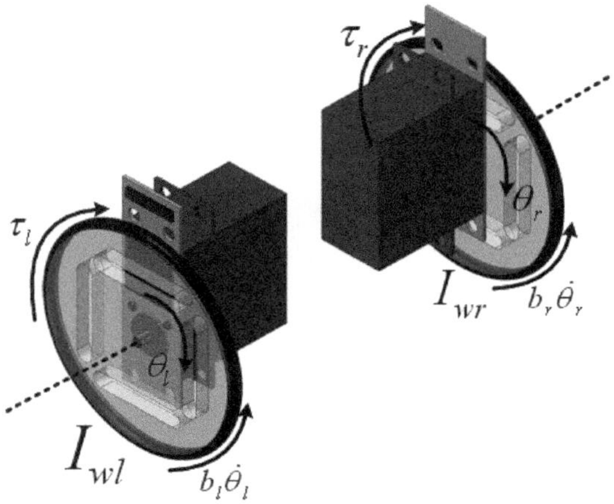

Fig. 5. Dual Servo Motor Drive Configuration

It is assumed that both the left and right servo motors are identical. The torque, $\tau_{l,r}$, of the motor is related to the armature current, $i_{l,r}$, by the constant factor K_t and is described as $\tau_{l,r} = K_t.i_{l,r}$. The input voltage source is described as $V_{l,r}$, which is used to drive the servo motors, where $R_{l,r}$ is internal resistance of the motor, $L_{l,r}$ is the internal inductance of the motor, and $e_{l,r}$ describes back electromagnetic field (EMF) of both the left and right electric servo motors. It is known that $e_{l,r} = K.\dot{\theta}_{l,r}$, where $K = K_e = K_t$ describe the electromotive force constants.

Symbol	Description
$I_{wl,r}$	Moment of inertia of the rotor
$b_{l,r}$	Damping ratio of the mechanical system
K	Electromotive force constant
$R_{l,r}$	Electric resistance
$L_{l,r}$	Electric inductance
$V_{l,r}$	Input voltage source

Table 2. Dual Servo Motor Configuration Value Definitions

The value definitions listed in Table 2 complement Fig. 5's representation of the mobile robot drive configuration.

Using Newton's laws of motion and Kirchhoff's circuit laws [14], the motion of the motors can be related to the electrical behaviour of the circuit.

$$I_{wl,r}\ddot{\theta}_{l,r} + b_{l,r}\dot{\theta}_{l,r} = Ki_{l,r} \tag{16}$$

$$L_{l,r}\frac{di_{l,r}}{dt} + R_{l,r}i_{l,r} = V_{l,r} - K\dot{\theta}_{l,r} \tag{17}$$

Equation (16) specifies the Newtonian derivation of motion of both motors and equation (17) specifies how the circuit behaves applying Kirchhoff's laws. Having derived equations (16) and (17), the next step is to relate the electrical circuit behaviour to the mechanical behaviour of rotating motors, and this is achieved using Laplace transforms and expressing equations (16) and (17) in terms of s as follows:

$$\frac{\dot{\theta}}{V_{l,r}} = \frac{K}{(I_{wl,r}s + b_{l,r})(L_{l,r}s + R_{l,r}) + K^2} \tag{18}$$

$$(L_{l,r}s + R_{l,r})I_{l,r}(s) = V_{l,r} - Ks\theta_{l,r}(s) \tag{19}$$

Using equations (18) and (19), the open-loop transfer function of this configuration can be derived by eliminating $\mathrm{I}_{l,r}(s)$ and relating the equation of motion to the circuit behaviour as follows:

$$\frac{\dot{\theta}}{V_{l,r}} = \frac{K}{(I_{wl,r}s + b_{l,r})(L_{l,r}s + R_{l,r}) + K^2} \tag{20}$$

Here, equation (20) equates the rotational speed of the motors as the systems output and the voltage applied to the motors as the systems input.

In summary, the constraints of the experimental vehicle have been derived in equations (7), (8), and (9), in both holonomic and nonholonomic forms. Using the system constraints, the behaviour of the nonholonomic manoeuvrable mobile robot is formed using the Lagrange-d'Alembert principle with Lagrangian multipliers to optimize the performance of the system, as specified in equations (10), (11), (12), (13), (14), and (15). Finally, using Newton's laws of motion and Kirchhoff's laws to evaluate circuits, a model is derived using Laplace transforms, to relate the behaviour of the systems motion to the electrical circuit.

3. Manoeuvrable mobile robot self-localization

The self-localization offers a different approach to solving for an object's position. The reason for the perpetuation of so many approaches is that no solution has yet offered an absolutely precise solution to position. In the introduction, three major approaches using the shaft encoder telemetry, using the visual odometry, and using the global position system were discussed, of which all three have inherent problems. This paper, hence, proposes an alternative approach, which is a hybrid model using the vehicle's dual shaft encoders and double compass configuration.

3.1 Implementing a dual-shaft encoder configuration

By using the quadrature shaft encoders that accumulate the distance travelled by the wheels, a form of position can be deduced by deriving the mobile robot's x, y Cartesian position and the manoeuvrable vehicle's orientation ϕ, with respect to time. The derivation starts by defining and considering $s(t)$ and $\varphi(t)$ to be function of time, which represents the velocity and orientation of the mobile robot, respectively. The velocity and orientation are derived from differentiating the position form as follows:

$$\frac{dx}{dt} = s(t).\cos(\phi(t)) \tag{21}$$

$$\frac{dy}{dt} = s(t).\sin(\phi(t)) \tag{22}$$

The change in orientation with respect to time is the angular velocity ω, which was defined in equation (4) and can be specified as follows:

$$\frac{d\phi}{dt} = \omega = \frac{\dot{\theta}_r - \dot{\theta}_l}{b} \tag{23}$$

When equation (23) is integrated, the mobile robot's angle orientation value $\phi(t)$ with respect to time is achieved. The mobile robot's initial angle of orientation $\phi(0)$ is written as ϕ_0 and is represented as follows:

$$\phi(t) = \frac{(\dot{\theta}_r - \dot{\theta}_l)t}{b} + \phi_0 \tag{24}$$

The velocity of the mobile robot is equal to the average speed of the two wheels and this can be incorporated into equations (21) and (22), as follows:

$$\frac{dx}{dt} = \frac{\dot{\theta}_r + \dot{\theta}_l}{2}\cos(\phi(t)) \tag{25}$$

$$\frac{dy}{dt} = \frac{\dot{\theta}_r + \dot{\theta}_l}{2}.\sin(\phi(t)) \tag{26}$$

The next step is to integrate equations (25) and (26) to the initial position of the mobile robot, as follows:

$$x(t) = x_0 + \frac{L(\dot{\theta}_r + \dot{\theta}_l)}{2(\dot{\theta}_r - \dot{\theta}_l)}\left(\sin\left(\frac{(\dot{\theta}_r - \dot{\theta}_l)t}{b} + \phi_0\right) - \sin(\phi_0)\right) \tag{27}$$

$$y(t) = y_0 + \frac{L(\dot{\theta}_r + \dot{\theta}_l)}{2(\dot{\theta}_r - \dot{\theta}_l)}\left(\cos\left(\frac{(\dot{\theta}_r - \dot{\theta}_l)t}{b} + \phi_0\right) - \cos(\phi_0)\right) \tag{28}$$

Equations (27) and (28) specify the mobile robot's position, where $x(0) = x_0$ and $y(0) = y_0$ are the mobile robot's initial positions. The next step is to represent equations (24), (27) and

(28) in terms of the distances that the left and right wheels have traversed, which are defined by d_R and d_L. This can be achieved by substituting $\dot{\theta}_r$ and $\dot{\theta}_l$ (in equations (24), (27) and (28)) for d_R and d_L, respectively, and also dropping the time constant t to achieve the following:

$$\theta = \frac{d_R - d_L}{2} + \theta_0 \tag{29}$$

$$x(t) = x_0 + \frac{L(d_R + d_L)}{2(d_R - d_L)}\left(\sin\left(\frac{(d_R - d_L)t}{b} + \phi_0\right) - \sin(\phi_0)\right) \tag{30}$$

$$y(t) = y_0 + \frac{L(d_R + d_L)}{2(d_R - d_L)}\left(\cos\left(\frac{(d_R - d_L)t}{b} + \phi_0\right) - \cos(\phi_0)\right) \tag{31}$$

By implementing equations (29), (30), and (31), we provide a solution to the relative position of a manoeuvrable mobile robot. This might offer a possible solution to the self-localization problem but is subject to accumulative drift of the position and orientation with no method of re-alignment. The accuracy of this method is subject to the sampling rate of the data accumulation, such that if small position or orientation changes are not recorded, then the position and orientation will be erroneous.

3.2 Numerical approach with a single compass configuration

Having derived a self-localization model using only the telemetry from the quadrature shaft encoders, the next step to evolve the model is to add a digital compass to input the manoeuvrable mobile robot's orientation. The ideal position on the vehicle for the digital compass would be at the midpoint between its centre of its mass and the intersection of the axis of symmetry with the mobile robot's driving wheel axis. In this case, the vehicle is configured such that there is no deviation between the two points, P_o and P_c.

When the manoeuvrable mobile robot starts a forward or backward rotation configuration, it induces two independent instantaneous centres of curvatures from the left and right wheel. The maximum arcs of the curvature lines are depicted in Fig. 9, showing the steady state changes to the configuration as it rotates.

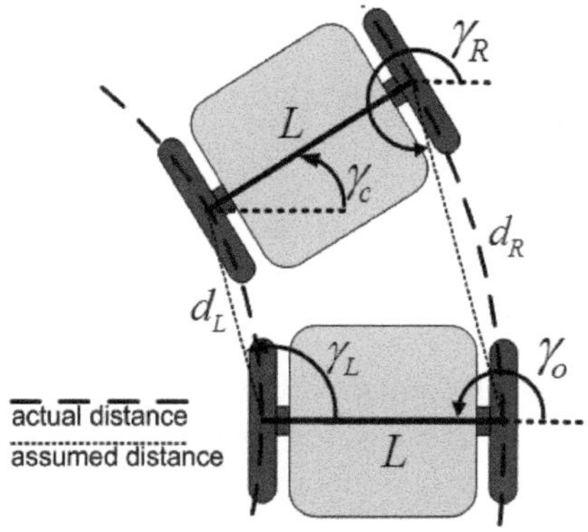

Fig. 6. Mobile Robot Manoeuvre Configuration

Using a steady state manoeuvre, it is assumed that the actual distance travelled in a rotation manoeuvre does not equal the distance used to model the calculation for position and orientation. This assumption is depicted in Fig. 6., clearly showing the actual and assumed distance travelled. The difference does not cause any consequence to the calculation model as the difference cannot be measured in the resolution of the quadrature shaft encoders.

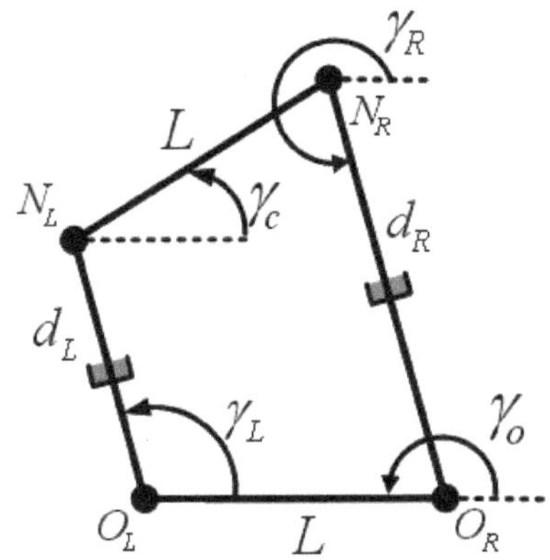

Fig. 7. Six-bar Linkage Manoeuvre Model with a Single Compass Configuration

Using the vector loop technique [15] to analyse the kinematic position of the linkage model in Fig. 7., the vector loop equation is written as follows:

$$\boldsymbol{I}_{N_L O_L} + \boldsymbol{I}_{N_R N_L} + \boldsymbol{I}_{O_R N_R} + \boldsymbol{I}_{O_L O_R} = 0 \tag{32}$$

Using the complex notation, equation (32) is written as follows:

$$\boldsymbol{d}_L.\boldsymbol{e}^{j.\gamma_L} + \mathbf{L}.\boldsymbol{e}^{j.\gamma_c} + \boldsymbol{d}_r.\boldsymbol{e}^{j.\gamma_R} + \mathbf{L}.\boldsymbol{e}^{j.\gamma_o} = 0 \tag{33}$$

Having derived the complex notation of the vector loop equation in equation (33), the next step is to substitute the Euler equations to the complex notations as follows:

$$\begin{aligned} &\boldsymbol{d}_L(\cos(\gamma_L) + j\sin(\gamma_L)) + \boldsymbol{L}(\cos(\gamma_c) + j\sin(\gamma_c)) + \\ &\boldsymbol{d}_r(\cos(\gamma_R) + j\sin(\gamma_R)) + \boldsymbol{L}(\cos(\gamma_o) + j\sin(\gamma_o)) = 0 \end{aligned} \tag{34}$$

Equation (34) is separated into its corresponding real and imaginary parts, considering $\gamma_o = 180°$, as follows:

$$\boldsymbol{d}_L \cos(\gamma_L) + \boldsymbol{L}\cos(\gamma_c) + \boldsymbol{d}_r \cos(\gamma_R) - \boldsymbol{L} = 0 \tag{35}$$

$$\boldsymbol{d}_L \sin(\gamma_L) + \boldsymbol{L}\sin(\gamma_c) + \boldsymbol{d}_r \sin(\gamma_R) = 0 \tag{36}$$

where, d_L, d_R, and L are all known constants, and γ_c specifies the new angle of orientation of the mobile robot. Having two equations (35) and (36), with two unknown angles γ_L and γ_R, when solving simultaneously for these two angles, it will return multiple independent values. This approach requires a large amount of computation to first find these angles and then deduce the relative position.

3.3 Double compass configuration methodology

Having described the issues associated with using a single compass to solve for the position above, it is clear that it would be preferable to have a model that eliminated having simultaneous solutions and led to a single solution. This would ideally mean that the angles γ_L and γ_R are known constants, and to achieve this condition requires additional telemetry from the vehicle. A model with dual compasses is proposed to resolve the angles γ_L and γ_R, as shown in Fig. 8.

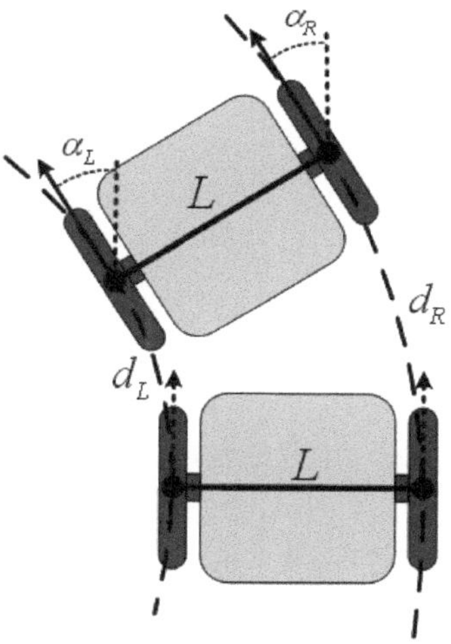

Fig. 8. Double Compass Manoeuvre Configuration

By introducing two compasses, which are placed directly above the rotating wheels, when a configuration change takes place, the difference is measured by α_L and α_R, which represent the change in orientation of the left and right mobile robot wheels, respectively. Using the same approach as a single compass configuration, the double compass manoeuvre configuration is modelled using a six-bar mechanism as shown in Fig. 9.

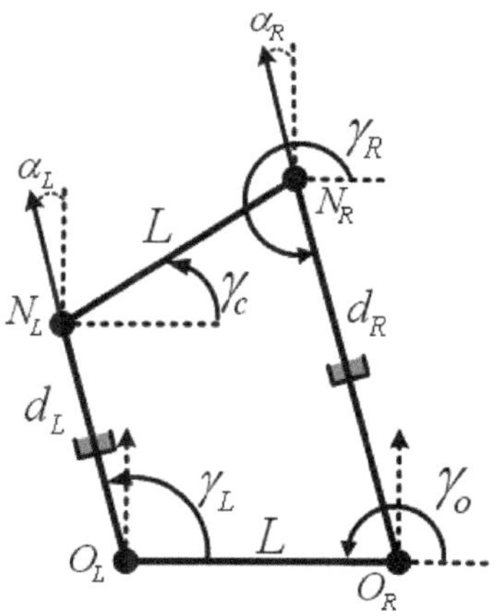

Fig. 9. Six-bar Linkage Manoeuvre Model with a Double Compass Configuration

Using an identical approach as the single compass configuration, the vector loop equations remain the same, equations (35) and (36), with the difference being the ability to define the angles γ_L and γ_R. For this manoeuvre model, presented in Fig. 9., the configuration can be calculated as follows:

$$\gamma_L = \beta_L + \alpha_L \tag{37}$$

$$\gamma_R = \beta_R + \alpha_R \tag{38}$$

where, β_L and β_R are the trigonometric angles used for calculating γ_L and γ_R for each pose of the mobile robot, based on the different configuration states of d_L and d_R. For the configuration state described by Fig. 13, $\beta_L = 90°$ and $\beta_R = 270°$. Having a constant value for the angles γ_L and γ_R from Equations (37) and (38), it allows either equation (35) or (36) to be used to derive the remaining angle γ_c, which is specified as follows:

$$\gamma_c = \cos^{-1}\left(\frac{-d_L\cos(\gamma_L) - d_r\cos(\gamma_R) + L}{L}\right) \tag{39}$$

$$\gamma_c = \sin^{-1}\left(\frac{-d_L\sin(\gamma_L) - d_r\sin(\gamma_R)}{L}\right) \tag{40}$$

where equations (39) and (40) solve the single compass solution to the position of the centrally positioned digital compass, indicated by γ_c.

Using the simultaneous solutions from comparing equations (35) and (36), and evaluating their difference to either equation (39) or (40), it is possible to derive the accuracy of the measurement. The simultaneous equation solution will be the best fit model for the hybrid model and any ambiguity can be considered resolution error or a factor of wheel slippage.

4. Mobile robot localization using visual odometry

4.1 Implementing visual odometry with an overhead camera

Using any standard camera at a resolution of 320 by 240 pixels, the system uses this device to capture images that are then used to compute the position and orientation of the manoeuvring mobile robot.

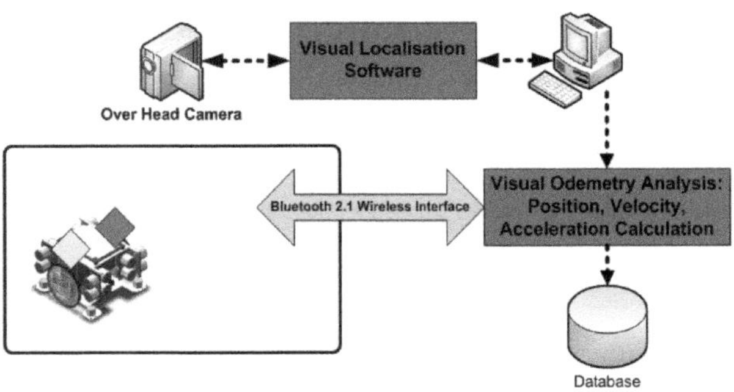

Fig. 10. Localization Configuration Setup

The camera is positioned above the manoeuvrable mobile robot at a fixed height. The camera is connected to an ordinary desktop computer that processes the images captured from the camera. The captured images are then processed to find the two markers on the mobile robot. The markers are positioned directly above the mobile robot's rolling wheels. The processing software then scans the current image identifying the marker positions. Having two markers allows the software to deduce the orientation of the manoeuvring mobile robot. When the software has deduced the position and orientation of the mobile robot, this telemetry is communicated to the mobile robot via a wireless Bluetooth signal.

4.2 A skip-list inspired searching algorithm

By using a systematic searching algorithm, the computation cost-time is dependent on the location of the markers. The closer the markers are to the origin (e.g. coordinates (0 pixels, 0 pixels) of the search, the faster the search for the markers will be performed. The systematic search algorithm is represented as follows:

```
For x-pixel = 0 To Image width
   For y-pixel  = 0 To Image height
          If current pixel (x-pixel, y-pixel) = Marker 1 Then
                   Marker 1 x position = x-pixel
                   Marker 1 y position = y-pixel
                   Flag = Flag + 1
          End if
          If current pixel (x-pixel, y-pixel) = Marker 2 Then
                   Marker 2 x position = x-pixel
                   Marker 2 y position = y-pixel
                   Flag = Flag + 1
          End if
          If Flag > 1 Then
                   x-pixel = Image width
                   y-pixel = Image height
          End if
   Next
Next
```

The search algorithm above shows how every vertical and horizontal pixel is scanned to identify the two defined markers on the mobile robot. When a marker is found, a flag is raised, and when the search has two raised flags, the search will end.

5. Results & analysis

5.1 Statistical results

To validate the double compass approach to self-localisation, an overhead camera visual odometry tracking system was set up, as presented in the previous section. The visual tracking results are used to compare the double compass tracking results.

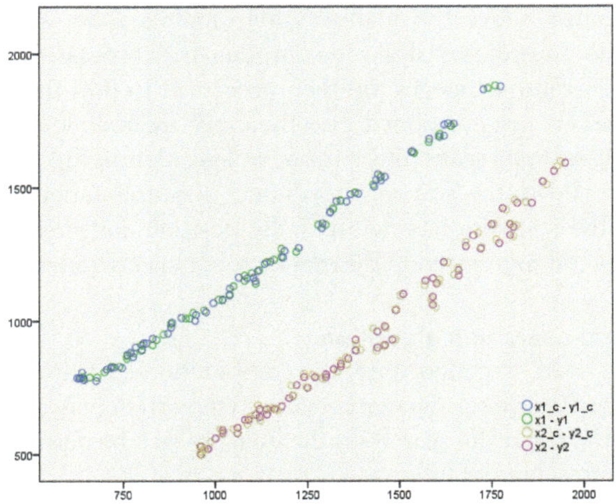

Fig. 11. Experimental data from linear path following

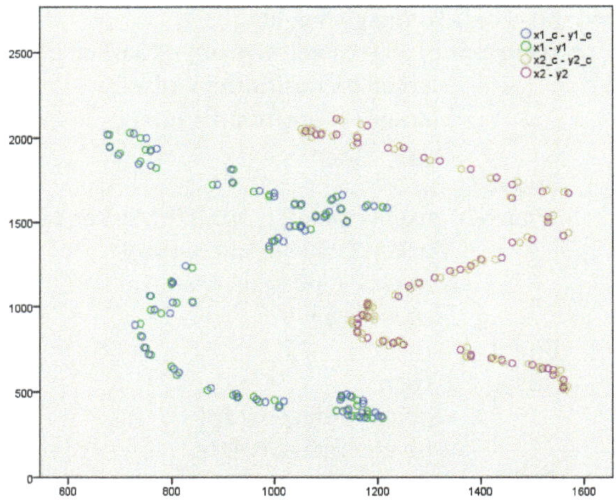

Fig. 12. Experimental data from sinusoidal path following

Using the configuration presented in Fig. 10. the mobile robot with a double compass configuration was set up to follow a linear path and a sinusoidal path. The experimental results showing the visual tracking results and the double compass estimation results are presented in Fig. 11. and Fig. 12.

The results presented in Table 3 show the statistical analyses of 362 samples recorded from the linear and sinusoidal manoeuvre experiments. Both left and right marker mean values are relatively low and, for more accuracy, because the data might be skewed, the median value is presented because it compensates for skewed data. The confidence intervals, which represent two standard deviations from the mean, equivalently present a low error rate.

Method	Right Marker	Left Marker
Mean	12.298302	11.254462
95% Confidence Interval for Mean (LB)	11.605941	10.616642
95% Confidence Interval for Mean (UB)	12.990663	11.892282
Median	12.806248	11.401754
Variance	22.284	18.911
Standard Deviation	4.7205636	4.3486990
Standard Error Mean	0.3508767	0.3232362
Minimum	1.0000	1.0000
Maximum	20.6155	19.3132
Range	19.6155	18.3132
Interquartile Range	7.0312	6.2556
Skewness	-0.263	-0.118

Table 3. Statistical analysis of experimental data

The histograms of the error distribution, shown in Fig. 13, present the distribution of error, which is the difference between the visual tracking position and double compass position.
The Q-Q plot depicted in Fig. 14. presents the performance of the observed values against the expected values. This demonstrates that the errors are approximately normally distributed centrally, with anomalies at both tail ends.
The boxplot presented in Fig. 15. visually shows the distance from 0 to the mean, which is 12.3mm and 11.3mm, for the right and left respectively. It also presents the interquartile range, which is 19.6mm and 18.3mm, for the right and left respectively, and minimum (1mm) and the maximum (20.6mm and 19.3mm) values.

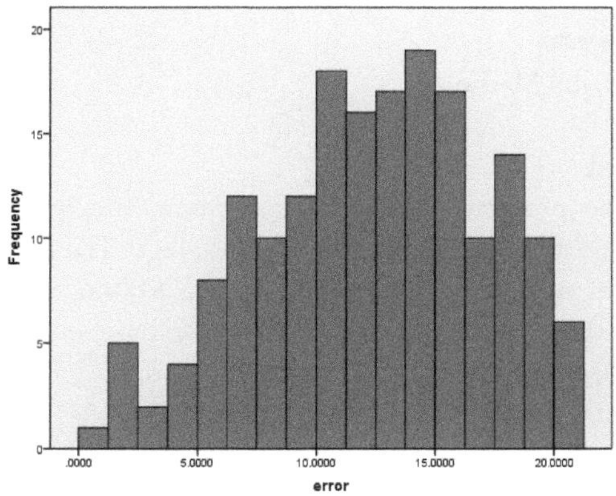

(a) Right Marker

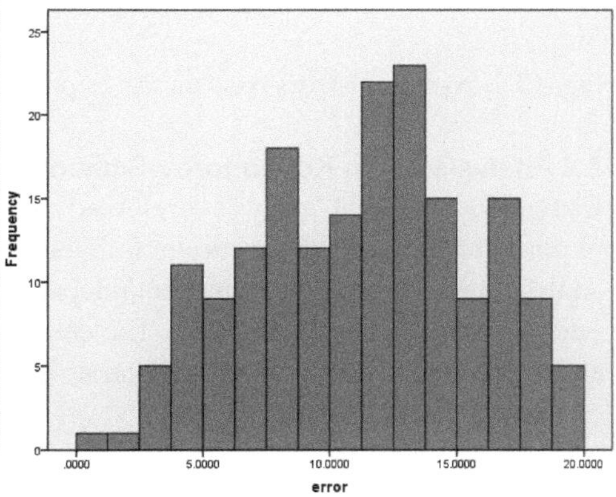

(b) Left Marker

Fig. 13. Histogram of the left and right marker error

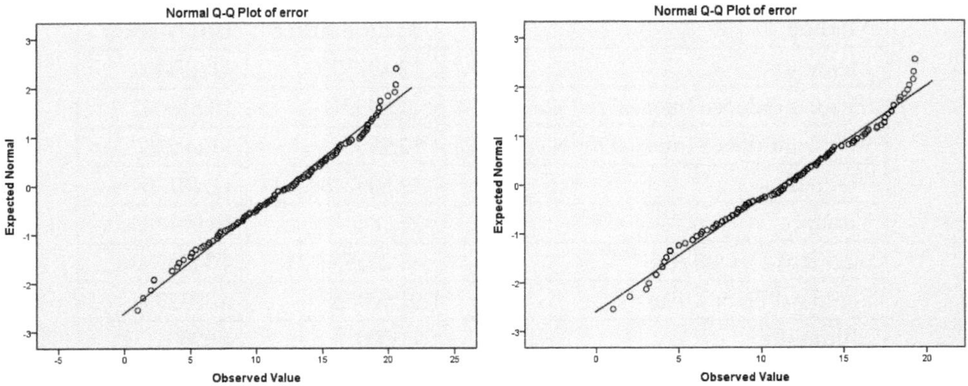

(a) Right Marker (b) Left Marker

Fig. 14. Normal Q-Q Plot of Error for the Right and Left Markers

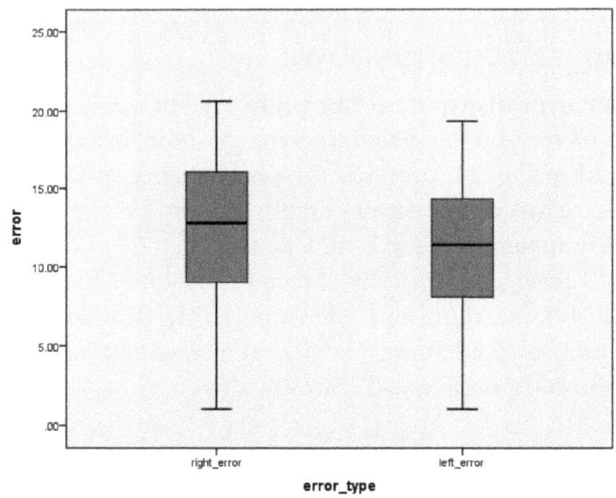

Fig. 15. Boxplot of Errors base on the Experimental Methods

5.2 Analysis using Kolmogorov-Smimov test

Using the statistical analysis presented in the previous section, a non-parametric test is required to validate the practical effectiveness of the double compass methodology. The ideal analysis test for a non-parametric independent one-sample set of data is the Kolmogorov-Smirnov test [16] for significance. The empirical distribution function F_n for n independent and identically distributed random variables observations X_i is defined as follows:

$$F_n(x) = \frac{1}{n}\sum_{i=1}^{n} I_{X_i \le x} \tag{41}$$

where $I_{X_i \le x}$ describes the indicator function, which is equal to 1 if $X_i \le x$ and is equal to 0 otherwise. The Kolmogrov-Smirnov statistic [16] for a cumulative distribution function $F(x)$ is as follows:

$$D_n = \sup_x | F_n(x) - F(x) | \tag{42}$$

where $\sup_x$ describes the supremum of the set of distances. For the analysis test to be effective to reject a null hypothesis, a relatively large number of data is required.
Under the null hypothesis that the sample originates from the hypothesized distribution $F(x)$, the distribution is specified as follows:

$$\sqrt{n}D_n \xrightarrow{n \to \infty} \sup_t | B(F(t)) | \tag{43}$$

where $B(t)$ describes the Brownian bridge [17]. If F is continuous, then under the null hypothesis $\sqrt{n}D_n$ converges to the Kolmogorov distribution, which does not depend on F. The analysis test is constructed by using the critical values of the Kolmogorov distribution. The null hypothesis is rejected at level α if $\sqrt{n}D_n > K_a$, where K_a is calculated from the following:

$$\Pr(K \le K_a) = 1 - \alpha \tag{44}$$

It should also be noted that the asymptotic power of this analysis test is 1.
For the experimental data presented in this paper, the null hypothesis is that the distribution of error is normal with a mean value of 11.78 and a standard deviation of 4.56. Based on significance level of 0.05, a significance of 0.663 is returned using the one-sample Kolmogorov-Smirnov test (44). The strength of the returned significance value allows us to retain the null hypothesis and say that the distribution of error is normal with a mean value of 11.78 and a standard deviation of 0.663.

6. Conclusions

The most fundamental part of implementing a successful maneuverable non-holonomic mobile robot is accurate self-localization telemetry. The biggest concern about using a self-localization technique is the amount of computation it will require to complete the task. Ideally, having an analytical solution to position offers the benefit of a single solution, where the numeric solution has many possibilities and requires more time and computation to derive.
In this paper, three different solutions have been presented for the position of the mobile robot. The first solution presented a method where only the quadrature shaft encoder telemetry is used to solve position. However, using an accumulative method leads to the drifting of the position and orientation result, caused by slippage of the wheels or even by the limiting resolution of the shaft encoders. The second solution presented the implementation of a hybrid model using both the quadrature shaft encoders and a single, centrally placed digital compass. Modeling the maneuver configuration of the mobile robot as a six-bar linkage mechanism and using vector loop equation method to analyze the mechanism for a numeric solution was derived. The final solution presented a model that implements a dual digital compass configuration to allow an analytical solution to the purposed six-bar linkage mechanism. An analytical model was derived to resolve the numeric nature of the single compass configuration, which will allow the swift resolution of position and orientation.

The benefit of the dual compass configuration is that it offers an analytical solution to the hybrid model that utilizes quadrature shaft encoders. This paper, nevertheless, presents a novel approach where visual odometry is not possible.

7. References

[1] Georgiou, E. *The KCLBOT Mobile Robot*. 2010; Available from: www.kclbot.com.

[2] *SRF05 Ultra-Sonic Ranger*. Robot Electronics; Available from: http://www.robot-electronics.co.uk/htm/srf05tech.htm

[3] *WC-132 WheelCommander Motion Controller*. Nu-Botics; Available from: http://www.nubotics.com/products/wc132/index.html

[4] *WW-01 WheelWatcher Encoder*. Nu-Botics; Available from: http://www.nubotics.com/products/ww01/index.html

[5] *CMPS03 Compass Module*. Robot Electronics; Available from: http://www.robot-electronics.co.uk/htm/cmps3tech.htm

[6] *GHI Chipworkx Module*. Available from: http://www.ghielectronics.com/catalog/product/123.

[7] Nister, D., Naroditsky, O., & Bergen, J, *Visual Odometry*, in *IEEE Computer Society Conference on Computer Vision and Pattern Recognition (CVPR)*. 2004.

[8] Frederico, C., Adelardo, M., & Pablo, A. , *Dynamic Stabilization of a Two-Wheeled Differentially Driven Nonholonomic Mobile Robot*, in *Simpósio Brasileiro de Automação Inteligente*. 2003. p. 620-624.

[9] Hofmeister, M., Liebsch, M., & Zell, A. , *Visual Self-Localization for Small Mobile Robots with Weighted Gradient Orientation Histograms.* , in *40th International Symposium on Robotics (ISR)*. 2009. p. 87-91.

[10] Haverinen, J., & Kemppainen, A. , *Global indoor self-localization based on the ambient magnetic field.* Robotics and Autonomous Systems, 2009: p. 1028-1035.

[11] Georgiou, E., Chhaniyara, S., Al-milli, S., Dai, J., & Althoefer, K. , *Experimental Study on Track-Terrain Interaction Dynamics in an Integrated Environment*, in *Proceedings of the Eleventh International Conference on Climbing and Walking Robots and the Support Technologies for Mobile Machines (CLAWAR)*. . 2008: Coimbra, Portugal.

[12] Kolmanovsky, I.a.M., N. , *Developments in nonholonomic control problems.* IEEE Control Systems Magazine, 1995: p. 20-36.

[13] Lew, A., Marsden, J., Ortiz, M. and West, M, *An Overview of Variational Integrators. In: Finite Element Methods: 1970's and Beyond. Theory and engineering applications of computational methods*, in *International Center for Numerical Methods in Engineering (CIMNE)*. 2004. p. 1-18.

[14] Alexander, J.a.M., J. , *On the kinematics of wheeled mobile robots.* The International Journal of Robotics Research, 1989. 8: p. 15-27.

[15] Acharyya, S.a.M., M. , *Performance of EAs for four-bar linkage synthesis.* Mechanism and Machine Theory, 2009. 44(9): p. 1784-1794.

[16] Zhang, G., Wang, X., Liang, Y., and Li, J. „ *Fast and Robust Spectrum Sensing via Kolmogorov-Smirnov Test.* IEEE TRANSACTIONS ON COMMUNICATIONS, 2010. 58(12): p. 3410-3416.

[17] Hu, L., Zhu, H.„ *Bounded Brownian bridge model for UWB indoor multipath channel.* IEEE International Symposium on Microwave, Antenna, Propagation and EMC Technologies for Wireless Communications„ 2005: p. 1411-1414.

4

Gaining Control Knowledge Through an Applied Mobile Robotics Course

Lluís Pacheco, Ningsu Luo, Inès Ferrer, Xavier Cufí and Roger Arbusé
University of Girona
Spain

1. Introduction

In this chapter the use of an open mobile robot platform as an innovative educational tool in order to promote and integrate control science knowledge is presented. The idea of including applied interdisciplinary concepts is an important objective in engineering education. Future work in Electrical and Computer Engineering education is pointing towards gaining the ability to understand the technical details of a wide variety of disciplines (Antsaklis et al., 1999, Murray et al., 2003). Moreover, experimental developments have helped to breathe life into theoretical concepts found in text books and have thereby greatly changed the educational experience of students (Hristu-Varsakelis and Brockett, 2002). Students react positively to realism, since the models used for such experiments are in general accurately described by some relatively simple differential equations. Within this framework, the challenge of using mobile robots becomes evident. Thus, realistic platforms that incorporate engineering standards and realistic constraints increase student skills and experience through engineering practices.

In this sense, mobile robot platforms can be used as educational tools to promote and integrate different curriculum subjects. Control system world and computer science share a mutual interest in robotics. Therefore, in the educational community, the Robotics Education Lab becomes a resource for supporting courses with an academic curriculum in a broad range of subjects. Some important institutions have developed various teaching activities by introducing the mobile robots as a necessary academic tool in a broad sense in order to extend the student's previously acquired knowledge (Carnegie Mellon University, 2011, Qu and Wu, 2006). In this context, many universities take advantage of mobile robot competitions in engineering education. This allows real world problem projects to be tackled, and fundamental concepts by increasing motivation and retention purposes to be reinforced. Thus, for example, FIRST (For Inspiration and Recognition of Science and Technology) mobile robot contest attracts young people to careers in engineering, technology and science. Robotics competition encourages students to apply knowledge gained throughout their engineering degree, it also offers all students a chance to serve as members of interdisciplinary engineering teams, and introduces both freshmen and sophomores to engineering concepts. Moreover, university curriculum is reinforced by the knowledge gained throughout applied experiences that embrace a wide spectrum of subjects (Wilczynski and Flowers, 2006). The educational and research objectives can also be achieved through the use of configurable, small, low-cost such as LEGO mobile robot kits (Valera, 2007).

In this Chapter, the program and acquired experience of an optional course named "Applied Mobile Robotics" is outlined. The main aim of the this educational course is to integrate different subjects such as electronics, programming, architecture, perception systems, communications, control and trajectory planning by using the educational open mobile robot platform PRIM (Pacheco et al., 2006). The course is addressed to the students of Computer Engineering around the time of their final academic year. As a practical approach, the majority of educational activities of the course are developed in our university labs. Section 1 introduces the teaching framework from a global point of view. Section 2 indicates which community of students is suitable for this teaching activity within the context of our university framework. The robotic platform used and the course program are presented from a wide scope within the multidisciplinary context. Section 3 presents the program related to the control educational content of the course in detail. Attention is paid particularly to describing the different experiments designed in order to fulfill the process of gaining knowledge. Section 4 briefly introduces the student opinions and future work.

2. Educational structure

Spanish educational framework arises from the present European convergence program (European Commission, 2008). In this way, the main challenge of Spain in 2010 was to design and adapt the university degrees studies. The objective was to improve the quality of Education processes by implementing new learning methodologies (Amante et al., 2007). In this work, knowledge integration is proposed by using an open mobile robot. The course is addressed to Computer Engineering students that can take this course to consolidate their curriculum studies. It is an optional subject within the degree of Computer Engineering at the Polytechnic School of University of Girona. However, the course *"Applied Mobile Robotics"* started as special subject in the summer of 2005 (Pacheco et al., 2009). It was proposed by a group of faculty members belonging to the teaching areas of Control and Computer Engineering and sharing a common background that involves the research on mobile robots. Hence, its purpose consists to consolidate, to integrate and to complement different teaching aspects such as electronics, programming, control and perception systems from a practical point of view by using an available robotic platform.
This section outlines the Computer Engineering curriculum and the course schedule. Especially attention is paid to relationships between different curriculum subjects and the *"Applied Mobile Robotics"* course. The mobile robot platform used is also described.

2.1 Course schedule and student curriculum

Students, who want to pursue the course, have previously acquired the basic skills in some fundamental areas such as electronics, programming, control and perception systems. In this way, an open mobile robot is used for linking multiple curriculum subjects from a practical point of view. Significant subjects of the actual university degree curriculum are depicted in Fig. 1.
The subjects of Fig. 1, which have blue background, have relationship with the *"Applied Mobile Robotics"* course. The main related subjects cover the following educational parts:

- *Programming* covers aspects such as: programming methodology and technology (first year); algorithms, data structure and programming projects (second year); user interfaces, multimedia, fundamentals of computing, programming languages and artificial intelligence (third year).

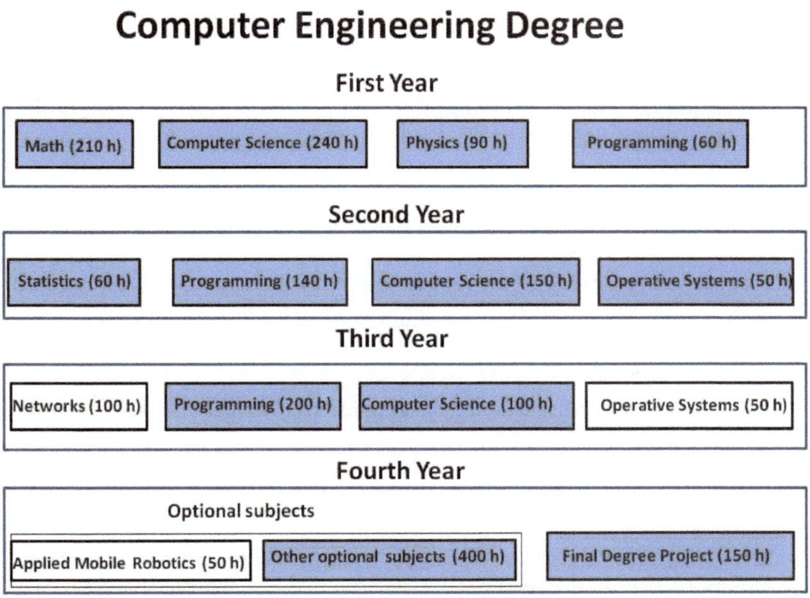

Fig. 1. Computer Engineering degree curriculum.

- *Computer Science* includes issues such as: computer structure and technology (first year); digital systems and microcomputers (second year); architecture of computers, industrial computer science and robotics (third year).

It is noted that topics such as computer vision and control, which are also highly related with the proposed course, are included within the optional subjects. Furthermore, some students are attracted to perform their *"final degree project"* by continuing and developing issues related with some practical aspects of the mobile robotics course. The students who pass the course successfully are awarded 5 credits on their degree which corresponds to 50 hours of teaching activities. It consists of theoretical and laboratory sessions, related to the following different topics: PSE (power system electronics), DPED (design & programming of electronic devices), MCS (modeling and control systems), TPC (trajectory planning and control) and CVS (computer vision systems). The theoretical parts are presented during 11 classroom hours, which are scheduled as follows:

1. Course and platform presentation. Introduction of power drivers (PSE).
2. Hardware description. Digital systems and PLD (programmable logic devices) (Wakerly, 1992). Theoretical introduction of DPED practices.
3. Hardware description. Digital systems and microprocessors (Barnett, 1994). Theoretical introduction of DPED practices.
4. Introducing the MCS block. System identification (Ljung, 1989). Discrete and continuous systems (Kuo, 1996).
5. Developing the MCS block. System reduction, observers and filters (Aström and Wittenmark, 1988).
6. Introducing the TPC block. Configuration space and trajectory following (Schilling, 1990). Fuzzy rules.
7. TPC & MCS integration. Trajectory following. Design of PID controllers.

8. TPC & MCS integration. Trajectory following. Introducing MPC (model predictive control) (Maciejowski, 2002).
9. Introducing the CVS block (Horn, 1998).
10. Practicing within CVS block, color and filters.
11. Final classroom, overall demos.

Laboratory parts are developed during 39 hours that are scheduled as follows:

1. DPED practices with PLD: 4 bits magnitude comparison, PWM generator, design of counters (encoders speed and position).
2. DPED practices with microcomputers: Serial port, I/O ports, A/D converters, PWM signal.
3. MCS practices with the robot PRIM and using MATLAB: excitation of the input system by using PRBS (pseudo random binary signals), the system model using the toolbox IDENT of MATLAB. Discrete and continuous models.
4. MCS practices using SIMULINK environment: reducing the system model and PID speed control.
5. MCS practices using the SISO tools of MATLAB: reduced SISO systems and PID speed control design. Testing of the controllers within SIMULINK environment.
6. MCS practices using the PRIM platform: on-robot testing of the obtained set of speed controllers.
7. DPED practices using PC and the robotic platform: introduction of the high level languages. Communication with the robot by using sockets that include the on-robot available functions (actuators and sensors).
8. TPC practices using the robotic platform: heuristic navigation using sonar sensors.
9. TPC practices using the robotic platform: the odometer system, trajectory following using PID and discontinuous control laws.
10. Time gap for finishing the previously proposed TPC practices.
11. MCS and TPC practices: introduction of MPC software simulation tools, trajectory following with different prediction horizons and trajectories.
12. CVS practices: introducing the software tools, binary images and segmentation.
13. CVS practices: color, filters and tracking of an object.

2.2 The educational open mobile platform

The robot structure, shown in Fig. 2.a, is made from aluminium. It consists of 4 levels where the different parts are placed. On the first level there are two differential driven wheels, controlled by two DC motors, and a third omni-directional wheel that gives a third contact point with the floor. On the second level there is an embedded PC computer, and on the third level specific hardware and the sonar sensors are placed. On the fourth level the machine vision system is placed. The system can be powered by 12V dc batteries or by an external power source through a 220V ac, and a switch selects either mode of operation. The 12V provided by the batteries or the external source of 220V are transformed into the wide range of dc voltages needed by the system. Moreover, the use of a 220V power supply allows an unlimited use of the platform during the teaching activities. The robot has the following sensorial system:

- Two encoders connected to the rotation axis of each dc motor.
- An array of sonars composed by 8 ultrasounds sensors.
- A machine vision system consisting of a monocular camera.

The meaningful hardware consists of the following electronic boards:

- The dc motor power drivers based on a MOSFET bridge that controls the energy supplied to the actuators.
- A set of PCB (printed circuits boards) based on PLD act as an interface between the embedded PC system, the encoders, and the dc motors. The interface between the PLD boards and the PC is carried out by the parallel port.
- A μc processor board controls the sonar sensors. Communication between this board and the embedded PC is made through a serial port. This board is also in charge of a radio control module that enables the tele-operation of the robot. The embedded PC is the core of the basic system, and it is where the high level decisions are made.

The PLD boards generate 23 khz PWM (pulse width modulation) signals for each motor and the consequent timing protection during the command changes. This protection system provides a delay during power connection, and at the moment of change of the rotation motor direction. A hardware ramp is also implemented in order to facilitate a better transition between command changes. The value of the speed is encoded in a byte, it can be generated from 0 to 127 advancing or reversing speed commands that are sent to the PLD boards through the parallel port. The PLD boards also measure the pulses provided by the encoders, during an adjustable period of time, giving the PC the speed of each wheel at every 25ms. The absolute position of each encoder is also measured by two absolute counters used in order to measure the position and orientation of the robot by the odometer system. The shaft encoders provide 500 counts/rev since encoders are placed at the motor axes; it means that the encoders provide 43,000 counts for each turn of the wheel due to the gear reduction. Moreover, the μc has control of the sonar sensors, so for each sensor a distance measure is obtained. The ultrasound sensor range is comprised of between 3cm and 5m. The data provided by these boards are gathered through the serial port in the central computer based on a VIA C3 EGBA 733/800 MHz CPU running under LINUX Debian OS. The rate of communication with these boards is 9600 b/s. Fig. 2.b shows the electronic and sensorial system blocks.

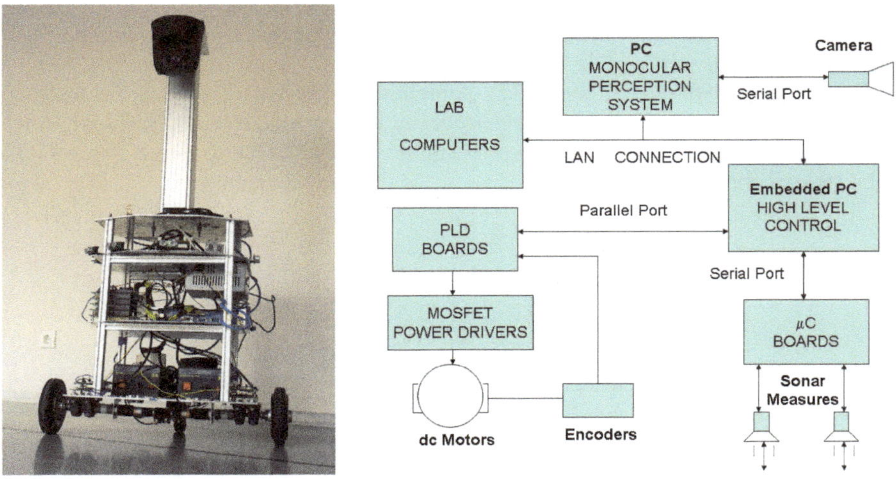

Fig. 2. (a) The open mobile educational platform PRIM.
(b) Sensorial and electronic system blocks.

The flexibility of the system allows different hardware configurations as a function of the desired application and consequently the ability to run different programs on the μc or PLD boards. The open platform philosophy is reinforced by the use of the similar μc and PLD boards that are used as teaching tools at our school. Furthermore, reinforcement of the teaching activities can be achieved through the integration of different subjects. The system's flexibility is increased with the possibility of connecting it to other computer systems through a local LAN. Hence, in this research a network server has been implemented, which allows the different lab groups to make their remote connection with the mobile robot during different practices. The use of a PC as a high level device is shown from two different points of view, consisting of PC high level programming and the TCP/IP protocol net knowledge that allow on-robot LINUX created functions to be run by the use of the laboratory LAN. Thus, lab personal computers become a flexible and easy tool that allows communication with the on-robot hardware real-time devices through the use of WIFI connections.

3. Gaining control knowledge

In this section it is presented the MCS and TPC learning activities that introduce the basic control theory. The control knowledge is mainly acquired through the following lab practices:

- Experimental identification. The model of the system.
- Designing and testing speed PID controllers.
- Trajectory following, controlling the robot position.
- Trajectory following by using MPC techniques as other suitable advanced control method.

It is noted that the compulsory curriculum of Computer Engineering has a lack of control science theory, while obviously students show experience in their programming skills. In this context, the control science is introduced from a general point of view. Moreover, students have a self-guide lab practices manual that includes the important theoretical aspects.

3.1 Experimental model and system identification

The model is obtained through the approach of a set of lineal transfer functions that include the nonlinearities of the whole system. The parametric identification process is based on black box models (Norton, 1986; Ljung, 1989). The nonholonomic system dealt with in this work is considered initially to be a MIMO (multiple input multiple output) system, as shown in Fig. 3, due to the dynamic influence between two DC motors. This MIMO system is composed of a set of SISO (single input single output) subsystems with coupled connection.

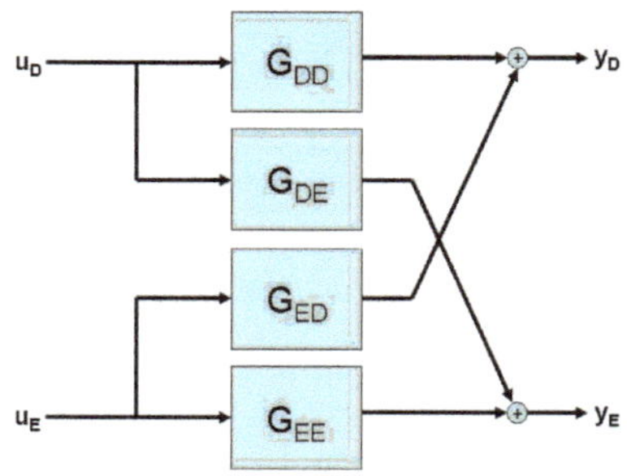

Fig. 3. The MIMO system structure

First lab control practical begins with experimental parametric model identification. Then, the model is obtained by sending different PRBS to the robot for the low, medium and high speeds, see Fig.4. Students select the adequate speed model and load the experimental data onto the MATLAB environment. Each data group has 5 different arrays containing time, right motor reference velocity, right motor measured velocity, left motor reference velocity, left motor measured velocity, respectively. Then, they obtain the time response of the inputs and outputs. The system identification is carried out by using the MATLAB toolbox *"ident"*, shown in Fig. 5, selecting a second-order ARX model and importing the necessary workspace data that should be identified.

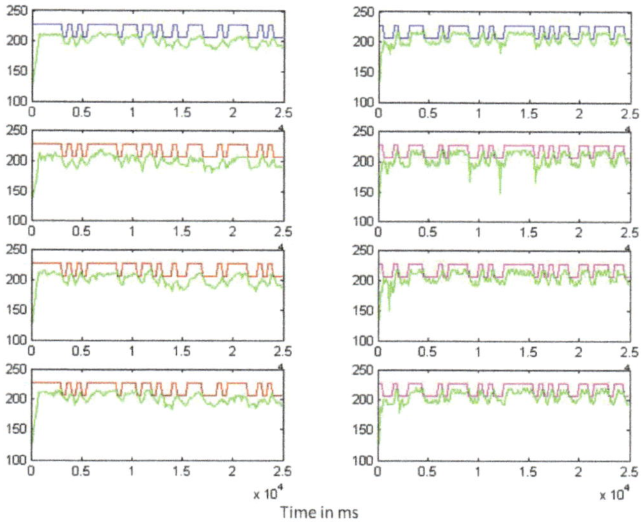

Fig. 4. Speed system outputs corresponding to the PRBS input signals

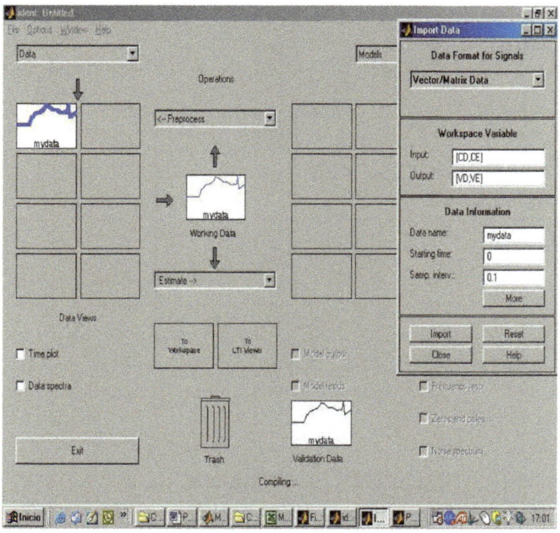

Fig. 5. System identification toolbox of MATLAB.

Tendency suppression and data filtering are suggested. The complete MIMO (multiple input multiple output) discrete-time model obtained will be used by students for validating the effectiveness of control. Table 1 shows a possible set of continuous transfer functions obtained for the three different speed lineal models used.

Linear Transfer Function	High velocities	Medium velocities	Low velocities
G_{DD}	$\frac{0.20s^2-3.15s+9.42}{s^2+6.55s+9.88}$	$\frac{0.20s^2+3.10s+8.44}{s^2+6.17s+9.14}$	$\frac{0.16s^2+2.26s+5.42}{s^2+5.21s+6.57}$
G_{ED}	$\frac{-0.04s^2-0.60s-0.32}{s^2+6.55s+9.88}$	$\frac{-0.02s^2-0.31s-0.03}{s^2+6.17s+9.14}$	$\frac{-0.02s^2-0.20s+0.41}{s^2+5.21s+6.57}$
G_{DE}	$\frac{-0.01s^2-0.08s-0.36}{s^2+6.55s+9.88}$	$\frac{0.01s^2+0.13s+0.20}{s^2+6.17s+9.14}$	$\frac{-0.01s^2-0.08s-0.17}{s^2+5.21s+6.57}$
G_{EE}	$\frac{0.31s^2+4.47s+8.97}{s^2+6.55s+9.88}$	$\frac{0.29s^2+4.11s+8.40}{s^2+6.17s+9.14}$	$\frac{0.25s^2+3.50s+6.31}{s^2+5.21s+6.57}$

Table 1. The second order WMR models

The second practice consists in finding a simplified robot model. The continuous-time model is introduced into the SIMULINK environment, as shown in Fig. 6. Several studies concerning about the importance of coupling terms are carried out by students through the analysis of stationary gains, while the order reduction is done by searching for dominant poles. The frequency response and BODE analysis can provide a dominant pole existence, which can reduce the model order of the system. Students learn from the obtained results that the MIMO model can be approximated by two SISO models.

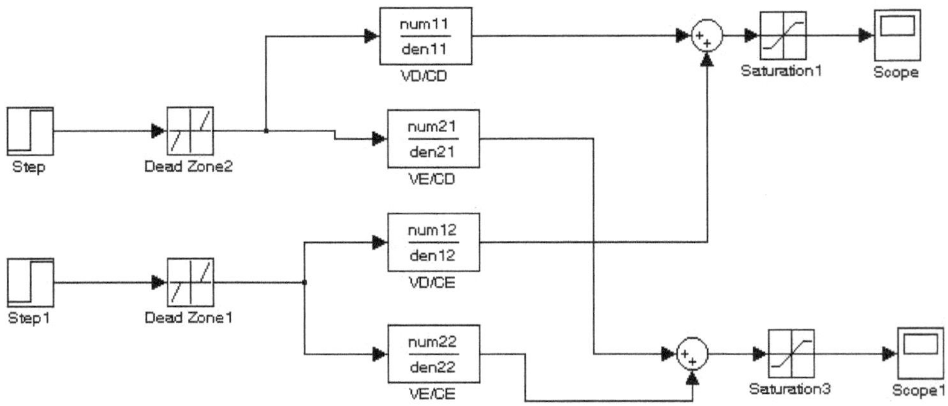

Fig. 6. MIMO robot model in SIMULINK environment.

Finally, the reduced SISO systems should be validated by considering the static gain and simulating reduced and complete robot models. Table 2 shows an example of the first order transfer functions obtained. Students can validate the results obtained by sending open loop speed commands to the robot.

Linear Transfer Function	High velocities	Medium velocities	Low velocities
G_{DD}	$\frac{0.95}{0.42s+1}$	$\frac{0.92}{0.41s+1}$	$\frac{0.82}{0.46s+1}$
G_{EE}	$\frac{0.91}{0.24s+1}$	$\frac{0.92}{0.27s+1}$	$\frac{0.96}{0.33s+1}$

Table 2. The reduced WMR model

3.2 Designing and testing the speed controllers

The third practice consists of obtaining a controller based on the simplified SISO models. The students will try to make the controller design by using pole placement techniques and the MATLAB design facilities. For instance, students can use the MATLAB "sisotool", as shown in Fig. 7, to design the controller by first importing the transfer functions corresponding to both robot wheels. Then, they depict the open loop frequency response without a compensator. Afterwards, students learn to add a PI controller to the system in order to achieve the desired response. The control performance is verified by using the complete MIMO model of the robot in the presence of several perturbations.

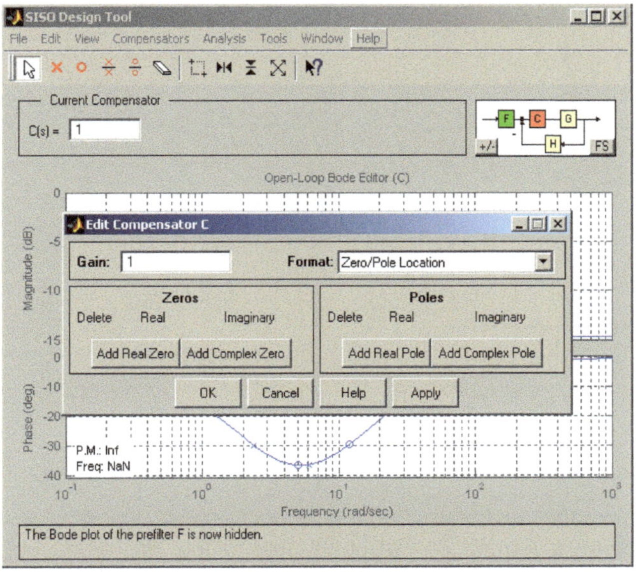

Fig. 7. "sisotool" design window.

On-robot speed control is an objective of the last MCS practice. Students transform firstly the controller from continuous to discrete time. Then, they use the robot remote network server to send the controller parameters to the robot in order to test the speed control performance.

3.3 Trajectory following using PID speed controllers

Once PID speed controllers are obtained, TPC block practices are used for performing trajectory tracking controllers. Student goal is following trajectories by using high level languages and the existing communication functions that allow sending speed commands to the robot and to receive the odometer data. The coordinates (x, y, θ) can be expressed:

$$\begin{aligned} x(k+1) &= x(k) + dS\cos(\theta(k)+d\theta) \\ y(k+1) &= y(k) + dS\sin(\theta(k)+d\theta) \\ \theta(k+1) &= \theta(k) + d\theta \end{aligned} \quad (1)$$

Fig. 8.a shows two consecutive robot coordinates. Fig. 8.b describes the positioning of robot as a function of the radius of left and right wheels (R_e, R_d), and the angular incremental positioning (θ_e,θ_d), with E being the distance between two wheels and dS the incremental displacement of the robot. The position and angular incremental displacements are expressed as:

$$dS = \frac{R_d d\theta_d + R_e d\theta_e}{2} \qquad d\theta = \frac{R_d d\theta_d - R_e d\theta_e}{E} \quad (2)$$

The incremental position of the robot can be obtained by the odometer system through the available encoder information from (1) and (2).

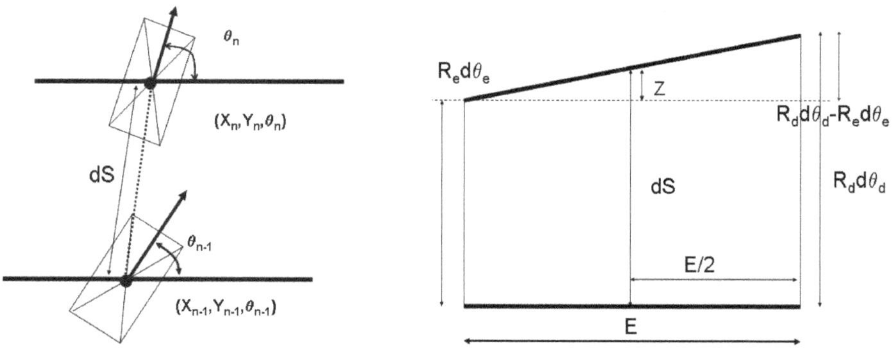

Fig. 8. (a) Robot coordinates at two consecutive instants of time.
(b) Robot position as function of the angular displacement of each wheel.

When students are able to obtain the robots coordinates, the path planning approach consists of following a sequence of straight lines. Thus, the trajectory tracking can be performed through either straight lines or turning actions (Reeds and Shepp, 1990). Trajectory tracking using discontinuous control laws can be implemented by using heuristic concepts related to experimental dynamic knowledge of the system. Therefore, when the error of path distance of the robot is greater than a heuristic threshold, the control law can minimize the distance to the path by performing a turning action. Otherwise the desired point distance should be minimized (Pacheco and Luo, 2006). Orientation and desired point distances can be considered as important data for achieving proper path-tracking. Their deviation errors are used in the control law as necessary information for going forward or turning. Students have to compute Cartesian distances between the robot and the trajectory to be followed. Fig. 9 depicts these concepts for two cases.

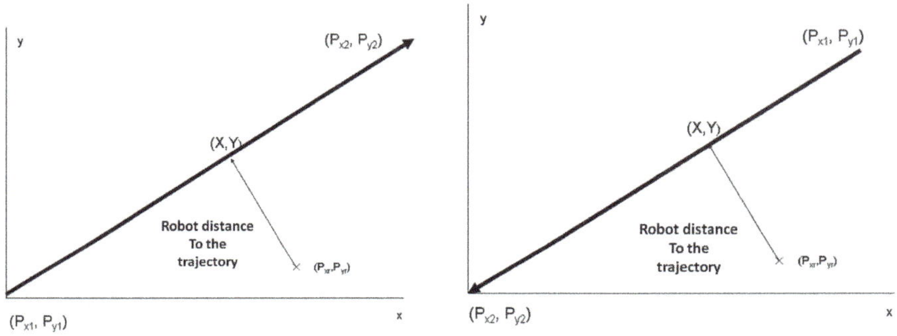

Fig. 9. (a) Straight line trajectory from (P_{x1}, P_{y1}) to (P_{x2}, P_{y2}).
(b) Straight line trajectory from (P_{x2}, P_{y2}) to (P_{x1}, P_{y1}).

The proposed control law is based on sending the same speed to both wheels when the robot distance to the trajectory is less than a heuristic threshold and different speed to each wheel in order to perform a turning action when such distance is greater. As can be observed in Fig. 9, the turning action must be different as a consequence of the robot position, up or down the trajectory, or the trajectory sense. Fig. 10 shows a trajectory following experience with two straight lines that have a 90° angle.

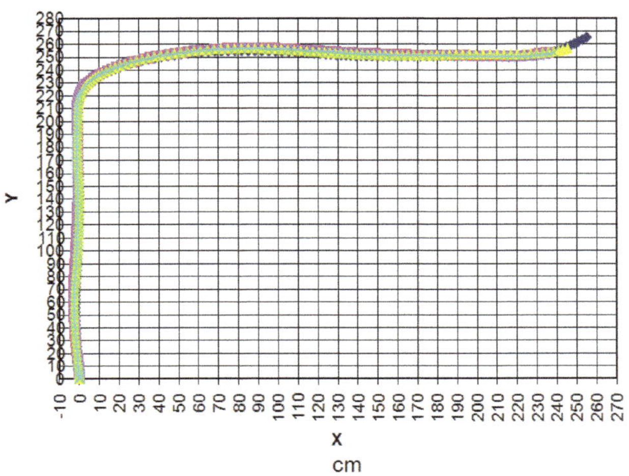

Fig. 10. Example of trajectory following experiences, using discontinuous control laws and PID speed controllers.

3.4 Trajectory following using MPC techniques

In this subsection the LMPC (local model predictive control) techniques based on the dynamics models obtained in the previous subsection are presented. LMPC strategies are simulated in the laboratory by using the methods and software developed by those of our teaching staff participating in the course (Pacheco and Luo, 2011).

3.4.1 The LMPC formulation

The main objective of highly precise motion tracking consists in minimizing the error between the robot and the desired path. Global path-planning becomes unfeasible since the sensorial system of some robots is just local. In this way, LMPC is proposed in order to use the available local perception data in the navigation strategies. Concretely, LMPC is based on minimizing a cost function related to the objectives for generating the optimal WMR inputs. Define the cost function as follows:

$$J(n,m) = \min_{\left\{U(k+i|k)_{i=0}^{i=m-1}\right\}} \left\{ \begin{array}{l} \left[X(k+n|k) - X_{ld}\right]^T P\left[X(k+n|k) - X_{ld}\right] \\ +\sum_{i=1}^{n-1}\left[X(k+i|k) - \overline{X_{ld}X_{l0}}\right]^T Q\left[X(k+i|k) - \overline{X_{ld}X_{l0}}\right] \\ +\sum_{i=1}^{n-1}\left[\theta(k+i|k) - \theta_{ld}\right]^T R\left[\theta(k+i|k) - \theta_{ld}\right] \\ +\sum_{i=0}^{m-1} U^T(k+i|k) S U(k+i|k) \end{array} \right\} \quad (3)$$

The first term of (3) refers to the attainment of the local desired coordinates, $X_{ld}=(x_d, y_d)$, where (x_d, y_d) denote the desired Cartesian coordinates. $X(k+n/k)$ represents the terminal value of the predicted output after the horizon of prediction n. The second one can be considered as an orientation term and is related to the distance between the predicted robot positions and the trajectory segment given by a straight line between the initial robot Cartesian coordinates $X_{l0}=(x_{l0}, y_{l0})$ from where the last perception was done and the desired local position, X_{ld}, to be achieved within the perceived field of view. This line orientation is denoted by θ_{ld} and denotes the desired orientation towards the local objective. $X(k+i/k)$ and $\theta(k+i/k)$ $(i=1,\ldots n-1)$ represents the predicted Cartesian and orientation values within the prediction horizon. The third term is the predicted orientation error. The last one is related to the power signals assigned to each DC motor and are denoted as U. The parameters P, Q, R and S are weighting parameters that express the importance of each term. The control horizon is designed by the parameter m. The system constraints are also considered:

$$\left\{ \begin{array}{ll} G_0 < |U(k)| \le G_1 & \alpha \in (0,1] \\ |X(K+n/k) - X_{ld}| \le \alpha |X(k) - X_{ld}| & \\ \text{or } |\theta(k+n/k) - \theta_{ld}| \le \alpha |\theta(k) - \theta_{ld}| & \end{array} \right\} \quad (4)$$

where $X(k)$ and $\theta(k)$ denote the current WMR coordinates and orientation, $X(k+n/k)$ and $\theta(k+n/k)$ denote the final predicted coordinates and orientation, respectively. The limitation of the input signal is taken into account in the first constraint, where G_0 and G_1 respectively denote the dead zone and saturation of the DC motors. The second and third terms are contractive constraints (Wang, 2007), which result in the convergence of coordinates or orientation to the objective, and should be accomplished at each control step.

3.4.2 The algorithms and simulated results

By using the basic ideas introduced in the previous subsection, the LMPC algorithms have the following steps:

1. Read the current position;
2. Minimize the cost function and to obtain a series of optimal input signals;
3. Choose the first obtained input signal as the command signal;
4. Go back to the step 1 in the next sampling period.

In order to reduce the set of possibilities, when optimal solution is searched for, some constraints over the DC motor inputs are taken into account:

- The signal increment is kept fixed within the prediction horizon.
- The input signals remain constant during the remaining interval of time.

The above considerations will result in the reduction of the computation time and the smooth behavior of the robot during the prediction horizon (Maciejowski, 2002). Thus, the set of available input is reduced to one value, as it is shown in Fig. 11.

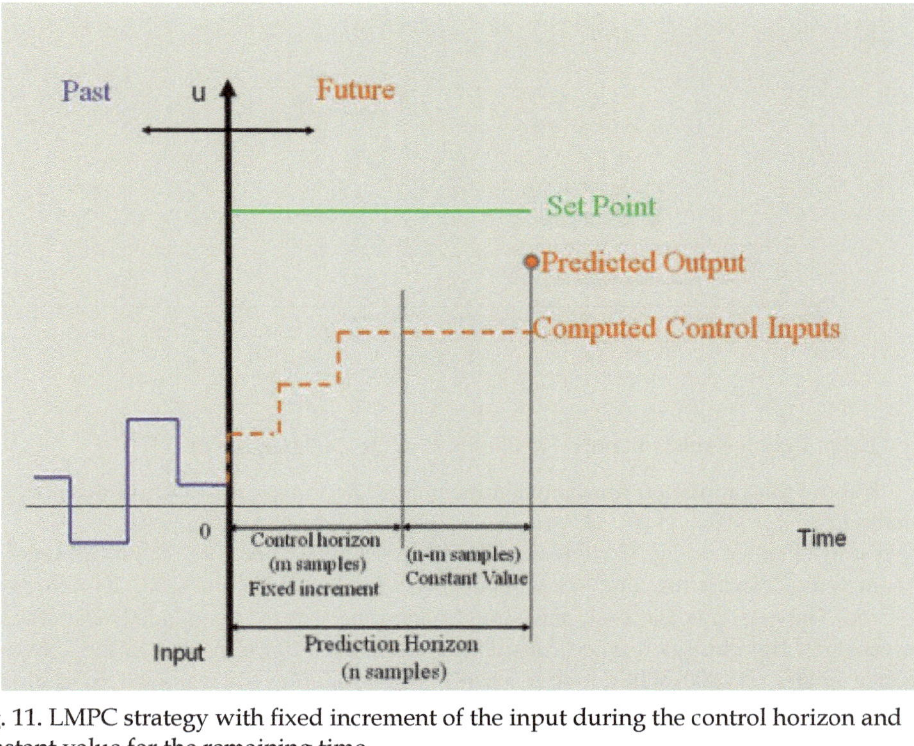

Fig. 11. LMPC strategy with fixed increment of the input during the control horizon and constant value for the remaining time

The LMPC simulation software provides to students a set of files and facilities to draw the results, and consequently the students can represent the files containing the cost function values, for different available set of inputs, corresponding to different moments of the simulation. Simulation environment allows testing the control law with different set up. Therefore, different control and prediction horizons can be tested. The control law parameters of equation (3) can be analyzed and modified according to the different horizons of prediction. Fig. 12 depicts the results obtained, m=3 and n=5, when a trajectory composed of 3 straight lines is followed. The lines are defined by the coordinates (0, 0), (0, 100), (50, 200) and (250, 200) where (x, y) are given in cm. It is noted that the sample time is 0.1s, so n=5 means 0.5s.

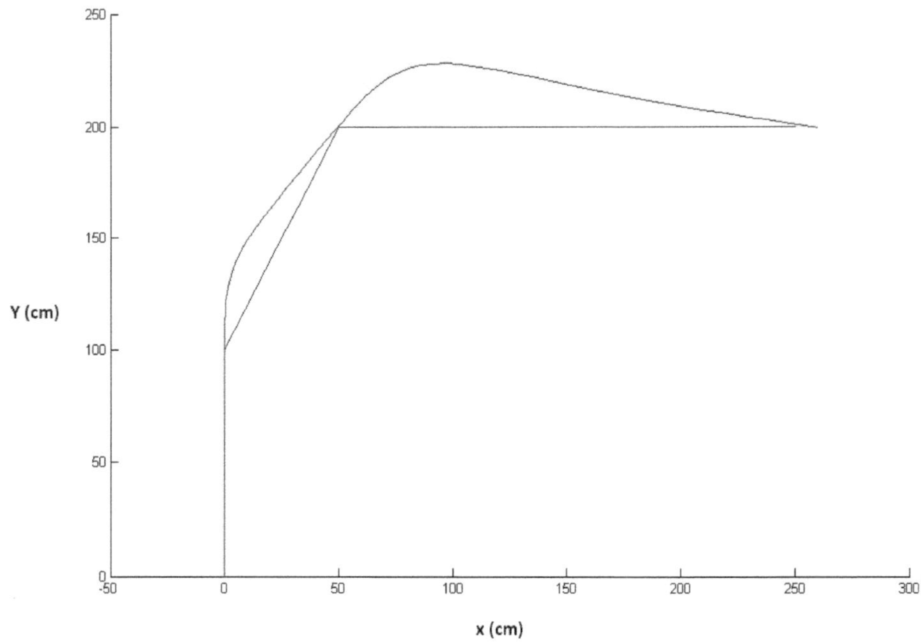

Fig. 12. Simulation results obtained (m=3, n=5, P=1, Q=1, R=1 and S=0).

Fig. 13 show the simulation results when the control horizon is set to 4 and the prediction horizon to 8.

The results obtained in Fig. 13 show that the robot is stopped near (50, 200) coordinates due to contractive constraints, and consequently the final coordinates (250, 200) cannot be achieved. The reason is that each new straight line trajectory is commanded when the final coordinate of trajectory is reached considering a tolerance threshold but in the above case the coordinates (50, 200) will not be reached even considering the tolerance threshold. The problem can be solved by modifying the parameter weights of the control law as it is shown in Fig. 14.

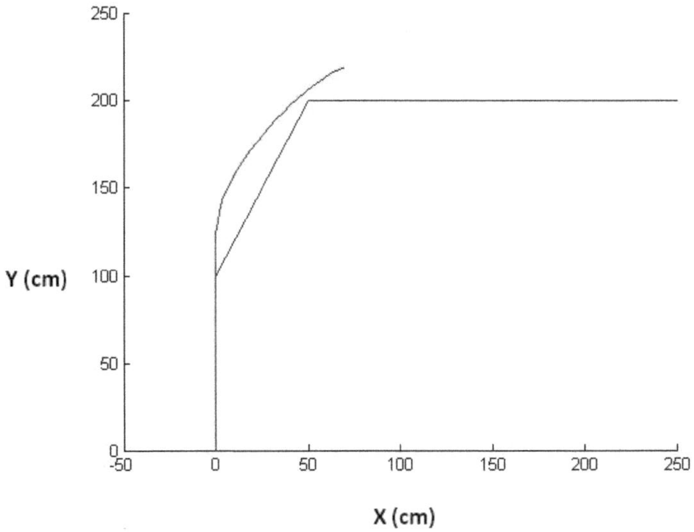

Fig. 13. Simulation results obtained (*m*=4, *n*=8, *P*=1, *Q*=1, *R*=1 and *S*=0).

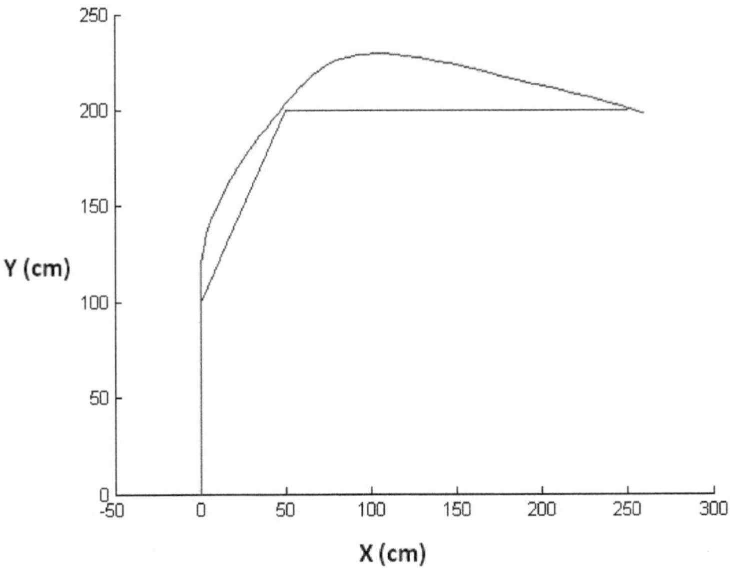

Fig. 14. Simulation results obtained (*m*=4, *n*=8, *P*=1, *Q*=2, R=1 and *S*=0).

4. Conclusions

The teaching experience gained from this course has shown the usefulness of the mobile robot as an important experimental platform for computer engineering education. The course framework allows a different level of knowledge learning according to the student skills. Various control techniques have been provided to students as a main issue in order to improve their knowledge. They can acquire a higher level of skills by integrating different subjects related to electronics, control system, computer science, communication, etc. Moreover, students have the opportunity to be familiar with basic control techniques as well as some advanced methods as MPC. It should be mentioned that the developed lab practices related to the theoretic counterparts have increased the motivation of the students and have achieved the multidisciplinary knowledge consolidation of the proposed objective of control education. In this context, according to the student's feedback on the course, the average opinion of 30 students of the course is 4 over 5 according to the different survey questions that the university provides at the end of the course. They underline the interesting and attractive educational and practical aspects as the relevant characteristics of the course. One of the suggestions is to increase the number of hours in order to perform better the theoretical analysis, lab practices and algorithm implementation. Furthermore, some students decide to do their compulsory final degree project by improving the robot performance in some practical aspects; e.g., nowadays the design of factorial experiments to perform the MPC cost function tuning. In the future, the teaching staff from different research areas will do their utmost to promote the development and consolidation of the course so that the quality of teaching can be further improved. In this process, the student's opinions are important enough to encourage this work. Therefore, the path for achieving renewed progress in robotics is through the integration of several fields of knowledge, such as computing, communications, and control sciences, so as to perform a higher level reasoning and use decision tools with strong theoretical base.

5. Acknowledgment

This work has been partially funded by the Commission of Science and Technology of Spain (CICYT) through the coordinated projects DPI2007-66796-C03-02 and DPI 2008-06699-C02-01. The authors are grateful to the support provided by Javi Cobos in implementing a useful on-robot network server as well as different functions that allow the dynamic interactions in the robot.

6. References

Amante, B., Romero, C., Piñuela, J.A. (2007). Aceptación de la metodología del aprendizaje colaborativo en diferentes ciclos de las carreras técnicas. Cuadernos de innovación educativa de las Enseñanzas Técnicas Universitarias. Vol. 1, No. 1, 2007, 65-74.

Antsaklis, P., Basar, T., Decarlo, R., McClamroch, N. H., Spong M., and Yurkovich, S. (1999). Report on NSF/CSS workshop on new directions in control engineering education. *IEEE Control Systems Magazine*, Vol. 19, No. 5, (Oct. 1999) 53-58.

Aström K.J and Wittenmark B., (1988). Computed Controlled Systems. Theory and Design. *Prentice-Hall International Ed.*, New Jersey, 1988.

Barnett, R. H. (1994). The 8051 Family of Microcontrollers. *Prentice Hall International Ed.*, New Jersey, 1994.

Carnegie Mellon University. *The Robotics Institute: Robotics Education in the School of Computer Science*. In URL: http://www.ri.cmu.edu, 2011.

European Commission. *Higher Education in Europe, inside the police areas of European Commission*, in URL:
http://ec.europa.eu/education/policies/higher/higher_en.html, 2008.

Horn, B. K. P. (1998). Robot Vision. *MIT press*, Ed. McGraw-Hill, ISBN 0-262-08159-8, London (England)

Hristu-Varsakelis, D., and Brockett, R. W. (2002). Experimenting with Hybrid Control. *IEEE Control Systems Magazine*, Vol. 22, No. 1, (Feb. 2002), 82-95.

Kuo, C. (1996). Automatic Control Systems. *Prentice-Hall International Editors*, New Jersey, 1996.

Ljung, L. (1989). System Identification: Theory for the User. Prentice-Hall Inter. Ed., New Jersey, 1989.

Maciejowski, J.M. (2002). Predictive Control with Constraints, *Ed. Prentice Hall*, ISBN 0-201-39823-0, Essex (England)

Murray, R. M., Aström, K. J., Boyd, S. P., Brockett, R. W., and Stein, G. (2003). Future Directions in Control in an Information-Rich World. *IEEE Control Systems Magazine*, Vol. 23, No. 2, (April 2003) 20-33.

Norton, J. (1986). An Introduction to Identification. *Academic Press*, ISBN 0125217307, London and New York, 1986

Pacheco, L., Batlle, J., Cufí, X., Arbusé, R. (2006). PRIM an Open Mobile Platform. Motivation, Present and Future Trends. *Inter. Conf. on Automation, Quality and Testing, Robotics*. Vol. 2, 231-236, 2006.

Pacheco, L. and Luo, N. (2006). Mobile robot experimental modelling and control strategies using sensor fusion. *Control Engineering and Applied Informatics*. Vol. 8, No.3, (2006), 47-55.

Pacheco, L., Luo, N., Arbusé, R., Ferrer, I., and Cufí, X. (2009). Interdisciplinary Knowledge Integration Through an Applied Mobile Robotics Course. *International Journal of Engineering Education*, Vol. 25, No. 4, (July 2009), 830-840.

Pacheco, L., Luo, N. (2011) Mobile robot local trajectory tracking with dynamic model predictive control techniques, *International Journal of Innovative Computing, Information and Control*, Vol.7, No.6, (June 2011), ISSN 1349-4198, 3457-3483

Qu, Z. and Wu XH. (2006). A new curriculum on planning and cooperative control of autonomous mobile robots. *International Journal of Engineering Education*, Vol. 22, No 4, (July, 2006), 804-814.

Reeds, J. A and Shepp, L. A. (1990). Optimal paths for a car that goes forwards and backwards. *Pacific Journal of Mathematics*, vol. 145, 1990.

Schilling, R. J. (1990). Fundamentals of Robotics. *Prentice-Hall International Ed.*, New Jersey, 1990.

Valera, A., Weiss, M., Valles, M., Diez, J. L. (2007). Control of mobile robots using mobile technologies. *International Journal of Engineering Education*, Vol. 23, No 3, (March 2007), 491-498.

Wakerly, J.F. (1992). Digital Design Principles and Practices. *Prentice-Hall International Ed.*, New Jersey, 1992.

Wan, J. (2007) Computational reliable approaches of contractive MPC for discrete-time systems, *PhD Thesis, University of Girona*

Wilczynski, V. and Flowers, W. (2006). FIRST Robotics Competition: University curriculum applications of mobile robots. *International Journal of Engineering Education*, Vol. 22, No. 4, (July, 2006), 792-803.

Part 2

Health-Care and Medical Robots

5

Walking Support and Power Assistance of a Wheelchair Typed Omnidirectional Mobile Robot with Admittance Control

Chi Zhu[1], Masashi Oda[1], Haoyong Yu[2], Hideomi Watanabe[3] and Yuling Yan[4]

[1]*Department of Systems Life Engineering, Maebashi Institute of Technology, Kamisadori 460-1, Maebashi, Gunma, 371-0816.*

[2]*Bioengineering Division, Faculty of Engineering, National University of Singapore*

[3]*Department of Health Sciences, Faculty of Medicine, Gunma University*

[4]*School of Engineering, Santa Clara University, CA*

[1,3]*Japan*

[2]*Singapore*

[4]*USA*

1. Introduction

Walking ability is essential for the elderly/the disabled's self-supported daily life. Currently the development of various robotic aids or devices for helping the elderly's walking has attracted many researcher's attentions. Guidecane ((Ulrich & Borenstein, 2001)), PAM-AID ((Lacey & Dawson-Howe, 1998)), Care-O-bot ((Graf et al., 2004)), PAMM ((S. Dubowsky, 2000)), and etc, are developed for intelligent walking support and other tasks for the elderly or the disabled. Mainly for the elderly's walking support, a robotic power-assisted walking support system is developed (Nemoto & M. Fujie, 1998). The system is to support the elderly needing help in standing up from the bed, walking around, sitting down for rehabilitation.

Meanwhile, a number of intelligent wheelchair-type aids are being designed for people who cannot walk and have extremely poor dexterity ((Yanco, 1998)-(Dicianno et al., 2007)). These devices are well suited for people who have little or no mobility, but they are not appropriate for the elderly with significant cognitive problems. Consequently, the facilities are generally reluctant to permit the elderly residents to use powered wheelchairs. Hence, the better solution is that the facility staff (the caregiver) pushes the wheelchair to move. Some technologies for the caregiver's power assistance are proposed ((Kakimoto et al., 1997)-(Miyata et al., 2008)).

Further, since the omnidirectional mobility is necessary to truly yield the user's (here refer to the elderly/disabled, or the caregiver) intentions of walking speed and direction, omnidirectional mobility is investigated in order to obtain a smooth motion and high payload capability, not with the special tires but with the conventional rubber or pneumatic tires. An omnidirectional and holonomic vehicle with two offset steered driving wheels and two free casters is proposed (Wada & Mori, 1996). An omnidirectional mobile robot platform using active dual-wheel caster mechanisms is developed (Han et al., 2000), in which, the kinematic model of the active dual-wheel caster mechanism is derived and a holonomic and omnidirectional motion of the mobile robot using two such assembly mechanisms is realized. An omnidirectional platform and vehicle design using active split offset casters (ASOC) is also developed (Yu et al., 2004). Its structure is very similar to (Han et al., 2000), but particular attention is paid to the system performance on uneven floors.

However, up to now, the wheelchair typed robot that can be used either for the elderly's walking support, or the disabled's walking rehabilitation, or the caregiver's power assistance has not been reported yet. In this study, we develop such kind of wheelchair typed omnidirectional robot that not only can support the elderly or the disabled's walking, but also can lessen the caregivers' burden, or saying, give the caregivers power assistance.

2. Overview of the developed robot

The developed new type of omnidirectional mobile robot for the elderly's walking support, the disabled's rehabilitation, and the caregiver's power assistance, is shown in Fig. 1. Its structure is illustrated in Fig. 2.

Fig. 1. The omnidirectional mobile robot in developing for the elderly's walking support, the disabled's walking rehabilitation, and the caregiver's power assistance

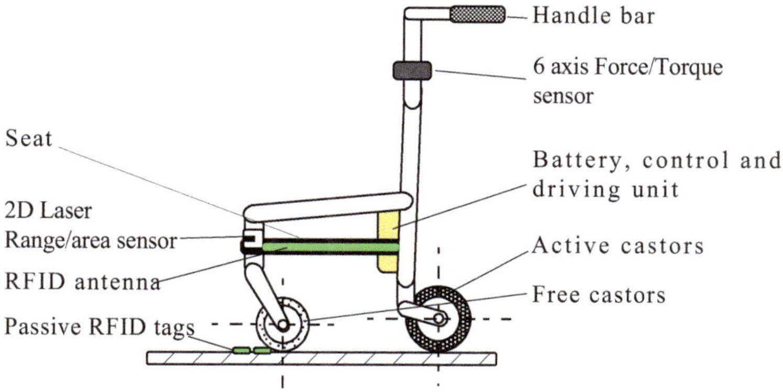

Fig. 2. The structure of the robot

2.1 The robot objective and functions

Different from the other intelligent aids, the robot has a handle and a seat. The user holds the handle with his/her hands during walking. Usually, the elderly's walking ability (the capable walking distance), can be greatly enhanced with the help of the intelligent walking support devices. However, because of the muscle weakness, the knee and waist of the elderly will be painful after walking some distances. At that time, the elderly hopes to have a seat for a rest. To meet such a very actual requirement, we embed a seat in the robot so that the elderly or the disabled can take a seat for rest when necessary. On the other hand, when the elderly or the disabled is sitting in the seat of the wheelchair type aids and wants to go to somewhere, the caregiver will have to push the wheelchair. Hence, the power assistance of the health caregivers is also necessary.

Consequently, the main the objectives of the robot are To meet the above two main objectives, the functions the robot should have:

1. omnidirectional mobility.
2. few load feeling when the user (the elderly/disabled /caregiver) holding the handle during walking or pushing the robot.
3. guidance to destinations and obstacle avoidance via sensors and pre-determined maps.
4. adaptation to the different motion abilities of the different users.

2.2 Robot structure

The robot shown in Fig.1, 2 is considered to be used either indoor or outdoor. It includes two active dual-wheel modules and two free casters that provides the robot omnidirectional mobility. It will be discussed in section 3 in detail.

A 6-axis force/torque sensor, mounted on the shaft of the handle, detects the forces and torques the user applies to the handle. With these force/torque signals, the embedded computer estimates the user's intention of walking velocity and walking direction, and the robot is controlled to yield the user's intention so that the user will feel few loads to fulfill the walking support and power assistance. The detail will be explained in section 4.

The 2D laser ranger/area finder is employed to detect the obstacles. The RFID system of the RFID antenna under the seat and the passive RFID tags laid under or near the places such as

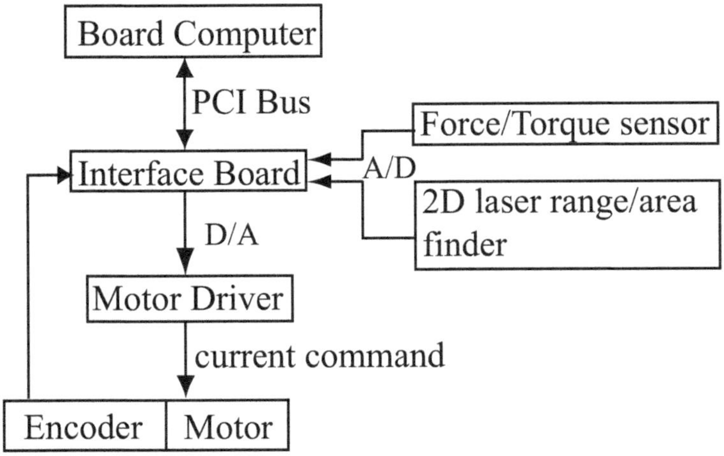

Fig. 3. The control system of the robot

the door, corner, elevator are used to identify the robot position and orientation. They are used for the destination guidance and obstacle avoidance. The problems of such obstacle avoidance and navigation are not discussed in this chapter.
The control system of the robot as shown in Fig. 3 consists of a board computer running RTLinux operation system, an multi-functional interface board, two motor drivers in which each motor driver can drive two motors.

3. Omnidirectional mobility

The wheels of the robot consists of two active dual-wheel modules and two free casters. The two active dual-wheel modules enable the robot omnidirectional mobility and the two free casters keep the robot balance. First we analyze the kinematics of one dual-wheel module, than investigate the omnidirectional mobility of the robot realized by two dual-wheel modules.

3.1 Kinematics of one dual-wheel module

As shown in Fig. 4, an active dual-wheel module consists of two independently driven coaxial wheels, which are separated at a distance $2d$ and connected via an offset link s to the robot at the joint O_i. The coordinate systems, valuables, and parameters are defined as follows.
$O_w X_w Y_w$: a world coordinate system.
$O_i X_i Y_i$: a moving coordinate system attached to the wheel module i at the offset link joint O_i.
x_i, y_i: coordinates of the origin O_i of the frame $O_i X_i Y_i$ with respect to the ground (the frame $O_w X_w Y_w$). Consequently, their derivations to time
$\dot{x}_i, \dot{y}_i$: velocities of the joint O_i respectively in the X_i, Y_i axes.
$\dot{x}_{wi}, \dot{y}_{wi}$: velocities of the joint O_i respectively in the X_w, Y_w axes.
ϕ_i: orientation of the module i with respect to the world frame.
$\boldsymbol{u_i}, \boldsymbol{u_{wi}}$: velocity vectors of the joint O_i respectively with respect to the frame $O_i X_i Y_i$ and ground $O_w X_w Y_w$, where, $\boldsymbol{u_i} = [\dot{x}_i, \dot{y}_i]^T$ and $\boldsymbol{u_{wi}} = [\dot{x}_{wi}, \dot{y}_{wi}]^T$.
ω_{li}, ω_{ri}: respectively the left and right wheel's angular velocities of the module i.

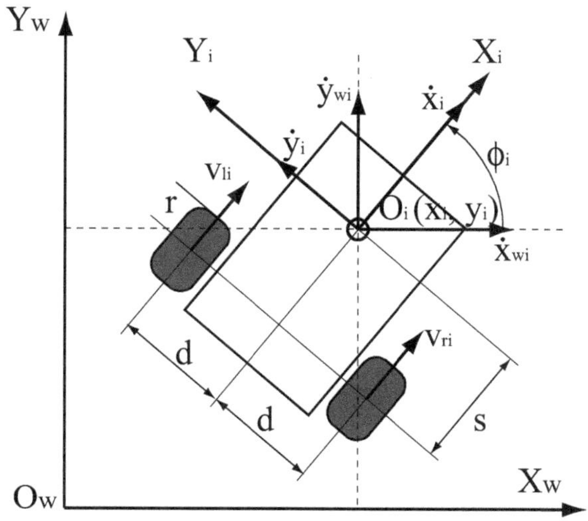

Fig. 4. Model of one active dual-wheel module

$\boldsymbol{\omega_i}$: angular velocity vector of the module i. $\boldsymbol{\omega_i} = [\omega_{li}, \omega_{ri}]^T$.
r: radius of the wheels.
The linear velocities of the joint O_i of the module i can be expressed as:

$$\boldsymbol{u_i} = \begin{bmatrix} \dot{x}_i \\ \dot{y}_i \end{bmatrix} = \frac{r}{2} \begin{bmatrix} 1 & 1 \\ \frac{s}{d} & -\frac{s}{d} \end{bmatrix} \cdot \begin{bmatrix} \omega_{ri} \\ \omega_{li} \end{bmatrix} = \boldsymbol{J_i} \cdot \boldsymbol{\omega_i} \tag{1}$$

where $\boldsymbol{J_i}$ is the Jacobian matrix of the module in the moving frame $O_i X_i Y_i$.
On the other hand, the velocity of the joint O_i expressed in the world system is given by

$$\boldsymbol{u_{wi}} = \begin{bmatrix} \dot{x}_{wi} \\ \dot{y}_{wi} \end{bmatrix} = \begin{bmatrix} \cos\phi_i & -\sin\phi_i \\ \sin\phi_i & \cos\phi_i \end{bmatrix} \cdot \boldsymbol{u_i} = \boldsymbol{R_i} \cdot \boldsymbol{u_i} \tag{2}$$

Therefore, the kinematics of the module between the velocity of the joint O_i and the two wheel rotation velocities is expressed as:

$$\boldsymbol{u_{wi}} = \boldsymbol{R_i} \cdot \boldsymbol{J_i} \cdot \boldsymbol{\omega_i} = \boldsymbol{J_{wi}} \cdot \boldsymbol{\omega_i} \tag{3}$$

where $\boldsymbol{J_{wi}}$ is the Jacobian matrix of the module in the world frame $O_w X_w Y_w$, and it is given as:

$$\boldsymbol{J_{wi}} = \frac{r}{2} \begin{bmatrix} \cos\phi_i - \frac{s}{d}\sin\phi_i & \cos\phi_i + \frac{s}{d}\sin\phi_i \\ \sin\phi_i + \frac{s}{d}\cos\phi_i & \sin\phi_i - \frac{s}{d}\cos\phi_i \end{bmatrix} \tag{4}$$

The above expressions indicate that the velocity of the module is determined by the parameters s, d, r, and ω_{ri}, ω_{li}. But since s, d, r are structurally determined for a robot, the only controllable values are angular velocities ω_{ri}, ω_{li} of the two driving wheels. The determinant of $\boldsymbol{J_{wi}}$ is :

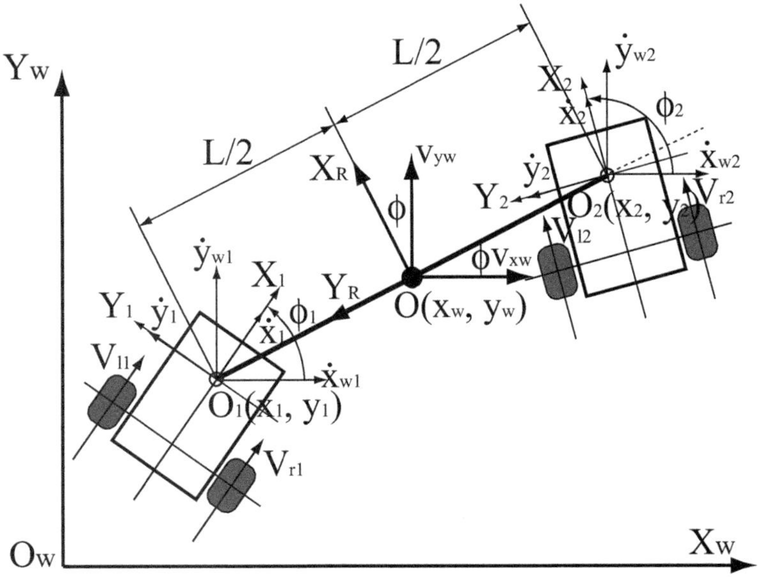

Fig. 5. Model of two active dual-wheel modules

$$\det J_{wi} = -\frac{r^2 s}{2d} \tag{5}$$

This means that the module has no singularity as long as the distance s is not zero. This fact implies that arbitrary and unique velocities at the joint point O_i can be achieved by appropriately controlling the rotational velocities of the two wheels. Note that one such a module only has 2DOF. In other words, omnidirectional mobility (3DOF) cannot be achieved. Consequently, at least two modules are necessary to realize omnidirectional mobility. In this study, two modules are used to achieve the omnidirectional mobility.

3.2 Omnidirectional mobility with two modules

As shown in Fig.5, two active dual-wheel modules consists of a basic part of the mobile robot that enable the robot to have omnidirectional mobility.

In Fig. 5, point O is the center of the robot platform. Its coordinates and velocities in the world frame $O_w X_w Y_w$ are respectively (x_w, y_w) and (v_{xw}, v_{yw}). The frame $OX_R Y_R$ is attached to the robot. ϕ is the orientation of the robot. The vector $\boldsymbol{V} = [v_{xw}, v_{yw}, \dot{\phi}]^T$ is the velocity vector of the robot. The distance between the two joint points of two modules is L. $\boldsymbol{\omega} = [\omega_{r1}, \omega_{l1}, \omega_{r2}, \omega_{l2}]^T$ is the angular velocity vector of the four driving wheels of the robot.

To obtain the rotational velocity $\dot{\phi}$ of the robot, it it convenient to transform $\dot{x}_{wi}, \dot{y}_{wi}$ in the world frame to the robot frame $OX_R Y_R$. The relationship is given by

$$\begin{bmatrix} \dot{x}_{Ri} \\ \dot{y}_{Ri} \end{bmatrix} = \begin{bmatrix} \cos\phi & -\sin\phi \\ \sin\phi & \cos\phi \end{bmatrix} \cdot \begin{bmatrix} \dot{x}_{wi} \\ \dot{y}_{wi} \end{bmatrix} \tag{6}$$

Therefore, $\dot{\phi}$ of the robot can be expressed as

$$\dot{\phi} = \frac{\dot{x}_{R2} - \dot{x}_{R1}}{L} \tag{7}$$

$$= \frac{1}{L}\left[(\dot{x}_{w2} - \dot{x}_{w1})\cos\phi + (\dot{y}_{w2} - \dot{y}_{w1})\sin\phi\right] \tag{8}$$

Hence, the velocity vector $\boldsymbol{V}$ can be expressed as:

$$\boldsymbol{V} = \begin{bmatrix} v_{xw} \\ v_{yw} \\ \dot{\phi} \end{bmatrix} = \frac{1}{2}\begin{bmatrix} 1 & 0 & 1 & 0 \\ 0 & 1 & 0 & 1 \\ -\frac{\cos\phi}{L} & -\frac{\sin\phi}{L} & \frac{\cos\phi}{L} & \frac{\sin\phi}{L} \end{bmatrix} \cdot \begin{bmatrix} \dot{x}_{w1} \\ \dot{y}_{w1} \\ \dot{x}_{w2} \\ \dot{y}_{w2} \end{bmatrix} \tag{9}$$

From (3), the velocity vector $[\dot{x}_{w1}, \dot{y}_{w1}, \dot{x}_{w2}, \dot{y}_{w2}]^T$ in the above expression is given by

$$\begin{bmatrix} \dot{x}_{w1} \\ \dot{y}_{w1} \\ \dot{x}_{w2} \\ \dot{y}_{w2} \end{bmatrix} = \frac{r}{2}\left[\begin{array}{c|c} \boldsymbol{A_1} & \boldsymbol{0_{2\times 2}} \\ \hline \boldsymbol{0_{2\times 2}} & \boldsymbol{A_2} \end{array}\right] \cdot \begin{bmatrix} \omega_{r1} \\ \omega_{l1} \\ \omega_{r2} \\ \omega_{l2} \end{bmatrix} = \boldsymbol{A} \cdot \begin{bmatrix} \omega_{r1} \\ \omega_{l1} \\ \omega_{r2} \\ \omega_{l2} \end{bmatrix} \tag{10}$$

where

$$\boldsymbol{A_i} = \begin{bmatrix} \cos\phi_i - \frac{s}{d}\sin\phi_i & \cos\phi_i + \frac{s}{d}\sin\phi_i \\ \sin\phi_i + \frac{s}{d}\cos\phi_i & \sin\phi_i - \frac{s}{d}\cos\phi_i \end{bmatrix} \tag{11}$$

Furthermore, the robot possessing omnidirectional mobility realized by two active dual-wheel modules has three degrees of freedom with a total of four actuators. There should be a constraint for this redundancy. The physically constant distance between the two joint points O_1 and O_2 is used to eliminate the redundancy. It leads the following velocity constraint:

$$\dot{y}_{R1} = \dot{y}_{R2} \tag{12}$$

From (6) it means

$$0 = \dot{x}_{w1}\sin\phi - \dot{y}_{w1}\cos\phi - \dot{x}_{w2}\sin\phi + \dot{y}_{w2}\cos\phi \tag{13}$$

By combining Eqs. (9), (10) and (13), we can get the following homogeneous forward kinematics of the robot:

$$\begin{bmatrix} \boldsymbol{V} \\ 0 \end{bmatrix} = \begin{bmatrix} v_{xw} \\ v_{yw} \\ \dot{\phi} \\ 0 \end{bmatrix} = \frac{1}{2}\begin{bmatrix} 1 & 0 & 1 & 0 \\ 0 & 1 & 0 & 1 \\ -\frac{\cos\phi}{L} & -\frac{\sin\phi}{L} & \frac{\cos\phi}{L} & \frac{\sin\phi}{L} \\ \sin\phi & -\cos\phi & -\sin\phi & \cos\phi \end{bmatrix} \cdot \begin{bmatrix} \dot{x}_{w1} \\ \dot{y}_{w1} \\ \dot{x}_{w2} \\ \dot{y}_{w2} \end{bmatrix}$$

$$= \boldsymbol{B} \cdot \begin{bmatrix} \dot{x}_{w1} \\ \dot{y}_{w1} \\ \dot{x}_{w2} \\ \dot{y}_{w2} \end{bmatrix} = \boldsymbol{B} \cdot \boldsymbol{A} \cdot \begin{bmatrix} \omega_{r1} \\ \omega_{l1} \\ \omega_{r2} \\ \omega_{l2} \end{bmatrix} = \boldsymbol{B} \cdot \boldsymbol{A} \cdot \boldsymbol{\omega} \tag{14}$$

Hence, the inverse kinematics of the robot, i.e., with the given desired velocity vector $\boldsymbol{V_d}$ of the robot, the necessary driving velocity vector $\boldsymbol{\omega_d}$ of the four wheels is given by

$$\boldsymbol{\omega_d} = \begin{bmatrix} \omega_{r1} \\ \omega_{l1} \\ \omega_{r2} \\ \omega_{l2} \end{bmatrix} = \boldsymbol{A}^{-1} \cdot \boldsymbol{B}^{-1} \cdot \begin{bmatrix} v_{xw} \\ v_{yw} \\ \dot{\phi} \\ 0 \end{bmatrix} = \boldsymbol{A}^{-1} \cdot \boldsymbol{B}^{-1} \cdot \begin{bmatrix} \boldsymbol{V_d} \\ 0 \end{bmatrix} \tag{15}$$

where,

$$\boldsymbol{A}^{-1} = \frac{1}{r} \left[\begin{array}{c|c} \boldsymbol{A_1}^{-1} & \boldsymbol{0} \\ \hline \boldsymbol{0} & \boldsymbol{A_2}^{-1} \end{array} \right] \tag{16}$$

$$\boldsymbol{A_i}^{-1} = \begin{bmatrix} \cos\phi_i - \frac{d}{s}\sin\phi_i & \sin\phi_i + \frac{d}{s}\cos\phi_i \\ \cos\phi_i + \frac{d}{s}\cos\phi_i & \sin\phi_i - \frac{d}{s}\cos\phi_i \end{bmatrix} \tag{17}$$

$$\boldsymbol{B}^{-1} = \begin{bmatrix} 1 & 0 & -L\cos\phi & \sin\phi \\ 0 & 1 & -L\sin\phi & -\cos\phi \\ 1 & 0 & L\cos\phi & -\sin\phi \\ 0 & 1 & L\sin\phi & \cos\phi \end{bmatrix} \tag{18}$$

Now by rewriting $\boldsymbol{B}^{-1}$ as:

$$\boldsymbol{B}^{*} = \begin{bmatrix} 1 & 0 & -L\cos\phi \\ 0 & 1 & -L\sin\phi \\ 1 & 0 & L\cos\phi \\ 0 & 1 & L\sin\phi \end{bmatrix} \tag{19}$$

We can get the following inverse kinematic equation of the robot:

$$\boldsymbol{\omega_d} = \begin{bmatrix} \omega_{r1} \\ \omega_{l1} \\ \omega_{r2} \\ \omega_{l2} \end{bmatrix} = \boldsymbol{A}^{-1} \cdot \boldsymbol{B}^{*} \cdot \begin{bmatrix} v_{xw} \\ v_{yw} \\ \dot{\phi} \end{bmatrix} = \boldsymbol{A}^{-1} \cdot \boldsymbol{B}^{*} \cdot \boldsymbol{V_d} \tag{20}$$

As discussed in section 4, the $\boldsymbol{V_d}$, the desired velocity of the robot, is obtained from the introduced admittance controller, and the $\boldsymbol{\omega_d}$, each wheel's speed command, is used for velocity control.

4. Admittance based interaction control for power assistance

Admittance of a mechanical system is defined as (Newman (1992))

$$\boldsymbol{G} = \frac{\boldsymbol{V}}{\boldsymbol{F}} \tag{21}$$

where, $\boldsymbol{V}$ is the velocity and $\boldsymbol{F}$ is the contact or applied force, both at the point of interaction. A large admittance corresponds to a rapid motion induced by applied forces; while a small admittance represents a slow reaction to act forces.
Here, admittance controller is introduced in order to fulfill the power assistance when the elderly/disabled holding the handle to walk or the caregiver holding the handle to push the robot while the elderly or the disabled sitting in the seat. Since user's (the elderly, the disabled, or the caregiver) walking speed doesn't change a lot, a large admittance implies relatively small forces needed to exert to the robot. This is the basic principle of power assistance. Note that almost similar approaches had been already used for the elderly's walking support (Nemoto & M. Fujie (1998); Yu et al. (2003)), since our robot can be used both by the elderly/disabled and the caregiver, the approach is extended to the caregiver's power assistance.
In this study, the admittance of the human-robot system can be defined as a transfer function with the user's applied forces and torques, $\boldsymbol{F}(s)$, as input and the robot's velocities, $\boldsymbol{V}(s)$, as the output. The time response $\boldsymbol{V_d}(t)$ of the admittance model is used as the desired velocity of the robot. Then, the desired driving speed $\boldsymbol{\omega_d}$ of each wheel is calculated from $\boldsymbol{V_d}(t)$ by the inverse kinematics equation (20), and the $\boldsymbol{\omega_d}$, as each wheel's speed command, is used for velocity control. The admittance based control process is shown in Fig.6, in which a digital LF (low-pass filter) cuts off the high frequency noises in the force/torque signals from the 6-axis force/torque sensor.
In the forward direction (X_R direction in Fig.5), the admittance can be expressed as

$$G_x(s) = \frac{V_x(s)}{F_x(s)} = \frac{1}{M_x s + D_x} \tag{22}$$

where, M_x and D_x are respectively the virtual mass and virtual damping of the system in forward (X_R) direction.

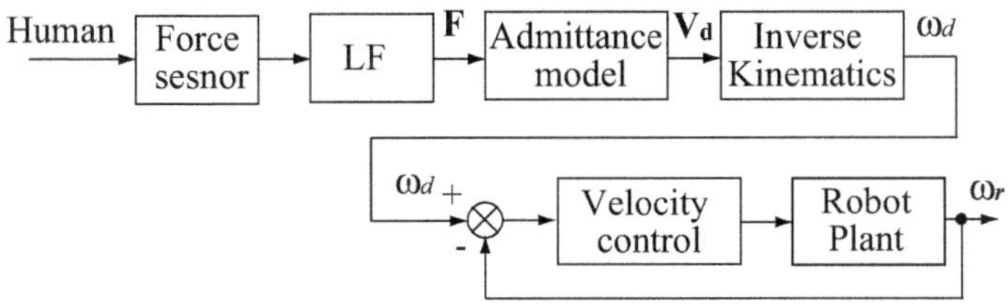

Fig. 6. Admittance based control

The time response $V_{xd}(t)$ for a step input F_x of the above transfer function is:

$$V_{xd}(t) = \frac{F_x}{D_x}(1 - e^{-t/\tau_x}) \tag{23}$$

where, τ_x is the time constant defined by $\tau_x = M_x/D_x$. The steady state velocity of the system is $V_{xs} = F_x/D_x$. This means that the forces to the robot exerted by the user determines the velocities of the system (user and machine). In other words, when the user's steady forward walking velocity is V_{xs} (this velocity usually doesn't change a lot for a user), then the necessary pushing force F_{xs}, or saying, the burden the user feels reacted from the robot, should be

$$F_{xs} = V_{xs} \cdot D_x \tag{24}$$

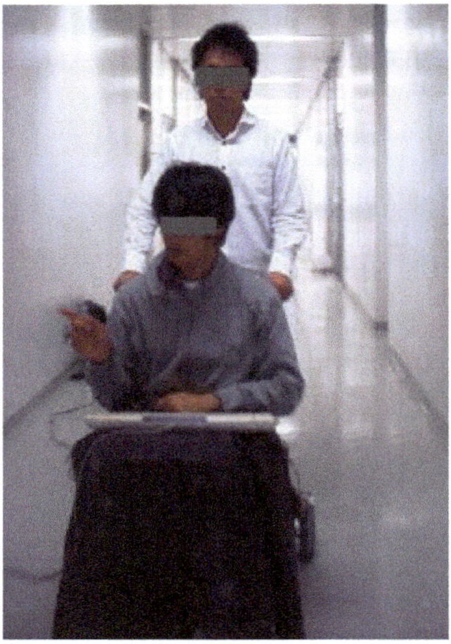

Fig. 7. A user pushes the robot while a person sitting in the seat.

Thus, by adjusting the virtual damping coefficient D_x, the user will have different burden feeling. And further by altering virtual mass coefficient M_x (therefore τ_x is changed), we can get the different dynamic corresponds of human-machine system. Our experiments will verify this.

5. Experiments and remarks

The omnidirectional mobility of the robot is experimentally implemented in (Zhu et al., 2010). Here, the developed admittance based interaction control is tested. The sampling and control period is 1 [kHz]. As shown in Fig. 7, a user is holding the robot handle to move the robot forward, or lateral, or tuning on the spot while a 65[kg] person sitting in the seat.

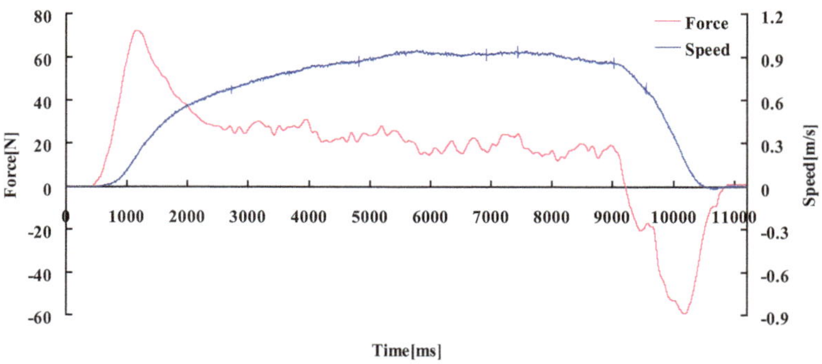

Fig. 8. Forward experimental result without assistance

5.1 Forward experiment without power assistance

First, we turn off the motor drivers and do the experiment without power assistance. In this case, the motors are in free state and the robot is just like a traditional manual-operated wheelchair. From the the result shown in Fig. 8, we can find:

1. to start to move the robot, a big pushing force as large as 70[N] is needed.
2. in steady state, the pushing force is about 24[N].
3. to stop the robot, a negative (pulling) force is needed.
4. the user's walking speed is about 0.9[m/s].

5.2 Forward experiments with power assistance as $D_x = 14$[N·s/m]

In this case, we try to halve the user's burden with power assistance to about 12[N]. Since the user's walking speed is about 0.9[m/s], according to eq.(24) we set $D_x = F_{xs}/V_{xs} = 14$[N·s/m].
Fig. 9 and 10 respectively show the experimental results with $\tau_x = 1.1$[s] ($M_x = \tau_x \cdot D_x = 15.4$[kg]) and $\tau_x = 0.5$[s] ($M_x = 7$[kg]). Both the results show that the pushing force is reduced to about 12[N] from about 24[N] when the steady walking speed is about 0.9[m/s] as we planned, and almost no pulling (negative) force is needed to stop the robot, which can be directly interpreted by eq.(23) or (24). The purpose of power assistance for the caregiver is achieved. In addition, Fig. 9 indicates that a slow and big pushing force occurs at the start to move the robot, while there is no such phenomenon in Fig. 10. This can be explained as follows. The parameter M_x, the virtual mass coefficient, is set to be 15.4[kg] in Fig.9 and M_x is 7.0[kg] in Fig.10. Since mass is a metric of inertia, the more the inertia of an object is, the slower the object starts to move. But a too small M_x will make the system too sensitive to the exerted force. This will lead to the user fearful to use the robot.

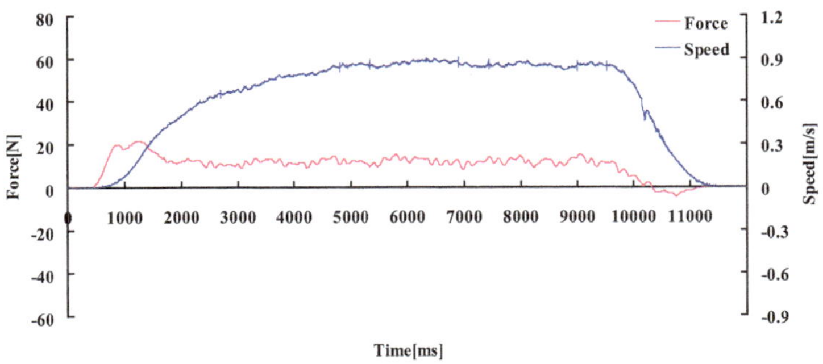

Fig. 9. Forward Experimental result with assistance (as $D_x = 14$ and $\tau_x = 1.1$)

5.3 Forward experiments with power assistance as D_x=9, 19[N·s/m]

The result of power assistance experiment with $D_x = 9$[N·s/m] and $\tau_x = 0.5$[s] ($M_x = 4.5$[kg]) is shown in Fig. 11. The result shows that the pushing force is about 7[N] when the walking speed is about 0.9[m/s]. The experiment result with $D_x = 19$[N·s/m] and $\tau_x = 0.5$[s] ($M_x = 9.5$[kg]) is shown in Fig. 12. The pushing force is about 17[N] as the walking speed is about 0.9[m/s]. Same as the above, these results demonstrate that the pushing force (the

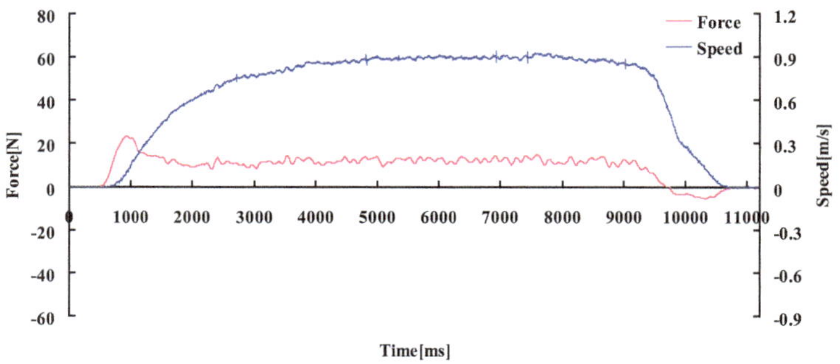

Fig. 10. Forward experimental result with assistance (as $D_x = 14$ and $\tau_x = 0.5$)

user's burden) is well controlled by the admittance based controller. In Fig. 10 there occurs a comparatively big force for starting to move the robot. As aforementioned, this is caused by a relatively big mass coefficient M_x. How to optimally select the two parameters is one of our future tasks.

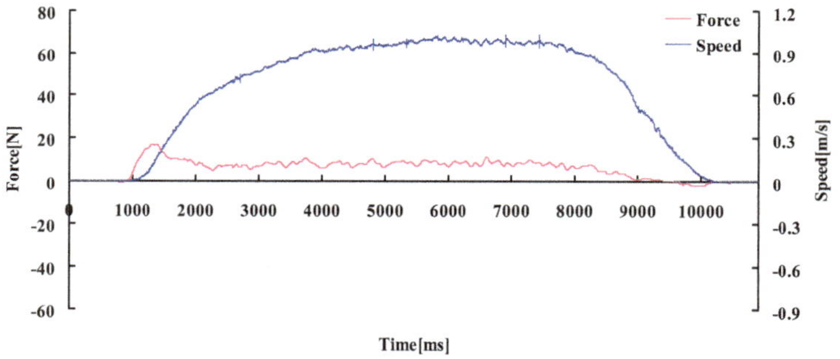

Fig. 11. Forward experimental result with assistance (as $D_x = 9$ and $\tau_x = 0.5$)

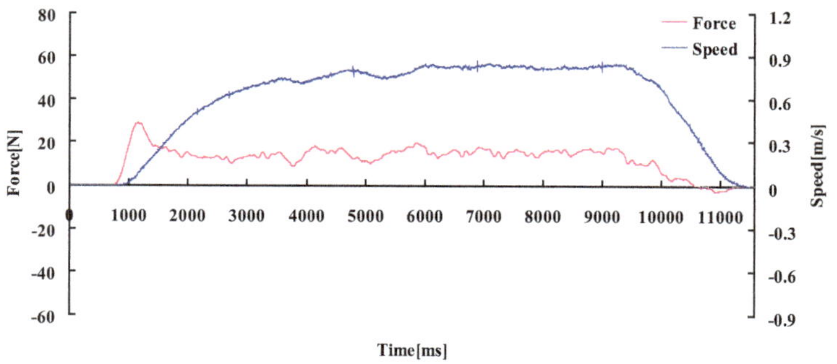

Fig. 12. Forward experimental result with assistance (as $D_x = 19$ and $\tau_x = 0.5$)

5.4 Lateral experiments without power assistance

We turn off the motor drivers and do the experiment without power assistance. In this case, the motors are in free state. From the the result shown in Fig. 13, we can find:

1. to start to move the robot, a big pushing force as large as 40[N] is needed.
2. in steady state, the pushing force is about 22[N].
3. to stop the robot, a negative (pulling) force is needed.
4. the user's walking speed is about 0.6[m/s].

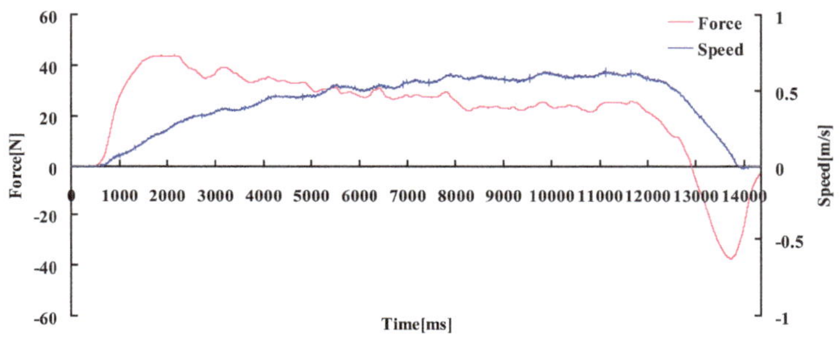

Fig. 13. Lateral experimental result without assistance

5.5 Lateral experiments with power assistance as $D_y = 18$[N·s/m]

In this case, we try to halve the user's burden with power assistance to about 11[N]. Since the user's walking speed is about 0.6[m/s], according to eq.(24) we set $D_y = F_{ys}/V_{ys} = 18$[N·s/m].

Fig. 14 shows the experimental results with $\tau_y = 0.5$[s] ($M_y = 9$[kg]). The results show that the pushing force is reduced to about 11[N] from about 22[N] when the steady walking speed is about 0.6[m/s] as we planned, and a very small pulling (negative) force appeared to stop the robot. The purpose of power assistance for the caregiver is achieved.

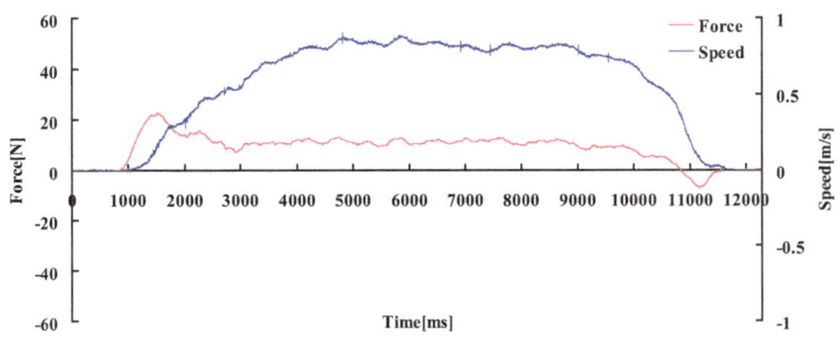

Fig. 14. Lateral experimental result with assistance (as $D_y = 14$ and $\tau_x = 0.5$)

5.6 Turning-on-the-spot experiment without power assistance

We switch off the motor drivers and do the experiment without power assistance. In this case, the motors are in free state. From the the result shown in Fig. 15, we can find:

1. to start to move the robot, a big pushing torque as large as 19[Nm] is needed.
2. in steady state, the pushing torque is about 10[Nm].
3. to stop the robot, a negative (pulling) torque is needed.
4. the user's walking speed is about 0.8[rad/s].

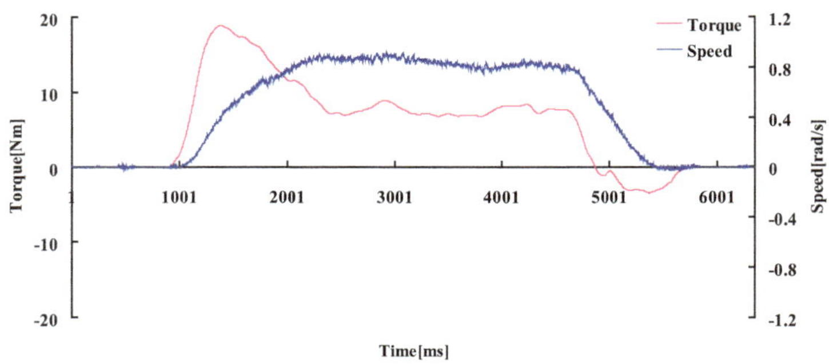

Fig. 15. Turning-on-the-spot experimental result without assistance

5.7 Turning-on-the-spot experiments with power assistance as $D_z = 6$[Nm·s/rad]

In this case, we also try to halve the user's burden with power assistance to about 5[Nm]. Since the user's walking speed is about 0.8[rad/s], according to eq.(24) we set $D_z = F_{zs}/V_{zs} = 6$[Nm·s/rad].

Fig. 16 shows the experimental results with $\tau_z = 0.5$[s] ($I_z = 9$[kg m^2]). The results show that the pushing torque is reduced to about 11[Nm] from about 22[Nm] when the steady walking speed is about 0.6[rad/s] as we planned, and no pulling (negative) torque is needed to stop the robot, which can be directly interpreted by eq.(23). The purpose of power assistance for the caregiver is achieved.

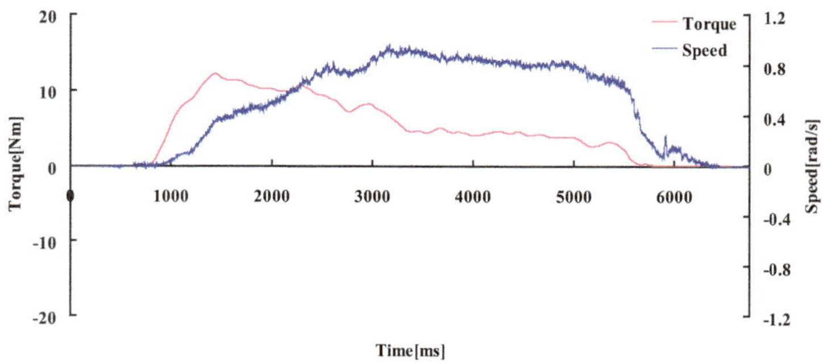

Fig. 16. Turning-on-the-spot experimental result with assistance (as $D_z = 6$ and $\tau_z = 0.5$)

6. Conclusions

In this research, we developed a new type of omnidirectional mobile robot that not only can fulfill the walking support or walking rehabilitation as the elderly or the disabled walks while holding the robot handle, but also can realize the power assistance for a caregiver when he/she pushes the robot to move while the elderly or the disabled is sitting in the seat. The omnidirectional mobility of the robot is analyzed, and an admittance based human-machine interaction controller is introduced for power assistance. The experimental results show that the pushing force is reduced and well controlled as we planned and the purpose of walking support and power assistance is achieved.

7. References

Dicianno, B. E., Spaeth, D. M. & Rory A. Cooper, e. a. (2007). Force control strategies while driving electric powered wheelchairs with isometric and movement-sensing joysticks, *IEEE Transactions on Neural Systems and Rehabilitation Engineering* Vol.15(No.1): 144–150.

Graf, B., Hans, M. & Schraft, R. D. (2004). Care-o-bot ii.development of a next generation robotic home assistant, *Autonomous Robots* Vol.16(No.2): 193–205.

Han, F., Yamada, T., Watanabe, K., Kiguchi, K. & Izumi, K. (2000). Construction of an omnidirectional mobile robot platform based on active dual-wheel caster mechanisms and development of a control simulator, *Journal of Intelligent and Robotics Systems* Vol.29(No.3): 257–275.

Kakimoto, A., Matsuda, H. & Sekiguchi, Y. (1997). Development of power-assisted attendant-propelled wheelchair, *Proc. of 19th Annual International Conference of the IEEE Engineering in Medicine and Biology Society*, pp. 1875–1876.

Lacey, G. & Dawson-Howe, K. M. (1998). The application of robotics to a mobility aid for the elderly blind, *Robotics and Autonomous Systems* Vol.23(No.4): 245–252.

Miyata, J., Kaida, Y. & Murakami, T. (2008). v–$\dot{\phi}$ coordinate-based power-assist control of electric wheelchair for a caregiver, *IEEE Transactions on Industrial Electronics* Vol.55(No.6): 2517–2524.

Nemoto, Y. & M. Fujie, e. a. (1998). Power-assisted walking support system for elderly, *Proc. of the 20th Annual International Conference of the IEEE Engineering in Medicine and Biology Society*, pp. 2693–2695.

Newman, W. (1992). Stability and performance limits of interaction controllers, *Transactions of the ASME, Journal of Dynamics Systems, Measurement, and Control* Vol.114(No.4): 563–570.

S. Dubowsky, e. a. (2000). Pamm-a robotic aid to the elderly for mobility assistance and monitoring: A "helping-hand" for the elderly, *Proc. of 2000 IEEE International Conference on Robotics and Automation*, pp. 570–576.

Ulrich, I. & Borenstein, J. (2001). The guidecane - applying mobile robot technologies to assist the visually impaired, *IEEE Transactions on System, Man, and Cybernetics-Part A: Systems and Humans* Vol. 31(No. 2): 131–136.

Wada, M. & Mori, S. (1996). Holonomic and omnidirectional vehicle with conventional tires, *Proc. of 1996 IEEE International Conference on Robotics and Automation*, pp. 3671–3676.

Yanco, H. A. (1998). Integrating robotic research: A survey of robotic wheelchair development, *AAAI Spring Symposium on Integrating Robotic Research*, Stanford University, CA.

Yu, H., Spenko, M. & Dubowsky, S. (2003). An adaptive shared control system for an intelligent mobility aid for the elderly, *Autonomous Robots* Vol.15(No.1): 53–66.

Yu, H., Spenko, M. & Dubowsky, S. (2004). Construction of an omnidirectional mobile robot platform based on active dual-wheel caster mechanisms and development of a control simulator, *Transactions of the ASME, Journal of Mechanical Design* Vol.126(No.8): 822–829.

Zhu, C., Oda, M., Suzuki, M., Luo, X., Watanabe, H. & Yan, Y. (2010). A new type of omnidirectional wheelchair robot for walking support and power assistance, *Proc. of 2010 IEEE International Conference on Intelligent Robots and Systems*, pp. 6028–6033.

A Control System for Robots and Wheelchairs: Its Application for People with Severe Motor Disability

Alonso A. Alonso, Ramón de la Rosa, Albano Carrera,
Alfonso Bahillo, Ramón Durán and Patricia Fernández
University of Valladolid
Spain

1. Introduction

There is a large number of people with disabilities that involve severe reduction of mobility, such as tetraplegia, brain stroke or vascular brain damage. These people usually have great impairments which prevent them from performing their normal daily activities. For our society, this fact means a great effort in cares and specialized attention that, in most cases, involves caregivers assistance dedicated almost exclusively to these patients.

Although only a small part of the disabled population can be included in these types of injuries, trying to solve or minimise the problems associated with severe motor disabilities is an especially difficult challenge for research and development in Rehabilitation Technologies (RT). The Laboratory of Electronics and Bioenginnering (LEB) of the University of Valladolid, to which the article authors belong, has been working since 1995 on the development of practical systems in this field.

Although, currently, it is impossible to find an effective medical solution for these disabilities, the improvement in patients' personal autonomy should be achieved through technical means. This approach focuses, primarily, on two actions based upon:

- The environment. This action consists in modifying the environment to facilitate mobility, interaction with the ambient and communication.
- The patient. In this case, the action consists in developing human-machine interfaces adapted to each type of disability and desired functionality. Additionally, it is necessary to implement the corresponding actuators or systems to be controlled, including prosthetics and orthotics.

As regards the adaptation of the environment, the implemented solutions have relied, primarily, on laws that promote accessibility, applied to architecture and urban planning. Thus, significant improvements were achieved to facilitate, mainly, the mobility of motor disabled people who use wheelchairs. In this regard, the buildings, the infrastructure planning and the public transport services have been especially taken into account. In order to achieve a suitable ambient, authorities subsidise the reforms in private homes where disabled people live, in some countries (Spain, 2006). Note that this first action, based on the transformation of the environment, also benefits a wider scope of population, such as the elderly or other people with milder motor disabilities.

In the case of adapting the environment, the action is mainly reflected in changes made on the urban and architectural physical space and, in any case, on the special furniture required and home automation elements that facilitate automated control of a home. In contrast, the aid techniques based on patients are more diverse as they rely on a multitude of particular applications. In this way, if functionality is will the topic to considered four main areas of work in developing patient-centric devices can be distinguished:

- Mobility
- Interaction with external technical elements.
- Telecare and telemonitoring
- Communication

In all electromechanical and electronic developments, the four work areas need a suitable interface between the patient and the technical means of control. The development of a suitable and customizable adapter is much more difficult in the case of severe motor disabilities, where there is no possibility of utilising the patient's limbs to control simple or standard devices.

The growing interest in all these work fields is established by the appearance of numerous different works and systems designed to make life easier for the user. A particularly interesting example of application is the control of an interface with a computer, since on that computer just about any control application can be developed. This fact allows to understand that the development of a human-machine interface is the core of the problem to be solved. In order for disabled people to function, one of the most studied tasks has been to facilitate the control of personal computers, (Gareth et al, 2000), (Kim, 2002), (Betke et al, 2002), (De Santis & Iacoviello, 2009) and (Barea et al, 2011). In these works different systems are developed to adapt the control of the mouse cursor using, firstly, an infrared pointer system and secondly a face tracking. Other works present as people with disabilities in the upper limbs can control peripherals (Azkoitia, 2007).

Based on the experience of the research group and the systems found in the scientific related literature, it has been determined that the human-machine interfaces developed must meet the following conditions (Maxwell,1995):

- Adaptation to the user. The interface should used, effectively, taking advantage of any remaining capability of the disabled, which depends on the type and degree of disability.
- Correct choice of the remaining capability to be used. It is important to identify which of the various remaining capabilities of the patient is the most adequate to act with the interface.
- Reliability of the interface. The error level in the interpretation of the commands generated by the patients should be minimised. This task is especially important for interfaces to control personal mobility systems, where potential failures may pose a real physical danger to the user.
- Capacity. The interfaces must be capable of generating a sufficient number of different commands to control the external system.
- Speed and computational cost. The human-machine interfaces should be able to work in real-time or with minimum delay.
- Comfort. This item implies easy usability, ergonomics and low proper interaction with other activities that the patient carries out at the same time.
- Cost. The interfaces to control the system must have a reasonable price for the potencial users.

As much as the work is focused on creating interfaces to control wheelchairs or other type of mobile systems, as robots, there are several papers that review the different existing systems. Their general conclusion is that improvements should be made on the control interfaces and monitoring technologies for automatic navigation (Fehr et al, 2000). The works found in the scientific literature show different approaches to the development of control interfaces in wheelchairs or surveillance robots for severely disabled people. The solutions proposed in the literature can be grouped into the following strategies:

- System based on voice recognition, (Mazo,1995). The operating principle of such interface systems is reduced to the interpretation of a set of predefined voice commands.
- Motion detection systems of the head or the eye, using cameras, (Úbeda et al, 2009) and (Perez et al, 2009). This type of detection systems is based on monitoring the user's face or a part of it. Thus, the movements made by the user are translated into control commands for the device of interest.
- Interfaces based on electromyography (EMG) and electrooculography (EOG) records generated by muscles during voluntary action, (Barea et al, 2002), (Frizera et al, 2006), (Ferreira et al, 2007) and (Freire Bastos et al, 2009). These techniques are based on the acquisition and analysis of electric logs generated by the muscles, which allow us to know the level of muscle activity and the duration thereof. The term EMG is utilised when unspecified muscle signals are used, while EOG indicates that EMG signals around the eye are used. Thus, combining the duration and intensity of the voluntary muscle contractions, various commands can be generated.
- Electroencephalography (EEG)-based interfaces - Brain Computer Interfaces (BCI), (Millán et al, 2004), (Ferreira et al, 2007), (Escolano et al, 2009), (Freire Bastos et al, 2009) and (Iturrate et al, 2009). This means using the signals of brain activity to control the system after processing and coding possible commands. Currently, there are two types of techniques in this category, the invasive and noninvasive (Millán & Carmena, 2010). The invasive strategy implies that the signals are collected inside the brain or, generally, on its surface. The noninvasive approach uses the signals recorded on the surface of the scalp.
- Systems based on the detection of residual muscle movements and not based on the registration of bioelectric signals, (Huo et al, 2008) and (Alonso et al, 2009). This is the kind of system developed in the research group of the authors and that will be explained in detail in this chapter. It is based on coding small residual muscle movements that the patient can perform and on a suitable sensor to detect such actions.
- Interfaces based on the detection of head position using inertial sensors, (Azkoitia, 2007) and (Freire Bastos et al, 2009). In this case, the objective is to detect the position and the movements of the head by placing an external element on it, so that the movement is collected. With this simple mechanism, a light processing system to manage the signals and encode the desired order can be used according to gesture made in each case.
- System based on control by sniffing, (Plotkin et al, 2010). The guidance system is based on the detection of inhalations and exhalations of the use and further processing to code orders according to a preset sequence of inhalations and exhalations.

- Autonomous navigation, (Alonso, 1999), (Levine et al, 1999), (Minguez, et al, 2007), (Angulo et al, 2007) and (Zeng et al, 2008). The autonomous navigation systems perform an automatic guide to the previously chosen destination. Thus, the user only interferes with the system for choosing the final point; it is the system that automatically takes the user to the desired position. The interface chosen, any of the above points, will be used only at the beginning of the operation or if it intends to stop the automatic guide option.

Additionally, hybrid systems that combine self-guided techniques with directed guidance have been developed. Thus, the device intelligently chooses the best possible option for the next move, after the environment analysis, (Hoppenot & Colle, 2002) and (Perrín et al, 2010). In other cases, the system knows different paths and it performs an automatic guidance after the order of destination, (Rebsamen et al, 2006).

For the development of assistance RT elements is of paramount importance, to have the opinion of the recipients of these developments, i.e., persons with disabilities. Therefore, the research group conducted a survey, among the population with severe motor disabilities, concerning the basic features that systems designed to enhance their autonomy and capabilities should have. The results of this study have been taken into account in the LEB works.

The system presented in this chapter is based on sensors capable of detecting winks performed voluntarily by the user. The devices with these interfaces are classified in the fifth group of the types presented above. To carry out this detection, several user interfaces, based mostly in a pair of glasses support have been developed to detect the movement of the skin near the eye. In this area, when there is a wink the orbicularis oculi muscle contractions and extensions produce a movement. The hardware needed for the treatment of signals received is simple, robust, inexpensive, compatible with commercial wheelchairs and implemented by microcontrollers.

This chapter is divided into different sections. Section 2 presents the objectives that the group aim to achieve by the use of the system. Then, section 3 presents and explains the architecture of each of the component parts and the different interfaces developed. Section 4 makes a comparative assessment of the advantages and disadvantages of the proposed solution and those contained in the related scientific literature, after the presentation of the conclusions reached from the interview survey. In addition, an explanation about the protocols and test which has been put into practice with the developed systems is included. Finally, Section 5 details the conclusions reached after the completion of all the work and subsequent analysis.

2. Objectives

This chapter presents the basic theoretical aspects of the interfaces and systems adapted to severe motor disabilities, including an extensive reference to the state of the art in this field. In addition, an interface development, especially suitable for this kind of disabilities, and two particular systems which use this interface are presented. These two systems have the aim, on the one hand, to manipulate a robot that acts as a surveillance or telecare device and, on the other hand, to guide an electric wheelchair. Therefore, the following objectives can be defined:

- Presentation of the parameters that an adapted interface to the severe motor disability must achieve.

- Review of related scientific literature to learn about the different approaches carried out by the research community in this field.
- Study the feasibility of assisted systems for the disabled taking into account the views of users.
- Presentation and operating principle of adapted interfaces developed in the research group of the authors.
- Presentation of the structure and principle of operation of two systems that use the adapted interfaces implemented. The first succeeds in establishing a surveillance and telecare system, using a mobile robot equipped with a webcam. The second facilitates patient mobility through the control of a modified electric wheelchair.
- Presentation of the results and experience in the use of the entire system and interfaces. Additionally, there is a discussion about the advantages of the interfaces based on residual voluntary actions of severe motor disabilities and compare its advantages over other systems found in the scientific literature for the same kind of disability.

The developed devices are described with sufficient level detail. Thus, a specialized reader can incorporate different tools and procedures to their own developments.

The comparison that will be made with similar systems will focus, primarily, on the study and adaptation to the parameters defined in the introduction. In addition, it should be noted that the adequacy of systems will be studied in relation to its use in patients with severe motor disabilities, excluding other types of disabilities like sensorial or cognitive ones and their specific needs.

3. Implementation

This section describes the working principle of a robot guidance system based on a user interface customized according to severe motor disabled patients. This system has been implemented in two different applications: the first one, a surveillance and telecare system using a mobile robot equipped with a webcam, and the second one, a system for patient mobility through control of a modified electric wheelchair.

The system diagram shown in Figure 1 describes the structure of the application developed for surveillance and telecare. This system includes a customizable interface, a processing module, a transmission channel, a reception module and, finally, the device to be controlled: a robot, an electric wheelchair or another convenient domotic system. A Scribbler robot type that represents the device to be controlled is included.

Throughout this section, each of the parts of the system presented in Figure 1 will be explained in detail. Additionally, other items and adaptations such as an electric wheelchair with a guide system based on an adapted interface joystick control will be described.

3.1 Adapted interfaces

The system presented is designed to collect patient information by detecting voluntary winks. Thus, several different adapted interfaces that are explained in detail in this section have been built in. Most of them required an additional support consisting of a pair of glasses and mechanical items for attachment and adjustment of the different sensors. Figure 2 shows three different types of the built interfaces.

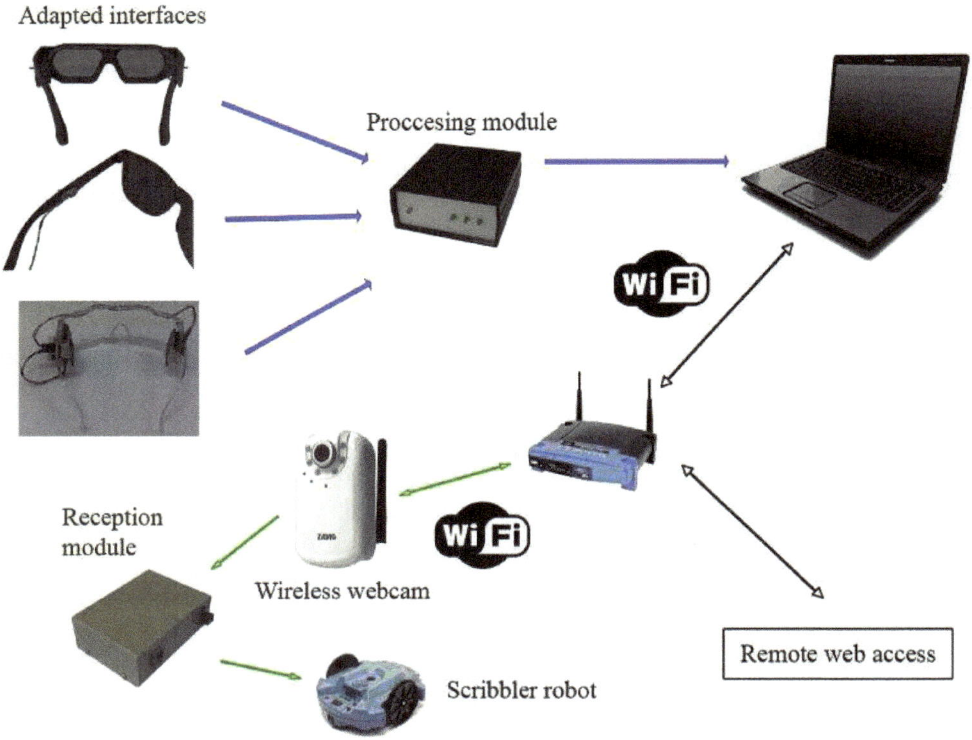

Fig. 1. Diagram of the developed system for telecare and surveillance using a Scribbler robot.

The sensors included in these adapted interfaces and their processing systems are detailed below:

- Optical sensors based on variations of light reflection on the mark that is placed on the skin above the *orbicularis oculi*. In this case, the integrated circuit CNY-70, (Vishay, 2008), has been employed. These devices consist of an infrared LED and a phototransistor whose range of vision is parallel and receives the signal reflected from the facing surface. The use of these sensors requires an additional black and white sticker on the user's skin (see Figure 3). If the sticker is properly placed on the surface of the *orbicularis oculi* muscle, the sensor on the glasses pin will detect the wink because the reflected light changes. This change is registered due to the color change that occurs with the movement of the sticker. Two sets of preprocessing systems for the acquired signals have been built following this screening procedure: the first one based on PIC16F84A microcontroller from Microchip Technology Inc., (Microchip, 2001), and the second based on the Arduino hardware platform that incorporates ATMega microcontrollers from Atmel Corporation, (Arduino, 2009) and (Atmel, 2009).
  The integrated circuit that has been utilised is very simple and only requires two resistors to bias the diode and phototransistor incorporated. The signal generated by the phototransistor is the information signal, and it discriminates a high or low level with the microcontroller.

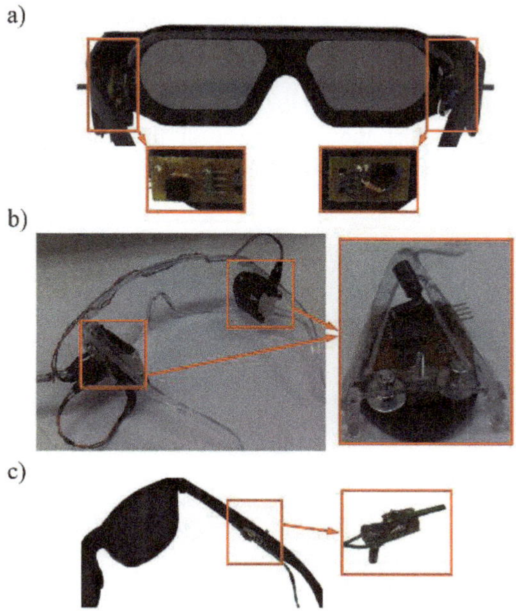

Fig. 2. Three different adapted interfaces built in: a) the highlighted rectangles indicate an optical sensor based on two integrated circuits CNY-70 and their mechanization on glasses. b) the rectangles indicate the circuits fixed on the patient's glasses, built using hardware sensors used in optical mice, c) the highlighted part indicates a customizable interface that uses vibration sensors to detect the winks.

- Optical sensors to detect movements of the *orbicularis oculi* based on optical mice system. Following the same philosophy as the above sensors, these devices are placed on the arm of glasses and detect the movement on the side of the eye. The advantage of this sensor is that it does not need any special attachments, simply placing it on the right place will make it work properly.

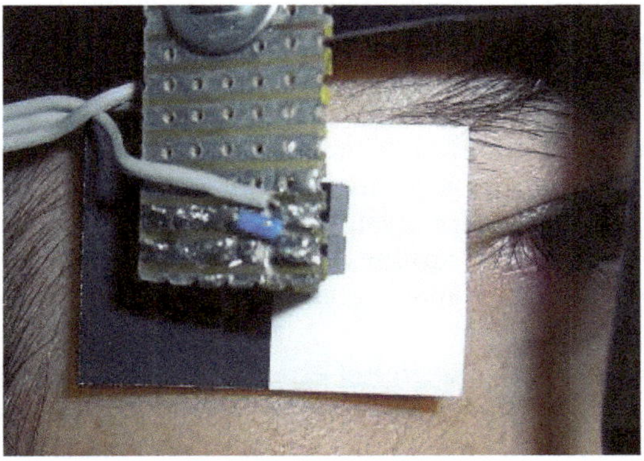

Fig. 3. Adapted interface based on the integrated circuit CNY-70 while it is being utilised by a user. Detail of the sticker on the skin for proper operation.

The sensors only need some parts of the mouse: the integrated optical camera and its peripherals, LED lighting and the light beam guidance prism (see Figure 4).
The system is controlled by the Arduino hardware platform. The interconnection between the optical devices and the platform is achieve through the same signals that a computer processes to detect the movement of the mouse. Thus, it is easy to set different thresholds in order to discriminate motion errors when capturing the information from the user.

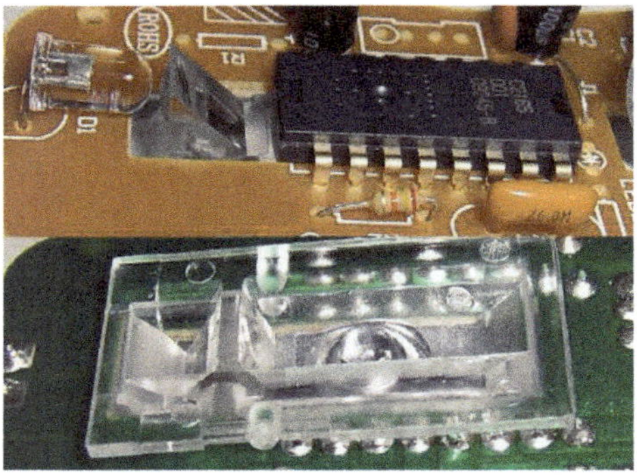

Fig. 4. Detail of the necessary components of the optical mouse for the adapted interface. The top of the image is the top view: LED, prism and integrated camera. The bottom of the image is the bottom view: prism and lens.

- Vibration sensors based on the piezoelectric effect. These sensors have been built in conjunction with a signal conditioner which adapts the signal to a PIC microcontroller, which selects the appropriate orders and discriminates the signals produced by involuntary movements. Again, the philosophy of these sensors is identical to the two mentioned above; it will detect the skin movement near the eye and possible sensor state changes when the eyes wink.
  In contrast with the previous circuits, in this case, the use of a circuit to condition the sensor generated signals is necessary. For this purpose, a simple operational amplifier inverter stage has been developed. This stage is based on the TL082 chip from Texas Instruments, Inc., (Texas, 1999). In addition to this amplifier device, a new stage for information signal pulse discrimination and bounce suppression has been developed. A NE555 integrated circuit from the same company, (Texas, 2002), has been utilised to give shape to this part of the system (see Figure 5).
- Electromyogram electrodes for acquisition of contraction signals of the *orbicularis oculi*. These sensors are based on the acquisition of the bioelectrical signals of each muscle. In this way, it would not be necessary to use the glasses structure interface for signal acquisition, simply, attaching surface electrodes to the skin and an acquisition and processing system, the signal could be obtained and discriminate voluntary winks from other involuntary actions.
  This research group has a high degree of experience in this kind of signals and also has its own neuromuscular training platform UVa-NTS (de la Rosa et al, 2010). This system utilises an amplification and conditioning module (see Figure 6). It consists of different stages: a step for electrostatic discharge, an instrumentation amplifier, a high-pass filter with programmable gain and a low-pass filtering.

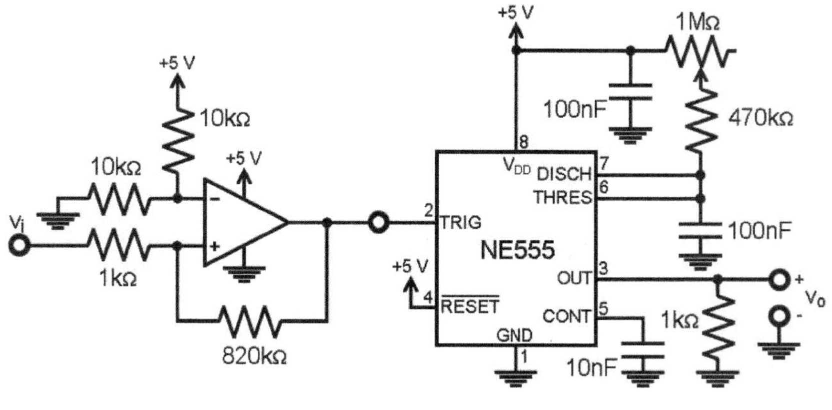

Fig. 5. Circuit developed for the adaptation of the vibration sensors output signal.

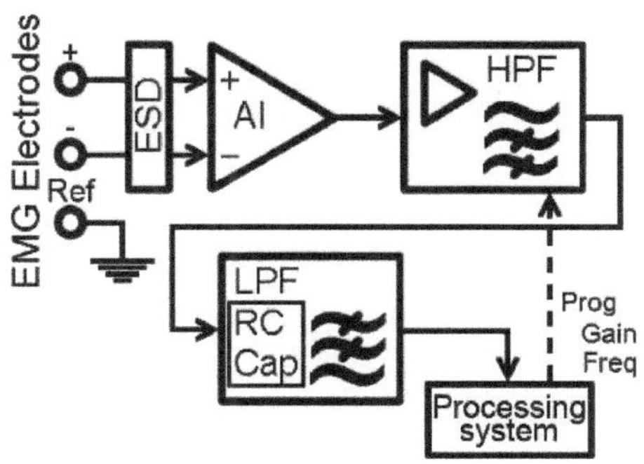

Fig. 6. Block diagram of the electrodes EMG signals acquisition and conditioning module.

In addition to these systems adapted to control devices by voluntary winks for people with major problems of mobility, a less suitable system has been built. It is an intermediate interface between the adapted one and the hardware processing system:

- Special keypad for disabled people. This keypad consists of four buttons that allow the user to define the same orders as those used in the adapted interface. The problem with this interface is a certain difficulty for the disabled person to control a robot or a wheelchair, as performing simultaneous button presses would be necessary, and the user should have greater flexibility and nimbleness with their hands and fingers.

3.2 Processing module

The main task of the implemented processing module is to identify and process the information received by different sensors of the adapted interfaces. This process is straightforward and easily implemented by microcontrollers or hardware platforms such as Arduino.

The processing module is an essential part for the proper functioning of the system and it will be responsible for the discrimination between voluntary or involuntary signal gestures. In addition, the processing system carries out the detection of possible complex orders as the combination of different voluntary signals. The flowchart presented in Figure 7 gives an idea of the philosophy behind the construction of the processing module and the orders discrimination.

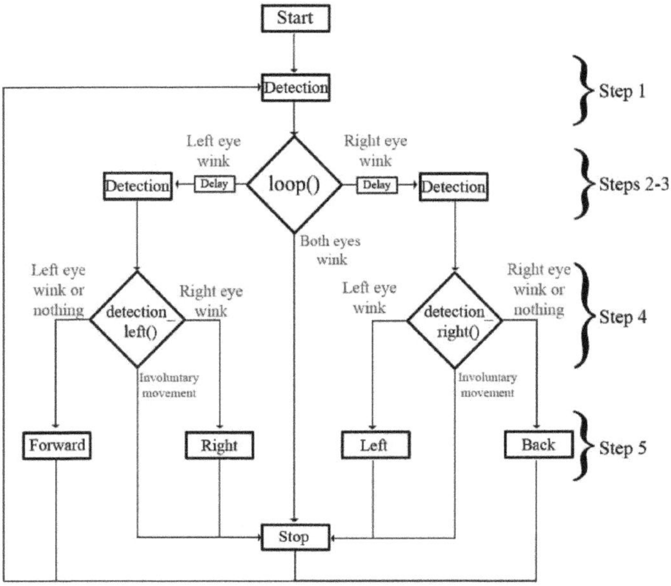

Fig. 7. Flowchart implemented in the processing modules with microcontrollers o hardware platforms.

In a first step, a loop-detection of a possible wink is performed. This initial blink can come from either or both eyes. Once the wink is characterized, the orders are discriminated. On the one hand, if the two eyes wink jointly the robot stop order is activated. Conversely, if the wink is coming from one of the eyes, after a moment's delay, a new similar detection loop to detect a possible new wink is accessed, which discriminates the simple or complex order. The complex commands consist of two consecutive winks made by the two eyes, regardless of the order; in this case, the activated order is 'robot turn right or left and stop'. The simple order consists of a single wink of an eye and it is used to command the movement forward and/or backward of the mobile device.
Following the implementation of this algorithm in the processing module, tests have been carried out with different adapted interfaces, to make sure their reliability and versatility. Switching between interfaces is automatic because the same connector is used. In the specific case of the optical mice-based interface, adapting the code to extract information from the generated sensors signals is necessary.
The received orders are interpreted and forwarded in the appropriate format to a communications device: laptop, notebook or PDA. This device transmits the necessary commands wirelessly to the device to be controlled. In this case, the processing module generates an audio signal of a certain frequency for each of the movements.

3.3 Transmission channel and webcams for telecare and surveillance applications

The Wi-Fi is the transmission channel and it is implemented using a router as an access point, allowing both an *ad-hoc* access (computer - webcam) and a remote Internet access. In

this way, the transmission channel will broadcast a two-way data flow using the video channel to receive images from the webcam and the audio channel to send commands.
Communication is simple for the adapted interface system, through the direct connection with the Wi-Fi device. Thus, the audio orders generated by the processing module are directly received through the communication channel, by the mobile robot with a webcam that has a wireless interface. The control orders are received by the device to be controlled through Internet and the wireless webcam interface. In this case, a router is used as the access point.
On the other hand, if the mobile device needs to be controlled remotely, a simple graphical interface based on a web page, programmed in HTML and JavaScript, has been developed. This graphical interface allows to send the same commands that the adapted one. Operating it is very easy because the screen has five buttons (stop, move forward and backward, turn left and right) that, when pressed with the mouse cursor it sends an audio signal for the execution of the chosen movement.
Hence, with this use of wireless communications and the Internet, both surveillance tasks for disabled people at home and telecare can be conducted. The first application allows the disabled person commanding the robot, with the adapted interface, to observe what occurs in different rooms of the home that he/she cannot reach. Moreover, the second application of remote assistance allows a caregiver, who cannot be physically with the patient, to have control and some information of the status of his/her patient, using the robot and the camera to display the environment of the disabled person.

3.4 Reception module and robot

The reception module is responsible for detecting the received order by decoding the information signal from the processing system through the communications channel. The received signal is demodulated to obtain the control command by a system based on Phase-locked loop (PLL). This detection system uses the 74HC4046 integrate circuit from Philips Semiconductor Company, (Philips, 1997), and the LM3914 from National Semiconductor Corp., (National, 1995). The first one, 74HC4046 is a PLL with and integrated voltage-controlled oscilator (VCO), while the sencond is a driver that activates a greater or smaller number of outputs according to an input analog voltage.
Thus, the control input of the LM3914 device is taken from the PLL VCO input and the signal amplitude is directly proportional to the detected frequency of the PLL input and it has been generated by the processing module. The control signal is converted to eight digital signals, where the number of high-voltage signals depends on the input amplitude. For the conversion of these digital signals to the three bits used to encode the commands of the robot, an encoder integrated circuit SN74HC148 Texas Instruments, Inc., (Texas, 2004), has been used. Figure 8 shows a block diagram of this module and the robot.

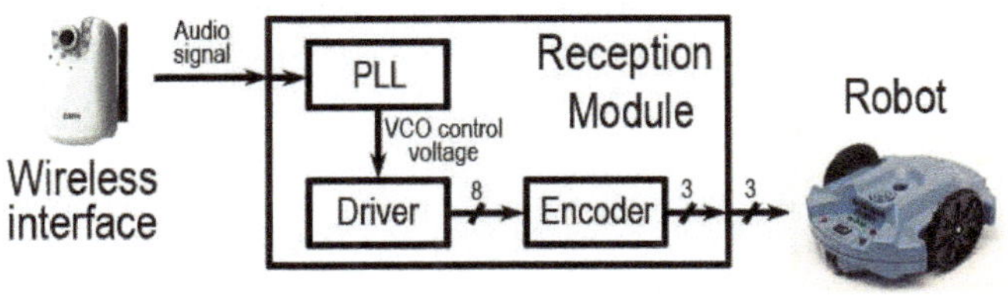

Fig. 8. Block diagram of reception module and robot.

The encoded robot orders, that consist of 3 bits, are summarized in Table 1. The five basic movements -move forward and backward, turn to one side or the other and stop- have been codified through five orders.

Order	Codification
Stop	000
Move forward	101
Move backward	010
Turn right	001
Turn left	100

Table 1. List of robot control commands and their encoding.

Through this combination of commands, three free combinations can be used for more complex robot movements, such as forward while turning slightly, or habilitate device standby option while controlling another system with the same adapted interface and similar orders.

Apart from other commercial systems such as the Scribbler robot, a robot with a custom design has been used. In the implemented robot, a video surveillance camera is incorporated (see Figure 9). It allows the applications described above, to control all the rooms at home by the disabled person and the caregiver`s remote tasks.

3.5 Adaptation to a commercial wheelchair

Once all the system was built and the optimal control of the robot was achieved using the developed interfaces, the equipment was adapted for use in guiding commercial wheelchairs. For this adaptation, a simpler design of the whole system has been implemented. In this design, the processing module is connected directly to the actuator system, the wheelchair. In any case, the amendments to include in the presented system are very easy to carried out.

Fig. 9. Custom design robot implemented with the surveillance system.

The first prototype built for this purpose was intended to replace the control signals generated by the original joystick. Therefore, it could only be used in those wheelchairs which incorporate an identical system: SHARK system with DK-REMA joystick and DK-PMA power module from the Dynamic Controls Company (Dynamic Controls, 2004, 2006). The joystick works easily; it incorporates four fixed coils and a mobile one that moves together with the joystick and induces specific signals in the rest of the coils. The detection of this signals generates the corresponding order for the movement of the wheelchair. With these data, as a result of a reverse engineering process developed, different signals for each type of movement have been generated by the P87LPC769 microcontroller, (Philips, 2002), (Figure 10). This figure shows three signals for each case: the clear line is the reference level, the upper and lower are the order signals and the middle one is the sync signal from the power module.

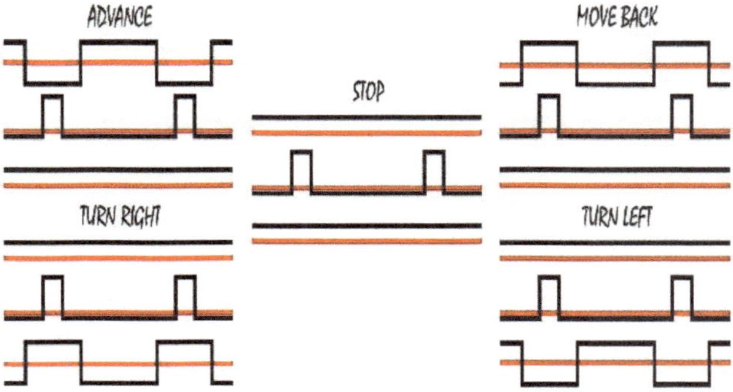

Fig. 10. Signals generated by the microcontroller for the five basic movements of the wheelchair.

These signals are faded directly into the circuit of the joystick, producing the desired movement in each case. The reverse engineering process concluded that the circuitry needed sine signals, but the system worked correctly with square ones because the embedded coils performed filtering functions, and finally, signals similar to sinusoidal ones were obtained. The wheelchair responds appropriately with the synthesized signals and its overall operation is excellent.

In this design, the problem encountered is that it is very delicate to introduce external changes in the internal joystick circuit. These changes can cause a considerable decrease of the device's lifetime. Moreover, the main problem lies in the specificity of the proposed solution, that is, each control system for each wheelchair would require a different model specifically designed.

To solve these two problems encountered when modifying the electronics, a mechanical system based on two 180° rotation servo motors was developed, (Figure 11). This device, controlled by the Arduino platform, is able to move the joystick easily, in every possible direction and ensures universal usability for all wheelchairs that incorporate this kind of guidance system.

Figure 11 shows the prototype built using two 180° servo motors. The joystick is positioned at the bottom of the device. The right image of Figure 11 shows, the top of the prototype, the control Arduino platform, along with a commands LED display and interfaces for connection: USB to a computer (optional connection), supply jack and RJ11 for the adapted interface. In this image also shows the servo motor responsible for setting the direction in which the joystick will be moved. This servo motor controls a central platform where the second motor is located. On the other hand, the left side of Figure 11 includes the interior view of the joystick control system where the second servo motor is responsible for setting off the final movement and speed by a rod, once the direction is positioned. It also allows the system to go to the rest position, i.e., stop the chair quickly and independent of the direction set firstly. With this system, all the 360° of possible positions for the joystick can be easily reached.

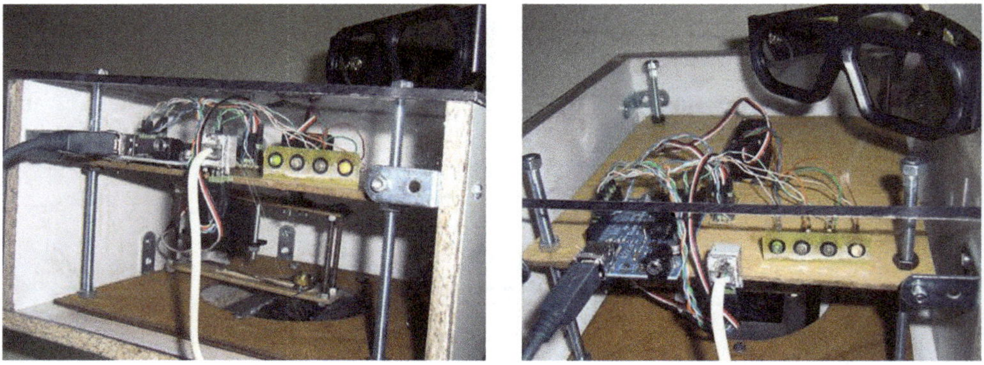

Fig. 11. Wheelchair joystick control prototype based on two servo motors, along with an adapted interface developed.

4. Discussion and results

First, there will be a brief analysis of the results obtained from surveys of a sample of the disabled community. The respondents were divided into two groups, face interviews and email consultations. The personal interviews were conducted in the Spinal Cord Injury Association, ASPAYM (Valladolid), and the Agency of Attention and Resources to the Handicapped, CEAPAT (Salamanca). The email consultation was carried out by two virtual network organizations: DISTEC and ListaMedular. Both organizations have been dedicated to the spread of technical assistance devices all over the world and have lots of users.

The statistical study of 40 disabled persons can be summarized to yield some conclusions of interest for guided assistance system designers. The respondents clearly associated these systems to patients with severe motor disabilities. Over 80% of them prefered the guidance systems that are not fully automatic, so that users can have some sort of continuous interaction with them. Finally, over 90% of the sample population declared to be willing to pay the cost of such systems.

Next, the experience gained from the use of the developments dealt with in this chapter will be presented. The LEB has been collaborating for several years with the National Paraplegics Hospital of Toledo (HNPT), putting the systems into practice and testing them. The HNPT is a Spain reference center in this field. This center is, probably, the most appropriate to test the validity of the developments in Rehabilitation Technologies (RT).

Regarding testing in the laboratory, some basic requirements to be met by adapted interfaces as presented in the introduction of the chapter are also rated here. These laboratory studies allow a comparison with other similar systems found in the literature. In relation to the interfaces found for handling external robotic systems, a comparative analysis of their advantages and disadvantages given the above basic requirements will be provided.

- System based on voice recognition (SVR). This type of system has some serious drawbacks. First, the complexity of pattern recognition systems that is neccesary to use. Second, the difficulty of speaking with other people while performing the guidance. Finally, the possibility of interference with environment talks in the control system. The main advantage is the ease of deployment of large numbers of orders.
- Motion detection systems of the head or the eye, using cameras (MDC). They have common problems like the need to install a PC with a camera, the importance of controlling the lighting of the scene and the interference caused by the use of the system during the patient's normal activity (speaking, freedom to look where interested, etc.).
- Interfaces based on EMG and EOG records generated in muscles during voluntary action (IEMGEOG). The problems in these cases are the need for electrodes and to ensure that they maintain a good contact, the need for a PC and the greater complexity of the system. A major advantage is that they can get analog control output, and after proper training the versatility of the device guidance can be improved.
- EEG based interfaces - BCI (BCI). They are very uncomfortable to use and difficult to apply each time they are used. They are slower and less reliable than other methods and, of course, need a PC. Furthermore, the interference is too great for any other activity that the patient is carring out. The advantage of the BCI is the potential to achieve a high level of control, long term, when the serious technical and biological constraints that currently exist can be overcome.
- Systems based on the detection of residual muscle movements and not based on the registration of bioelectric signals (IRMNBS). These are systems with high reliability because they depend on voluntary actions easily detectable with simple sensors. Moreover, the complexity required is low and limited to record the sensors' output. On the other hand, if both the action and the muscle that performs it are chosen properly, the system allows the user great freedom to carry out daily activities. In this case, sensors like the tracking tongue should be excluded because the ability to talk while performing the guidance is affected.
- Interfaces based on the detection of head position using inertial sensors (IHPIS). They are more robust than previous ones and possibly more suitable for marketing. A PC is not necessary but it requires head movements that interfere with the normal activity of the patient.
- System based on control by sniffing (SNIF). Such systems may interfere with the ability to speak while being used. Furthermore, in a situation in which the user breathes with difficulty, due to fatigue or illness, their performance would be affected.
- Autonomous navigation aids for wheelchair users (ANW). It is a good suplement to the above guidance interfaces. These systems are complex and often require installation of some kind of hardware in the environment and in their own wheelchair. They also need the existence of an adapted interface to select the destination. These systems are justified only for extreme cases of immobility and when the subject reaction rate is

incompatible with the dynamic of the guide. In these cases, the only solution is to choose a destination in advance and let the system do the rest.

A summarized comparative analysis between all adapted guidance systems found in the scientific related literature is presented in Table 2. Thus, there has been an evaluation of the parameters presented in the chapter introduction, using six progressive degrees, from very inappropriate (- - -) to very suitable (+ + +).

	AU	CCRC	RI	C	SCC	CF	CT
SVR	+ +	+ +	+	+ + +	-	- -	+
MDC	+ +	-	-	-	- -	- -	- -
IEMGEOG	+ +	+	+	+	-	-	-
BCI	+	-	- -	-	- - -	- - -	- -
IRMNBS	+ +	+ +	+	+	+ +	+ +	+
IHPIS	+ +	+	+	-	+ +	-	+
SNIF	++	++	+	-	+ +	- -	+
ANW	+	+	+	- -	- - -	+	- - -

Table 2. Comparative analysis between the adapted system interfaces found in the literature and the parameters presented in the introduction. Abbreviations for each of the systems and the parameters have been used. Thus, for columns: AU (adaptation to the user), CCRC (correct choice of the remaining capability to be used), RI (reliability of the interface), C (capacity), SCC (speed and computational cost), CF (comfort) and CT (cost); for rows: SVR (System based on voice recognition), MDC (Motion detection systems of the head or the eye, using cameras), IEMGEOG (interfaces based on EMG and EOG records generated by muscles during voluntary action), BCI (EEG-based interfaces - BCI), IRMNBS (systems based on the detection of residual muscle movements and not based on the registration of bioelectric signals), IHPIS (interfaces based on the detection of head position using inertial sensors), SNIF (system based on control by sniffing) and ANW (autonomous navigation aids for wheelchair users).

Taking into account the previous disccussion and the type of disability to which the developments are directed, the LEB chose the most appropriate option to carry out the adapted interfaces. The system presented in this chapter is based on sensors capable of detecting winks performed voluntarily by the user (a particular IRMNBS system). The hardware needed for the treatment of signals received is simple, robust, inexpensive, compatible with commercial wheelchairs and implemented by microcontrollers. The main advantages of this system are listed below:

- Simplicity of the hardware.
- Light software implemented on microcontrollers.
- More reliable than other systems.
- Low cost and easy adaptation to different wheelchair models.
- Minimal interference with patient daily activities. The necessary conventional biological flashes are not detected, as it should. The winks that activate the system need not be as intense as for having to close the eyes, in order not to lose sight at any time, due to the interface.

The results obtained from practical trials of the devices have responded to the expectations of reliability, speed and convenience for the user. The tests performed with the robotic system monitoring and telecare have been successful in the controlled environment of the LEB, (Figure 12). Additionally, the trials were prepared in private homes equipped with an Internet connection with a Wi-Fi router.

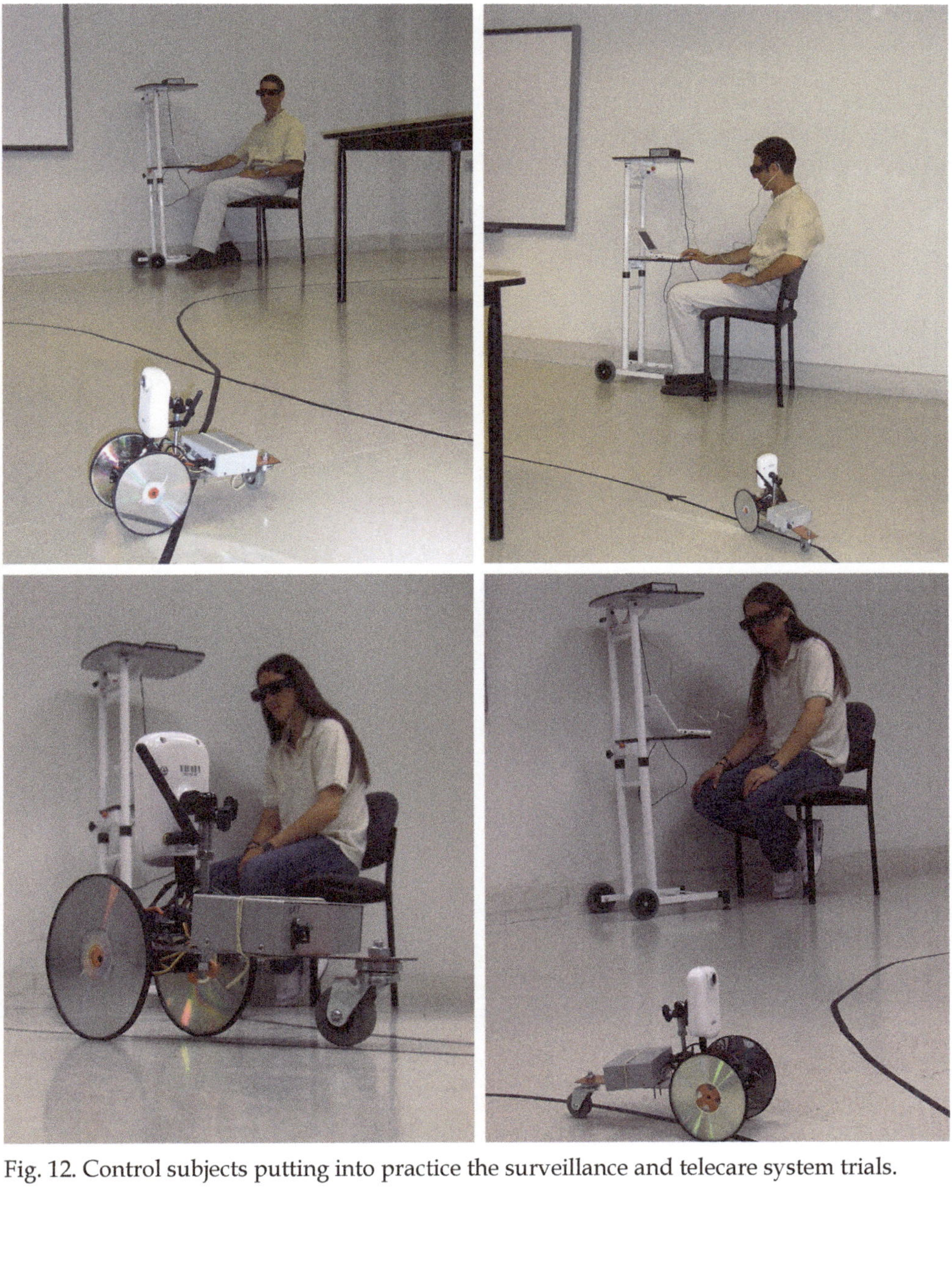

Fig. 12. Control subjects putting into practice the surveillance and telecare system trials.

Regarding the direct control of the electric wheelchair with the adapted interface developed, the data were collected from control subjects, Figure 13, pending further trials in the HNPT research center. In this case, indeed, it is not important that the subjects who test the system are disabled, since the residual capability used to control the adapted interface, i.e., the eye winks, is similar in both populations.

The trials consisted of training and two definitive tests that indicate the progression of learning. Tests are based on a predetermined circuit, Figure 14, and the complete protocol is summarized as follows:

- Previous 5 minutes free training.
- First test scoring errors and the time to complete the circuit.
- 5 minutes training after a break.
- Second test with new errors and time registration.

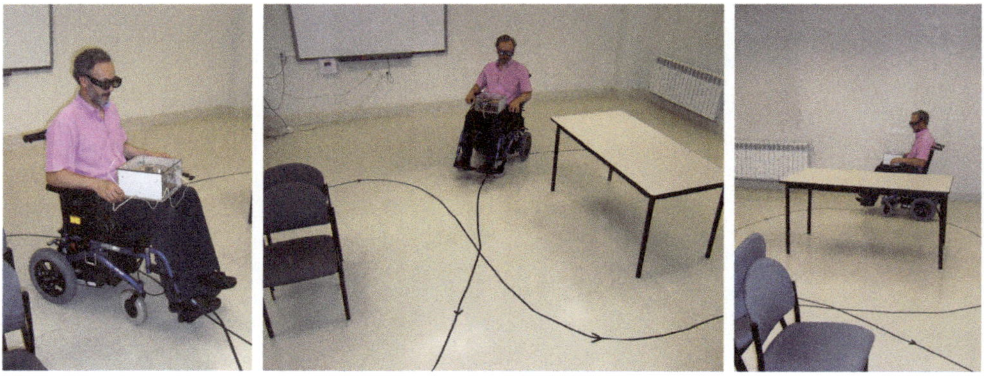

Fig. 13. Control subject testing the adapted wheelchair control system.

The conditions for conducting the trial are based on an optimized circuit for an estimated duration of one minute. For safety reasons, for the users without prior training, the circuit did not contain physical barriers. The recorded values can make an assessment of the usability of the system, user satisfaction and speed of learning.

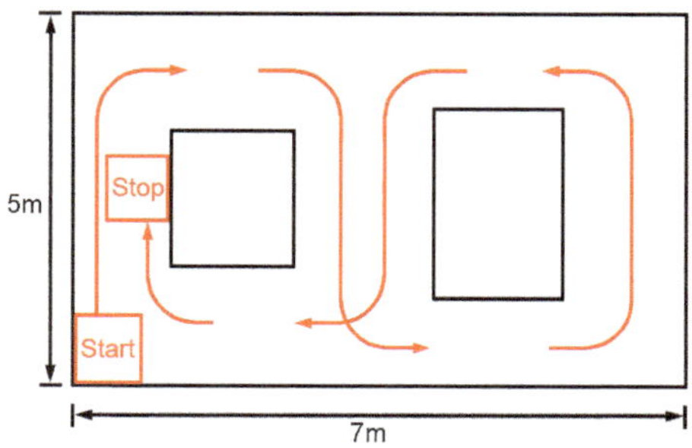

Fig. 14. Designed test circuit for both surveillance and telecare system and adapted wheelchair mobility.

Both systems showed an annoying delay in the response for the action ordered by the user through the interface. This effect is caused by how the interpreting the control commands has been programmed on the microcontroller. This delay was introduced to facilitate beginners to execute the commands performed by a successive wink of each eye. However, this is not a serious problem, since it is simple to modify the code to speed up the response. In this case, the functioning mode would be customized for users who have been properly trained.

5. Conclusions

The design and development of human-machine interfaces for people with severe motor disabilities are activities of great interest to improve the quality of life and independence of such persons. The use of adapted interfaces to their residual abilities serves the disabled to train their remaining capabilities and also provides an assessment tool of the patients for medical specialists.

In this chapter, a brief collection of techniques and interfaces developed by different researchers for this type of patients has been made. The interfaces are generally used to manage domotics and robotic elements with different functions such as mobility, environment control and communication. In the introduction, some basic requirements that the very low mobility disabled-interfaces should meet have been defined. Such requirements can be used to assess the goodness of a given interface or to compare different human-machine interfaces. The control interfaces based on slight voluntary actions, such as winks, are worthy of a high score because they meet most of the conditions.

Once the human-machine interface is developed, a control for two different robotic systems, with good results of satisfaction for users, has been applied. Both systems have been described with sufficient detail to be implemented by those readers with appropriate technical training. The first of these systems -environment surveillance and telecare of the patient system- can be easily modified to use their two-way audio channel, and thus can also perform the communication function. The second system -a mobility system developed- is based on the control of a conventional electric wheelchair, for which only a cheap adaptation between the joystick and the interface of winks is needed. This circumstance significantly increases the accessibility of the system for patients. Both systems can be programmed for a faster response, as users are learning to manage the interface.

6. Acknowledgments

This work has been funded by the *research Excellence Group GR72* project of the Regional Administration, *Junta de Castilla y León*, Valladolid, Spain.

7. References

Alonso, A. (1999). Diseño de un sistema de silla de ruedas autoguiada en entornos controlados. *International Simposium on Biomechanic Methods (SIBVA '99)*, Valladolid (Spain), November, 1999.

Alonso, A., de la Rosa, R., del Val, L., Jimenez, M.I. & Franco, S. (2009). A Robot Controlled by Blinking for Ambient Assisted Living, In: *Distributed Computing, Artificial Intelligence, Bioinformatics, Soft Computing, and Ambient Assisted Living. Lecture Notes in Computer Science*, Omatu, S., Rocha, M.P., Bravo, J., Fernández, F., Corchado, E.,

Bustillo, A. & Corchado, J.M., pp.(839-842), Springer-Verlag, 3-642-02480-7, Berlin (Germany).

Angulo, C., Minguez, J., Díaz, M. & Cabestany, J. (2007). Ongoing Research on Adaptive Smart Assistive Systems for Disabled People in Autonomous Movement. *Proceedings of II International Congress on Domotics, Robotics and Remote-Assistance for All - DRT4all2007*, 84-8473-258-4, Madrid (Spain), April, 2007.

Arduino. (2009). Arduino Duemilanove. In: *ArduinoBoardDuemilanove*, 29 June 2011, Available from: http://www.arduino.cc/en/Main/ArduinoBoardDuemilanove.

Atmel Corporation. (2009). 8-bit AVR Microcontroller with 4/8/16/32 bytes in-system programmable flash: ATmega48PA, ATmega88PA, ATmega168PA, Atmega328P. *Datasheet*. San Jose (CA,USA), October 2009.

Azkoitia, J.M., Eizmendi, G., Manterota, I., Zabaleta, H. & Pérez, M. (2007) Non-Invasive, Wireless and Universal Interface for the Control of Peripheral Devices by Means of Head Movements. *Proceedings of II International Congress on Domotics, Robotics and Remote-Assistance for All - DRT4all2007*, 84-8473-258-4, Madrid (Spain), April, 2007.

Barea, R., Boquete, L., Rodriguez-Ascariz, J.M., Ortega, S. &López, E. (2011) Sensory System for Implementing a Human-Computer Interface Based on Electrooculography. *Sensors*, Vol.11, No 1, January 2011, pp. (310-328), 1424-8220.

Barea, R., Boquete, L., Mazo, M. & López, E. (2002). System for assisted mobility using eye movements based on electrooculography. *IEEE transactions on neural systems and rehabilitation engineering*, Vol. 10, No 4, December 2002, pp. (209-218), 1534-4320.

Betke, M., Gips, P. & Fleming, P. (2002). The Camera Mouse: Visual Tracking of Body Features to Provide Computer Access for People With Severe Disabilities. *IEEE transactions on neural systems and rehabilitation engineering*, Vol 10, No 1, March 2002, pp. (1-10), 1534-4320.

De la Rosa, R., Alonso, A., Carrera, A., Durán, r. & Fernández, P. (2010) Man-Machine Interface System for Neuromuscular Training and Evaluation Based on EMG and MMG Signals. *Sensors*, Vol. 10, No 12, December 2010, pp. (11100-11125), 1424-8220.

De Santis, A. & Iacoviello, D. (2009). Robust real time eye tracking for computer interface for disabled people. *Computer Methods and Programs in Biomedicine*, Vol 96, No 1, October 2009, pp. (1-11), 0169-2607.

Dynamic Controls. (2004). SHARK DK-REMA Remotes. *Installation Manual*. Christchurch, (New Zealand), June, 2004.

Dynamic Controls. (2006). DK-PMA SHARK Power Module. *Installation Manual*. Christchurch, (New Zealand), June, 2006.

Escolano, C., Antelis, J. & Minguez, J. (2009). Human brain-teleoperated robot between remote places, *Proceedings of the 2009 IEEE International conference on robotics and automation*, 1-4244-2789-5, Kobe (Japan), May, 2009.

Fehr, L., Langbein, W.E. & Skaar, S.B. (2000). Adequacy of power wheelchair control interfaces for persons with severe disabilities: a clinical survey. *Journal of Rehabilitation Research and Development*, Vol 37, No 3, May - June 2000, pp. (353-360), 0898-2732.

Ferreira, A., Cardoso Celeste, W., Freire Bastos-Filho, T., Sarcinelli-Filho, M., Auat Cheein, F. & Carelli, R. (2007). Development of interfaces for impaired people based on EMG and EEG. *Proceedings of II International Congress on Domotics, Robotics and Remote-Assistance for All - DRT4all2007*, 84-8473-258-4, Madrid (Spain), April, 2007.

Freire Bastos, T., Ferreira, A., Cardoso Celeste, W., Cruz Calieri, D., Sarcinelli Filho, M. & de la Cruz, C. (2009). Silla de ruedas robótica multiaccionada inteligente con capacidad de comunicación. *Proceedings of III International Congress on Domotics, Robotics and Remote-Assistance for All - DRT4all2009*, 978-84-88934-39-0, Barcelona (Spain), May, 2009.

Frizera, A., Cardoso, W., Ruiz, V., FreireBastos, T. & Sarcinelli, M. (2006). Human-machine interface based on electrobiological signals for mobile vehicles. *Proceedings of the 2006 IEEE International Symposium on Industrial Electronics*, 1-4244-0496-7, Montreal (Quebec, Canada), July, 2006.

Gareth Evans, D., Drew, R. & Blenkon, P. (2000). Controlling mouse pointer position using an infrared head-operated joystick. *IEEE Transactions on rehabilitation engineering*, Vol 8, No 1, March 2000, pp.(107-116), 1063-6528.

Hoppenot, P. & Colle, E. (2002). Mobile robot command by man-machine co-operation - Application to disabled and elderly people assistance. *Journal of Intelligent and Robotic Systems*, Vol 34, No 3, July 2002, pp.(235-252), 0921-0296.

Huo, X., Wang, J. & Ghovanloo, M. (2008). Tracking tongue drive system as a new interface to control powered wheelchairs, *Proceedings of the Rehabilitation Engineering and Assistive Technology Society of North America (RESNA) Conference*,Washington DC (USA), June, 2008.

Iturrate, I., Escolano, C., Antelis J. & Minguez, J. (2009). Dispositivos robóticos de rehabilitación basados en Interfaces cerebro-ordenador: silla de ruedas y robot para teleoperación, *Proceedings of III International Congress on Domotics, Robotics and Remote-Assistance for All - DRT4all2009*, 978-84-88934-39-0, Barcelona (Spain), May, 2009.

Kim, Y.W. (2002) Development of Headset-Type Computer Mouse Using Gyro Sensors for the Handicapped. *Electronics letters*, Vol. 38, No 22, October 2002, pp.(1313-1314), 0013-5194.

Levine, S.P., Bell, D.A., Jaros, L.A., Simpson, R.C., Koren, Y. & Borenstein, J. (1999). The NavChair Assistive Wheelchair Navigation System. *IEEE Transactions on rehabilitation engineering*, Vol. 7, No 4, December 1999, pp. (443-451), 1063-6528.

Mazo, M., Rodríguez, F.J., Lázaro, J.L., Ureña, J., Gracía, J.C., Santiso, E., Revenga, P. & García, J.J. (1995). Wheelchair for physically disabled people with voice, ultrasonic and infrared sensor control. *Autonomous Robots*, Vol 2, No 3, September 1995, pp.(203-224), 0929-5593.

Maxwell K.J. (1995). Human-Computer Interface Design Issues, In: *The biomedical Engineering Handbook*. Bronzino, J.D. pp. (2263-2277), CRC press - IEEE press, 0-8493-8346-3, Salem (MA, USA).

Microchip Technology Incorporated. (2001). PIC16F84A: 18-pin enhanced FLASH/EEPROM, 8-bit microcontroller. *Datasheet*. Chandler, (AZ, USA), January 2001.

Millán, J. del R., Renkens, F., Mouriño, J. & GerstneR, W. (2004). Noninvasive brain-acturated control o a mobile robot by human EEG. *IEEE Transactions on biomedical engineering*, Vol 51, No 6, June 2004, pp.(1026-1033), 0018-9294.

Millán, J. del R. & Carmena, J.M. (2010). Invasive or noninvasive: understanding brain-machine interface technology. *IEEE Engineering in Medicine and Biology Magazine*, Vol 29, No 1, January 2010, pp.(16-22), 0739-5175.

Minguez, J., Montesano, L., Díaz, M. & Canalis, C. (2007). Intelligent robotic mobility system for cognitive disabled children. *Proceedings of II International Congress on Domotics, Robotics and Remote-Assistance for All - DRT4all2007*, 84-8473-258-4, Madrid (Spain), April, 2007.

National Semiconductor Corporation. (1995). LM3914 Dot/Bar Display Driver. *Datasheet*. Arlington (TX, USA), February, 1995.

Perez, E., Soria, C., Mut, V., Nasisi, O. & Freire Bastos, T. (2009). Interfaz basada en visión aplicada a una silla de ruedas robótica. *Proceedings of III International Congress on Domotics, Robotics and Remote-Assistance for All - DRT4all2009*, 978-84-88934-39-0, Barcelona (Spain), May, 2009.

Perrín, X., Chavarriaga, R., Colas, F., Siegwart, R. & Millán, J. del R. (2010). Brain-coupled interaction for semi-autonomous navigation of an assistive robot. *Robotics and Autonomous Systems*, Vol 58, No 12, December 2010, pp.(1246-1255), 0921-8890.

Philips Semiconductors, Incorporated. 1997. 74HC/HCT4046A Phase-locked-loop with VCO. *Datasheet*. Eindhoven (The Netherlands), November, 1997.

Philips Semiconductors, Incorporated. 2002. P87LPC769 - Low power, low price, low pin count (20 pin) microcontroller with 4 kB OTP 8-bit A/D, and DAC. *Datasheet*. Eindhoven (The Netherlands), March, 2002.

Plotkin, A., Sela, L., Weissbrod, A., Kahana, R., Haviv, L., Yeshurun, Y., Soroker, N. & Sobel, N. (2010). Sniffing enables communication and environmental control for the severely disabled. *Proceedings of the National Academy of Sciences of the United States of America*, Vol. 107, No. 32, August 2010, pp. (14413-14418), 0027-8424.

Rebsamen, B., Burdet, E., Guan, C., Zhang, H., Teo, C.L., Zeng, Q., Ang, M. & Laugier, C. (2006). A Brain-controlled wheelchair based on P300 and path guidance, *Proceedings of the first IEEE/RAS-EMBS International conference on biomedical robotics and biomechatronics, 2006. BioRob 2006.*, 1-4244-0040-6, Pisa (Italy), February, 2006.

Spain. Ley 39/2006, de 14 de diciembre, de Promoción de la Autonomía Personal y Atención a las personas en situación de dependencia. Disposición adicional tercera. Ayudas económicas para facilitar la autonomía personal. *Boletín Oficial del Estado (BOE) – Viernes, 15 de diciembre de 2006*, No 299, pp. (44142-44156), Madrid (Spain), December, 2006.

Texas Instruments, Incorporated. 1999. TL082, TL082A, TL082B, TL082Y JFET-Input operational amplifiers. *Datasheet*. Dallas (TX,USA), February, 1999.

Texas Instruments, Incorporated. 2002. NE555, SA555, SE555 precision timers. *Datasheet*. Dallas (TX,USA), February, 2002.

Texas Instruments, Incorporated. 2004. SN54HC148, SN74HC148 – 8-line to 3-line priority encoders. *Datasheet*. Dallas, (TX,USA), April, 2004.

Úbeda, A., Azorín, J.M., Iáñez, E. & Sabater, J.M. (2009). Interfaz de seguimiento ocular basado en visión artificial para personas con discapacidad, *Proceedings of III International Congress on Domotics, Robotics and Remote-Assistance for All - DRT4all2009*, 978-84-88934-39-0, Barcelona (Spain), May, 2009.

Vishay Intertechnology, Incorporated. (2008). CNY 70 – Reflective Optical Sensor with Transistor Output. *Datasheet*. Shelton (CT, USA), July, 2008.

Zeng, Q., Teo, C.L., Rebsamen, B. & Burdet, E. (2008). A collaborative wheelchair system. *IEEE Transactions on Neural Systems and Rehabilitation Engineering*, Vol 16, No 2, April 2008, pp. (161-170), 1534-4320.

7

A Micro Mobile Robot with Suction Cups in the Abdominal Cavity for NOTES

Chika Hiroki and Wenwei Yu
Graduate School of Engineering, Chiba University
Japan

1. Introduction

NOTES (Natural Orifice Translumenal Endoscopic Surgery), in which forceps are put through a natural orifice, such as the mouth, anus or vagina, and a hole is cut at the site to reach intra-abdominal cavity. Because this surgery is able to minimize incision size and the amount of pain, thus greatly improve the quality of life of patients. Although the NOTES approach may hold tremendous potential, there are difficulties that should be overcome before this technique is introduced into clinical care. The most serious one is that since the distance from the surgeon's fingertip to the targeted site is generally longer than that of the usual endoscopic operations, manipulation of forceps is much more difficult, which brings more burdens on the surgeons; meanwhile, there are few surgical devices that could be specifically used for NOTES.

The aim of this study is to develop surgical devices that could facilitate high manipulability and high functionality (cut, hold tissues, hold camera) for NOTES.

The biggest issue when developing an device for NOTES support use is that it has to show both flexibility and rigidity. On one hand, in order to pass a long pathway (i.e., the esophagus) to reach a site (i.e., the stomach) it should be flexible enough. On the other hand, after reaching its target site (i.e., the abdominal cavity), it should show sufficient rigidity which could stay at the site steadily and perform its tasks, such as holding a camera for inspection, and/or a soft forceps in operations.

The first type expanded the traditional flexible endoscope for single port access surgery (Xu. et. al., 2009), which has built-in camera, forceps, and electric scalpel, all folded in a small cylinder before and during insertion, then deployed after reaching its targeted site. This type of device owns sufficient flexibility, however, since the fulcrum of the manipulation (a point to provide support of manipulation) is outside the port, as the distance between the fulcrum and targeted site increases, its rigidity of system will be reduced by its inherent flexibility, and force will be even more difficult to transmitted to the endeffector usually located at the detail portion of the device.

The robot type goes to another extreme. The robot moves around the targeted site, after being inserted through the port. The fulcrum of manipulation is usually near the endeffector, thus the Robot Type usually has good manipulability. It has been reported that a wheeled surgical robot system could move on the surface of liver (Rentschler & Reid, 2009). However, the mobile mechanism could not provide mobility to cover whole abdominal cavity for NOTES support usage. Moreover, not all the surface of inner organs are suitable as the movement plane. Sudden involuntary movement of inner organs in the abdominal cavity would be irresistible disturbance to robots. For the use of surgery support in the abdominal cavity, magnetic based mechanism has been employed (Lehman. et. al., 2008a, 2008b, 2008c, 2009; Dumpert, 2009). Generally, adhered to an outer magnet, a movable magnetic object inside the body cavity could be moved through moving the outer magnet. However, high magnet field needed (0.6T-2T) unavoidably gives strong constraint to surgical tools and devices, which increases cost of surgery. Moreover, the safety issues concerning long-term exposure to strong magnet field should also be carefully investigated.

Suction cup has been shown as an effective way to adhere to living organisms or biological objects, since it gives less damage to attaching surface. Suction effect to avian livers has been investigated in (Horie. et al, 2007), for developing a micro switchable sucker system. HeartLander (Patronik. et al, 2005a, 2005b, 2009), which was designed for moving on epicardium, moved using suction cups and a wire driven mechanism. It is recognized as the most successful mobile robotic system working in human body cavity. However, neither the mobility of the system, nor the adhesion to intraperitoneal environment were tested. The aim of this study was to construct a robotic system that can improve the manipulability and functionality for NOTES, using vacuum suction cups, moving on the peritoneum (the abdominal wall). The advantages of the robotic system, as well as the design details will be discussed in section 2.

2. Construction of the robot

2.1 Basic considerations and specification of the robot

- Stable movement and surgery support actions.

Since the surface of the intraperitoneal organs, such as the large intestine and the stomach, is unstable due to peristalsis and other factors, in order to achieve stable movements and surgery support actions, we had to consider the other possibilities. In this study, we proposed using the peritoneum (the smooth serous membrane that lines the abdominal cavity) as the surface for moving in the abdominal cavity.

- Less damage to the internal environment and fewer electrical devices.

The effort to produce less damage to the internal environment (according to the issue discussed above, the peritoneum) and use fewer electrical devices is significantly important considering the robot's medical application. In this study, we employed vacuum suction as the means to attach to the peritoneum and devised a cable-driven mechanism to realize the relative movements of the two suction cups.

- Degree of freedom (D.O.F.) necessary for mobility and surgery support.

In order to move on the peritoneum and guide the forceps to the targeted sites, 3- D.O.F. movements, i.e., moving forward/backward, turning left/right and moving up/down, should be realized. Moving up/down is important because (i) when one of the suction cups is detached from the peritoneum for the relative movement, anti-gravity control is necessary; (ii) the peritoneum is not a flat plane – especially after being filled with gas (one procedure during laparoscopic surgery), the peritoneum turns to be a dome-like three-dimensional (3-D) landscape; (iii) the elasticity of the peritoneum will increase the difference in elevation; and (iv) the robot should be able to clear the ramp when moving from the overtube to abdominal cavity.

• Size and weight constraints.

While realizing the functions discussed above, there are design constraints that should be met, i.e., the robot should be small and light enough. The constraints come from the size of the overtube for endoscopic operation, and robot's load-bearing ability that depends on the sucking force, and the weight of the robot body, camera and flexible forceps. Considering the inner diameter of the overtube, the diameter of our robot should be limited to less than 16.5 [mm]. However, since the purpose of this study is to investigate the feasibility of the robot, the diametral size of the robot was set to 34 [mm], according to the machining accuracy and parts currently available.

2.2 The robot with suction cups developed

Different with the other robots and devices using suction cups, the robot developed in our previous study moves in the abdominal cavity, hanging upside down from the abdominal wall and supports NOTES. The first prototype is shown in Fig. 1.

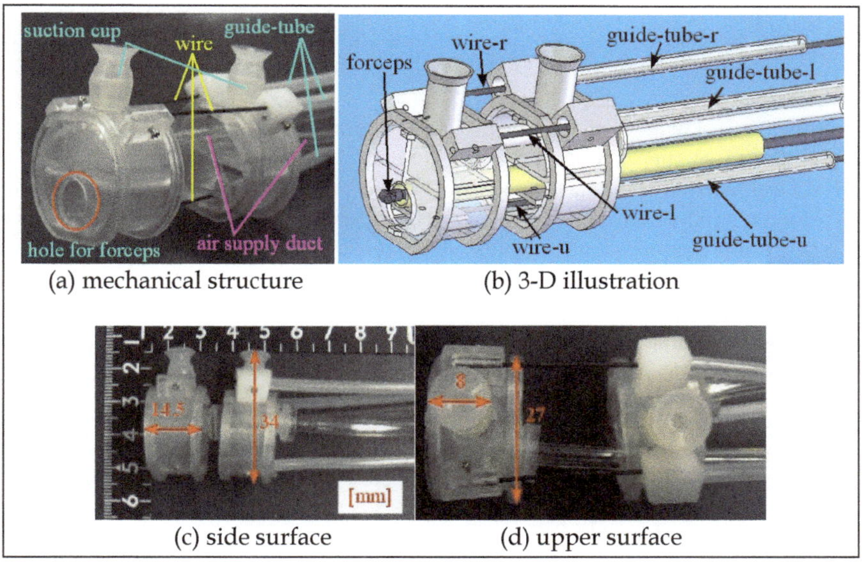

(a) mechanical structure (b) 3-D illustration

(c) side surface (d) upper surface

Fig. 1. Construction of the robot

Three pairs of wires and guide-tubes were equipped to realize relative movement of the two suction cups, thus realizing the 3-D.O.F. movements for the robot. The three pairs of wires and guide-tubes, together with two air supply ducts, were contained in two housings (front and rear housings). Wires were fixed to the front housing and guide-tubes were fixed to the rear housing. In the front housing, there's a hole for loading and guiding forceps. The diametral size of robot is 34 mm, bigger than the inner diameter of the overtube; after carefully re-inspecting the current design of the robot, we could have an optimistic view that the problem can be solved in the next version of prototype. The wire-u and the guide-tube-u are equipped to enable the vertical movement of the robot (shown in Fig. 1). We adopted Ti–Ni alloy as wires for control of the front housing.

2.3 Control system

Fig. 2 shows general view of the robot's control system.

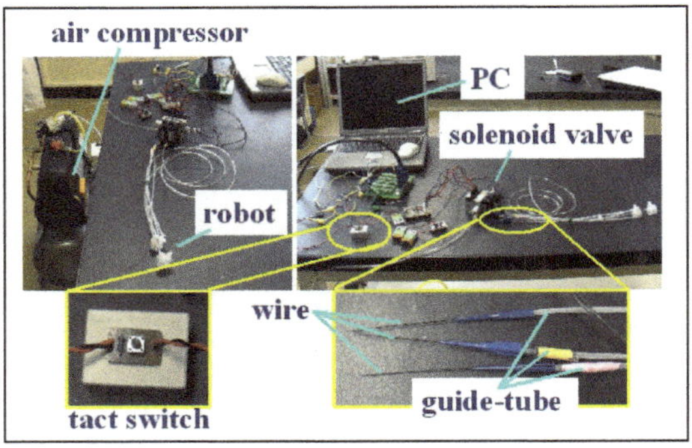

Fig. 2. Robot's control system

The block diagram for the robot control is shown in Fig. 3. Movement of three degrees of freedom could be realized by controlling the state transition between adsorption and release of the suction cups, as well as the relative displacement of the two housings (containing a suction cup in each).

The suction control was achieved by using a digital solenoid valve [ZX1103-K16LZ-EC; SMC] that regulates the vacuum pressure generated by a compressor [0.75LP-75; Hitachi Industrial Equipment Systems], which generates air pressure up to 0.69 MPa. The valve can switch between adsorption and release of suction cups according to the command sent to it from output ports of a personal computer. Concretely, the negative pressure supplied from the compressor is controlled by an input voltage (ON: 12 v or OFF: 0 v), which corresponds deaeration (for adsorption state) or insufflation (for release state) of the air to the suction cups. Moreover, the adsorption state could be detected by the pressure sensor in the valve, and sent back to PC for control.

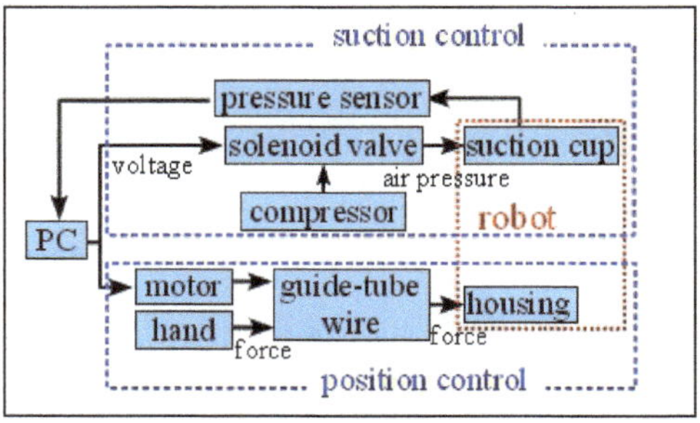

Fig. 3. Block diagram of the robot control

2.4 Phase diagrams of movements

1. Moving forward/backward

The phase diagram of moving forward is shown in Fig. 4(a), where a hatched circle means an adsorbed suction cup, whereas an open circle means a released one. It is obvious that, the motion could be divided into 5 phases, starting from an initial phase 0 (both suction cups absorb), releasing one suction cup and pushing forward (shown by an arrow marked with F) the wires (for moving the front suction cup) or the guide-tubes (for moving the rear suction cup), sequentially and repeatedly. Also the phase diagram of moving backward could be acquired by simply reversing the direction of force (arrows).

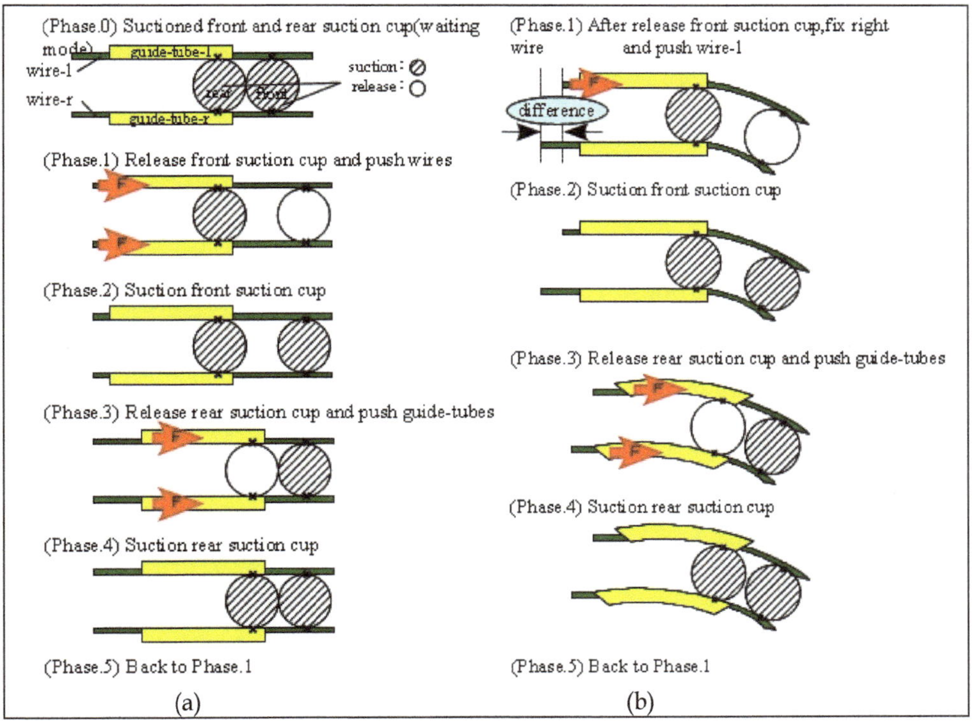

Fig. 4. Phase diagram (a) moving forward (b) turning right

2. Turning left/right

Turning motion is realized by making difference between the stretching-out length of left and right wires. The phase diagram of turning right is shown in Fig. 4(b). Different with moving forward motion, in phase 1, in order to move the front housing right, the left wire was pushed out, while fixing the right wire length, thus, the left wire will bend towards right, and the front housing will turn right. Turning left could be achieved by making stretching-out length of right wire longer during phase 1. Except this making difference between the left and right stretching-out wire length, the suction control of turning motion is basically the same as that of moving forward.

3. Moving up/down

Vertical movement is also possible by making difference between the stretching-out length of upper wires (those two wires close to suction cups) and lower wire (one single wire) (Fig. 5).

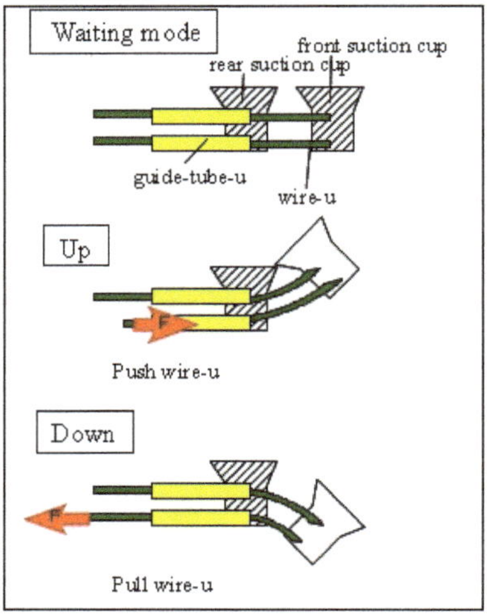

Fig. 5. Phase diagram of moving up/down

2.5 Experiment with the first prototype

The motions described in the last section were verified using the first prototype, by manual operation. The robot was operated to move on a piece of transparent film pasted onto a flat, level frame of a laparoscope operation simulation unit (Fig. 6).

It was confirmed that all the motions could be realized by manual operation. The forces needed to operate the robot will be shown in the next section. Moreover, even with a 50g load, the robot could move forward. Fig. 7 shows the pictures of each phase of robot with 50g load, in moving forward motion.

Fig. 6. Laparoscope operation simulation unit

3. Automatic control system

As shown before, with manual operation, the robot prototype could realize the movement designed, hanging upside down from a transparent film. However, the operation needs the manipulation of three wires and three guide-tubes independently, and sometimes several of them simultaneously. Therefore, it's complicated and difficult to operate the robot by one surgeon. Thus, it is necessary to develop an automation system that could operate the wires and guide-tubes automatically, as well as a user interface that could receive and translation

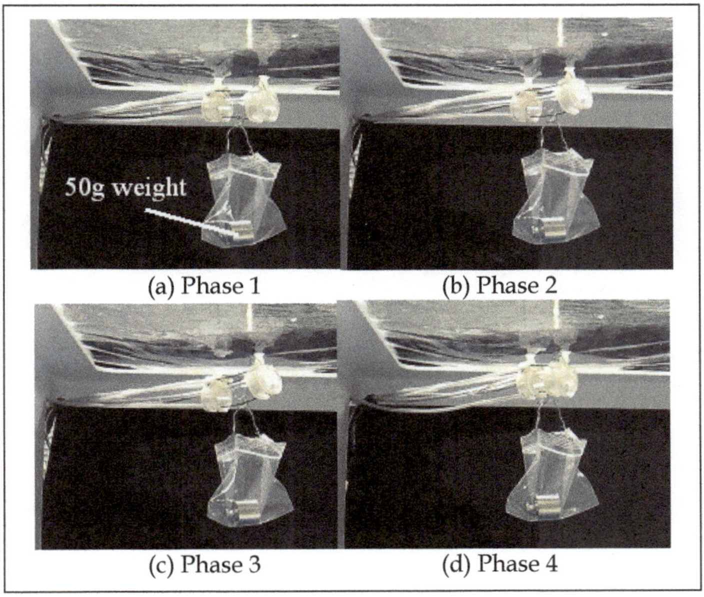

Fig. 7. Phase views of moving forward motion with a 50g weight

the instruction from surgeons to the automation system. This section describes the such hardware and software for such a controller: a WGL controller (Wire and Guide-tube Length controller).

3.1 Developing the WGL controller

In order to adjust the length of wires and guide-tubes, the manipulation part should be able to 1) realize a push and pull linear motion with the stroke longer than 300mm, and with suitable axial forces; 2) guarantee independent linear motion of paired wire and guide-tube to be coaxial; 3) hold the wires and guide-tubes stably and fixedly without breaking the wires and guide-tubes. Besides, control algorithm should be developed to realize effective automatic manipulation.

In this section, we described our efforts to realize the WGL controller. We measured the force required to manipulate the wires and guide-tubes in a manual operation. Secondly, we developed a linear motion mechanism by using motors and timing belts. Thirdly, we designed a griping mechanism that is fixed to the timing belt and able to hold wires and guide-tubes. Finally, we proposed control algorithms using sensor information.

3.1.1 Measurement of the force required to manipulate wire and guide-tube

In order to develop the device which controls the length of wires and guide-tubes, it's necessary to measure how much force is actually required for operation. Then, we measured force required for push/pull of each wire and each guide-tube by using the force measurement system (Fig. 8(a)). We measured the forces of 4 operations of moving forward/backward and turning left/right 3 times each. Moreover, the force gauge as shown in Fig. 8(b) was used for the measurement (ZPS-DPU-50N; IMADA).

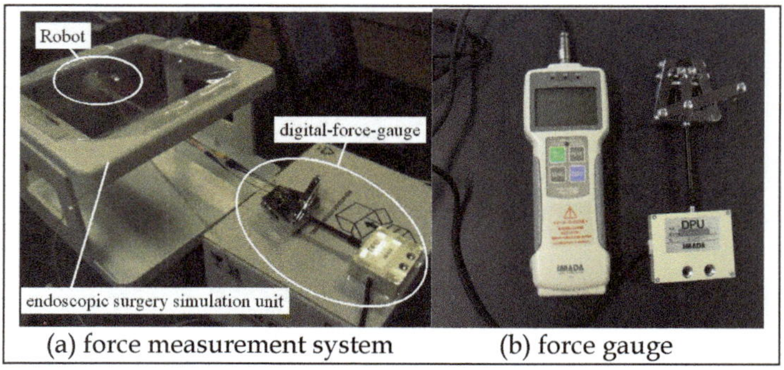

(a) force measurement system (b) force gauge

Fig. 8. Experiment setup for force measurement

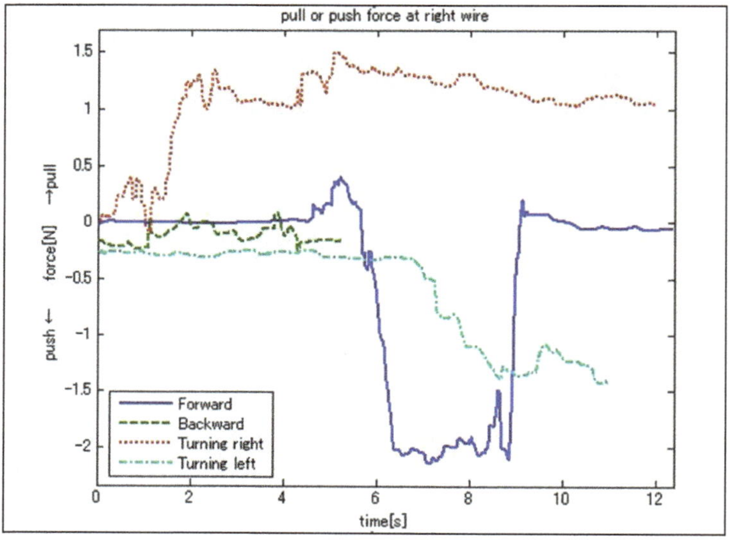

Fig. 9. Force required to manipulate wire-r in phase 1.

Fig. 9 shows the output of the force gauge and the output of force which the wire R undergoes at the time of each operation . The following formulas were used in order to represent graphically.

$$F_i = \begin{cases} F_{pull}(|F_{max}| > |F_{min}|) \\ F_{push}(|F_{max}| < |F_{min}|) \end{cases} \tag{1}$$

$$F = \frac{1}{3}\sum_{i=1}^{n} F_i \quad \text{(i:trial number)} \tag{2}$$

Fig. 10 shows the force measured in forward, backward and turning right and left motions, respectively. Fig. 10 showed that force required for operation was 5N at the maximum. From the above result, we determined that we connect the actuator which outputs the force more than 5N to each 6 operating portions.

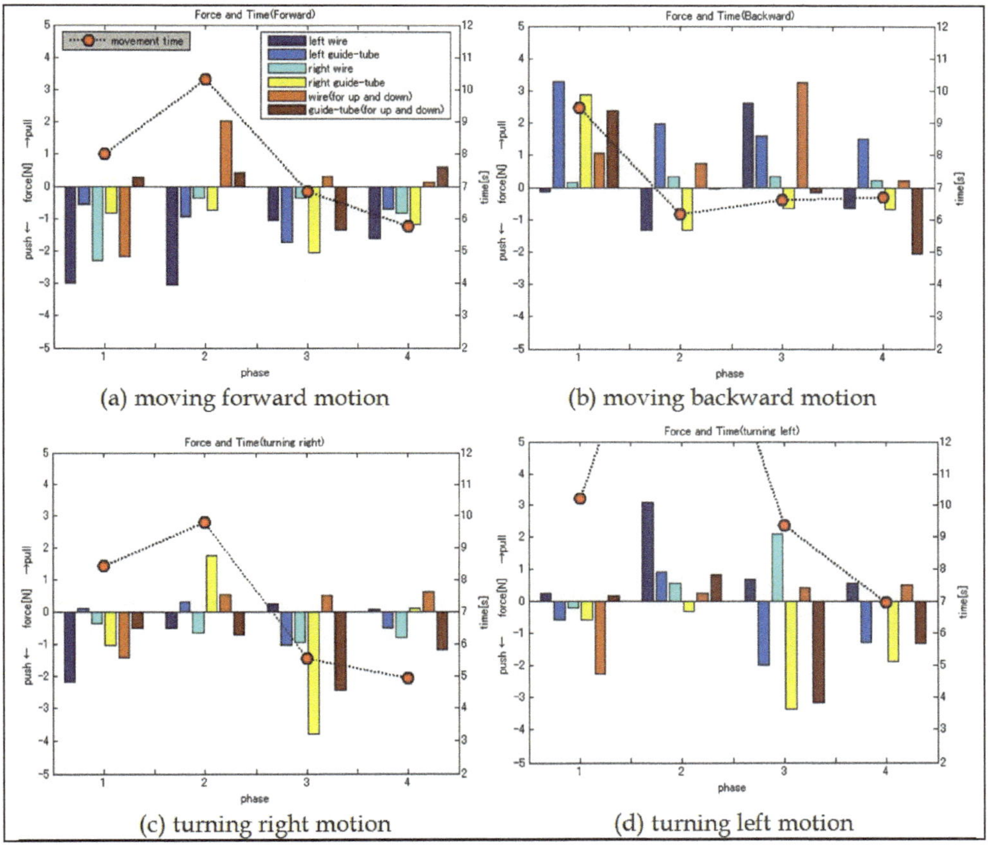

Fig. 10. Force and time required for each motion

3.1.2 A mechanism to realize the linear motions for manipulating wire and guide-tube pairs

As shown in the last section, the robot motion could be realized by pulling or pushing the wire and guide-tubes to change the relation length between them. The requirements to the mechanism is: linear motion, with big stroke and a certain level of force (around 5 N) .

In this research, stepping motor with timing belt and pulley were employed as the main mechanism. The stepping motor selected is shown in Fig. 11, and the specification is shown in Table 1.

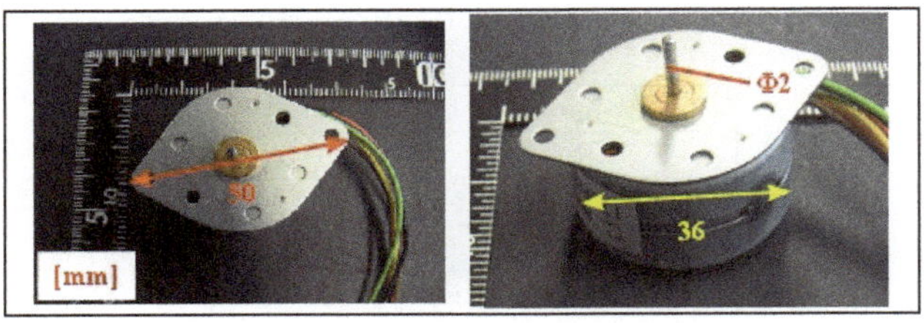

Fig. 11. Selected stepping motor (35L048B-2U,Servo Supplies Ltd)

Step angle	7.5 [deg]
Hold torque	2.5 [Ncm]
Rotor inertia	0.04 [kgcm2]
Resistance	64 [Ω]
Current	0.18 [A]
Voltage	12 [V]
Inductance	40 [mH]
Mass	88 [g]

Table 1. Stepping motor's specification

The reason for selecting the stepping motor from various motors is that a stepping motor has the following features:

- Easy to control, as compared with other motors.
- Sufficiently high positioning accuracy.
- Position detection without an additional encoder.

The selected timing belt and pulley are shown in Table 2, Fig. 12 and 13.

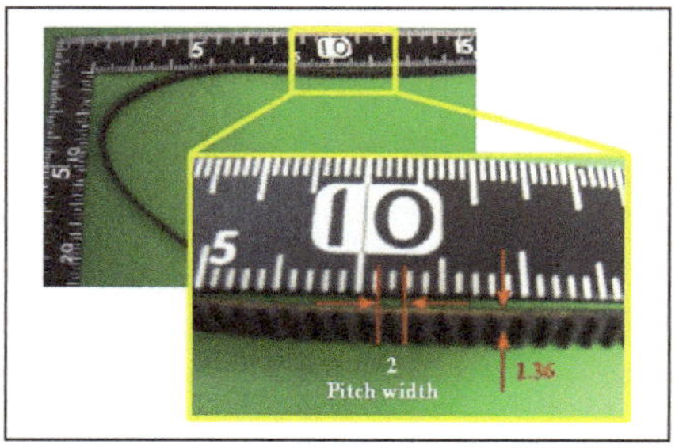

Fig. 12. Selected timing belt

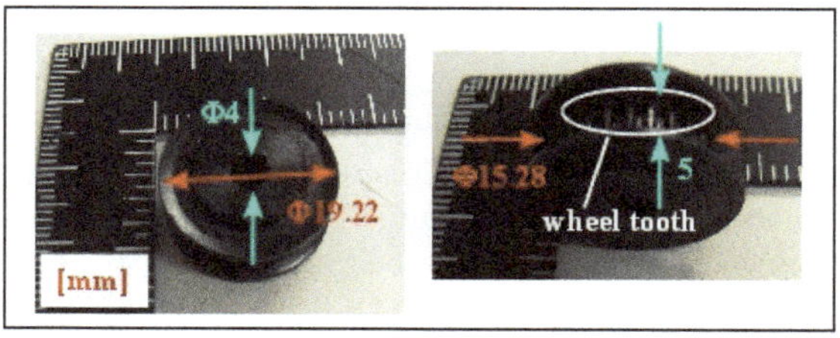

Fig. 13. Selected timing pulley

Number of teeth	24
Pitch	2.0 [mm]
Inside diameter	20 [mm]
Belt width	9 [mm]
Timing belt's length	800 [mm]

Table 2. Specification of selected timing belt and pulley

Moreover, the linear distance per rotation [mm/step] (positioning accuracy) could be determined by the following formula 3.

$$x = \frac{\theta \times P}{360} \tag{3}$$

Where, x stands for the linear moving distance, θ represents rotation angle, and P is the product of pulley's pitch and number of teeth. Therefore, considering the specification of selected motor and pulley (see Table 1 and 2), the positioning accuracy of the combined system is± 1 [mm].

The mechanical system designed to realize the linear motion of wires and guide-tubes is shown in Fig. 14. The dimension of this device is summarized as follows: full length 480 [mm], height 100 [mm], and stroke of the wire or tube 376 [mm].

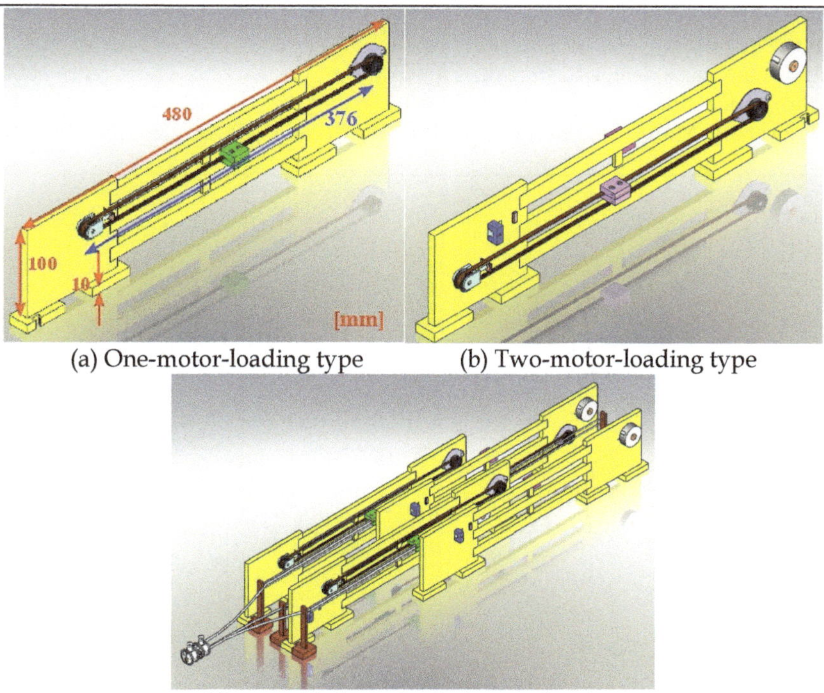

(a) One-motor-loading type (b) Two-motor-loading type

(c) assembled manipulation mechanism

Fig. 14. Designed manipulation mechanism

In order to guarantee independent linear motion of paired wire and guide-tube to be coaxial, special consideration is given to the setting of motor-loading stands, and the mounting position of the gripping mechanism.

Two types of motor-loading stands were designed. In the one-motor-loading type (Fig. 14(a)), one motor was loaded on only one side, while in the two-motor-loading type, motors were load on both sides, one for each side. The grip mechanism could be mounted either on the upper side or the lower side of the timing belt. Two gripping parts, one mounted on the upper side, another one mounted on the lower side, from two different stands will be put to face each other for manipulating one wire and guide-tube pair.

The motor loading position (height from the horizontal plane) was decided so that the paired gripping parts have the same height. Through adjusting the positions of paired stands, paired wire and guide-tube can be manipulated in a coaxial manner.

3.1.3 Gripping mechanism

In order to transmit the force from the linear mechanism to the wires and guide-tubes, a gripping mechanism that is able to hold the wires and guide-tubes steadily and attach to the timing belt should be designed. Moreover, because the timing belt is made of rubber, deflection is likely to happen. Thus, it's necessary to consider the following factors in the development of the gripping mechanism:

- To have teeth that have the same pitch with, in turn, mesh with timing belt;
- To have self-support against the deflection due to weight of the part;
- To hold the wires and guide-tubes to avoid the wires and guide-tubes' fleeing away from the gripping.

Fig. 15 shows the parts designed to realize the gripping mechanism. These parts have a Φ3.0 mm hole for guide-tube and a Φ0.8 mm hole for wire, and two Φ2.5mm up and down for long sliding shaft. Moreover, the teeth in the parts could enable gripping mechanism to

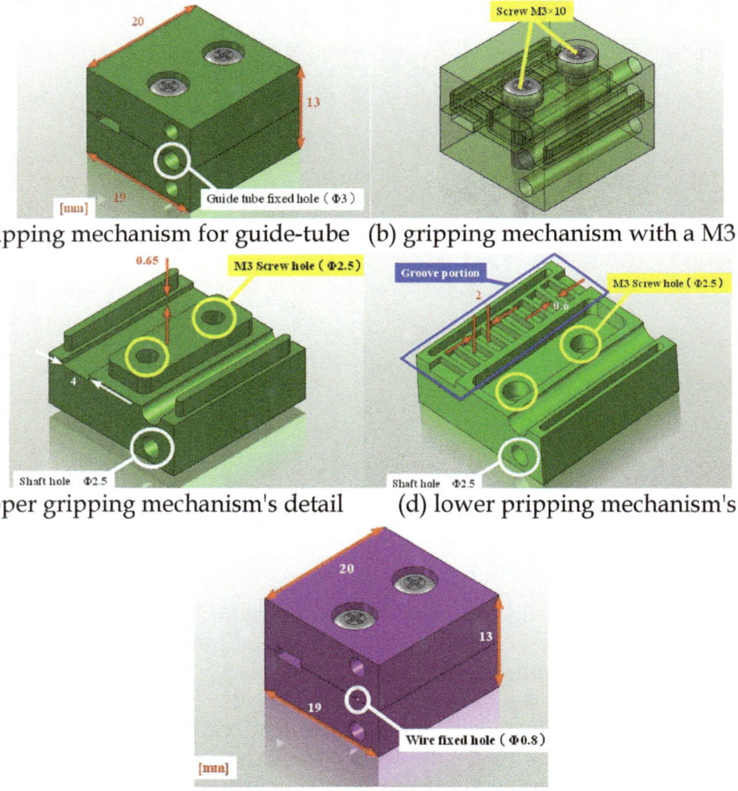

(a) gripping mechanism for guide-tube (b) gripping mechanism with a M3 screw

(c) upper gripping mechanism's detail (d) lower pripping mechanism's detail

(e) gripping mechanism for wire

Fig. 15. Parts for realizing the gripping mechanism

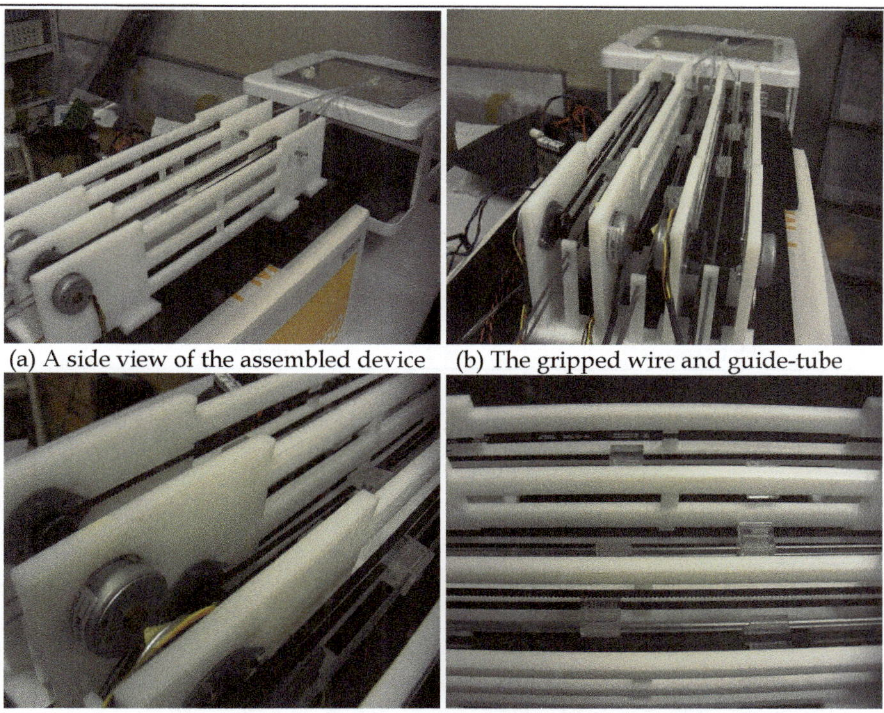

(a) A side view of the assembled device (b) The gripped wire and guide-tube

(c) Motor and timing belt fixture (d) Gripping mechanism and timing belt

Fig. 16. Assembled WGL controller

mesh with the timing belt. And, two screws shown in Fig15(b) could prevent the wire and guide-tube from fleeing away from the gripping mechanism. Moreover, in Fig. 15, a green part is the gripping mechanism for the guide-tube, and a purple part is the gripping mechanism for the wire.The assembled WGL controlling device is shown in Fig. 16.

3.1.4 Sensors for automatic control

As shown in Fig. 3, the robot control system could be divided into two independent control parts: suction control of suction cups and position control. Detection of the adsorption state is realizable by using the pressure sensor of the solenoid valve. The position of two housings could detected using a magnetometric sensor. Magnetometric sensor used: trakSTRAR [Model130 and Ascension Technology Corporation] is shown in Fig. 17, and the specification is shown in Table 3. Although the range of detection is a sphere with a radius of 30 [cm], it should be sufficient to cover the abdominal cavity. The block diagram of the control system using this magnetometric sensor is shown in Fig. 18.

Size	Φ1.5x7.7 [mm]
Sampling rate	20-255 [Hz] (Default 80[Hz])
Axis	6 Axis (Position 3 Axis and Direction 3 Axis)
Measurement position range	x:20-66/y:± 28/z:± 30 [cm]
Measurement angle range	azimuth and roll: ± 180 [deg] elevation: ± 90 [deg]
Static accuracy	position: 1.4 [mm RMS] direction: 0.5 [deg RMS]
Static resolution	position: 0.5[mm/305mm] direction: 0.1[deg/305mm]
Output	coordinate value of x, y and z, azimuth, rotation number
Interface	USB1.1/2.0 or RS-232
Data format	Binary
Correspondence	WindowsAPI, driver

Table 3. Magnetometric sensor's specification

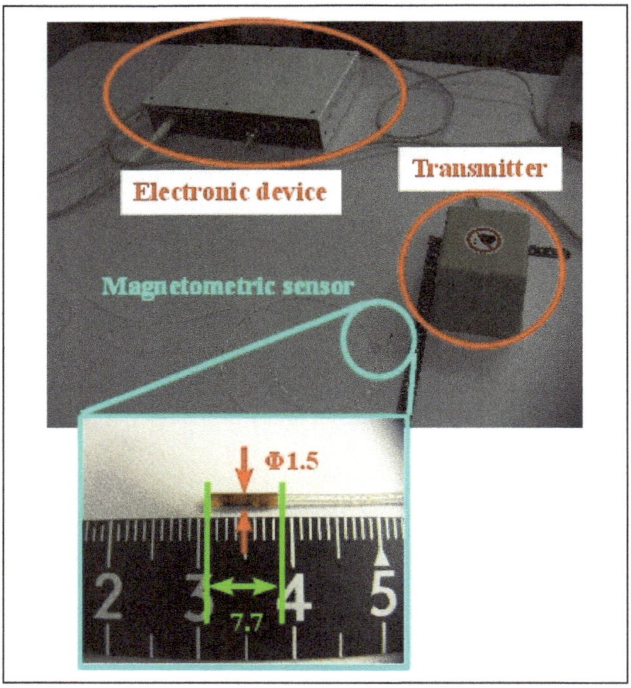

Fig. 17. Magnetometric sensor (Model130, Ascension Technology Corporation)

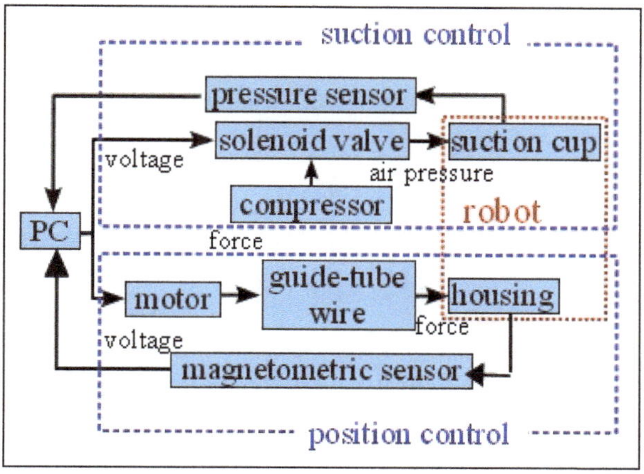

Fig. 18. Block diagram including the magnetometric sensor

3.2 Control algorithm

In this paper, only the control algorithm for moving forward motion is explained in detail, because the other motions could be realized by modifying the algorithm. As shown in the phase transition diagram of the moving forward motion in Fig. 4(a), the sensors to detect the end of four phases differ from each other. Position sensing is required for the end of Phase1 and 3, and adsorption sensing is required for the end of Phase2 and 4. The automatic control algorithm could be considered for each phase independently.

- Phase 1

Phase1 is for the movement of the front suction cup (see Fig. 4(a)). Fig. 19 is an illustration of the movement, where x_{0f} expresses the initial position of the front suction cup, x_{0r} expresses the initial position of the rear suction cup, x_f means the position of the present front suction cup and x_A stands for the target position of the front suction cup. The distance from the initial position of the front suction cup to the target position is shown with the following formula 4:

$$D_f=|x_A - x_{0f}| \tag{4}$$

Let the tolerance of the target position be ± a, then an end judge function of Phase1 could be defined by formula 5:

$$g=x_f - x_{0f} - D_f \tag{5}$$

According to the value of the judge function g, the rotation of the stepping motors for Wire L, R, U (left, right, up-down wire respectively) were determined. That is, the robot would keep moving until $|g| < a$. In the case $g < 0$, the motor would rotate to move forward, while in the case $g > 0$, i.e., the case that the front suction cup goes over its target, the motor would rotate to move backward.

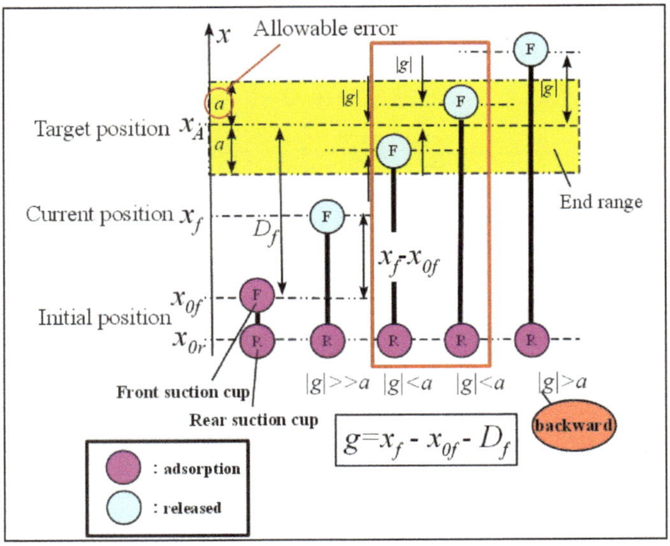

Fig. 19. An illustration for the moving forward motion in Phase 1

- Phase 2

The purpose of Phase 2 is to make the front suction cup adsorbed to its moving surface. Although through the solenoid valve, the deaeration (for adsorption state) or insufflation (for release state) could be controlled, the deaeration could not gurantee the adsorption of the front suction cup, i.e., only if the the suction cup is brought close enough to the moving surface, and the plane of the suction cup surface is almost parallel to the moving surface, then the deaeration could result in adsorption state. Then, the vertical up/down motion illustrated in Fig. 5 becomes important. This motion would be repeated until the front suction cup adsorbs to the moving surface. For each Wire U operation, the adsorption state would be confirmed by investigating the output of the pressure sensor of solenoid valve (formula 6).

$$V_{output}=\begin{cases}0V:Adsorption\\6V:Release\end{cases} \tag{6}$$

- Phase 3

In Phase 3, which is illustrated in Fig. 20, the rear housing (containing the rear suction cup) is moved towards the front housing. In Fig. 20, x_{0f} and x_{0r} express the position (initial position) of the rear suction cup at the end of Phase4 (or Phase0), x_r is the position of the present rear suction cup, x_{1f} is the position of the front suction cup at the end of Phase1 and x_B is the position of the target rear suction cup. Moreover, as in the Fig. 19, *a* stands for the tolerance and D_f is defined by formula 4, i.e., the distance of front suction cup was supposed to move. Since, if the distance for rear suction cup is set to D_f, the rear suction cup may contact the front suction cup after this phase, thus there would be no enough margin to move the rear suction cup in Phase 4, the expected moving distance for the rear suction cup is set to $D_f/2$. Therefore, the end judge function of Phase3 is defined in formula 7.:

$$h=x_{1f}-x_r-\frac{D_f}{2} \tag{7}$$

According to the value of the judge function h, the rotation of the stepping motors for Guide-tube L, R, U (left, right, up-down wire respectively) were determined. That is, the robot would keep moving until | h | < a. In the case h > 0, the motor would rotate to move forward, while in the case h < 0, i.e., the case that the rear suction cup goes over its target, the motor would rotate to move backward.

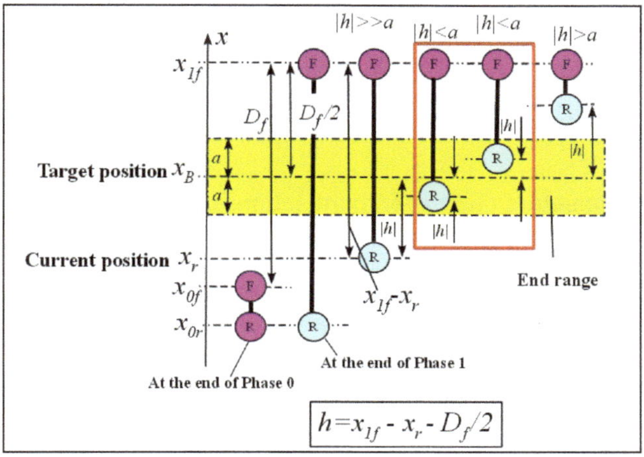

Fig. 20. An illustration for the moving forward motion in Phase 3

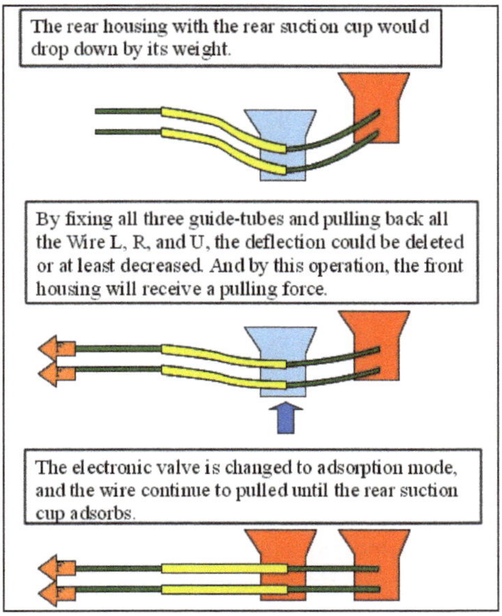

Fig. 21. The lift-up operation of the rear suction cup

- Phase 4

The purpose of Phase 4 is to adsorb the rear suction cup to the moving surface. Since, when released at the end of Phase3, the rear housing with the rear suction cup would drop down by its weight, there is a deflection of the wires and guide-tubes in the beginning of the Phase

4. Basically, the operation that could lift the rear suction cup up to the moving surface is required. However, by using Wire U, only the front suction cup could be lifted up towards the moving surface, as shown in Fig. 5. In this study, we employed a motion shown in Fig. 21 to achieve the same effect. That is, by fixing all three guide-tubes and pulling back all the Wire L, R, and U, the deflection could be deleted or at least decreased. Certainly, by this operation, the front housing will receive a pulling force, which would affect the adsorption of the front suction cup to the moving surface. Although the experiment results would show that the feasibility of this operation, in the near future, a new mechanism should be designed improve lift-up of the rear suction cup.
Except the lift-up operation, the other sensing and operation are just the same as the Phase 2.

3.3 The effect of parameters and experiment setting-up

It is clear that, the two parameter D_f and a, would influence the behavior of the robot. The parameter a determines the accuracy of the robot. In the experiment, the a was set as 1.5 [mm], decided according to the magnetometric sensor accuracy (1.4 [mm]), and linear motion accuracy (1.0[mm]). Due to the property of the robot, the D_f, which determines the pace of the robot motion, also affects force required for wire operation. Because the relative distance between suction cups will become large when D_f is increased, thus bending of the wire also becomes large. Thereby, at the phase of the front suction cup adsorption (Phase 2), bigger force would be needed.
On the other hand, by setting up a smaller D_f, the deflection could be reduced, and force needed could be kept within an allowable range. However, this would result in a slow moving speed, and a bigger influence from measurement error of magnetometric sensor. Thus, an optimal D_f has to be decided by trial and error. In the experiment, the D_f was set to 10, 15, 20, 25, 30, and 50 [mm], and movement speed was calculated for each value.
In order to verify the capacity of the developed automatic control algorithm, operation is verified using a laparoscope operation simulation unit (Fig. 6).

4. Results of the automatic control for moving-forward motion and discussion

In the experiment, each motion (from Phase 1-4) was taken as one step, and 3 consecutive steps were measured and recorded as a trial. During each trial, if there is a falling down from moving surface, or a deadlock due to the shortage of the torque happened, then the trial was considered as a failure.
When calculating moving speed, since the robot moves on a x - y plane (ref. Fig. 23(a)), moving distance was calculated using the square root of the squared sum of distance on x and y axis. The moving speed for each step and each trial (3 Steps) were calculated.
Table 4. shows the moving speed [mm/s] in a single trial for each D_f value, and the value in brackets shows the amount of moving distance [mm] for each case.
For detailed explanation, the case of D_f=20 [mm] is taken as an example. In Fig. 24, the output voltage of pressure sensor is shown, where 0V expresses an adsorption state, 6[V] shows a release state, and the upper and lower part of graphs depict the output of sensors for the front and rear suction cup, respectively. This graph expressed the adsorption state of the suction cup of each rear and front part. The relationship between the phase and output voltages is shown as follows.

$$\begin{aligned} &\text{Phase 0: } Vf = 0,\ Vr = 0 \\ &\text{Phase 1: } Vf = 6,\ Vr = 0 \\ &\text{Phase 2: } Vf = 6,\ Vr = 0 \\ &\text{Phase 3: } Vf = 0,\ Vr = 6 \\ &\text{Phase 4: } Vf = 0,\ Vr = 6 \end{aligned} \tag{8}$$

Where, Vf means the front output voltage, and Vr means the rear output voltage. From Fig. 24, it is clear that, the robot could move 3 Steps without falling-down from the moving surface.

The change of the coordinate of the front and rear suction cup is shown the Fig. 25(a) and Fig. 25(b), respectively. The x, y and z coordinates (see Fig. 23) at the starting point were set to 0. From the figure, it is clear that both suction cups seldomly moved in the y and z direction, but moved mostly in x direction. Moreover, it was observed the front suction cup moved in x direction about 20 [mm] (value of D_f) in Phase1 of each step, and the rear suction cup moved more than 10 [mm] ($D_f/2$) only in Phase3 of each step. This shows that the robot is automatically manipulated, exactly following the control algorithms designed. Moreover, Fig. 26 shows the representative situations for each phase in the moving forward motion. The moving speed for the trial in the case of D_f=20 [mm] was 1.85 [mm/s].

Because the difference of the speed for each value of D_f was not remarkable, the adsorption sequence of each value of D_f was also investigated. By increasing the value of D_f, deflection of the wire becomes large and the time required for adsorption operation becomes long. Thus, a trade-off relation exists between the value of D_f and the adsorption time. Then, with each value of D_f, we conducted the experiment that repeats adsorption operation (Phase 2, 4) and investigated about repeatability. Moreover, we also investigated about whether adsorption time changes by increasing the value of D_f. In the experiment, the suction cup's adsorption state was detected per motor's a full revolution (7.5deg), and a number of motor rotation required by adsorption was measured. The greater the number of motor rotation, the longer adsorption time. The period until suction cup's adsorption is detected is set as one trial, and it is repeated 10 trials. Then, the difference (repeatability of adsorption) of each trial and the difference of the number of rotations by the value of D_f were compared.

The result of Phase 2 is Table 5. The value of Table 5 shows value of D_f or the number of motor rotation required by adsorption in each trial. From Table 5, in $D_f \leq 30$, there was no difference in the adsorption time for each value of D_f, and the repeatability of the adsorption operation in each trial. However, in the first trial of D_f =50, the number of rotation and time required by adsorption became twice other values of D_f. Thus, if D_f becomes very large, deflection of the wire has various influences and has big influence on adsorption time or the reproducibility of adsorption.

Next, the rear suction cup's adsorption operation in Phase 4 of forward motion was investigated. In Phase 4, the relative distance during suction cups is adjusted by moving the rear suction cup after the front suction cup moves according to set D_f. For this reason, we have to set the value of D_f and relative distance. Therefore, in each of D_f =30(almost no influence of deflection of the wire) and D_f =50(some influences of deflection of the wire), it investigated by changing relative distance with 10, 15, 20, 25, and 30 [mm]. Each relative distance and the number of motor rotation of the value of D_f are shown in Table 6. From

Table 6, below in 25 [mm], the differences of adsorption time and the reproducibility of trial don't have relative distance. Therefore, it is considered that there is almost no influence of deflection of the wire. On the other hand, in the relative distance 30 [mm], only in the case of D_f =50, the increase of number of rotation and adsorption time was confirmed. Moreover, this increase was verified in the relative distance 30 [mm] and all the trial of D_f =50 [mm]. From this result, it is considered that deflection of the wire by D_f =50 of Phase1 influenced not only the front but adsorption operation of the rear.

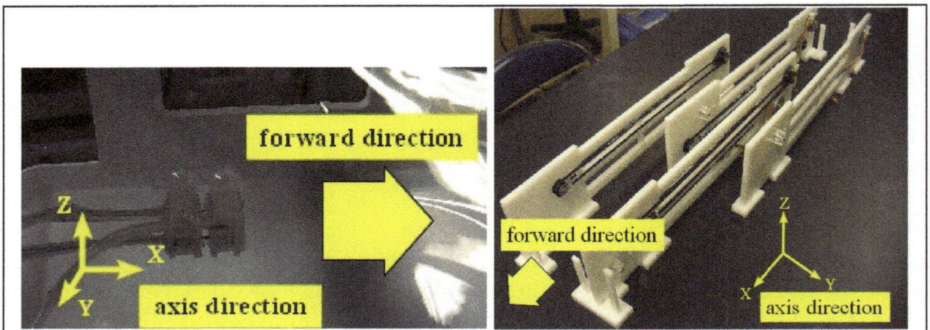

(a) axis direction on the physical simulator (b) axis direction on the WGL controller

Fig. 23. Robot's move and axis direction

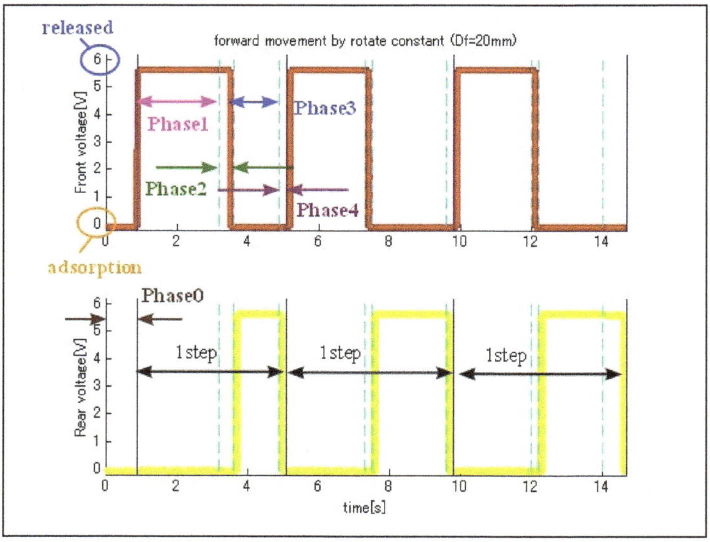

Fig. 24. The output of the adsorption switch in the unit moving distance 20 [mm]

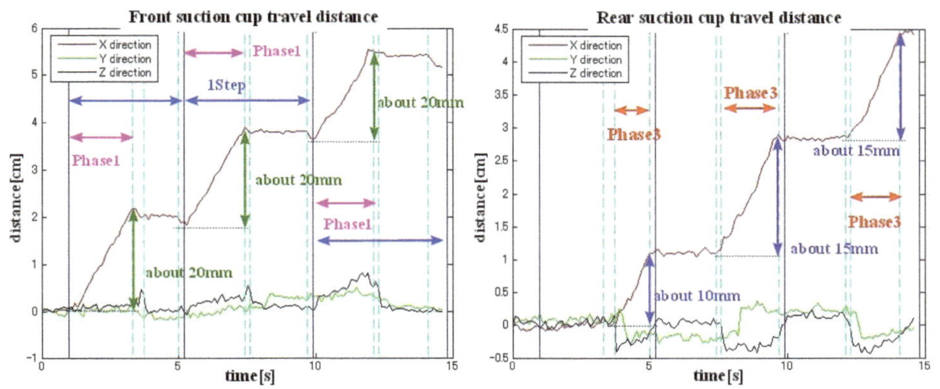

(a) front suction cup (b) rear suction cup

Fig. 25. Change of travel distance of each suction cups (D_f=20)

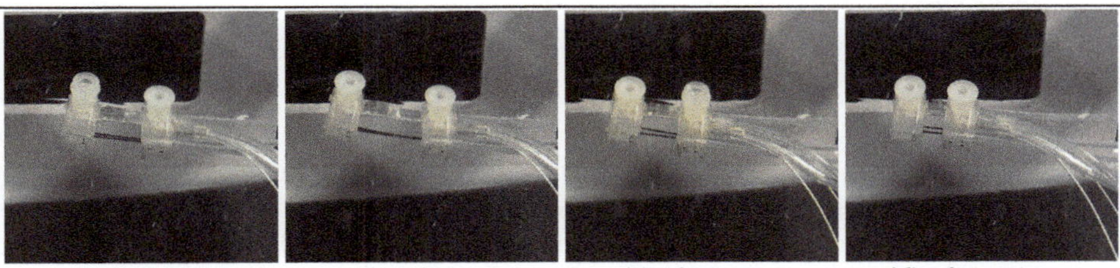

(a) phase 1 (b) phase 2 (c) phase 3 (d) phase 4

Fig. 26. Representative situations for each phase in the moving forward motion in case of D_f = 20 [mm]

D_f [mm]	Step 1-3	Step 1	Step 2	Step 3
10	1.86(31.11)	2.54 (11.41)	1.70 (10.73)	1.65 (9.09)
15	1.81(40.56)	1.90 (13.52)	1.93 (14.54)	2.12 (15.53)
20	1.85(51.46)	2.36 (19.65)	2.12 (19.84)	1.59 (15.49)
25	1.73(63.55)	2.44 (24.33)	1.87 (22.77)	1.29 (18.28)
30	1.94(79.78)	2.70 (29.05)	1.70 (25.83)	1.77 (25.92)
50	1.95(127.99)	2.16 (44.69)	1.55 (38.13)	2.37 (47.11)

Table 4. The speed of moving-forward motion

D_f [mm]	1st trial	2nd trial	3rd trial	4-10 trial's average
10	2	2	2	2
15	2	2	2	2
20	2	2	2	2
25	2	2	2	2
30	2	2	2	2
50	4	2	2	2

Table 5. The number of motor rotation of each value of D_f in Phase2

relative distance[mm]	D_f [mm]	1st trial	2nd trial	3rd trial	4-10 trial's average
10	30	2	2	2	2
10	50	2	2	2	2
15	30	2	2	2	2
15	50	2	2	2	2
20	30	2	2	2	2
20	50	2	2	2	2
25	30	2	2	2	2
25	50	2	2	2	2
30	30	2	2	2	2
30	50	3	3	3	3

Table 6. The number of motor rotation of each value of D_f in Phase4

5. Conclusion

In this paper, we described a NOTES support robot which uses suction cups and a wire driven mechanism. The robot has 3 pairs of wire and guide-tube, so it is difficult to manipulate for surgeons in operation. To realize automatic control for the robot, we developed a WGL controller, which adjusts the relative length of the wire and guide-tube pairs, and the control algorithms for it. In the experiment, it was shown that, the moving forward motion could be realized by the automatic control system.

The moving speed was also measured. From the result of Table 4, even if the value of D_f was changed, there was no great change in the total movement speed (Step1-3), and the average moving speed was 1.86 [mm/s].

However, the moving speed 1.86 [mm/s] is not fast enough for clinical application, and improvement in speed is needed. Also, in this study, we investigated the moving forward motion. The control algorithms for other motions should be developed and verified. Furthermore, as Chapter 2 described, the robot size must be less than the over tube's inner diameter of 17mm (made by TOP Corporation). Moreover, in order that the robot may correspond to various operations, it is necessary to develop in consideration of usage in laparoscopic surgery. The inner diameter of the port used by laparoscopic surgery is 12mm (made by Applied Medical Resources Corporation). Therefore, as first aim, robot size less than the inner diameter of the over tube is realized, and after that, robot size less than the inner diameter of a port is realized. Finally, we have to test the whole robotic system in in-vivo experiment.

6. References

Amy, L.; Kyle, B.; J, Dumpert.; N, Wood, A, Visty, ME, Rentschler.; SR, Platt.; SM, Farritor. & D. Oleynikov. (2008). Surgery with cooperative robots, *Computer aided surgery*, Vol. 13, No. 2, pp. 95-105

Amy, L.; Nathan. W.; Jason. D.; Dmitry. O & Shane, F. (2008). Robotic natural orifice translumenal endoscopic surgery, *Proceedings of IEEE International Conference on Robotics and Automation (ICRA)*, pp. 2969-2974, May, 2008

Amy, L.; Nathan, W.; Jason, D.; Dmitry, O. & Shane, F. (2008). Dexterous miniature in vivo robot for notes, *Proceedings of the 2nd IEEE/RAS-EMBS International Conference on Biomedical Robotics and Biomechatronics*, pp. 244-249, October, 2008

Amy, L.; Jason, D.; Nathan, W.; Lee, R.; Abigail, V.; Shane, F.; Brandon, V. & Dmitry, O. (2009). Natural orifice cholecystectomy using a miniature robot, *Surgical Endoscopy*, Vol. 23, No. 2, pp. 260-266

H, Zhang.; J, Gonzalez-Gomez.; S, Chen.; W, Wang.; R, Liu.; D, Li. & J, Zhang. (2007). A novel modular climbing caterpillar using low-frequency vibrating passive suckers, *Proceedings of IEEE/ASME international conference on Advanced intelligent mechatronics*, pp. 1-6, September, 2007

J, Hazey.; V, Narula.; D, Renton.; K, Reavis.; C, Paul.; K, Hinshaw.; P, Muscarella.; E, Ellison. & W, Melvin. (2008). Natural-orifice transgastric endoscopic peritoneoscopy in humans: initial clinical trial, *Surgical Endoscopy*, Vol. 22, No. 1, pp. 16-20

Kai, X.; Roger, G.; Jienan, D.; Peter, A.; Dennis, Fowler. & Nabil, S. (2009). System design of an insertable robotic effector platform for single port access (spa) surgery, *Proceedings of IEEE/RSJ International Conference on Intelligent Robots and Systems*, pp. 5546-5552, 2009

M, Rentschler.; J, Dumpert.; S, Platt.; S, Farritor. & D, Oleynikov. (2007). Natural orifice surgery with an endoluminal mobile robot. *Surgical endoscopy*, Vol. 21, No. 7, pp. 1212-1215

M, Bessler.; P, Stevens.; L, Milone.; M, Parikh. & D. Fowler. (2007). Transvaginal laparoscopically assisted endoscopic cholecystectomy: a hybrid approach to natural orifice surgery, *Gastrointestinal Endoscopy*, Vol. 66, No. 6, pp. 1243-1245

Naoki, S.; Maki, H.; Satoshi, I.; Morimasa, T.; Hajime, K. & Makoto, H. (2010). The function which an oral type operation robot system should have, and its development, *The 19th Society of Computer Aided Surgery conference special edition*, Vol. 12, pp. 180-181

Satoshi, O.; Junichi. T. & Wenwei, Y. (2009). Development of a micro mobile robot in the abdominal cavity, *Proceedings of IEEE/RSJ International Conference on Intelligent Robots and Systems*, pp. 4707-4711, October, 2009

Satoshi, O.; Chika, H. & Wenwei, Y. (2010). Design and manipulation of a suction-based micro robot for moving in the abdominal cavity, *Advanced Robotics*, Vol. 24, No. 12, pp. 1741-1761

Satoshi, O.; Chika, H. & Wenwei, Y. (2010). Development of a Control System for Micro Mobile Robotwith Suction Cups in the Abdominal Cavity, *The 19th Society of Computer Aided Surgery conference special edition*, Vol. 12, pp. 470-471

Toshiaki, H.; Satoshi, S. & Satoshi, K. (2007). Micro switchable sucker for fixable and mobile mechanism of medical mems, *Proceedings of IEEE International Conference on Micro Electro Mechanical Systems (MEMS)*, pp. 691-694, 2007

W, Tierney.; D, Adler.; J, Conway.; D, Diehl.; F, Farraye.; S, Kantsevoy.; V, Kaul.; S, Kethu.; R, Kwon.; P, Mamula.; M, Pedrosa & S, Rodriguez. (2009). Overtube use in gastrointestinal endoscopy, *Gastrointest. Endosc.* 70, pp. 828–834

Yoshiyuki, T.; Tomoya, N.; Emi, S.; Norihito, W.; Kazuhiro, S. & Takashi, Y. (2010). Development of multiple degrees of freedom active forceps for endoscopic submucosal dissection, *The 19th Society of Computer Aided Surgery conference special edition,* Vol. 12, pp. 356-357

8

Influence of the Size Factor of a Mobile Robot Moving Toward a Human on Subjective Acceptable Distance

Yutaka Hiroi[1] and Akinori Ito[2]
[1]Osaka Institute of Technology,
[2]Tohoku University
Japan

1. Introduction

Service robots working around humans are expected to become widespread in the next decade. There have been numerous works for developing autonomous mobile robots, starting as early as the 1980s. For example, Crowley developed the Intelligent Mobile Platform (IMP) which moved around a known domain according to given commands (Crowley, 1985). The issue in the earlier works was how to navigate a robot in a room. HelpMate (Evans et al., 1989) was a mobile platform intended to be used in hospitals for carrying medical records, meal trays, medications, etc. In the 1990s, robots were developed which were equipped with manipulators and executed tasks such as moving objects. Bishoff (1997) developed a mobile robot called HERMES, which is an upper-body humanoid equipped with two arms with hands and an omni-directional vehicle. HERMES recognizes objects around it using stereo vision, and executes tasks such as moving an object from one place to another. Recently, service robots that can execute more complicated tasks using three-dimensional distance sensors and more powerful actuators have been actively developed (Borst et al., 2009; Graf et al., 2009; Droeschel et al., 2011). Along with the development of such service robots, service robot contests have been held such as RoboCup@Home League (RoboCup Federation, 2011), in which mobile service robots compete for accuracy, robustness and safety of task execution in home-like environments. We have also developed an experimental care service robot called IRIS (Hiroi et al., 2003). This robot understood a patient's commands through spoken dialogue and face recognition, and performed several care tasks such as carrying bottles or opening/closing curtains in a real environment. The other feature of IRIS was its safety; IRIS was equipped with various devices for physical safety, such as arms with torque limiters (Jeong et al., 2004).
Safety is the most important issue for this kind of robot, and there have been many studies on keeping a robot safe for humans. Here, we consider two kinds of "safety." The first one is the physical safety of avoiding collisions between a robot and humans; physical safety is the most important requirement for a mobile robot working around humans. The other is mental safety, which means ensuring that the robot does not frighten people around it. Mental safety is as important as physical safety; if a robot's appearance or behavior is frightening, it will not be accepted by people even if it is physically harmless.

There have been many researches for improving the physical safety of robots. For example, sensors are commonly used for avoiding collisions with humans (Prassler et al., 2002; Burgard, 1998), and shock absorbers are deployed around a robot to reduce the risk of injury in case of a collision with a human (Jeong et al., 2005). Heinzman and Zelinsky (2003) proposed a scheme that restricts the torque of a manipulator to a pre-defined limit for safety against collision. As mentioned above, IRIS had a similar kind of torque limiter (Jeong, 2004). Furthermore, a method for evaluating the physical safety of a robot has been proposed (Ikuta et al., 2003).
Compared with physical safety, there have been few studies on improving mental safety. The purpose of the present work was to investigate the relationship between a robot's physical properties—especially the size of the robot—and the psychological threat that humans feel from the robot.

2. Mental safety of mobile robots

In this section, we briefly review previous works that investigated issues related to the mental safety of robots, and describe the objective of our work.

2.1 Previous works

Ikeura et al. (1995) investigated the human response to an approaching mobile robot through subjective tests as well as objective analysis using skin resistance. They used a small robot (250×180×170 mm) moving on a desk. The robot was set at a distance of 700 mm from the subject, and moved along rails toward the seated subject at various velocities and accelerations. The robot approached to a distance of 400 mm from the subject. A subjective evaluation suggested that humans fear the robot's velocity, while they are surprised by its acceleration. Ikeura et al.'s work is interesting, but their robot was too small to generalize their conclusion to real service robots.
Nakashima and Sato (1999) investigated the relationship between a mobile robot's velocity and anxiety. They used HelpMate (Evans et al., 1989) as a mobile robot, and measured the distance between the robot and subject at which the subject did not feel anxiety or threat when the robot moved toward the subject. They changed the velocity with which the robot moved toward the subject, and investigated the relationship between the velocity and the distance. They used 21 university students aged from 22 to 28 as subjects, and five velocities of 0.2, 0.4, 0.6, 0.8 and 1.0 m/s. They examined two postures of the subject: standing and seated. The experimental results showed that the distance was proportional to the velocity, and that the distance was longer when the subject was seated.
Walters et al. (2005) carried out an experiment similar to that of Nakashima and Sato, using a mobile robot called PeopleBot. They discussed personal factors such as gender on the impression on the robot. As these studies used commercially available robots, they could not change the size of the robot.

2.2 Size does matter

Factors of a robot other than velocity also affect the psychological threat to humans around it. The size of a robot seems to have a great psychological effect. The size of a robot is determined by its width, depth and height. When a robot is approaching a subject from in front of the subject, the width and height are the factors that ought to be considered. In this chapter, we consider only the height of a robot because we cannot vary the width greatly due to stability (a thin shape makes the robot unstable) and environmental (a very wide robot cannot pass through a door) restrictions. Thus, we define the height of a robot as the "robot size." The heights of robots used in conventional experiments have been around 1200 mm.

In this study, we investigated the psychological effects by varying the size of a robot. Although other factors such as the robot's color or materials also affect the impression of the robot, we assume that the effects of those factors are independent of the effects of the robot's size. Next, we define "subjective acceptable distance" as the minimum distance at which a subject does not feel any anxiety or threat. The concept of subjective acceptable distance is identical to that measured by Nakashima and Sato (1999). They defined this distance as "personal space" (Sommer, 1959). However, we decided to avoid the word "personal space" and used "subjective acceptable distance" instead because personal space seems to be a much broader concept compared with the distance we are trying to measure.

We measured subjective acceptable distances using robots of various sizes in order to investigate the relationship between robot size and subjective acceptable distance. Next, we determined whether or not changing the size of a robot affects the anxiety or threat perceived by a subject. We also asked the subjects to answer questionnaires to investigate differences in impression on the robots of different sizes.

3. Experimental conditions

3.1 Robot size

To decide the sizes of robots to be examined in the experiment, we considered the sizes of existing robots. Robots around 1200 mm tall are used in many works such as the general-purpose mobile humanoid Robovie (Ishiguro et al., 2001), a mobile robot for hospital work HOSPI (Sakai et al., 2005) and a mobile robot for health care (Kouno & Kanda, 1998). As a small robot, the assistive mobile robot AMOS was 700 mm tall (Takahashi et al., 2004). AMOS is not a humanoid but a cubic-shaped vehicle with a manipulator and camera. As a large robot, HERMES was 1850 mm tall (Bischoff, 1997). A robot smaller than AMOS could not easily carry objects in an office, for example, while a robot larger than HERMES would have difficulty in moving through a door. We therefore decided to examine three sizes around 1200 mm: 600, 1200 and 1800 mm.

3.2 Velocity of the robot

Next, we decided the velocity of the robots in the experiment. Nakashima and Sato (1999) examined five velocities in their experiment: 200, 400, 600, 800 and 1000 mm/s. They concluded that 800 and 1000 mm/s were too fast and caused great anxiety to the subjects. On the other hand, a velocity as slow as 200 mm/s caused no anxiety at all for some subjects. Considering their results, we set the velocity of our robot to 400 mm/s, which was an intermediate level in Nakashima's experiment.

3.3 Posture of the subjects

During experiments, subjects can either stand or sit on a chair. Nakashima et al. (1999) reported that the subjective acceptable distance became larger when the subject was seated. To investigate the relationship between this effect and the robot size, we conducted our experiment for both conditions of the subject standing or seated.

4. Experimental setup

Figure 1 shows the base of the robots used in the experiments. The base included two driving wheels and two castors, and was 450 mm wide, 390 mm deep and 250 mm high, and weighed 15.0 kg. The body of the robot could be changed by replacing the aluminum frame on its base. A sheet of white paper was glued to the front of the frame so that the robot looked like a white parallelepiped. We prepared three frames, 600 mm, 1200 mm and 1800 mm in height, as shown in Fig. 2.

Nineteen male subjects aged from 19 to 22 years old participated in the experiment. The mobile robot was first positioned at 3 m from the nearest part of the subject, as shown in Fig. 3. The subject started and stopped the robot using a push switch. After starting the robot to move toward himself, he stopped the robot when he did not want the robot to move any nearer toward him.

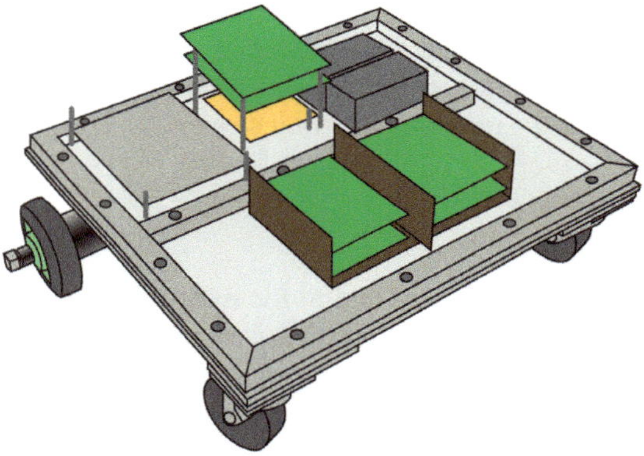

Fig. 1. Overview of the base of the mobile robot

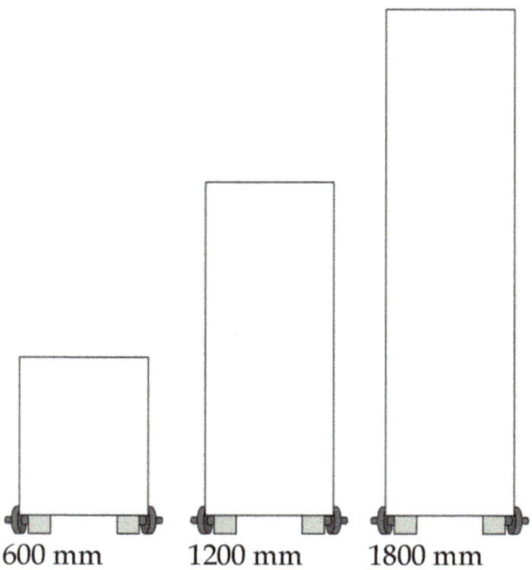

Fig. 2. External view of the robots

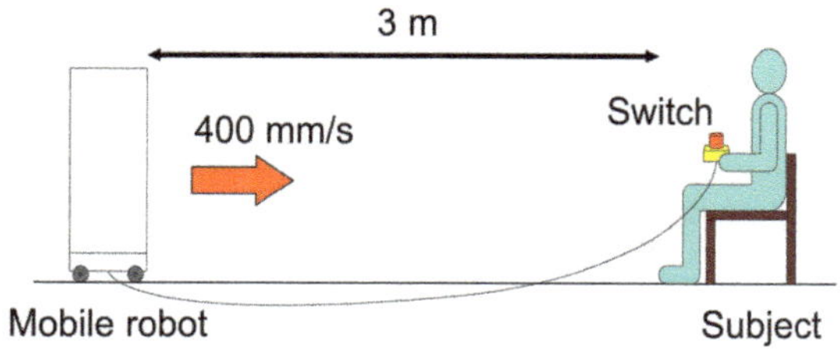

Fig. 3. Arrangement of the robot and subject

We allowed the subjects to practice using the switch to ensure the safety of the experiment. Before the experiment, we gave all the subjects the following instructions:

Just after pushing the switch, the robot will immediately start to move toward you from a distance of 3 m at a speed of 400 mm/s. If you feel any anxiety or fear and do not want the robot to come any nearer, please push the switch again to stop the robot immediately. If you feel the distance between you and the halted robot is nearer or further than the distance you intended, please let us know. In case of emergency such as if the robot does not stop, please avoid the robot by yourself. This experiment will be conducted in two postures, seated and standing, using three robots. Please keep looking at the robot throughout the experiment. After the experiment, we will ask you to fill in a questionnaire.

We randomized the order of the experiment (robot size and posture) to cancel out the order effect. If a subject reported that the distance was different from his intention, the experiment was repeated. The measurement was conducted only once for each condition, except failure of measurement. Nakashima and Sato (1999) measured the subjective acceptable distances many times for the same condition, and reported that the variance of distances obtained by multiple measurements was sufficiently smaller than the change caused by other factors. In view of their result, we decided that we did not need to conduct multiple measurements for one condition.

As a result, no subject asked to measure the distance again. There was no operation accident involving the switch, and no collision between the robot and the subject either. The robot remained in good order throughout the experiment. Therefore, the measurement was done just once for one subject and one condition.

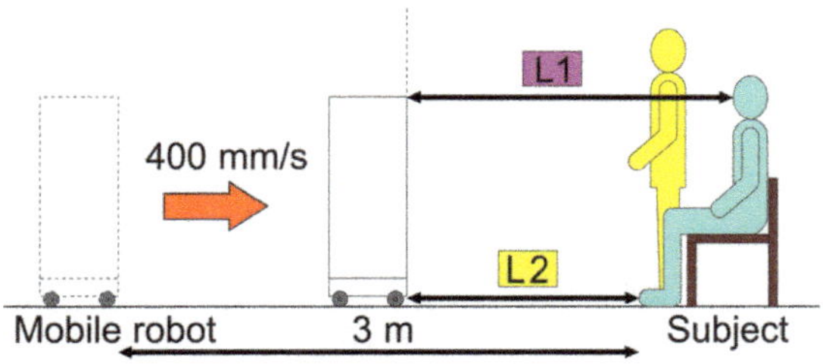

Fig. 4. Definition of distances between robot and subject

After stopping the robot, we measured two distances between the robot and the subject, as shown in Fig. 4. L1 is the distance between the front of the robot and the seated subject's eyes, and L2 is that between the front of the robot and the toes of the subject.

After the experiment, we asked the subjects to answer the following questions:

- Sort the six conditions (three robot sizes by two postures) in order of anxiety.
- Did you feel any differences between the two postures (standing and seated)? If you did, please describe them.

- Other suggestions (if any)

After a subject answered the questionnaire, we assigned scores to the conditions according to the order the subject gave. For example, if a subject answered that he felt the greatest anxiety for the (1800 mm, standing) condition, we gave a score of "6" to that condition (the larger the score, the more frightening the condition). Then we summed up the scores for a condition given by all subjects to calculate the final score for that condition.

5. Experimental results and discussion

5.1 Subjective acceptable distance and subjects' posture

Figure 5 shows the average subjective acceptable distances (L1 and L2) with respect to the three robot sizes. From the figure, L2 seems to change according to robot size. However, L1 for 1200 mm and 1800 mm does not look different. To validate these data, we conducted ANOVA using a randomized block design for both L1 and L2 to determine whether the subjective acceptable distance was affected by robot size. The results showed significant differences in both L1 and L2 ($p<0.001$). Next, we conducted Dunnett's test to find out whether the subjective acceptable distances at 600 mm or 1800 mm were different from that at 1200 mm. The results showed significant differences for L2s of 600 mm and 1800 mm ($p<0.05$), and an L1 of 600 mm ($p<0.01$). However, there was no significant difference between L1s of 1200 mm and 1800 mm. These results suggest that the subjective acceptable distance is greater for larger robots when the subject is standing. However, the subjective acceptable distance does not increase when the robot is larger than 1200 mm and the subject is seated.

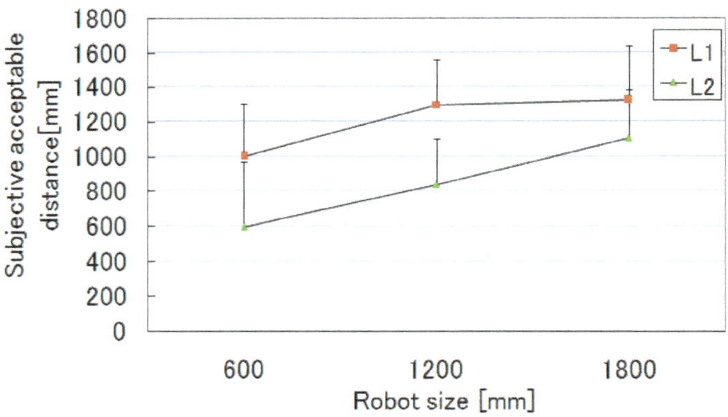

Fig. 5. Relationship between robot size and subjective acceptable distance (error bars show standard deviation)

The average height of a seated subject's eyes from the floor was 1186 mm, which was comparable with the medium robot size of 1200 mm. When watching an object higher than the observer's eyes, the object's height within the observer's view does not change with the distance between the observer and the object, which means that one of the important cues for distance perception is lost (Gary, 2002). As the effect of cues for perceiving distance is additive (Cutting & Vishton, 1995), losing one of the cues may affect the observer's perception of distance.

When a subject was standing, the average height of the eyes was 1601 mm, which was larger than the small and medium robot sizes (600 and 1200 mm). This fact might have caused the significant differences of acceptable distances when a subject was standing.

If this conjecture is correct, the subjective acceptable distance does not increase for robots larger than 1800 mm even when the subject is standing. However, a robot taller than 1800 mm is not suitable for working in a typical environment such as a home, office or hospital, because it cannot go through a door. Therefore, we did not consider robots taller than 1800 mm.

5.2 Effect of posture on subjective acceptable distance

Nakashima et al. (1999) reported that the subjective acceptable distance was larger when subjects were seated than when standing. To confirm this relationship, we conducted a paired t-test to compare L1 and L2 for each robot size. As a result, we observed significant differences between L1 and L2 for all robot sizes ($p<0.001$ for 600 and 1200 mm, $p<0.01$ for 1800 mm). This result supports Nakashima's conclusion that the distance was larger when a subject was seated.

5.3 Questionnaire results

The results of the questionnaires are summarized in Fig. 6. The x-axis is the robot size, and the y-axis denotes the score of the condition. Each condition is denoted as a symbol from A to E, as shown in Table 1.

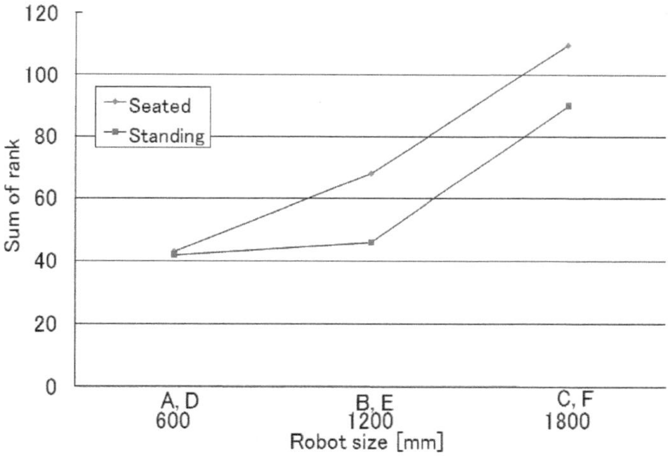

Fig. 6. Comparison of perceived psychological threat between seated posture and standing posture

We conducted Friedman's test upon the result, and obtained a significant difference between conditions ($p<0.001$). Therefore, the anxiety felt by the subjects differed from condition to condition. Next, we conducted a Steel-Dwass multiple comparison test to investigate if there were differences in anxiety between two specific conditions. Table 2 shows the results.
These results can be analyzed as follows.

1. Subjects felt the maximum anxiety for the 1800 mm robot regardless of their posture.
2. Different postures did not affect anxiety for the 600 mm robot. When the robot was larger than 600 mm, seated subjects felt more anxiety than when they were standing.
3. When subjects were seated, subjects felt more anxiety for the larger robot.

Compared with the results shown in Fig. 5, result 1 is consistent with the subjective acceptable distance. However, when the subject was seated, the subjective acceptable distances for the 1200 mm and 1800 mm robots were not different, whereas the anxiety was larger for the 1800 mm robot. This result suggests that the subjective acceptable distance does not simply reflect the subject's anxiety about the robot.

Condition	Robot size (mm)	Subject's posture
A	600	
B	1200	Seated
C	1800	
D	600	
E	1200	Standing
F	1800	

Table 1. Symbols for the conditions

Cond.	**Cond. B**	**Cond. C**	**Cond. D**	**Cond. E**	**Cond. F**
Cond. A	$p<0.05$	$p<0.001$	NS	NS	$p<0.001$
Cond. B		$p<0.001$	$p<0.05$	$p<0.05$	$p<0.01$
Cond. C			$p<0.001$	$p<0.001$	$p<0.05$
Cond. D				NS	$p<0.001$
Cond. E					$p<0.001$

Table 2. Results of Steel-Dwass test

Next, let us consider result 2. In section 4.1, we observed that the subjective acceptable distance of a seated subject became longer than that of a standing subject regardless of the subject's posture. This result could be interpreted to mean that a subject feels more anxiety when standing; however, result 2 from the questionnaire suggests that there was no difference in anxiety between the two postures for the 600 mm robot. To investigate the reason for this inconsistency, we focus on the second question of the questionnaire that asks the subject to describe the difference in feeling between the two postures.

Nine out of the 19 subjects felt anxiety for the 600 mm robot. The comments given by those nine subjects are shown in Table 3; many of them pointed out the relationship between the robot and the subject's sight.

- I was anxious when the robot went out of my sight
- When standing, I felt anxious because I couldn't see the robot (anxious)
- It went out of sight when it got nearer
- The lower robot almost vanished from my sight and I felt anxious
- I felt anxious when it went out of sight
- I felt somewhat uncomfortable because it was below my sight
- I felt it approached toward my feet
- I was more scared than when standing because I had to look down
- I didn't feel any threat, but it was uncomfortable to look down

Table 3. Descriptions of the reasons why the subjects felt more anxiety for the 600 mm robot when standing

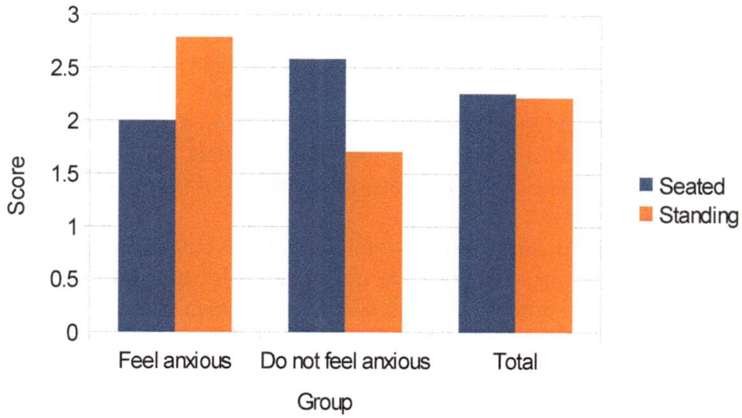

Fig. 7. Effect of anxiety about small robot (600 mm) on the difference in perceived threat for different postures

Figure 7 shows the average score (order of anxiety, larger score for more anxiety) for the 600 mm robot for the two postures. "Feel anxious" is the average score for the subjects who felt anxiety for the 600 mm robot (9 subjects), and "Do not feel anxious" is that for the others (10 subjects). From this result, the anxiety score for the "feel anxious" group and the other group showed a different tendency for the two postures, and therefore the total average score for these two groups is similar (2.26 and 2.21). The existence of these two groups of subjects (subjects who were concerned when the robot went below their sight and those who did not care) seems to be the reason why there was no difference in score for the 600 mm robot.

Next, we investigate result 3. In section 4.1, we observed no difference between the subjective acceptable distances for the 1200 mm and 1800 mm robots when the subjects were seated. Result 3 looks inconsistent with this result. In fact, 13 out of the 19 subjects described anxiety for robots taller than their eye height. Therefore, we can conclude that the subjective acceptable distance does not become larger when the robot is taller than the subject's eye height, but it does not mean that the anxieties for larger robots are the same. The invariance of the subjective acceptable distance for larger robots is not because the larger robots cause the same impression of anxiety but is caused by the difficulty of estimating the distance to the robot.

In summary, the robot with 1800 mm height caused the most anxiety to the subjects and the subjective acceptable distance for that robot was the longest. In this case, note that a shorter subjective acceptable distance does not mean less anxiety. Thus, robots of 1800 mm height or more are not suitable for service robots working around humans. Humans will allow a smaller robot to get nearer, but some people feel anxiety for the robot's behavior when they are standing.

6. A method of reducing the threat and anxiety by using preliminary announcement

6.1 Preliminary announcement of a robot's behavior

The experimental results supported the conjecture that the subjective acceptable distance becomes smaller for a robot shorter than 1200 mm and vice versa. Therefore, we can design the size of a robot considering the assumed distance between the robot and users. However, in reality, it is not usually possible to change the size of a robot, particularly when using a commercial robot. In this section, we discuss ways of reducing a person's psychological threat and anxiety about the robot without changing the robot's size.

In the above experiment, a subject could stop the robot at will when he did not want the robot to come any nearer. However, in real situations where a robot is working around humans, a person by the robot cannot stop it even if he/she does not want the robot to move any closer. Therefore, the distance kept between the human and the robot can be larger than that measured in the experiment.

One reason for keeping at a distance from a robot is that it is difficult to predict the robot's motion. As a robot is an artificial being, it is not possible for a human to infer a robot's behavior from common sense among humans; we cannot predict when a robot will begin to move and stop.

One possible method for reducing the threat or anxiety about robots derived from this uncertainty of behavior is to explicitly announce what the robot will do next. If humans around the robot know the direction in which it will move, they will not be so anxious even when in close proximity.

Existing works on this concept include announcing the robot's velocity and moving direction using a laser beam (Matsumaru et al., 2006) or LCD projector (Matsumaru, 2006). In the former method, the robot draws the trajectory along which it is going to move using a laser beam pointer. The direction of the laser beam is controlled by a mirror, and the beam is projected on and swept over the floor. They evaluated the effect of such an announcement by questionnaires conducted in an exhibition hall. As a result, they reported that half of the respondents answered that they could easily understand the direction and velocity of the robot, and received the following comments:

- The shape drawn by the laser beam should be an arrow, rather than a line
- The method of showing the velocity could be improved
- This method can be applied to industrial robots
- The method should be combined with another type of method, such as a speech-based method
- Children might have difficulty understanding the meaning of the beam

As shown, Matsumaru et al. received many opinions on ways of improving the display method, even though they explained to the respondents the background and purpose of their research, proposed announcement method, and overview of the robot. A preliminary announcement must be easy to understand for someone looking at the robot for the first time. These opinions reveal that the laser-beam-based announcement method needs further improvement.

The method based on video projector (Matsumaru, 2006) projects icons such as an arrow, turning signs, "STOP" sign and "BACK" sign onto the floor. A video projector can present more information than a laser beam. The direction of the robot is presented by the direction of the arrow, and the velocity is expressed as the thickness and color of the arrow. When the robot is going to rotate, the color of the turning sign changes. When the robot is going to stop or go backward, the color of "STOP" or "BACK" is changed. Matsumaru evaluated the announcement method using the same criteria as used for evaluating the laser-beam method. The subjective evaluation showed that the projector method was easier to understand than the laser-beam method for both direction and velocity. The problems of the projector method are summarized as follows:

- As the projected image depends on both the lighting condition and the floor, it cannot be used in a bright environment or on ground that is not sufficiently flat; for example, the method is difficult to use outdoors.
- As the information is projected in front of the robot, humans behind the robot cannot see the projected information.
- Humans around the robot need to pay attention to the floor rather than the robot itself.

Therefore, we need to develop a method that is not affected by the environmental conditions such as lighting or floor condition, is able to present the robot's behavior in all directions, and attracts human attention to the robot's body.

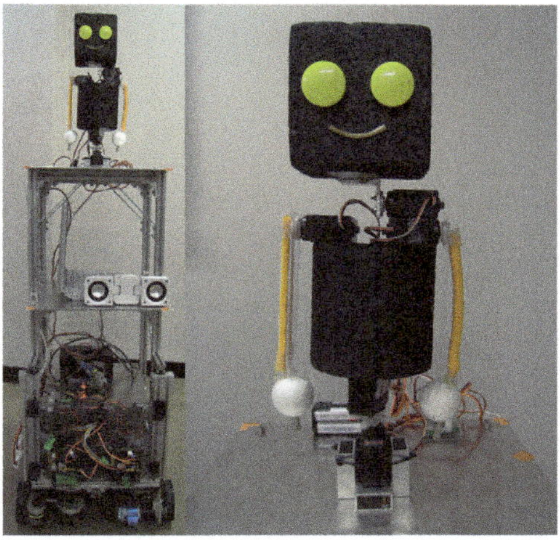

Fig. 8. Overview of the experimental robot avatar

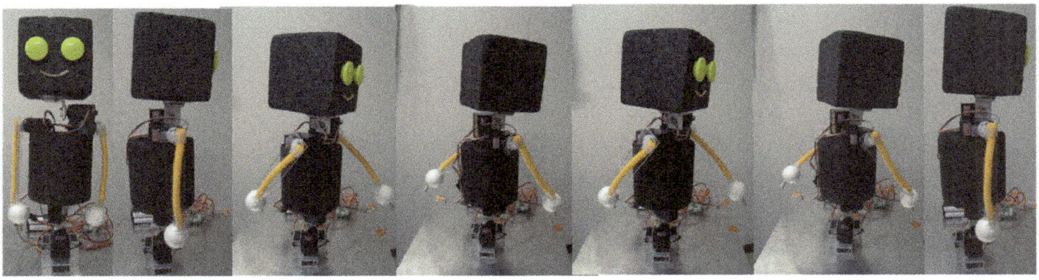

Fig. 9. Motion of the robot avatar

6.2 Preliminary announcement using a robot avatar

We propose a preliminary announcement method using a robot avatar. A robot avatar is a small robot mounted on a large robot for communication (Hiroi et al., 2005). An example of a robot avatar is shown in Fig. 8. The proposed method announces the (bigger) robot's behavior through the robot avatar's motion. There are many advantages of using a robot avatar for preliminary announcement of robot motion:

- The preliminary announcement can be intuitive.
- The motion of the robot avatar can be prominent if the avatar is mounted at around a human's eye height.

- Its visibility is more robust to environmental changes than the projection-based announcement method.
- If the large robot were to make announcements, the motion could be dangerous because the robot's arm could collide with a human. A robot avatar is safer because it is smaller and lighter.
- It is easy to make a robot avatar look friendly.

Figure 8 shows the robot avatar we are developing. This robot avatar can swing its arms and change direction in order to make the announcement motion. As an announcement motion, the robot avatar swings its arms and changes direction towards which the larger robot is about to move (Fig. 9). We are verifying the effectiveness of this method through subjective evaluations.

7. Conclusions

We investigated the effect of robot size on subjective acceptable distance using three robots of different sizes. The results showed that the acceptable distance becomes smaller when the robot is smaller than 1200 mm, and vice versa. However, the relationship between robot size and distance was nonlinear, and the acceptable distance saturated when a subject was seated. One possible reason for this saturation could be a relationship between a robot's height and the eye height of the subject. This experiment showed a mutual effect between robot size and posture of a subject. Therefore, the size of a robot should be designed considering the likely posture of robot users.

We also conducted a survey on the impression of the robots, and found that robots taller than 1800 mm caused too much anxiety to be used safely in practice.

This work is the first investigation of the relationship between robot size and its effect on human impression, considering the real use of service robots. The results of this work will be useful for designing actual robots working around humans.

The experimental results obtained for square-shaped robots can be directly applied to mobile carriers (Kouno & Kanda, 1998; Sakai et al., 2005). A smaller robot is desired if we wish to minimize the acceptable distance between a robot and a human; however, the robot should be larger for carrying larger payloads. Considering these conditions, a robot size of 1200 mm could be the best among the examined three robot sizes.

In future, we plan to evaluate various factors affecting the impression of a robot, including the age and gender of subjects (Walters et al., 2005; Mutlu et al., 2006), human factors such as knowledge and rapport with the robot (Nomura et al., 2007), as well as the robot's color (Goetz et al., 2003) and velocity patterns (Ikeura et al., 1995; Nakashima & Sato, 1999). The effectiveness of using a robot avatar for preliminary announcement should also be tested.

8. References

Bischoff, R. (1997). HERMES - A Humanoid Mobile Manipulator for Service Tasks. *Proceedings of International Conference on Field and Service Robotics*, Canberra, pp. 508-515

Crowley, J. L. (1985). Navigation for an Intelligent Mobile Robot. *IEEE Journal of Robotics and Automation*, Vol. 1, No. 1, pp. 31-41

Evans, J., Krishnamurthy, B., Ponga, W., Croston, R., Weiman, C., & Engelberger, G. (1989). HelpMate™: A robotic materials transport system. *Robotics and Autonomous Systems*, Vol. 5, No. 3, pp. 251-256

Gary, H. (2002). Perception as Unconscious Inference. In: *Perception and the Physical World: Psychological and Philosophical Issues in Perception*, ed. by Dieter Heyer and Rainer Mausfeld, pp. 115-143, New York: Wiley

Goetz, J., Kiesler, S., & Powers, A. (2003). Matching robot appearance and behaviors to tasks to improve human-robot cooperation. *Proceedings of 12th IEEE International Workshop on Robot and Human Interactive Communication (RO-MAN 2003)*, USA, pp. 55-60

Heinzman, J., & Zelinsky, A. (2003). Quantitative Safety Guarantees for Physical Human-Robot Interaction. *International Journal of Robotics Research*, Vol. 22, No. 7–8, pp. 479-504

Hiroi, Y., Nakano, E., Takahashi, T., Makino, S., Ito, A., Kotani, K., Takatsu, N., & Ohmi, T. (2003). A Patient Care Service Robot System Based on a State Transition Architecture. *Proceedings of the 2nd International Conference on Mechatronics and Information Technology*, China, pp. 231-236

Hiroi, Y., Nakano, E., Takahashi, T., Ito, A., Kotani, K., & Takatsu, N. (2005). A New Design Concept of Robotic Interface for the Improvement of User Familiarity. *Proceedings of SPIE*, Vol. 6042, doi:10.1117/12.664685

Ikeura, R., Otsuka, H., & Inooka, H. (1995). Study on emotional evaluation of robot motions based on galvanic skin reflex. *The Japanese Journal of Ergonomics*, Vol. 31, No. 5, pp. 355-358

Ikuta, K., Ishii, H., & Nokata, M. (2003). Safety Evaluation Method of Design and Control for Human-Care Robots. *International Journal of Robotics Research,* Vol. 22, No. 5, pp. 281-297

Ishiguro, H., Ono, T., Imai, M., Maeda, T., Kanda, T., & Nakatsu, R. (2001). Robovie: an interactive humanoid robot. *International Journal of Industrial Robotics*, Vol. 28, No. 6, pp. 498-503

James E. Cutting and Peter M. Vishton. (1995). Perceiving layout and knowing distances: The integration, relative potency, and contextual use of different information about depth. In: *Handbook of perception and cognition, Vol 5; Perception of space and motion,* W. Epstein & S. Rogers (eds.), pp. 69-117, CA: Academic Press, San Diego.

Jeong, S. H., Takahashi, T., & Nakano, E. (2004). A safety service manipulator system: the reduction of harmful force by a controllable torque limiter. *Proceedings of IEEE/RSJ International Conference on Intelligent Robots and Systems, 2004. (IROS 2004)*, Japan, pp. 162-167

Jeong, S. H., Takahashi, T., Shoji, M., & Nakano, E. (2005). Harmful Force Reduction of a Manipulator by Using a Collision Detecting System with a Shock Absorbing Function. *Journal of the Robotics Society of Japan*, Vol. 23, No. 8, pp. 31-38

Kouno, T., & Kanda, S. (1998). Robot for Carrying Food Trays to the Aged and Disabled. *Journal of the Robotics Society of Japan*, Vol. 16, No. 3, pp. 317-320

Matsumaru, T. (2006). Mobile Robot with Preliminary-Announcement and Display Function of Following Motion using Projection Equipment. *Proceedings of the 15th IEEE International Symposium on Robot and Human Interactive Communication (RO-MAN 06)*, United Kingdom, pp. 443-450

Matsumaru, T., Kusada, T., & Iwase, K. (2006). Mobile Robot with Preliminary-Announcement Function of Following Motion using Light-ray. *Proceedings of the 2006 IEEE/RSJ International Conference on Intelligent Robots and Systems (IROS 2006)*, China, pp. 1516-1523

Mutlu, B., Osman, S., Forlizzi, J., Hodgins, J., & Kiesler, S. (2006). Task Structure and User Attributes as Elements of Human-Robot Interaction Design. *Proceedings of the 15th IEEE International Symposium on Robot and Human Interactive Communication (RO-MAN 2006)*, United Kingdom, pp. 74-79

Nakashima, K., & Sato, H. (1999). Personal distance against mobile robot. *The Japanese Journal of Ergonomics*, Vol. 35, No. 2, pp. 87-95

Nomura, T., Shintani, T., Fujii, K., & Hokabe, K. (2007). Experimental Investigation of Relationships between Anxiety, Negative Attitudes, and Allowable Distance of Robots. *Proceedings of the Second IASTED International Conference on Human-Computer Interaction*, France, pp. 13-18

Prassler, E., Bank, D., & Kluge, B. (2002). Key Technologies in Robot Assistants: Motion Coordination Between a Human and a Mobile Robot. *Transactions on Control, Automation and Systems Engineering*, Vol. 4, No. 1, pp. 56-61

RoboCup Federation. (2011) RoboCup @Home. Web available. http://www.robocup.org/robocup-home/.

Sakai, T., Nakajima, H., Nishimura, D., Uematsu, H., & Kitano, Y. (2005). Autonomous Mobile Robot System for Delivery in Hospital. *Technical Report of Matsushita Electric Works*, Vol. 53, No. 2, pp. 62-67

Sommer, R. (1959). Studies in Personal Space. *Sociometry*, Vol. 22, No. 3, pp. 247-260

Takahashi, Y., Komeda, T., & Koyama, H. (2004). Development of the assistive mobile robot system: AMOS - to aid in the daily life of the physically handicapped. *Advanced Robotics*, Vol. 18, No. 5, pp. 473-496

Walters, M. L., Dautenhahn, K., Boekhorst, R. te, Koay, K. L., Kaouri, C., Woods, S., Nehaniv, C., Lee, D., & Werry, I. (2005). The influence of subjects' personality traits on personal spatial zones in a human-robot interaction experiment. *Proceedings of the 14th IEEE International Workshop on Robots and Human Interactive Communication (RO-MAN 2005)*, USA, pp. 347-352

9

Mobile Platform with Leg-Wheel Mechanism for Practical Use

Shuro Nakajima
The Department of Advanced Robotics, Chiba Institute of Technology
Japan

1. Introduction

Robots that can move over rough terrain with active body leveling are now in strong demand. Such a robot can have a variety of uses, such as carrying packages, assisting people who have difficulty in walking, and safety monitoring outdoors.

Many robots capable of moving over rough terrain exist as research tools; however, few are suitable for practical use. These robots can be generally classified into the three categories.

1) Legged robots: These have excellent mobility with high stability. The mobility of legged robots has been extensively studied: for example, ASV (Song and Waldron 1989), the TITAN series (Hirose et al. 1985), DANTE II (Bares and Wettergreen 1997), the hexapod robot (Delcomyn and Nelson 2000), Tekken2 (Kimura et al. 2007), and Hyperion3 (Yoneda 2007).

2) Wheeled robots: These are most commonly selected for traversing continuous surfaces that include rough terrain. Because of their stability, maneuverability, and simple controls, wheels are the most frequently used mechanism for exploration rovers. Examples of wheeled mobile robots are Micro5 (Kubota et al. 2003), Rocky7 (Volpe et al. 1997), Shrimp (Siegwart et al. 2002), CRAB (Thueer et al. 2006), and Zaurus (Sato et al. 2007). These have passive linkage mechanisms. SpaceCat (Lauria et al. 1998) and Nanokhod (Winnendael et al. 1999) have active linkage mechanisms. The high-grip stair climber (Yoneda et al. 2009) is a crawler-type robot.

3) Leg-wheel robots: These attempt to combine the advantages of both legs and wheels in various configurations. Work Partner (Halme et al. 2003), Roller Walker (Endo and Hirose 2000), Zero Carrier (Yuan and Hirose 2004), Hylos (Grand et al. 2004), and PAW (J.A. Smith et al. 2006) are equipped with wheels placed at the ends of their legs; the Chariot series (Nakajima and Nakano 2008a,b, 2009a-c,Fig. 1), RoboTrac (Six and Kecskem'ethy 1999), and a wheel chair robot(Morales et al.2006) have separate wheels and legs; Whegs (Quinn et al. 2003; Daltorio et al. 2009) and Epi.q-1(Quaglia et al.2010) have four wheels composed of rotating legs or wheels; and Wheeleg (Lacagnina et al. 2003) has two front legs and two rear wheels.

Although a legged robot is highly mobile on rough terrain, the mechanism is complex and more energy is required for walking. On the other hand, while wheeled robots are usually the best solution for continuous terrain, most cannot travel over discontinuous terrain. Generally speaking, a hybrid mechanism like Fig. 1 provides the strengths of both wheels and legs, although such mechanisms tend to be complex. Chariot 3 is equipped with four legs of three degrees of freedom and two independent wheels. On the other hand, Whegs is not complex,

(a) (b)

Fig. 1. A leg-wheel robot. (a)Chariot 3. (b)Chari-Bee is a demonstration robot of Aichi EXPO, 2005.

but the posture of its body cannot be easily controlled. PAW has both wheel and leg modes with a simple mechanism and can control its posture in wheel mode by adjusting each leg tip position; however, PAW cannot get over a step statically while maintaining a horizontal posture. The evaluation point for the mechanism in this chapter is to maintain the horizontal posture of the body on rough terrain statically, because a person or object should be carried stably on the robot.

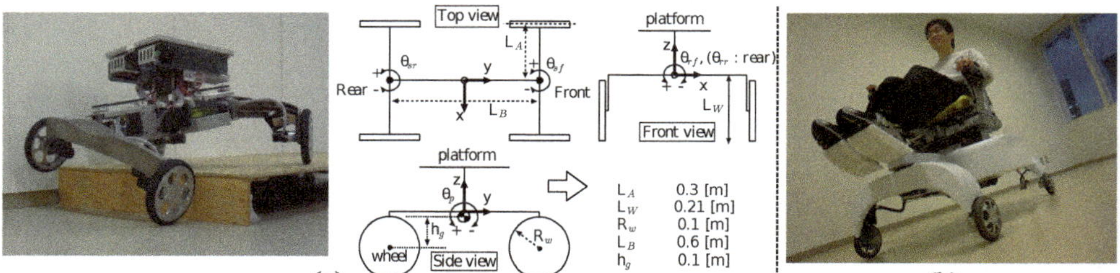

(a) (b)

Fig. 2. RT-Mover series. (a)A middle size type. (b)A personal mobility vehicle type.

The proposed robots, the RT-Mover series (Nakajima 2011a, b)(Fig. 2), have a simple mechanism and adequate mobility with a stable posture position for the following target environments: 1. an indoor environment including a step and an uneven ground surface; 2. an artificial outdoor environment with an uneven ground surface and a bump; and 3. natural terrain such as a path in a forest. RT-Mover's mechanism is different from that of conventional mobile robots: four wheels are mounted at the tip of every leg but it has only four active wheels and only five other active shafts. With an emphasis on minimizing the number of drive shafts, the mechanism is designed for a four-wheeled mobile body that is widely used in practical locomotive machinery. RT-Mover can move like a wheeled robot and also walk over a step like a legged robot, despite the simplicity of the mechanism.

The robot can move on discontinuous, rough terrain while maintaining its platform in a horizontal position. Therefore, it can be used as a stable carrier for an object (Fig. 2(a)) or a person (Fig. 2(b)).

In this chapter, the mechanical design concept for RT-Mover is discussed, and strategies for moving on rough terrain are proposed. The kinematics, stability, and control of the robot are also described in detail. The performance of the proposed locomotion is evaluated through simulations and experiments.

2. RT-mover

2.1 Mechanical concept

The target of this chapter is a practical mobile robot that carries objects or people or is used as a mobile bogie for a service robot. It is necessary to keep objects, the upper half of the onboard parts, and the boarding seat (hereinafter, the platform) horizontal in order to carry people and objects. We also aim to develop a practical robot with high cost performance by reducing the number of drive shafts to the extent possible and employing a simple mechanism.

Table 1 shows the practical use status of robots with various locomotion mechanisms. Wheeled robots have been used for practical applications. On the other hand, robots with complex mechanisms are not currently suitable for practical use because they are difficult to control, operate, and maintain, and they are expensive because of their many actuators.

Type	Status
Leg	No examples of practical use to date
Wheel	Some examples of practical use (e.g., cleaning robots)
Crawler	Few examples of practical use (e.g., in leisure and construction fields)
Hybrid mechanism	No examples of practical use to date

Table 1. Practical use status of mobile robots with various locomotion mechanisms

A mobile robot used for general purposes should have high speed on a paved road and good ability of locomotion over rough terrain. Although there is no mechanism superior to a wheel for high speed and energy efficiency, a wheeled robot can move only on continuous rough terrain. To move on discontinuous terrain, a leg mechanism is better than a wheel mechanism. Therefore, to perform all the essential functions of mobile robots in rough terrain, both wheel and leg mechanisms are needed.

Table 2 shows the strengths and limitations of leg-wheel robots. Moreover, the target application environment such as an urban environment is basically a paved surface upon which transportation by wheel is possible, with several steps that necessitate a leg function in order to reach the destination.

Strengths	Ability of locomotion on rough terrain is good because of the use of a leg mechanism.
	High speed movement is possible because of the use of a wheel mechanism.
	Ability of locomotion can be enhanced by using leg and wheel mechanisms cooperatively.
Limitations	There is a danger of collision between the legs of the robot and a person near the robot if the leg's motion space is wide.
	The number of actuators (required for legs) increases, so the cost also rises.
	Reliability and maintainability worsen because of the complexity of the leg mechanism.

Table 2. Strengths and limitations of leg-wheel robots

Therefore, it is advantageous to reduce the complexity of the leg mechanism to a minimum and to limit each leg's motion space.

We take a four-wheeled mobile body often used in practice as the starting point in considering the mechanism of the proposed robot, and from there we develop the proposed mechanism. When seeking a high degree of ability of locomotion on rough terrain in a highly used wheel mode, it is clear that each of the four wheels should generate good driving force. In addition, when driving on rough terrain, each of the wheels of a four-wheeled mobile body is required to be driven independently, since each wheel travels a different route. Accordingly, this discussion is based on a four-wheeled mobile body that drives each wheel independently.

2.2 Mechanical design

Cost, reliability, and maintainability are important for practical mobile bodies. These factors can be evaluated from the number of drive shafts to a certain extent. In other words, using fewer drive shafts tends to lower the cost and simplify the mechanism, which in turn leads to increased reliability and maintainability. The above is evident if a comparison is made between practical transport machinery, such as automobiles and trains, and the mobile robot currently being developed.

Since the objective for the robot developed in this chapter is to add the minimum leg functions necessary, a mechanism that minimizes the number of added drive shafts is designed. Specifically, after listing possible mechanisms for each function, the mechanism with the minimum number of drive shafts is chosen for the proposed robot with consideration of the possible combinations.

2.2.1 Steering function

Practical mechanisms to achieve a steering function for a four-wheeled mobile body are shown in Fig. 3(a). Needless to say, there are other mechanisms to achieve steering, but we aim to realize a practical mobile robot. Accordingly, the following discussion targets highly practical representative mechanisms. The mechanism of 1-1 is the Ackermann steering system, an automobile steering system. That of 1-3 is a mechanism that rotates the center of a shaft to steer. It is possible to attach a steering mechanism to the rear wheels of both 1-1 and 1-3. However, this case is the same as 1-1 and 1-3 and is omitted from following discussion. Those of 1-2 and 1-4 are 4-Wheel Steering (4WS) systems in which the four wheels can be directed.

2.2.2 Suspension function in wheel mode

The wheels are required to have active vertical travel according to terrain in order to keep the platform horizontal when moving on gently varying irregular terrain. Systems to fulfill this requirement are shown in Fig. 3(b). The mechanism of 2-1 provides an up-down function to each wheel. The necessary number of drive shafts is four, and the platform can be kept horizontal on rough terrain by the vertical movement of each wheel according to the ground surface.

The mechanism of 2-2 is a bogie mechanism for the left and right wheel that uses a rotating shaft as a drive shaft. The horizontal degree of the roll direction of the platform can be maintained by the active control of the rotating shaft of front and rear bogie mechanism in response to the ground surface. The horizontal degree of the pitch direction of the platform is maintained by attaching an independent pitch control shaft. The number of necessary drive shafts is three.

The mechanism of 2-3 is a bogie mechanism for the front and back wheels.

2.2.3 Suspension mechanism in leg mode

The following factors are necessary for traveling in leg mode over steps and other such obstacles, which cannot be traversed in wheel mode, while maintaining the platform in a horizontal plane: 1. lifting, moving forward, and landing for each leg; 2. supporting the body with three other legs.

The up-and-down function of each leg is achieved by a wheel mode suspension function. The mechanisms to move each wheel back and forth are shown in Fig. 3(c). The mechanism of 3-1 is a method to affix a back-and-forth drive function to each wheel, and the number of required drive shafts is four. That of 3-2 is the same as steering mechanism 1-4, with the wheels moving back-and-forth through rotation of the center shaft. Both 3-1 and 3-2 can locate the center of gravity of the body in the support polygon and thus support the body by shifting to the front and rear of the landing points of the three legs.

2.2.4 Mechanism of proposed robot

The realistically possible combinations of each aforementioned mechanism are examined, and the combinations with the minimum number of drive shafts are determined.

The combinations of each mechanism are listed in Table 3. Impossible or unavailable combinations are indicated by "-".

As mechanism 3-2 is the same as that of 1-4, it is also a steering mechanism. Therefore, 3-2 cannot be combined with 1-1, 1-2, and 1-3 because this would result in unnecessarily duplicating steering functions in equipping the drive shaft. The mechanism by which to move the wheels back and forth is duplicated in the combination of 1-4 and 3-1, so they are not combined. Also, the simultaneous employment of 1-3 and 2-3, or 1-4 and 2-3, results in a physical contradiction because the distance between front and rear wheels is altered, and these are fixed by the bogie mechanism.

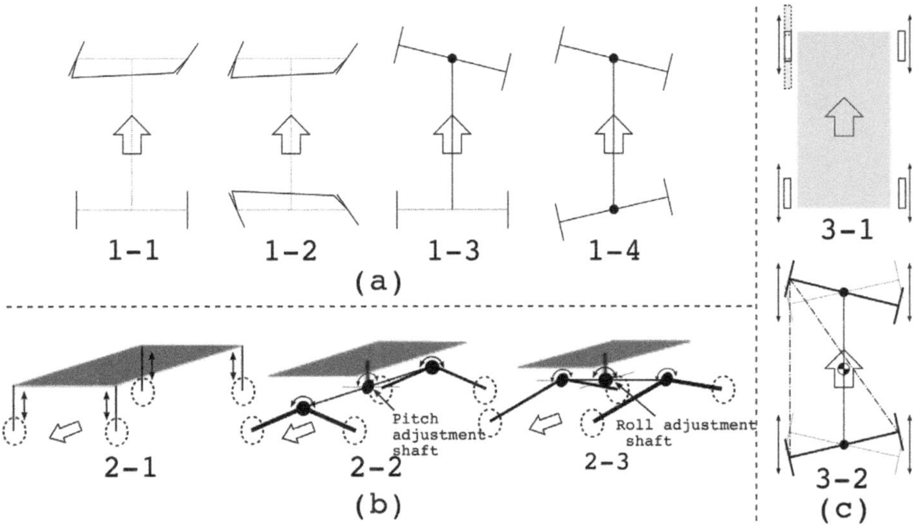

Fig. 3. Mechanism for each function. (a)Steering mechanism. (b)Suspension mechanism for wheel mode. (c)Suspension mechanism for leg mode (Top view).

Steering	Suspension (wheel mode)	Suspension (leg mode)	Number of drive shafts
1-1	2-1	3-1 / 3-2	9 / -
	2-2	3-1 / 3-2	8 / -
	2-3	3-1 / 3-2	8 / -
1-2	2-1	3-1 / 3-2	10/ -
	2-2	3-1 / 3-2	9 / -
	2-3	3-1 / 3-2	9 / -
1-3	2-1	3-1 / 3-2	7 / -
	2-2	3-1 / 3-2	6 / -
	2-3	3-1 / 3-2	- / -
1-4	2-1	3-1 / 3-2	- / 6
	2-2	3-1 / 3-2	- / 5
	2-3	3-1 / 3-2	- / -

Table 3. Combinations of each mechanism

The reason the combination of 1-3, 2-1, and 3-1 results in seven drive shafts is because the front wheels are able to move back and forth by the steering function of 1-3, and only the rear wheels are required to be moved by mechanism 3-1.
As shown in Table 3, the combination of 1-4, 2-2, and 3-2 gives five drive shafts, which is a minimum. Taken together, the mechanism of the proposed robot consists of four drive wheels and five drive shafts, as shown in Fig. 4. The mechanism that has an intersection of the roll-adjustment axis and the steering axis at the center of the steering arm is called the leg-like axle.

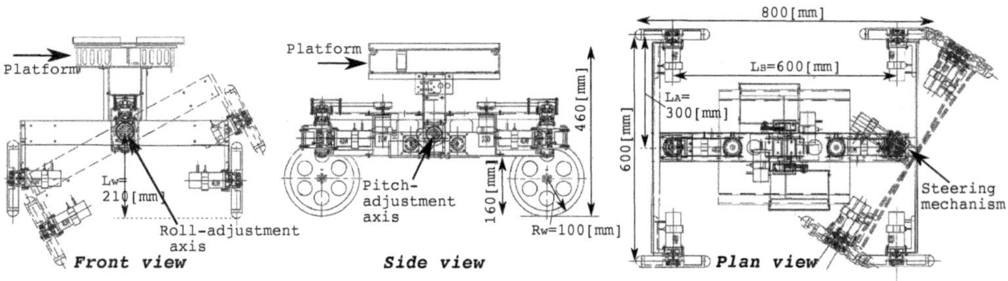

Fig. 4. Assembly drawing of RT-Mover

2.2.5 Consideration of major dimensions

The size of the robot(Fig. 2(a)) is suitable for transporting small objects or serving as a mobile bogie of a service robot. The target is to traverse steps about 0.07 [m] high in wheel mode and 0.15 [m] high in leg mode, on the assumption that a robot of this size would be used in indoor offices, factories, and public facilities. Although depending on the coefficient friction with the floor surface, the wheel mechanism allows the robot to traverse a floor surface with irregularities having height up to about 2/3 of the wheel radius. Accordingly, the wheel radius for the robot being developed is configured to 0.1 [m].
The specifications of the robot are listed in Table 4.

Dimensions	Length 0.8[m]; Width 0.63[m] (Tread 0.6[m]); Height 0.46[m]; Height to bottom 0.16[m]
Wheel	Radius:0.1[m]; Width:0.03[m]
Weight	28[kg] (including the platform at 6.5[kg] and batteries at 5.4[kg])
Motor (DC servo)	23[W] (front and rear steering θ_{sf}, θ_{sr}: $\times 2$)
	40[W] (front and rear roll θ_{rf}, θ_{rr}: $\times 2$; pitch θ_p: $\times 2$(double motor); each wheel: $\times 4$)
Gear ratio	100 (steering θ_{sf}, θ_{sr}, and roll θ_{rf}, θ_{rr})
	250 (pitch θ_p)
	50 (each wheel)
Sensor	Encoder (each motor)
	Current sensor (each motor)
	Posture angle sensor (roll and pitch of platform)
Angle limit	± 30 (steering θ_{sf}, θ_{sr}, roll θ_{rf}, θ_{rr}, and pitch θ_p)
Minimum rotation radius	0.52[m]
Max speed	0.63[m/s]
Power supply	24[V] lead accumulator

Table 4. Main specifications

The length of the steering arm (tread) was 0.6 [m], and the maximum angle for steering, roll-adjustment axis, and pitch-adjustment axis was ± 30 [deg]. When rotating the roll-adjustment axis through 30[deg] such that the wheel on one side is in contact with the ground, the other wheel attached to a 0.6[m] steering arm can rise 0.3[m]. Therefore, the movement range is sufficient for this initial step. Likewise, moving 0.3[m] in the front and rear directions is possible by moving the steering from 0[deg] to 30[deg], and holes of 0.3[m] can be crossed (Fig. 7(c)). Also, the radius of rotation is 0.52[m] if the front and rear steering angles are turned a full 30[deg] in antiphase. With regards to locomotion on a slope, back-and-forth movement and traversal of a slope of up to 30[deg] is possible.

3. Control method in wheel mode

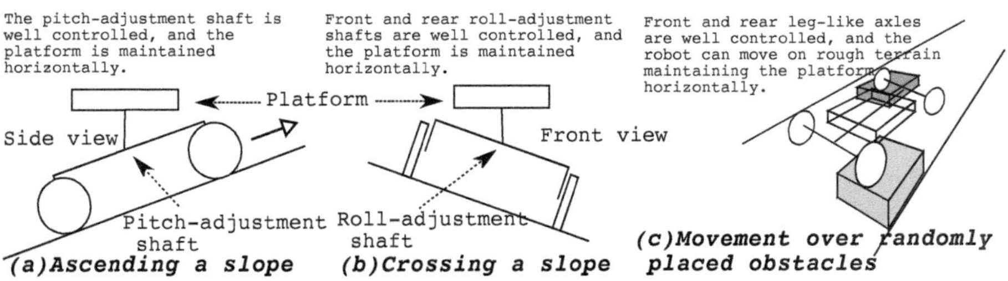

Fig. 5. Wheel mode locomotion

The movement method in wheel mode is shown in Fig. 5. RT-Mover can move on continuous rough terrain while maintaining the platform in a horizontal plane by applying eq. (1) to the pitch-adjustment shaft θ_p, and to the front and rear roll-adjustment shafts θ_{rf}, θ_{rr}.

$$T_d = K(\theta_d - \theta) + D(\dot{\theta}_d - \dot{\theta}) = -K\theta - D\dot{\theta}, \tag{1}$$

where T_d is the target torque, θ is the posture angle of the platform, θ_d: target posture angle of the platform (= 0), K is the angle gain, and D is the angular velocity gain.

3.1 Assessment of ability of locomotion in wheel mode

The situations of (a) and (b) shown in Fig. 5 partially appear in (c). Therefore, Fig. 5(c) alone is evaluated by simulation. The conditions employed in the simulation are as follows. Each gain value is obtained experimentally. For the initial step of the study, the velocity is set such that movement in the static state is possible. Since high speeds are a characteristic of wheel driving systems, traveling states with dynamic behavior will be studied in the near future.

1. K_p =800[N·m], D_p =15[N·m·s], $K_{rf} = K_{rr}$ =250[N·m], $D_{rf} = D_{rr}$ =10[N·m·s].
2. The speeds of all the wheels are maintained at a constant 0.2[m/s].
3. The front and rear steering angles are maintained at 0.
4. The wheels and steering are controlled using a proportional derivative (PD) controller.
5. The coefficient of friction between wheel and road is 0.7, and there is no friction on the shafts of the robot.
6. Open Dynamics Engine (ODE) is used for simulation.

Figure 6 shows a simulation of moving from point A to B in Fig. 6(a) over randomly placed obstacles. In (b) and (c) we see that each adjustment shaft is controlled appropriately, as well as that the platform's posture angle remains horizontal to within $\pm$0.8[deg]. This shows that in wheel mode the platform can move over rough terrain with obstacles about 2/3 the size of the wheel radius as shown in the figure.

When a wheel hits an obstacle, the steering shaft is turned because of the reaction force from the obstacle. If the robot is required to move exactly straight, it is necessary to adjust the corresponding wheel speed according to both the rotation angle of the steering shaft and that of the roll-adjustment shaft. This is a subject for future study.

4. Gait strategy and control method in leg mode

The leg mode locomotion method is shown in Fig. 7. As an initial step, evaluations are performed by simulations and experiments taking Fig. 7 (a) and (b) as examples.

In the future, a method for integrating external sensor information with the robot system will be studied because, for example, such information is necessary to recognize a downward step before descending in Fig. 7(b). At the current stage, road shapes are known in advance.

4.1 Step-up gait

Using the control method in eq. (1), RT-Mover can move over rough terrain where its wheels can be in continuous contact with the ground. However, with steps higher than the wheel radius or gaps larger than the wheel diameter, the ground contact points of wheels need to be altered by lifting the wheels. The case of lifting a wheel onto a step which the robot cannot climb in wheel mode is shown in Fig. 8. Assuming that static stability is maintained, a wheel is lifted like a leg while the body is constantly supported on at least three wheels.

Figure 9 shows the flow of the step-up gait. Before and after an upward step, the robot runs in wheel mode (Fig. 9(a) and (l)). When a front wheel reaches the step, the rear steering is rotated so that the margin of static stability during leg motion increases (Fig. 9(b)). Since RT-Mover cannot adjust the position of its center of gravity due to having a small number of degrees of

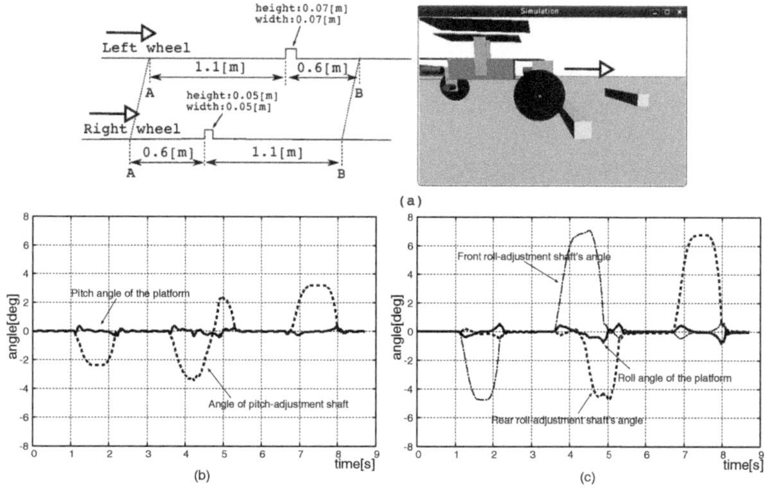

Fig. 6. Simulation of moving over randomly placed obstacles. (a) Shape of the road and a scene from the simulation. (b) Platform's pitch and the angle of the pitch-adjustment shaft. (c) Data on platform's roll and the angles of the front and rear roll-adjustment shafts for the movement from point A to B in (a).

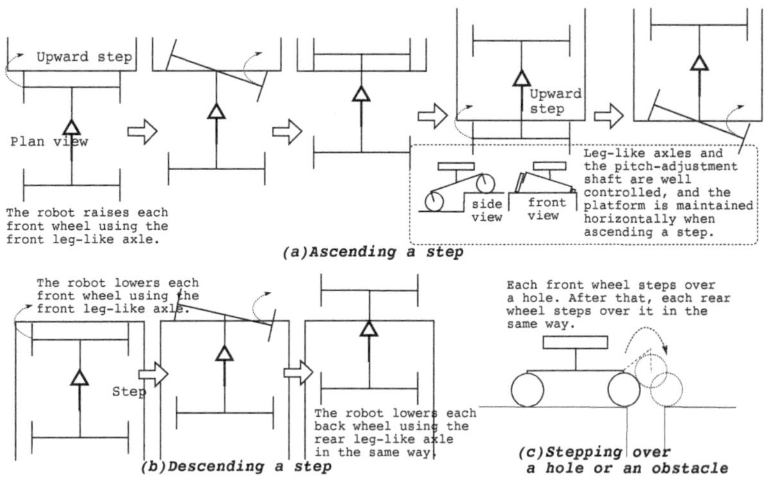

Fig. 7. Leg mode locomotion

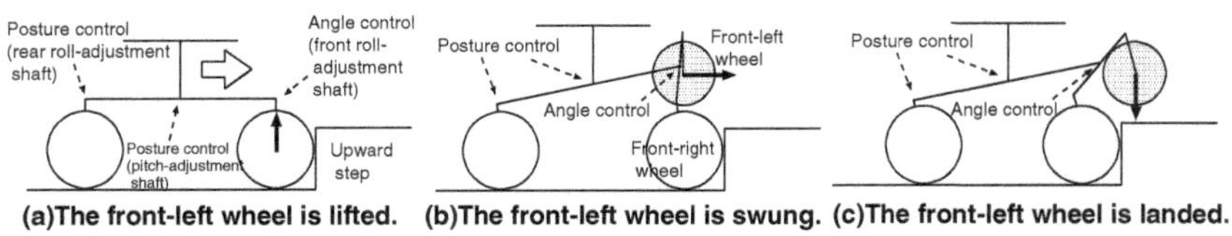

Fig. 8. Leg motion of front-left wheel

freedom, the positions of supporting wheels are adjusted by rotating a steering shaft in order to maintain static stability. Since the leg-side steering shaft is used for moving the lifted wheel forward, static stability is increased by rotating the support-side steering shaft to the limit (-30[deg]). In order to lift a leg, the front roll-adjustment shaft is switched from posture control (eq. (1)) to angle control, and the leg is lifted to the desired height (Fig. 8(a)). Meanwhile, to prevent the platform from inclining, the rear roll-adjustment shaft and pitch-adjustment shaft continue to use posture control. After lifting, the angle of the front roll-adjustment shaft is kept constant, and the lifted wheel is moved forward onto the step (Fig. 8(b) and Fig. 9(c)). Then, the lifted wheel is lowered, and when landing is detected, leg motion of the front-left wheel ends (Fig. 8(c)). As can be seen in Fig. 18(a), since the sign of the roll angle of the platform changes from negative to positive at (A) when the wheel lands, this timing can be used for detection. Next, the front-right wheel becomes the lifted leg (Fig. 9(d) and (e)).

After the front wheels have gone up the step, the robot changes its yaw angle relative to the step to ensure static stability when the rear wheels go up (this is considered in detail in a later section). The robot moves forward keeping the rear steering angle at 30[deg] until its yaw angle reaches the desired value (Fig. 9(f)). After that, it moves forward in wheel mode while maintaining the desired yaw angle (Fig. 9(g) and (h)). When a rear wheel reaches the step, the rear wheels are lifted onto the step in the same way as the front wheels (Fig. 9(i) and (j)). The rear roll-adjustment shaft is controlled using angle control, and the front one by posture control. Finally, the robot changes its yaw angle back to 0 (Fig. 9(k)).

Since the left-right order does not affect this movement, each wheel is lifted in turn in the order front-left, front-right, rear-left, and rear-right.

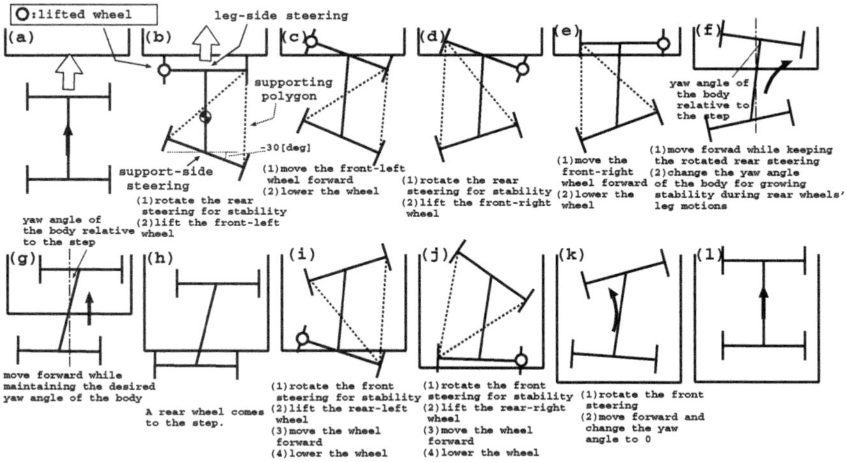

Fig. 9. Flow of processes in the step-up gait

4.2 Step-down gait

Figure 10 shows the flow of the step-down gait. When all wheels support the body, the robot is controlled under wheel mode.

When a front wheel encounters a downward step, the robot changes the yaw angle of its body relative to the step for ensuring static stability. So, after the robot reaches the step for the first time (Fig. 10(a)), it moves backward keeping the rear steering angle at -30[deg] until

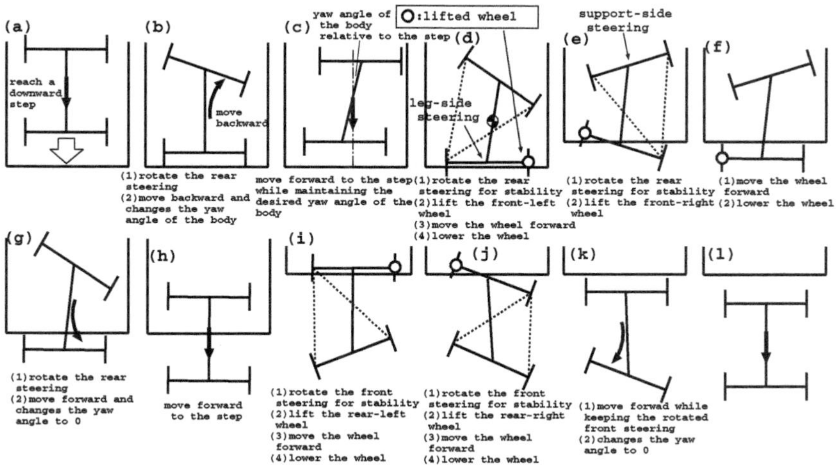

Fig. 10. Flow of processes in the step-down gait

the yaw angle of its body relative to the step acquires the desired value (Fig. 10(b)). Then, it moves forward to the step maintaining this yaw angle (Fig. 10(c)). When a front wheel reaches the step, the rear steering is rotated so that the margin of static stability during leg motion increases (Fig. 10(d)). First, the front-left leg is lifted (Fig. 10(d)), then the front right-leg is lifted (Fig. 10(e) and (f)). After both front wheels have completed the leg motion, the robot changes its yaw angle back to 0[deg] relative to the step (Fig. 10(g)). The yaw angle at this time is not important, because the static stability is sufficient during the rear wheels' leg motion. After coming to the step (Fig. 10(h)), the rear wheels are lifted down in the same way as the front wheels (Fig. 10(i) and (j)). Finally, the robot again changes the yaw angle of its body to 0 (Fig. 10(k)). The roll-adjustment shaft on the leg side is controlled using angle control, and that on the support side uses posture control.

5. Inverse kinematics

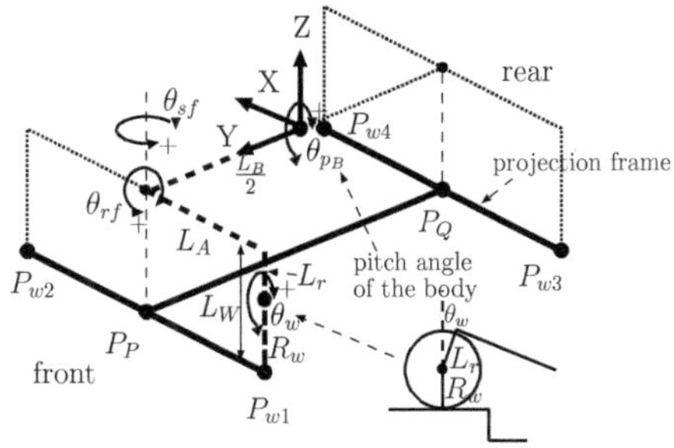

Fig. 11. Frame model for analysis

In this section, the target angle of each joint shaft to achieve the configured leg tip trajectory and the target angle of each wheel when in wheel mode are obtained. In other words, in leg

mode, for example, the inverse kinematics to achieve the trajectory by lifting the transfer leg vertically, swinging it forward, and setting it down vertically to land is described.
A "projection frame" is introduced (Fig. 11), which comprises projecting line segments connecting the wheel supporting points (front arm $P_{w1}P_{w2}$, and rear arm $P_{w3}P_{w4}$) and a line segment connecting the centers of the arms (body P_PP_Q) to a horizontal plane. Here, the inverse kinematics are discussed using this projection frame. We use a right-handed coordinate system with the center of the projection frame as the origin. The direction of travel is defined as Y and the vertical axis as Z. Then, the following matrix ${}^0T_{w_{fl}}$ maps to coordinates with the front-left leg at the origin of the body-centered coordinate system:

$$
{}^0T_{w_{fl}} = \begin{pmatrix} 1 & 0 & 0 & 0 \\ 0 & C\theta_{p_B} & -S\theta_{p_B} & 0 \\ 0 & S\theta_{p_B} & C\theta_{p_B} & 0 \\ 0 & 0 & 0 & 1 \end{pmatrix} \begin{pmatrix} 1 & 0 & 0 & 0 \\ 0 & 1 & 0 & \frac{L_B}{2} \\ 0 & 0 & 1 & 0 \\ 0 & 0 & 0 & 1 \end{pmatrix} \begin{pmatrix} C\theta_{sf} & -S\theta_{sf} & 0 & 0 \\ S\theta_{sf} & C\theta_{sf} & 0 & 0 \\ 0 & 0 & 1 & 0 \\ 0 & 0 & 0 & 1 \end{pmatrix} \begin{pmatrix} C\theta_{rf} & 0 & S\theta_{rf} & 0 \\ 0 & 1 & 0 & 0 \\ -S\theta_{rf} & 0 & C\theta_{rf} & 0 \\ 0 & 0 & 0 & 1 \end{pmatrix} \cdot
$$

$$
\begin{pmatrix} 1 & 0 & 0 & -L_A \\ 0 & 1 & 0 & 0 \\ 0 & 0 & 1 & 0 \\ 0 & 0 & 0 & 1 \end{pmatrix} \begin{pmatrix} 1 & 0 & 0 & 0 \\ 0 & 1 & 0 & 0 \\ 0 & 0 & 1 & -L_r \\ 0 & 0 & 0 & 1 \end{pmatrix} \begin{pmatrix} 1 & 0 & 0 & 0 \\ 0 & C\theta_w & -S\theta_w & 0 \\ 0 & S\theta_w & C\theta_w & 0 \\ 0 & 0 & 0 & 1 \end{pmatrix} \begin{pmatrix} 1 & 0 & 0 & 0 \\ 0 & 1 & 0 & 0 \\ 0 & 0 & 1 & -R_w \\ 0 & 0 & 0 & 1 \end{pmatrix} \tag{2}
$$

5.1 Lifting and landing phases

When lifting or landing the front-right wheel, an angular velocity value will be determined for the front roll-adjustment shaft. In order to avoid contacting a lateral surface of the step, the lifted wheel is moved up and down without moving back or forth. The posture control given by eq. (1) is applied to the pitch adjustment and rear roll-adjustment shafts, and the rotation of the supporting front-left wheel is stopped. In order to widen the supporting polygon, the rear steering shaft is rotated to its steering limit. The control parameters of the front steering shaft, the rear-left wheel, and the rear-right wheel are determined by the value set for the front roll-adjustment shaft.
The derivation of these three control parameters is described in an absolute coordinate system with its origin at the supporting position of the front-left wheel $P_{w1}(t)$, as shown in Fig. 12(a). In Fig. 12(a), the position of the front-right wheel $P_{w2}(t)$ and $P_{w2}(t+\Delta t)$, when rotating the front roll-adjustment shaft for a small amount of time Δt, are calculated. The angular velocity of the front steering shaft, $\dot{\theta}_{sf}(t)$, and the velocities of the rear-left and rear-right wheels, $\mathbf{V_{w3}}(t)$ and $\mathbf{V_{w4}}(t)$, are also derived. Since the wheel is moved up and down without moving in the Y direction, the Y coordinate of P_P is constant.
The distance between $P_{w1}(t)$ and $P_P(t)$ is $A_f(t)$; since this is half the distance between $P_{w1}(t)$ and $P_{w2}(t)$, it may be derived from eq. (2). According to eq. (2), $A_f(t)$ depends on the front steering $\theta_{sf}(t)$, the front roll axis $\theta_{rf}(t)$, and the body pitch angle $\theta_{p_B}(t)$. The value of A_f after a small incremental movement of $\theta_{rf}(t)$ is $A_f(t+\Delta t)$. Because an analytic solution is difficult, θ_{sf} and θ_{p_B} are approximated as not varying over time Δt. Since the Y axis moves in a fixed up and down path, the Y coordinate of P_{w2} is fixed and is given below:

$$
(P_{w2x}(t), P_{w2y}(t)) = (2A_f(t)\cos(\theta_{leg}(t)+\theta_B(t)), 2A_f(t)\sin(\theta_{leg}(t)+\theta_B(t))), \tag{3}
$$

$$
(P_{w2x}(t+\Delta t), P_{w2y}(t+\Delta t)) = (\sqrt{4A_f(t+\Delta t)^2 - P_{w2y}(t)^2}, P_{w2y}(t)), \tag{4}
$$

$\theta_{leg}(t)$ and $\theta_B(t)$ are obtained by eqs. (6), (7), and (10), when their initial values are given.
The velocity of P_P and the small angle $\Delta\theta_o(t)$ are given by

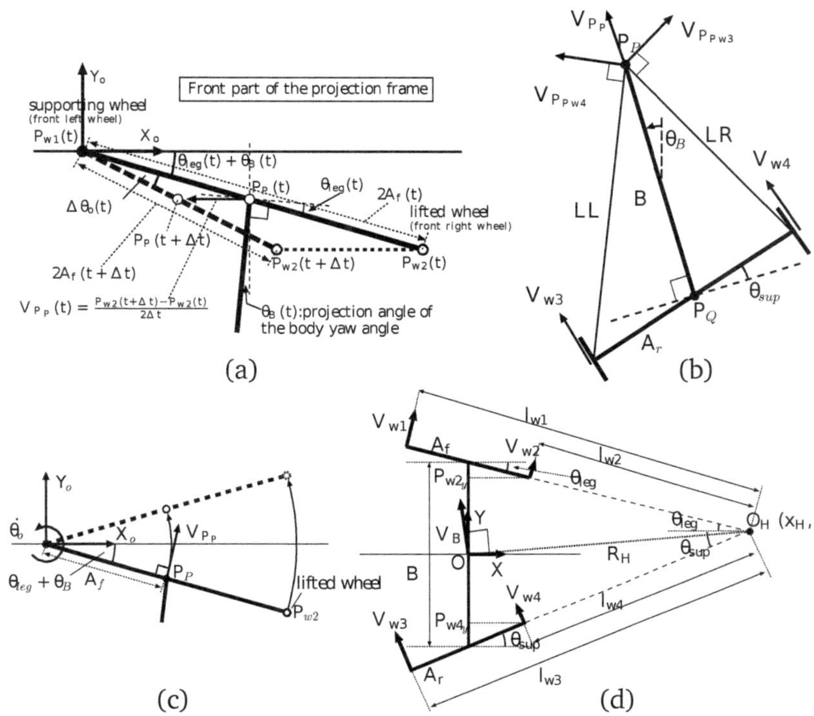

Fig. 12. Calculation model. (a) For the trajectory of a leg tip when raising and lowering a wheel. (b) For V_{w_3} and V_{w_4}. (c) For swing phase. (d) For wheel mode.

$$\mathbf{V_{P_P}}(t) = (V_{P_{px}}(t), V_{P_{py}}(t)) = (\frac{P_{w2x}(t+\Delta t) - P_{w2x}(t)}{2\Delta t}, 0), \tag{5}$$

$$\Delta\theta_o(t) = -\tan^{-1}\frac{P_{w2y}(t)}{P_{w2x}(t)} - \tan^{-1}\frac{P_{w2y}(t+\Delta t)}{P_{w2x}(t+\Delta t)}. \tag{6}$$

$\Delta\theta_o(t)$ is the sum of the changes in the projected front steering angle $\theta_{leg}(t)$ and the body yaw angle $\theta_B(t)$:

$$\Delta\theta_o(t) = \Delta\theta_{leg}(t) + \Delta\theta_B(t). \tag{7}$$

From these variables, the angular velocity of the projected front steering shaft $\dot{\theta}_{leg}(t)$ and the angular velocity of the front steering $\dot{\theta}_{sf}(t)$ are determined by calculating $\dot{\theta}_B(t)$ and using the relationship between $\dot{\theta}_{leg}(t)$ and $\dot{\theta}_{sf}(t)$, which is determined topologically from the relations below.

$$\theta_{leg}(t) = \theta_{sf}(t)\cos\theta_{p_B}(t) + \theta_{rf}(t)\sin\theta_{p_B}(t), \tag{8}$$

$$\therefore \dot{\theta}_{sf}(t) = \frac{\dot{\theta}_{leg}(t) - \dot{\theta}_{rf}(t)\sin\theta_{p_B}(t) + \dot{\theta}_{P_B}(t)(\theta_{sf}(t)\sin\theta_{p_B}(t) - \theta_{rf}(t)\cos\theta_{p_B}(t))}{\cos\theta_{p_B}(t)}, \tag{9}$$

where θ_{p_B} is obtained from attitude sensor information on the platform and the pitch adjustment angle.

The angular velocity of the body rotation $\dot{\theta}_B$ is

$$\dot{\theta}_B(t) = \frac{V_{P_{Qx}}(t) - V_{P_{Px}}(t)}{B(t)}, \tag{10}$$

where B is the length of the projection body and $V_{P_{Qx}}$ is the x element of the velocity of P_Q (Fig. 12(b)). $B(t)$ is the length between ${}^0P_P$ and ${}^0P_Q$, where ${}^0P_P$ and ${}^0P_Q$ are the positions of P_P and P_Q in the body-centered coordinate system. The velocity of P_Q, $\mathbf{V_{P_Q}}$, is given by

$$\mathbf{V_{P_Q}}(t) = ({}^0\mathbf{P_Q}(t) - {}^0\mathbf{P_Q}(t - \Delta t) - \Delta\mathbf{O_o})/\Delta t, \tag{11}$$

where $\Delta\mathbf{O_o}$ is the movement of the origin of the body-centered coordinate system relative to the absolute coordinate system:

$$\Delta\mathbf{O_o} = {}^0\mathbf{P_{w1}}(t) - {}^0\mathbf{P_{w1}}(t - \Delta t). \tag{12}$$

The angular velocity of the front steering shaft $\dot{\theta_{sf}}$, which is one of the three control parameters, is determined by eqs. (6), (7), (9), and (10).

5.1.1 How to derive velocities of rear-left and rear-right wheel

Here, we derive the velocities of the rear-left and rear-right wheels, $\mathbf{V_{w3}}(t)$ and $\mathbf{V_{w4}}(t)$. The velocity generated at point P_P when stopping the right-back wheel ($\mathbf{V_{w4}} = 0$) and moving left-back wheel at $\mathbf{V_{w3}}$ is $\mathbf{V_{P_{Pw3}}}$ shown in Fig. 12(b). If we define $\mathbf{V_{P_{Pw4}}}$ similarly, then the velocity of $P_P(t)$ is

$$\mathbf{V_{P_P}}(t) = \mathbf{V_{P_{Pw3}}}(t) + \mathbf{V_{P_{Pw4}}}(t). \tag{13}$$

The relationships between $\mathbf{V_{w3}}$ and $\mathbf{V_{P_{Pw3}}}$, and between $\mathbf{V_{w4}}$ and $\mathbf{V_{P_{Pw4}}}$ are

$$\mathbf{V_{w3}}(t) = \frac{2A_r(t)}{LR(t)}\mathbf{V_{P_{Pw3}}}(t), \tag{14}$$

$$\mathbf{V_{w4}}(t) = \frac{2A_r(t)}{LL(t)}\mathbf{V_{P_{Pw4}}}(t), \tag{15}$$

where $LR(t)$ and $LL(t)$ are obtained from $B(t)$, $\theta_{sup}(t)$, and the distance $A_r(t)$ between $P_{w3}(t)$ and $P_Q(t)$ in Fig. 12(b).

The velocities of the rear-left wheel and the rear-right wheel are determined by eqs. (5), (13), (14), and (15).

5.2 Swing phase

Figure 12(c) shows a model of the swing phase, where the origin of the absolute coordinate system is the front-left wheel and the lifted wheel is the front-right wheel. The trajectory is set such that point P_P draws a circular path around the front-left wheel. The angular velocity of the front steering shaft and the velocities of the rear wheels are determined so that they produce $\mathbf{V_{P_P}}$. Setting a command value for $\dot{\theta}_o$, we obtain

$$|\mathbf{V_{P_P}}(t)| = A_f(t)|\dot{\theta}_o|, \tag{16}$$

$$\mathbf{V_{P_P}}(t) = (-|\mathbf{V_{P_P}}(t)|\sin(\theta_{leg}(t) + \theta_B(t)), |\mathbf{V_{P_P}}(t)|\cos(\theta_{leg}(t) + \theta_B(t))). \tag{17}$$

With the velocity of point P_P determined, as in the lifting and landing phases, the three control parameters, the angular velocity of the front steering shaft and the velocities of the rear wheels, can be obtained.

5.3 Wheel mode

In Fig. 9(g) and (h), for example, the robot moves with all four wheels supporting the body. Since the velocity of the body center, $\mathbf{V_B}$, and the angles of the front and rear steering axes in the projection frame, θ_{leg} and θ_{sup}, are given as parameters, the desired wheel velocities with no slipping, $\mathbf{V_{w1}} \sim \mathbf{V_{w4}}$, are derived. Since each wheel rotates about O_H, $\mathbf{V_{wi}}$ is given by $\mathbf{V_{wi}}(t) = l_{wi}(t)\mathbf{V_B}(t)/R_H(t)(i = 1 \sim 4)$ where $R_H(t)$ is the turning radius. Except under conditions, such as $\theta_{leg} = \theta_{sup}$, where the front and rear steering angles are equal and the turning radius becomes infinite, the topology in Fig. 12(d) leads to

$$O_H(t) = (x_H(t), y_H(t)) = (\frac{B(t)}{\tan\theta_{sup}(t) - \tan\theta_{leg}(t)}, \frac{B(t)}{2}\frac{\tan\theta_{sup}(t) + \tan\theta_{leg}(t)}{\tan\theta_{sup}(t) - \tan\theta_{leg}(t)}) \tag{18}$$

and $R_H(t) = \sqrt{x_H(t)^2 + y_H(t)^2}$. Variables such as l_{w1} are obtained in the form $l_{w1}(t) = |(x_H(t) - P_{w1x}(t))/\cos\theta_{leg}(t)|$.
However, when $\theta_{leg}(t) = \theta_{sup}(t)$, we have $\mathbf{V_{wi}} = \mathbf{V_B}(i = 1 \sim 4)$.

6. Stability in leg mode

In this section, whether the robot can maintain static stability while moving over a target step of 0.15[m] is analyzed for the gait strategy given above. Static state locomotion is considered as an initial step. In general, statically stable locomotion can be achieved if the center of gravity is located inside the support polygon. Here, the stability during movement of the proposed robot in leg mode is specifically investigated. For example, the best range of body yaw angle shown in Fig. 9(g) to climb a step while maintaining stability is derived.
Figure 13(a) shows the static stability when lifting the front-left wheel. Static stability is positive if the center of gravity is in the supporting polygon. Since RT-Mover employs a mechanism with a small number of driving shafts, it cannot move its center of gravity without altering the position of the supporting wheels. In addition, the supporting point of the front-right wheel in Fig. 13(a) cannot move since the lifted wheel is needed to move forward. Thus, the rear steering is used so that the center of gravity stays within the supporting polygon. As shown in Fig. 13(b), if the body inclines backward when going up a step, the center of gravity is displaced backward by $h_g \sin\theta_{p_B}$, where θ_{p_B} is the body pitch angle.
Figure 14(A) shows four phases during the step-up gait. Out of the four phases in which a wheel is lifted during the step-up gait, only those shown in Fig. 14(A-c) and (A-d) cause static instability, because the center of gravity is displaced backward due to the backward inclination of the body and the stability margin consequently decreases. Here, the front steering is rotated up to the limit of ±30[deg] in the direction that increases stability. First, the rear-left wheel is lifted (Fig. 14(A-c)), moved forward, and then lowered. Next, the rear-right wheel is lifted, moved forward, and lowered. Therefore, the rear steering angle when the rear-right wheel is lifted depends on the rear steering angle when the rear-left wheel is lifted. It can be seen in Fig. 14(A-c) and (A-d) that the less the lifted rear-left wheel goes forward, the more static stability the robot has at the beginning of lifting the rear-right wheel. Hence, the rear-left

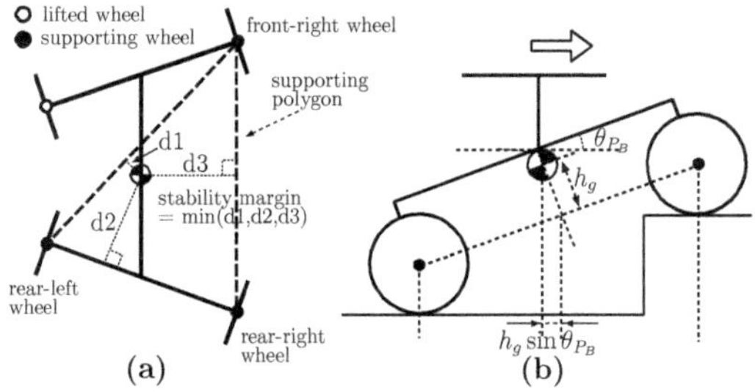

Fig. 13. Stability margin

wheel must be advanced by the minimum distance required for going up the step. Since the lifted wheel can be placed on the step from the state shown in Fig. 14(A-c) by advancing it a distance equal to its radius, θ_A is set at $\tan^{-1}(R_w^{'}/(2A_r))$, where $R_w^{'} = R_w + 0.02$[m](margin).

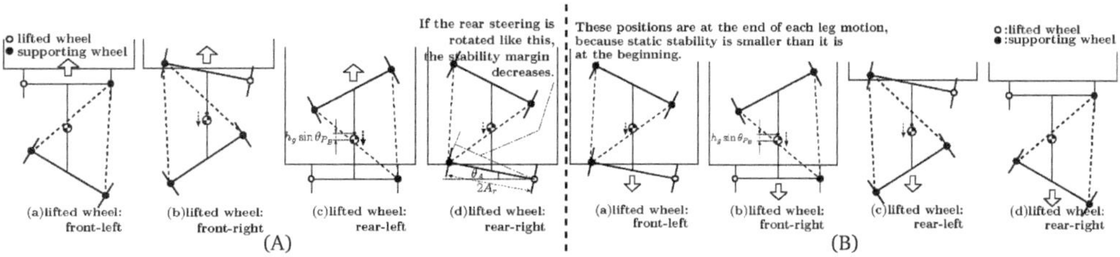

Fig. 14. Four phases during the gait. (A)The step-up gait. (B)The step-down gait.

Since the rear-left wheel is already on the step when lifting the rear-right wheel, the body pitch angle is smaller in (A-d) than in (A-c).

Figure 15 shows the results of numerical calculations of the margin of static stability (the minimum distance between the center of gravity and the supporting polygon) on a 0.15[m] high step. 0.15[m] is the maximum targeted height for the middle size type of RT-Mover.

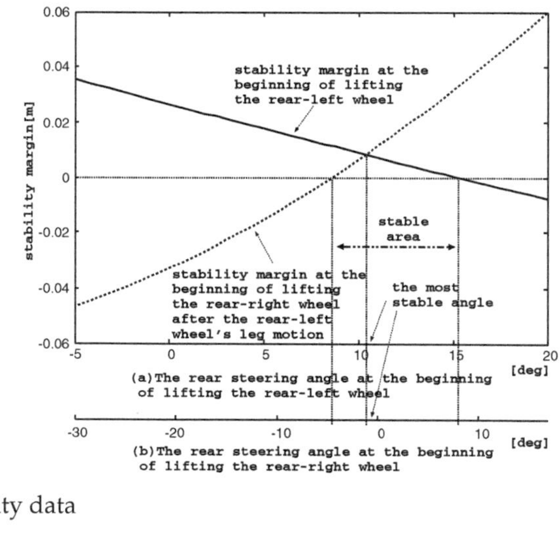

Fig. 15. Static stability data

A positive value of static stability indicates that the robot is stable, and a negative one indicates that it is unstable. Figure 15(a) shows that it is possible to go up a 0.15[m] step while maintaining static stability by setting the rear steering angle to be between 8 and 15.5[deg] when lifting the rear-left leg. The most stable angle is 11[deg], so the yaw angle of the robot becomes 11[deg] in Fig. 9(g).

When descending a step, the four phases in Fig. 14(A) occur in reverse order as shown in Fig. 14(B). The positions shown in Fig. 14(B) are at the end of each leg motion, because static stability is smaller than it is at the beginning. Out of the four phases, only those shown in Fig. 14(B-a) and (B-b) cause static instability due to an inclination of the center of gravity. Because the stability of Fig. 14(B-b) is determined by the condition of Fig. 14(B-a) and Fig. 14(B-a) corresponds to Fig. 14(A-d), Fig. 15(b) can be used for discussing the stability margin for the step-down gait. Figure 15(b) shows that it is possible to go down a 0.15[m] step while maintaining static stability by setting the front steering angle to be between -4.5 and 8[deg] when landing the front-left leg. The most stable angle is -1[deg].

For the maximum stable angle, the yaw angle of the robot shown in Fig. 10(c) is configured to a value calculated by (A) + (B) + (C). Here, (A) is the maximum stable angle of Fig. 15(b), (B) is the change in front steering angle generated by swinging front-left wheel ($\theta_b - \theta_a$ in Fig. 16), and (C) is the change in the front steering angle generated by the front-left wheel landing (Fig. 16 (c)).

As (A)=-1[deg], (B)=12[deg], and (C)=4[deg] for the robot, the yaw angle of the body is determined to be 15[deg] in Fig. 10(c).

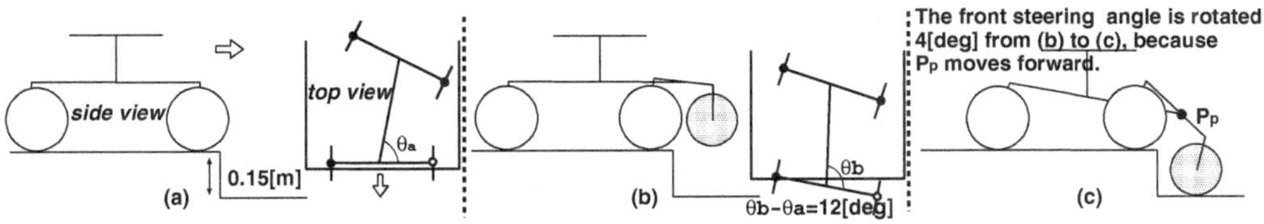

Fig. 16. Change of the front steering angle when moving the front-left wheel forward and lowering it

7. Assessment of ability of locomotion in leg mode

7.1 Step-up gait

The proposed step-up gait was evaluated through a simulation and an experiment. The conditions of the simulation are the following. The upward step height is 0.15[m], the height when lifting a wheel is 0.16[m], the distance that the lifted wheel is moved forward is 0.12[m], the yaw angle of the body relative to the step in Fig. 9(g) is 11[deg], the angular velocity of a roll-adjustment shaft when lifting the wheel is 0.2[rad/s], $\dot{\theta_0}$ in Fig. 12(c) is 0.2[rad/s], the angular velocity of a roll-adjustment shaft when landing the wheel is 0.1[rad/s], and the forward velocity of the body in wheel mode is 0.1[m/s]. In this chapter, the road shape is assumed to be known in advance. The robot starts 0.2[m] from the step, as shown in Fig. 17. The configured values are given a margin of 0.01[m] when lifting a wheel onto a step of height 0.15[m] and a margin of 0.02[m] when extending the wheel by the wheel radius of 0.1 [m]. The configured value of each process velocity in leg mode is obtained experimentally from a velocity that gives static leg motion. There are plans to address high-speed leg processes for both step-up and step-down gaits in the future.

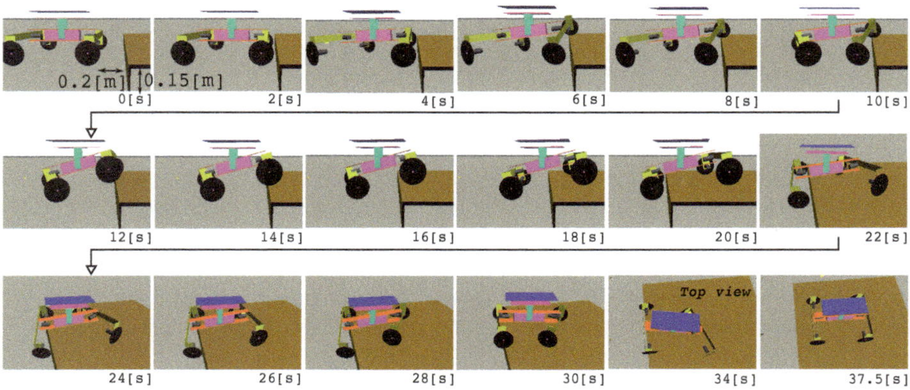

Fig. 17. Snapshots of the step-up gait simulation

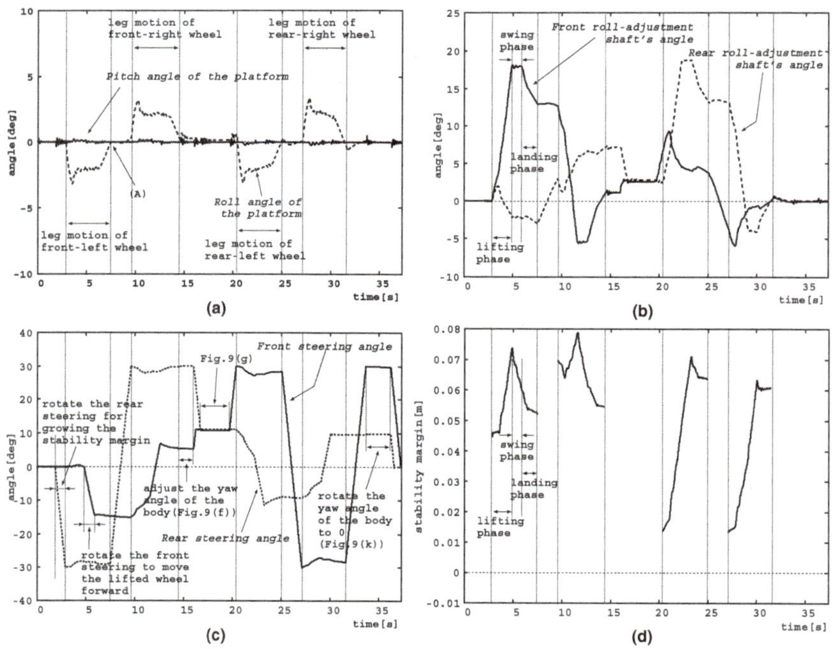

Fig. 18. Simulation data for the step-up gait. (a) Posture angles of the platform. (b) Front and rear roll-adjustment shaft's angles. (c) Front and rear steering angles. (d) Static stability during each leg motion.

Figure 18 shows the posture of the platform, the angles of the front and rear roll-adjustment shafts, the front and rear steering angles, and the static stability during each leg motion. Figure 18(a) shows that the pitch posture angle of the platform is almost kept horizontal. The roll angle of the platform is kept horizontal to within ± 3[deg]. At 2.8 $\sim$ 7.5[s], 9.6 $\sim$ 14.5[s], 20.4 $\sim$ 25.0[s], and 27.1 $\sim$ 31.6[s], the roll angle is larger than at other times because the twisting force around the body, caused by the roll-adjustment shaft that produces the torque for lifting a wheel, disturbs the posture control of the other roll-adjustment shaft. The timings given are those during each leg motion.

Figure 18(b) shows the transition of angles of the front and rear roll-adjustment shafts. From 2.8[s] to 7.5[s], the front-left wheel is lifted. First, the wheel is lifted until the front roll-adjustment shaft is rotated at 18[deg] (2.8[s] to 4.9[s]). From 4.9[s] to 5.9[s], the front steering is rotated until it reaches -14.5[deg] so that the wheel moves forward 0.12[m] (Fig. 18(c)). Then the wheel moves downward from 5.9[s] to 7.5[s]. Since the roll angle of the platform changes from negative to positive at 7.5[s]((A) in Fig. 18(a)), the landing of the wheel can be detected. The other legs behave similarly.

Figure 18(c) shows the transition of angles of the front and rear steering shafts. From 2.8[s] to 7.5[s], the front wheels are lifted. While the front-left wheel is lifted, the rear steering shaft rotates to its steering limit of -30[deg] (1.8[s] to 7.5[s]) so that the static stability increases. After lifting the front-left wheel, the wheel is moved forward until the front steering angle becomes -14.5[deg] (4.9[s] to 5.9[s]). While the front-right wheel is lifted, the rear steering shaft is maintained at the steering limit of 30[deg] (9.6[s] to 14.5[s]) so that the static stability increases. The rear steering shaft is also maintained at 30[deg] (14.5[s] to 15.9[s]) after the front wheels are lifted, thereby adjusting the yaw angle of the body relative to the step to 11[deg] for lifting the rear wheels. Rear wheels are lifted between 20.4[s] and 31.6[s]. While the rear-left wheel is lifted, the wheel is moved forward 0.12[m] until the rear steering shaft reaches an angle of -10.8[deg] (22.1[s] to 23.1[s]). The front steering shaft is rotated to $\pm$ 30[deg] in order to ensure static stability.

Figure 18(d) shows the data for static stability only during leg motion, because static stability is large enough during wheel mode. The figure shows that the static stability is maintained. When lifting the front-left wheel, the static stability increases, because the center of gravity of the robot moves backward according to the body pitch (2.8[s] to 4.9[s]). In the swing phase of the front-left wheel, static stability decreases, because the position of the front-right wheel with respect to the body changes and the supporting polygon becomes smaller (4.9[s] to 5.9[s]). Finally, in its landing phase, static stability decreases, because the center of gravity of the robot moves forward due to the body pitch (5.9[s] to 7.5[s]).

Figure 19 shows scenes from a step-up gait experiment and the experimental data. The conditions of the experiment are the same as those of the simulation except the D gains for each shaft are set experimentally. The actual robot can also move up onto the 0.15[m]-high step, and the features of the experimental data are almost the same as those of the simulation data. However, it takes about 2.5[s] longer to perform the movement in the experiment than in the simulation. The main reason is that the detection of the landing of each wheel is delayed due to a difference in the posture of the platform between the simulation and the experiment. The inclination of the pitch angle of the platform is larger in the experiment than in the simulation, because of the backlash of the pitch-adjustment shaft and the friction acting on it in the actual robot. Thus, the proposed step-up gait was proved to be effective.

7.2 Step-down gait

The proposed step-down gait was evaluated using a simulation and an experiment. Due to space limitations, only the result of simulation is shown. The conditions of the simulation are the following. The downward step height is 0.15[m], the height when lifting a wheel is 0.02[m], the length the lifted wheel is moved forward is 0.12[m], the yaw angle of the body in Fig. 10(c) is 15[deg], the angular velocity of a roll-adjustment shaft when lifting a wheel is 0.2[rad/s], $\dot{\theta}_0$ in Fig. 12(c) is 0.2[rad/s], the angular velocity of a roll-adjustment shaft when landing a wheel is 0.1[rad/s], the forward velocity of the body in wheel mode is 0.1[m/s], and the road shape is known in advance. The robot starts at a position 0.2[m] from the step, as shown in Fig. 20. The configured value allows a margin of 0.02[m] in the height by which to lift the wheel and in the length by which to swing the lifted wheel forward. The configured value of each process velocity in leg mode is obtained experimentally from a velocity that gives static leg motion.

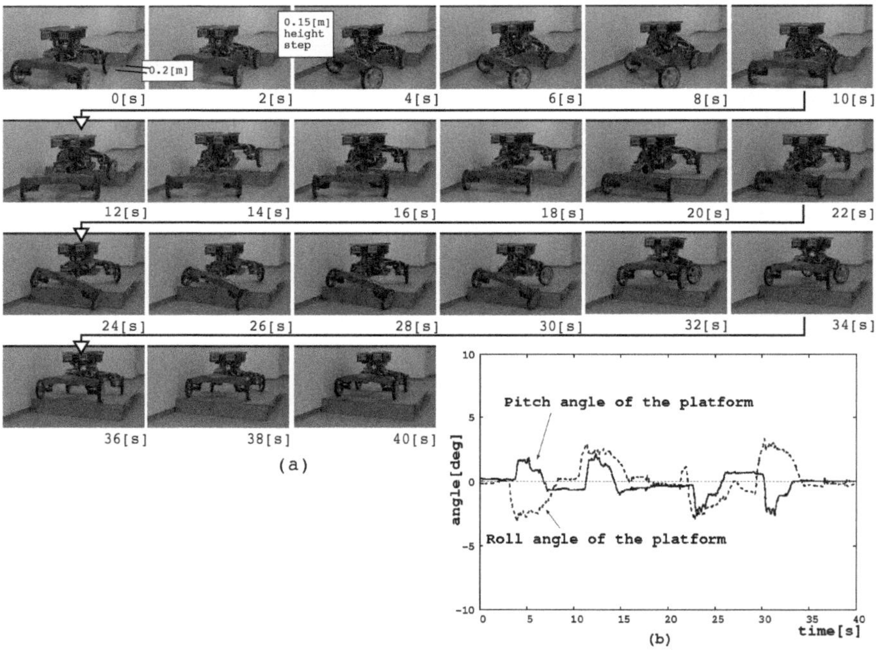

Fig. 19. Experimental data for the step-up gait. (a) Experimental scenes. (b) Posture angles of the platform.

Figure 20 shows snapshots of the step-down gait simulation. It can be seen that the step-down gait presented in Fig. 10 is performed stably.

8. A personal mobility vehicle, RT-mover P-type

RT-Mover P-type (Fig. 21) that is one of RT-Mover series is introduced. This robot can carry a person even if on the targetted rough terrain. The specifications of it are listed in Table 5. When rotating the roll-adjustment axis through 30[deg] such that the wheel on one side is in contact with the ground, the other wheel attached to a 0.65[m] steering arm can rise

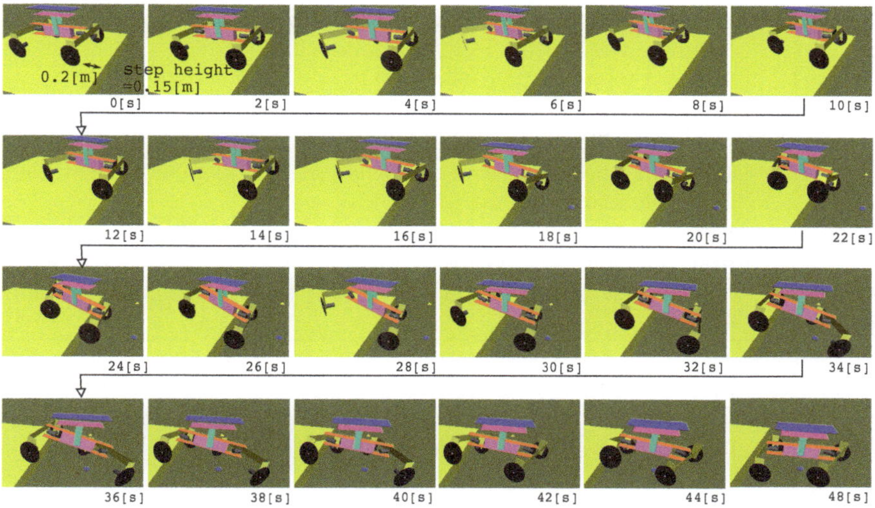

Fig. 20. Snapshots of the step-down gait simulation

0.325[m]. Therefore, the movement range is sufficient for the targeted terrain. Likewise, moving 0.325[m] in the front and rear directions is possible by moving the steering from 0[deg] to 30[deg], and holes of 0.325[m] can be crossed. With regards to locomotion on a slope, back-and-forth movement and traversal of a slope of up to 30[deg] is possible.

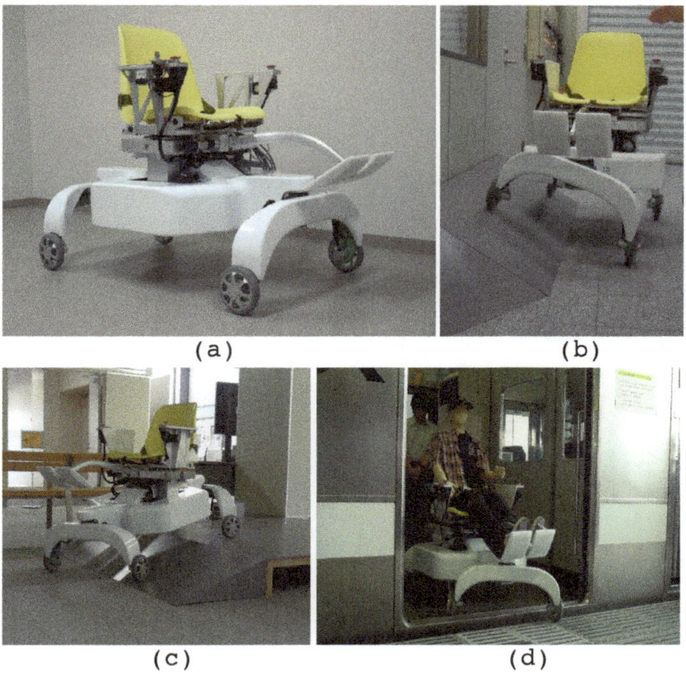

Fig. 21. (a)RT-Mover P-type. (b)On a bank. (c)On a slope. (d)Getting off a train.

Dimensions	Length 1.15[m](excluding footrest); Width 0.70[m] (Tread 0.60[m]); Height to seat 0.58[m]; Height to bottom 0.17[m]
Wheel	Radius:0.15[m]; Width:0.03[m]
Weight	80[kg] (including batteries at 20[kg])
Motor	maxon brushless motor 100[W] $\times$ 9
Gear ratio	100 (each wheel, front and rear steering); 820 (pitch-adjustment shaft); 2400 (roll-adjustment shaft)
Sensor	Encoder (each motor); Current sensor (each motor); Posture angle sensor (roll and pitch of platform)
Angle limit	$\pm$30[deg] (steering, roll-adjustment shaft, and pitch-adjustment shaft)
Max speed	4.5[km/s]
Power supply	48[V] lead accumulator

Table 5. Main specifications of P-type

In fact, additional motors are attached to the robot, for example, for adjusting footrest mechanism. Those are, however, not essential functions for moving on rough terrain, so they are not discussed here.

9. Assessment of ability of locomotion of P-type

Evaluations were performed through experiments taking a step-up gait and a step-down gait as examples. The above-mentioned methodology is also used for these gaits. At the current stage, road shapes are known in advance.

Fig.22 shows data of the step-up walking experiment over a 0.15[m]-high step. The robot can get over a 0.15[m] step with a person riding on it while maintaing the horizontal position of its platform within $\pm$5[deg]. The main conditions are the followings. The angular velocity of a roll-adjustment shaft when lifting and landing the wheel are 0.2[rad/s] and 0.1[rad/s]

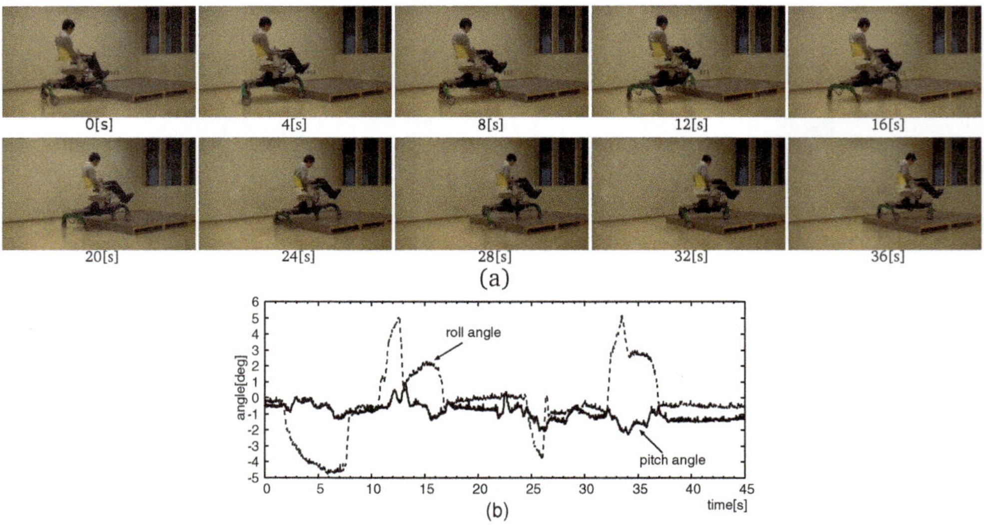

Fig. 22. Experimental result of the step-up gait. (a)Snapshots. (b)Posture angles of the platform.

respectively, that of a steering shaft to put forward the lifted leg is 0.2[rad/s], and the forward velocity of the body in wheel mode is 0.12[m/s]. The configured value of each process velocity in leg mode is obtained experimentally from a velocity that gives static leg motion. There are plans to address high-speed leg processes in near future.

Fig.23 shows data of the step-down walking experiment down a 0.15[m]-high step. The main conditions are basically the same as the step-up gait experiment. The robot can decline a 0.15[m] step with a person riding on it while maintaing the horizontal position of its platform within ± 4.5[deg].

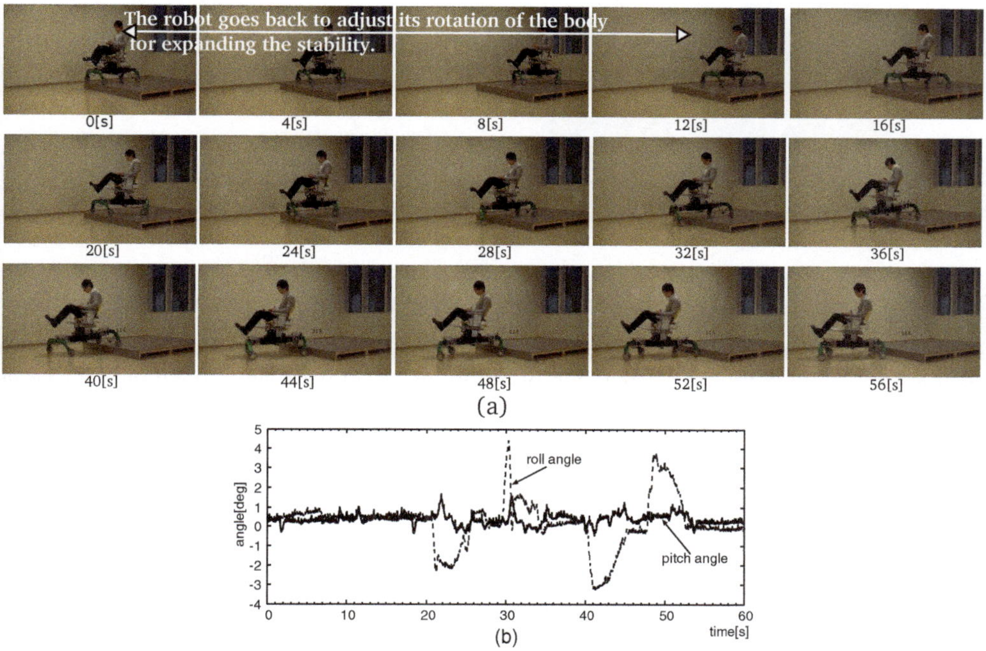

Fig. 23. Experimental result of the step-down gait. (a)Snapshots. (b)Posture angles of the platform.

10. Conclusions

We have developed some mobile platforms with leg-wheel mechanism for practical use, including a real size personal mobility vehicle (Fig. 24(a)). They are RT-Movers that have both of a wheel mode and a leg mode in a simple mechanism. They have four drivable wheels and two leg-like axles. The wheels are mounted on one side of the leg-like axles at the front and rear of the body. The mechanism is realized with few drive shafts to achieve the minimum necessary leg functions, taking a four-wheel model as the base.

The mechanical design concept was discussed and strategies for moving on rough terrain were proposed. The kinematics, stability, and control method of RT-Mover were also described in detail. Some typical cases of wheel mode and leg mode locomotion were selected, and the robot's ability of locomotion on rough terrain was assessed through simulations and

experiments. In every case, the robot was able to move while maintaining the horizontal position of its platform.
We are undertaking joint research with a railway company to develop a personal mobility robot for outdoor use, including on rough terrain. Good coordination between the personal mobility robot and the railway system may also lead to a new type of transportation system (see Fig. 24(b)).

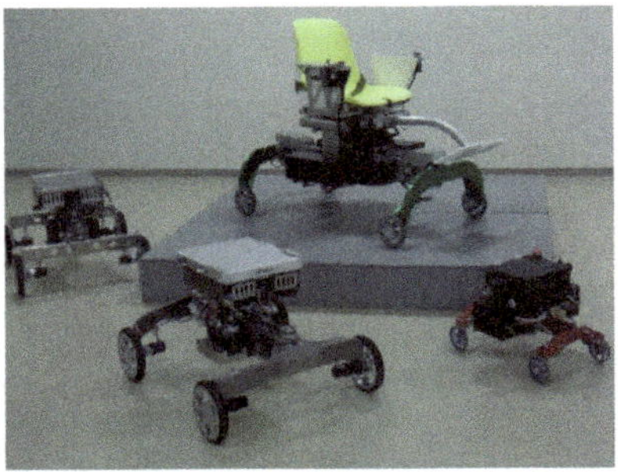

(a)

A future transportation system image of seamless connection between railway system and personal mobility vehicles

(b)

Fig. 24. Snapshots. (a) RT-Mover series. (b) A future transportation system image.

Since this research has just started, there is much work that should be done in the future, for example: 1. allowing for the perception of rough terrain rather than moving over obstacles whose position is known in advance; 2. adapting control methods for moving on different types of rough terrain; 3. dynamic control on rough terrain for high-speed locomotion.

11. References

Bares J. and Wettergreen D., (1997). Lessons from the development and deployment of Dante II, *Proceedings of the 1997 Field and Service Robotics Conference*, pp.72-79.

Daltorio K. A., et al., (2009). Mini-Whegs Climbs Steep Surfaces Using Insect-inspired Attachment Mechanisms, *The International Journal of Robotics Research*, 28(2): 285-302.

Delcomyn F. and Nelson M. E., (2000). Architectures for a biomimetic hexapod robot, *Robotics and Autonomous Systems*, 30: 5-15.

Endo G. and Hirose S., (2000). Study on roller-walker (multi-mode steering control and self-contained locomotion), *Journal of Robotics and Mechatronics*, 12(5): 559-566.

Grand C., et al., (2004). Stability and Traction Optimization of a Reconfigurable Wheel-Legged Robot, *The International Journal of Robotics Research*, 23(10-11): 1041-1058.

Halme A., et al., (2003). WorkPartner: Interactive Human-Like Service Robot for Outdoor Applications, *The International Journal of Robotics Research*, 22(7-8): 627-640.

Hirose S., et al., (1985). The Gait Control System of the Quadruped Walking Vehicle, *Journal of the Robotics Society of Japan*, 3(4): 304-323.

Kimura H., et al., (2007). Adaptive Dynamic Walking of a Quadruped Robot on Natural Ground Based on Biological Concepts, *The International Journal of Robotics Research*, 26(5): 475-490.

Kubota T., et al., (2003). Small, light-weight rover "Micro5" for lunar exploration, *Acta Astronautica*, 52: 447-453.

Lacagnina M., et al., (2003). Kinematics, dynamics and control of a hybrid robot Wheeleg, *Robotics and Autonomous Systems*, 45: 161-180.

Lauria M., et al., (1998). Design and control of an innovative micro-rover, *Proceedings of the Fifth ESA Workshop on Advanced Space Technologies for Robotics and Automation,* The Netherlands, 1998.

Morales R., et al., (2006). Kinematic Model of a New Staircase Climbing Wheelchair and its Experimental Validation, *The International Journal of Robotics Research,* 25(9): 825-841.

Nakajima S., (2011). RT-Mover: a rough terrain mobile robot with a simple leg-wheel hybrid mechanism, *The International Journal of Robotics Research,* doi:10.1177/0278364911405697 .

Nakajima S., (2011). Development of a Personal Mobility Robot for Rough Terrain, *Proceedings of the 14th CLAWAR,* accepted.

Nakajima S. and Nakano E., (2008a). Adaptive Gait for Large Rough Terrain of a Leg-wheel Robot (First Report: Gait Strategy), *Journal of Robotics and Mechatronics,* 20(5): 801-805.

Nakajima S. and Nakano E., (2008b). Adaptive Gait for Large Rough Terrain of a Leg-wheel Robot (Second Report:Step-Up Gait), *Journal of Robotics and Mechatronics,* 20(6): 913-920.

Nakajima S. and Nakano E., (2009a). Adaptive Gait for Large Rough Terrain of a Leg-wheel Robot (Third Report: Step-Down Gait), *Journal of Robotics and Mechatronics,* 21(1): 12-19.

Nakajima S. and Nakano E., (2009b). Adaptive Gait for Large Rough Terrain of a Leg-wheel Robot (Fourth Report: Step-Over Gait), *Journal of Robotics and Mechatronics,* 21(2): 285-292.

Nakajima S. and Nakano E., (2009c). Adaptive Gait for Large Rough Terrain of a Leg-wheel Robot (Fifth Report: Integrated Gait), *Journal of Robotics and Mechatronics,* 21(3): 419-426.

Quaglia G., et al., (2010). The Epi.q-1 Hybrid Mobile Robot, *The International Journal of Robotics Research,* 29(1): 81-91.

Quinn R. D., et al., (2003). Parallel Complementary Strategies for Implementing Biological Principles into Mobile Robots, *The International Journal of Robotics Research,* 22(3): 169-186.

Sato M., et al., (2007). An Environmental Adaptive Control System of a Wheel Type Mobile Robot for the Rough Terrain Movement, *Proceedings of the 2007 IEEE/RSJ International Conference on Intelligent Robots and Systems,* pp.3962-3967.

Siegwart R., et al., (2002). Innovative design for wheeled locomotion in rough terrain, *Robotics and Autonomous Systems,* 40: 151-162.

Six K. and Kecskem'ethy A., (1999). Steering properties of a combined wheeled and legged striding excavator, *Proceedings of the 10th World Congress on the Theory of Machines and Mechanisms,* pp.135-140.

Smith J. A., et al., (2006). PAW: a Hybrid Wheeled-Leg Robot, *Proceedings of the 2006 IEEE International Conference on Robotics and Automation,* pp.4043-4048.

Song S. M. and Waldron K. J., (1989). Machines That Walk: The Adaptive Suspension Vehicle, *MIT Press.*

Thueer T., et al., (2006). CRAB-Exploration rover with advanced obstacle negotiation capabilities, *Proceedings of the 9th ESA Workshop on Advanced Space Technologies for Robotics and Automation,* pp.1-8.

Volpe R., et al., (1997). Rocky 7: A next generation Mars rover prototype, *Journal of Advanced Robotics,* 11(4): 341-358.

Winnendael M. V., et al., (1999). Nanokhod micro-rover heading towards Mars, *Proceedings of the Fifth International Symposium on Artificial Intelligence, Robotics and Automation in Space*, pp.69-76.

Yoneda K., (2007). Light Weight Quadruped with Nine Actuators, *Journal of Robotics and Mechatronics*, 19(2): 160-165.

Yoneda K., et al., (2009). High-grip Stair Climber with Powder-filled Belts, *The International Journal of Robotics Research*, 28(1): 81-89.

Yuan J. and Hirose S., (2004). Research on Leg-wheel Hybrid Stair Climbing Robot, Zero Carrier, *Proceedings of the 2004 IEEE International Conference on Robotics and Biomimetics*, pp.1-6.

Part 3

Hardware – State of the Art

10

Construction of a Vertical Displacement Service Robot with Vacuum Cups

Nicolae Alexandrescu[1], Tudor Cătălin Apostolescu[2], Despina Duminică[1], Constantin Udrea[1], Georgeta Ionaşcu[1] and Lucian Bogatu[1]
[1]"POLITEHNICA" University of Bucharest
[2]"TITU MAIORESCU" University of Bucharest
Romania

1. Introduction

Based on the results of interdisciplinary fields like mechanics, electronics and informatics the autonomous mobile robots are gaining more and more attention. The methods used for movement, actuation and none the last the ability to operate in unknown and dynamic environments give them a great complexity and a certain level of intelligence.
The human model is still a challenge for mobility and robot movement. Robot mobility is satisfactorily achieved, in some cases even better with other types of locomotion means, but technical solutions that rigorously reproduce human walking are not yet identified.
Feasible solutions can be found in the situation when human mobility reaches its limits (movement on planes with high inclination angles, including 90° - vertical and 180° - parallel to the horizontal plane - ceiling type). Autonomous robot mobility on planes characterized by high inclinations is less tackled in specialty literature, some encountered solutions being objects of patents. The research in this field has a tremendous potential especially in providing new ideas and/or knowledge development.
The field of professional and domestic services, especially the wall cleaning of high buildings, is one of the areas that are expected to obtain a strong benefit from the use of robotic systems able to displace on vertical surfaces.
The advantages of the technologies that use climbing and walking robots consist mainly of two aspects:

- automatic cleaning of high buildings, improvement of technological level and of productivity of service industry in the field of building maintenance;
- cleaning robots can be used on various types of building, avoiding thus high costs involved by permanent gondola-type systems (open platform-car or baskets) for individual buildings.

The most common attachment principle is the vacuum adhesion (Cepolina et al., 2003), (Miyake et al., 2007), (Novotny&Horak, 2009), where the robot carries an onboard pump to create a vacuum inside the cups, which are pressed against the wall. This system enables the robots to adhere on any type of material, with low energy consumption. Vacuum adhesion is suitable for usage on smooth surfaces, because the roughness can influence a leakage loss in the vacuum chamber.

The mobile robots endowed with platforms and legs with cups are widely spread in practical applications due to high relative forces of locomotion, mobility and good suspension. The disadvantage of increased overall size less disturbs in applications of cleaning and inspection of large vitrified surfaces covering the buildings, (Sun.et al., 2007).
A new generation of cleaning robots based on all-pneumatic technology is on study (Belforte et al., 2005).
In this context, the authors developed an original solution of a cleaning robot with vertical displacement and vacuum attachment system (Alexandrescu, 2010a, 2010b), (Apostolescu, 2010). The novelty of the approach consists of the robot capability to move on vertical surfaces, which involves basic studies enlarging the horizon of knowledge related to: displacement cinematic structures, robot leg anchoring solutions, actuating solutions, as well as control system of such robots.

2. Robot construction

The robot structure (Fig. 1) contains two triangular platforms with relative mobility and reduced overall size. It is an original feature, not being found till now in the structure of the robots moving in the vertical plane. The fixing of the robot on the vertical surface is achieved by vacuum cups.

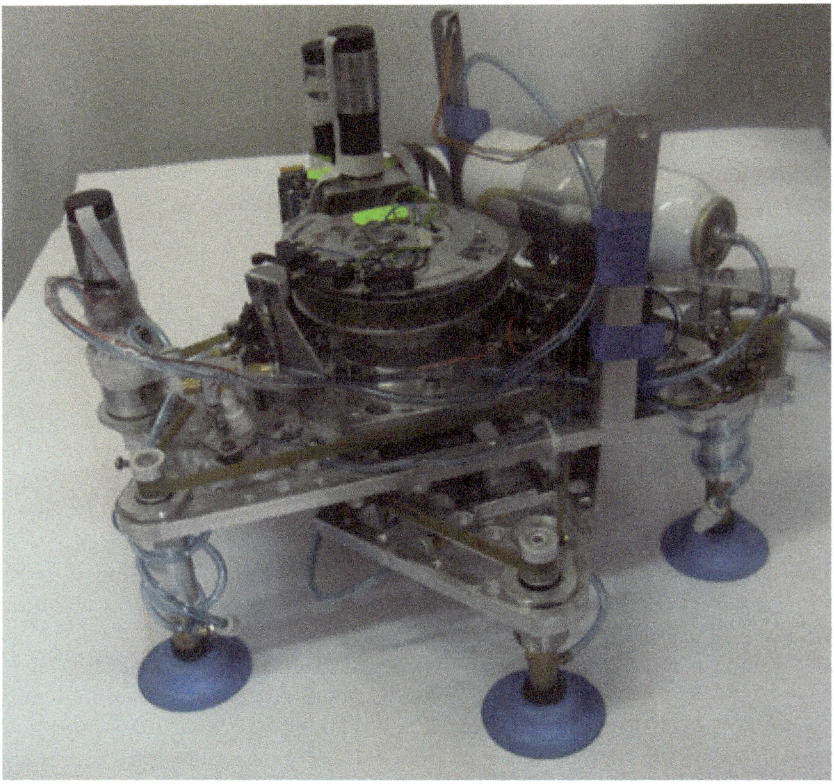

Fig. 1. The autonomous robot with vertical displacement

In order to obtain an autonomous robot, the electric actuation of all degrees of freedom was chosen, as well as for the depressurization needed by the cups. An original driving system was introduced for the moving of robot legs. The system uses screw mechanisms synchronized by a toothed belt transmission. The system developed for the relative translation of the platforms contains a ball guideway of reduced size and good guiding accuracy. The use of a rack mechanism allows a compact and fast actuation. The rotation of the robot is achieved by modifying the relative angular position of the two platforms.
The kinematic scheme of the robot is presented in Fig. 2. Although the robot attaches itself to the glass surface by the means of vacuum cups, a significant miniaturization was achieved. The interior platform - PLE - 7 is fixed on three vacuum suction cups 9. The raising and lowering of the platform is controlled using a screw mechanism 8 actuated by the motor-reduction gear M_1R_1. The toothed driving belt 4 allows the synchronous movement of the three suction cups. Similarly, for the exterior platform - PLE - 3 we have suction cups 1, the mechanism 2 and the motor-reduction gear M_2R_2 .

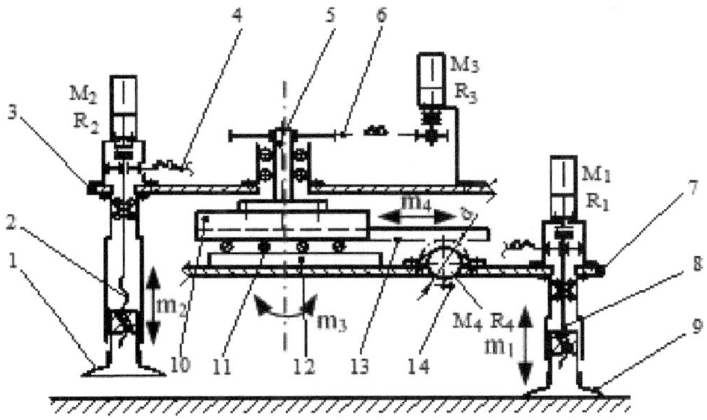

Fig. 2. The kinematic scheme

The driving toothed belt 6 and the motor-reduction gear M_3R_3 - movement m_3 allow the orienting rotation of the robot. The guiding part 10 is fixed on the shaft 5.
Translation between platforms - movement m_4 is achieved by the means of motor-reduction gear M_4R_4 fixed on the interior platform 7. A mechanism consisting of the gear 14 and the rack 13 solidary with the guiding part is used. This type of mechanism was preferred instead of a screw mechanism which would be difficult to miniaturize. A rolling guideway 11 is also used.
Figure 3 presents the 3D model of the robot placed in two positions: on a horizontal surface (Fig. 3a) and on a vertical surface (Fig. 3b).
The main parameters of the robot are:

- triangular platform side length L = 247 mm, corresponding to a stroke S of about 100…110 mm;
- full cycle for a translation step: 200…220 mm;
- duration of a cycle: 8s;
- vacuum of Δp=0.57 bar;
- diameter of the vacuum cups: 50 mm;
- normal detachment force: 86 N;
- lateral detachment force: 110 N;
- cup raising and lowering speed: approx. 6mm/s.

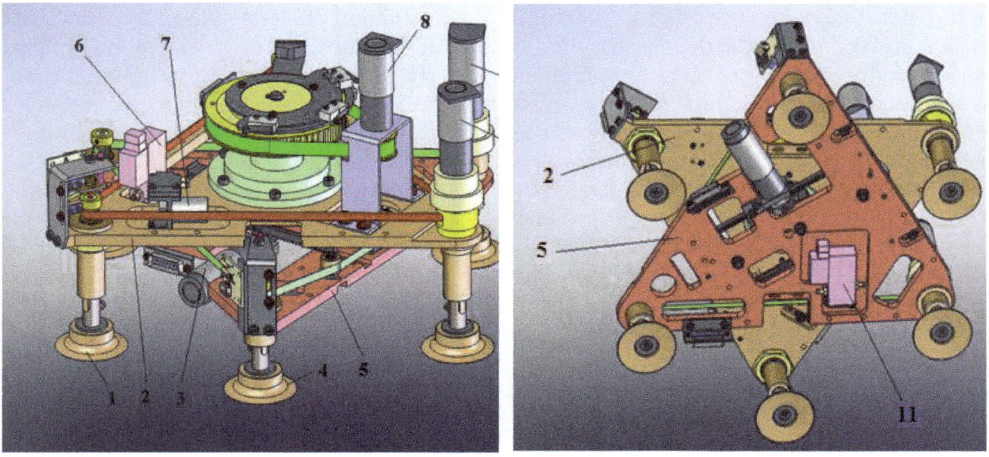

a) b)

Fig. 3. The 3D model of the robot: a) on a horizontal surface; b) on a vertical surface

The pneumatic diagram used for reducing the pressure inside the cups of the robot legs, in order to ensure the contact force when vacuumed suction cups adhere on the surface to be cleaned, is presented in Fig. 4: PV - vacuum micro pump (NMP 015 B – KNF Neuberger); Ac - tank; EM1, EM2 - electromagnets for the electro valve operating; V1, V2, V3 - vacuum cups for the interior platform; V4, V5, V6 - vacuum cups for the exterior platform; D - depressurization; A - atmospheric pressure.

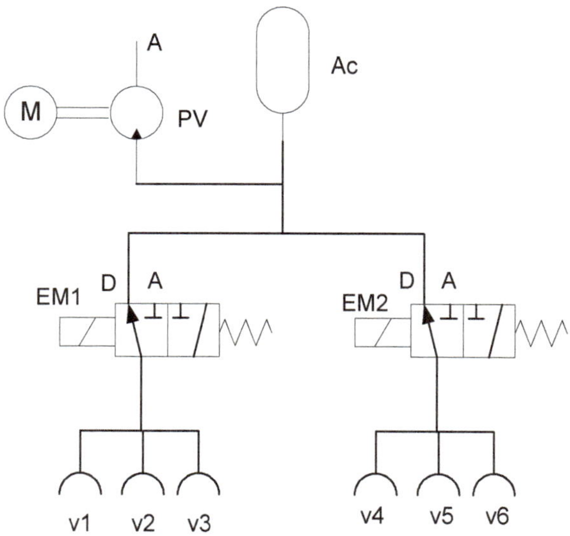

Fig. 4. Pneumatic diagram

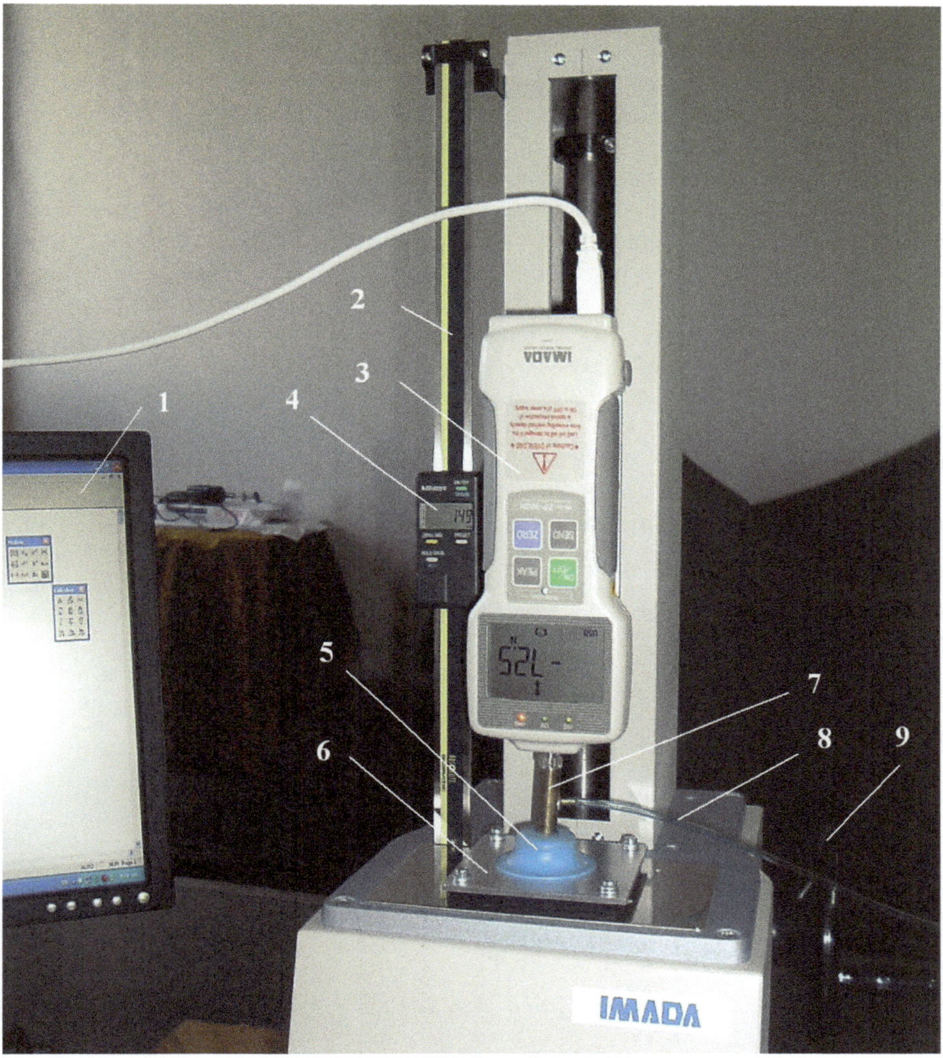

Fig. 5. Test stand used in order to establish the cup characteristic in the presence of normal forces F_x: 1 - computer, 2 - incremental rule, 3 - dynamometer, 4 - electronic vernier, 5 - vacuum cup, 6 - working surface, 7 - rod, 8 - table, 9 - hand wheel for obtaining the displacement w_x.

3. Research of the vacuum attachment system of the robot

An important part of the research concerned the fixing of the robot on the vacuum cups (Alexandrescu, 2010b). In order to establish their bearing capacity, the cups were subjected to external normal, lateral and combined loads. Tests were performed for different depressions. The influence of the different supporting materials was also studied, as well as the behaviour of the cup in presence of different liquids on the surface.

ESS-50 vacuum cups (FESTO) with an outer diameter of 50 mm were used. Glass (R_a=0.02µm), polished aluminium (R_a=0.59µm;) and textolite (R_a=1.1µm) were used as supporting surface. Different functional conditions were simulated. The roughness was measured using the roughness tester Surftest SJ 201P (Mitutoyo). Different working conditions were generated: cleaning of dry surfaces, cleaning of watered surfaces and cleaning using window detergent solution.

Fig. 5 presents the experimental stand used for the study of the effects of normal forces F_X.

Figure 6 presents the results obtained for dry glass surfaces. The positive values of the force correspond to the tendency of wrenching/detachment of the cup. The negative values of the force are conventionally assigned to the compression tendency of the cup.

Tests were performed also on dry aluminium surfaces and on dry textolite surfaces. The obtained diagrams are similar. Only small differences are noticed among the values corresponding to the three materials. For reduced deformations, glass and textolite have almost the same behaviour.

In fact, values of the roughness below 1µm do not influence the detachment force. Only the aluminium surface requires smaller loads.

In the range of negative deformations (cup compression), rigidity of the elastic membrane is the only factor that influences the characteristic, thus the three diagrams are very similar.

The study performed on watered surfaces showed almost insignificant decrease of the cup performance, due to the fact that water was practically eliminated from the contact area during cup fixing.

A most significant reduction of maximum normal load was noticed when detergent for window cleaning was used, because it remained in the contact area even after the cup was fixed on the supporting surface.

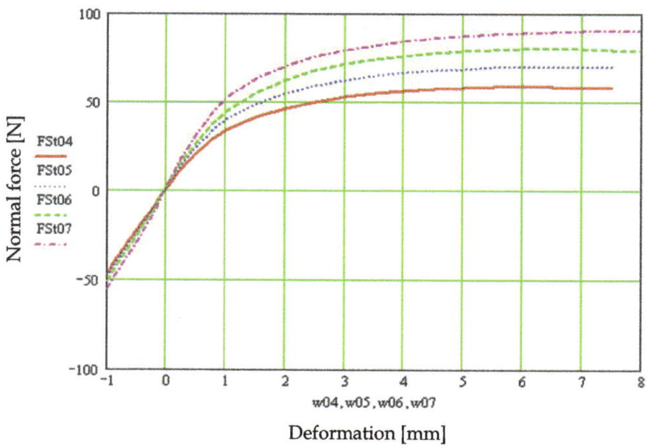

Fig. 6. Experimental diagrams obtained for dry glass, normal force; FSt04 – depression of 0.4bar; FSt05 – depression of 0.5bar; FSt06 – depression of 0.6bar; FSt07 – depression of 0.7bar.

Figure 7 presents the diagrams obtained for glass surfaces in the three situations. A decrease of about 6% of the cup capacity appears in the presence of detergent.

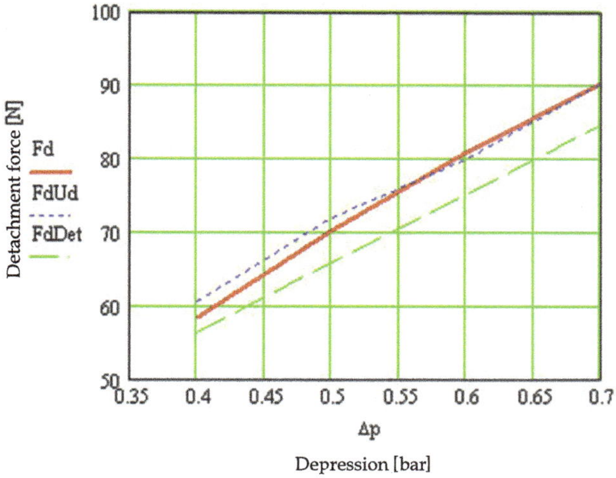

Fig. 7. Comparation among the maximum values of the detachment force: Fd - dry glass; FdUd - watered glass; FdDet - glass with detergent for window cleaning.

The test stand presented in Figure 8 was developed for the study of cup fixing on vertical planes.
In the presence of lateral forces, it was noticed that a slip of the cup appeared, leading to the re-positioning of the cup on the supporting surface. The slip stopped and the cup stabilized itself after 5...10s.
In order to deliver a correct information regarding the characteristic of the cup, after its stabilization, the deformation presented on the diagrams is obtained as difference between the total deformation obtained by progressive loading and the lateral slips.
The obtained results are presented in Figure 9. It can be noticed that the maximum supported loads are 50%-76% larger in the case of lateral forces that in the case of normal forces, an important advantage for a vertical robot. Similar results were obtained in the cases of aluminium and textolite.
Figure 10 presents a comparison among the three working surfaces for a depression of Δp=0.7bar. It can be noticed that the behaviour of the cup is practically identical.
The last part of the experimental research concerned the study of the cup behavior in presence of combined forces. The tests were performed for a maximum value of the normal force of F_n=49N, generated by the compression f=9mm of the spring 17 (Fig. 8). This force corresponds to approximately 60% of the normal detachment force previously established.
The diagrams corresponding to the tests on dry glass, in the presence of combined forces, are shown in Figure 11.
The same values of the depression as in the previous experiments were used for tests.
A comparison between the behaviour of the cup subjected to combined forces and subjected only to a normal force is presented in Figure 12. Experiments showed that, when lateral and normal forces acted together, the values of lateral forces were 36...45% smaller than in the absence of normal forces. Tests performed on aluminium and textolite surfaces led to similar results.
As previously, the research covered also the situation of fixing on watered glass. In the presence of lateral forces, the values of the cup slip increased significantly, e.g. for a depression of 0.6bar the slip exceeded 2mm. The obtained diagram is presented in Figure 13.

If the cup was supported on glass surfaces washed with detergent for window cleaning, the values of maximum lateral forces decreased with another 10...15%.
The experimental results obtained for aluminium and textolite were comparable to the results obtained in the case of glass supporting surfaces.

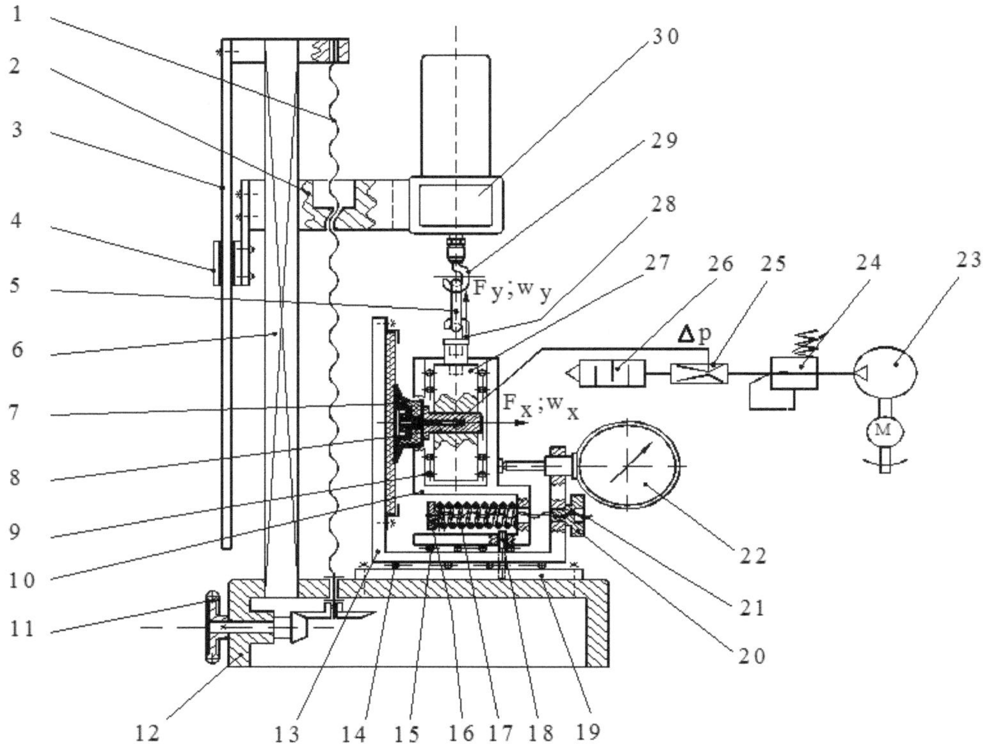

Fig. 8. Scheme of the experimental stand developed for testing the cups subjected to lateral forces F_y: 1 – screw for the displacement w_y; 2 -guide yoke; 3 -Vernier rod; 4 – digital vernier for w_y; 5 – link for transmitting the force F_y; 6 – guiding riser; 7 – cup; 8 – cup fixing body; 9, 14, 15 – ball guides; 10, 13, 27 – mobile elements; 11 – hand wheel w_y; 12 – dynamometer body; 16 – pan; 17 -spring for generating the force F_x; 18 – rod for positioning the element II; 19 – base; 20 - hand wheel w_x; 21 - screw for generating the force F_x; 22 – dial gauge; 23 – compressor; 24 – pressure regulation; 25 – ejector; 26 – noise damper; 27 – element I; 28 – stand hook; 29 – dynamometer hook.

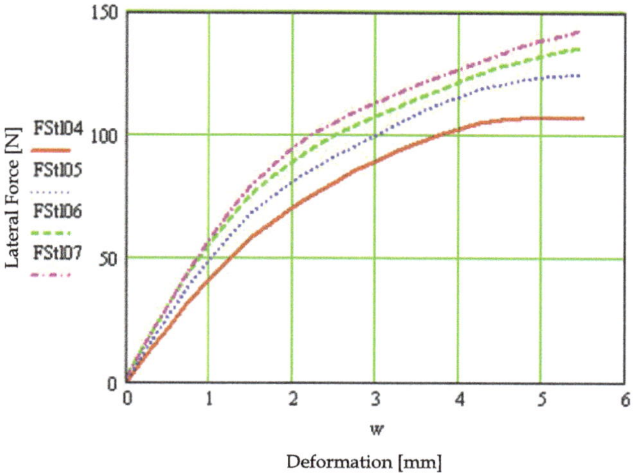

Fig. 9. Experimental diagrams for dry glass, lateral force; FSt04 - depression of 0.4bar; FSt05 - depression of 0.5bar; FSt06 - depression of 0.6bar; FSt07 - depression of 0.7bar.

Due to the fact that design conditions imposed a maximum robot leg slip of 0.5mm, the experimental research showed that the lateral forces corresponding to the vertical fixing of the robot must not exceed 25…30N. The values of the allowed lateral forces were 50…60% smaller than in the case of fixing on dry glass.

The results contributed significantly to the development of a robot able to sustain its own weight on vertical surfaces in order to perform cleaning tasks.

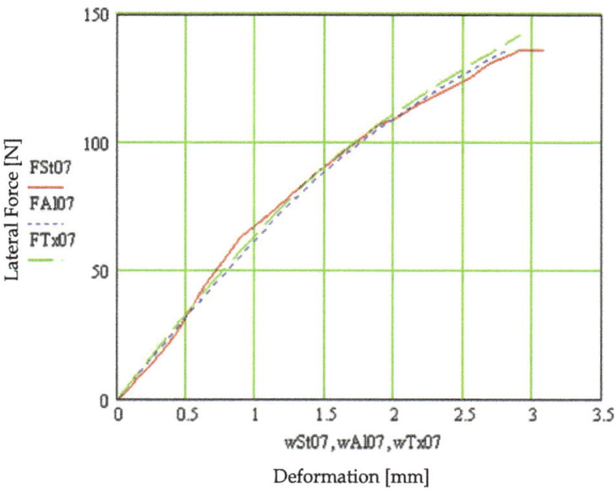

Fig. 10. Cup behavior in the presence of lateral force; depression of 0.7bar, different working surfaces: FSt07 - glass; FAl07 - aluminium; FTx07 - textolite.

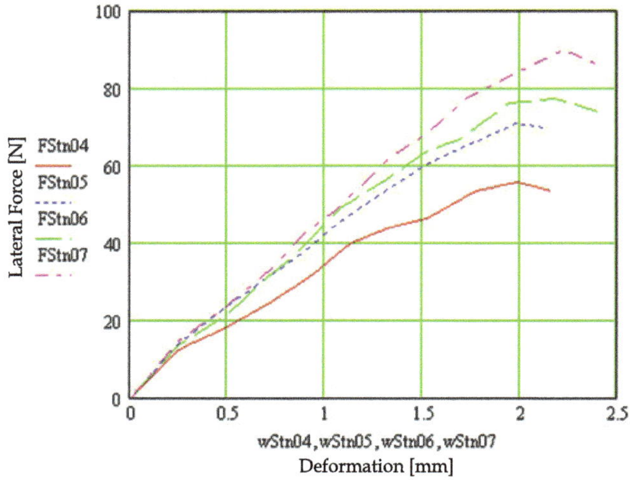

Fig. 11. Results of tests performed for combined forces (lateral force and normal force of 49N), for dry glass; FStn04 - depression of 0.4bar; FStn05 - depression of 0.5bar; FStn06 - depression of 0.6bar; FStn07 - depression of 0.7bar.

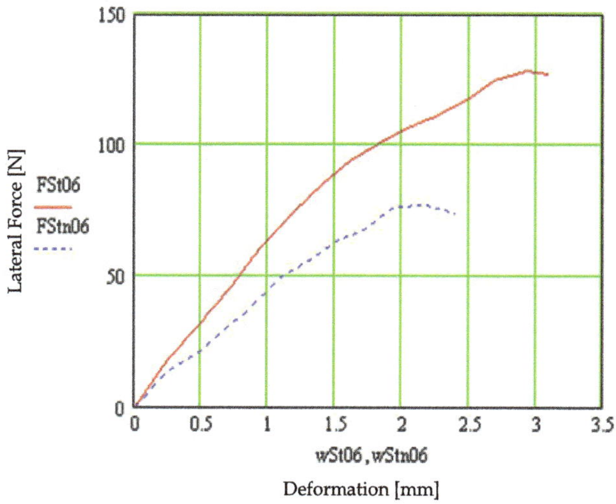

Fig. 12. Comparison between the behavior of the cup subjected to combined forces (FStn06) and normal force (FSt06) for a depression of 0.6 bar

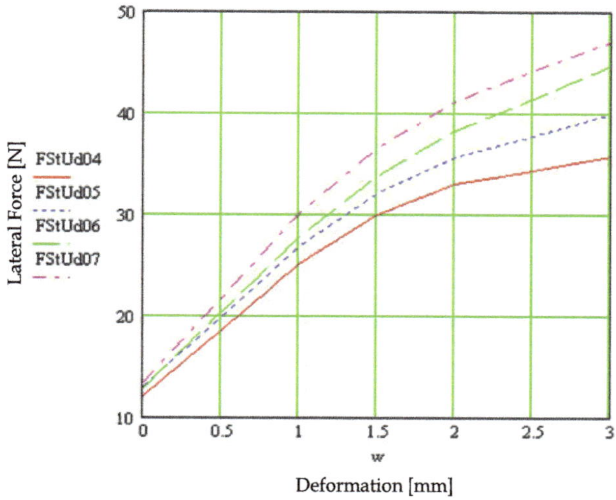

Fig. 13. Experimental diagrams for watered glass, lateral force; FStUd04 - depression of 0.4bar; FStUd05 - depression of 0.5bar; FStUd06 - depression of 0.6bar; FStUd07 - depression of 0.7bar.

4. Displacement kinematics

The value of the triangular platform side L = 247 mm was adopted in order to obtain a stroke S of about 100...110 mm. A translation of 200...220 mm of the robot is obtained for a full translation cycle (attachment on PLI suction cups - PLE translation - attachment on PLE suction cups - PLI translation), allowing the tracking of a glass surface of 1500 mm in about seven cycles. If a cycle duration of 8s is considered, the whole window size is travelled in less than one minute, which is convenient.

Figure 14 presents the constructive solution adopted for the robot displacement. The vacuum suction cup 1 (supplied from the vacuum miniature pump by the nozzle 16) is embedded in the sliding body 15 and can displace relatively to the guideway 3. The parallel key 2 restraints rotation.

The used mechanism consists of the shaft-screw 13 and the nut - interior thread manufactured in the body 15. The toothed belt 11 and the belt gear 6 solidary with the shaft allow the driving. Radial bearings 4 and stroke limiter with microswitch 7 can also be noticed.

The microswitches - two for each leg - are fixed on the corner brackets 8, adjustable relatively to the arm 5 solidary with the plate 12 (PLI). The positioning of the suction cup is discerned by the disk 10 driven by the rod 14 in contact with the inner part of the slider. The spring 9 helps maintaining this contact.

The displacements of the robot result as a combination of the following categories of movements: one-step translation, rotation and pseudo-circular movement (Alexandrescu, 2010a).

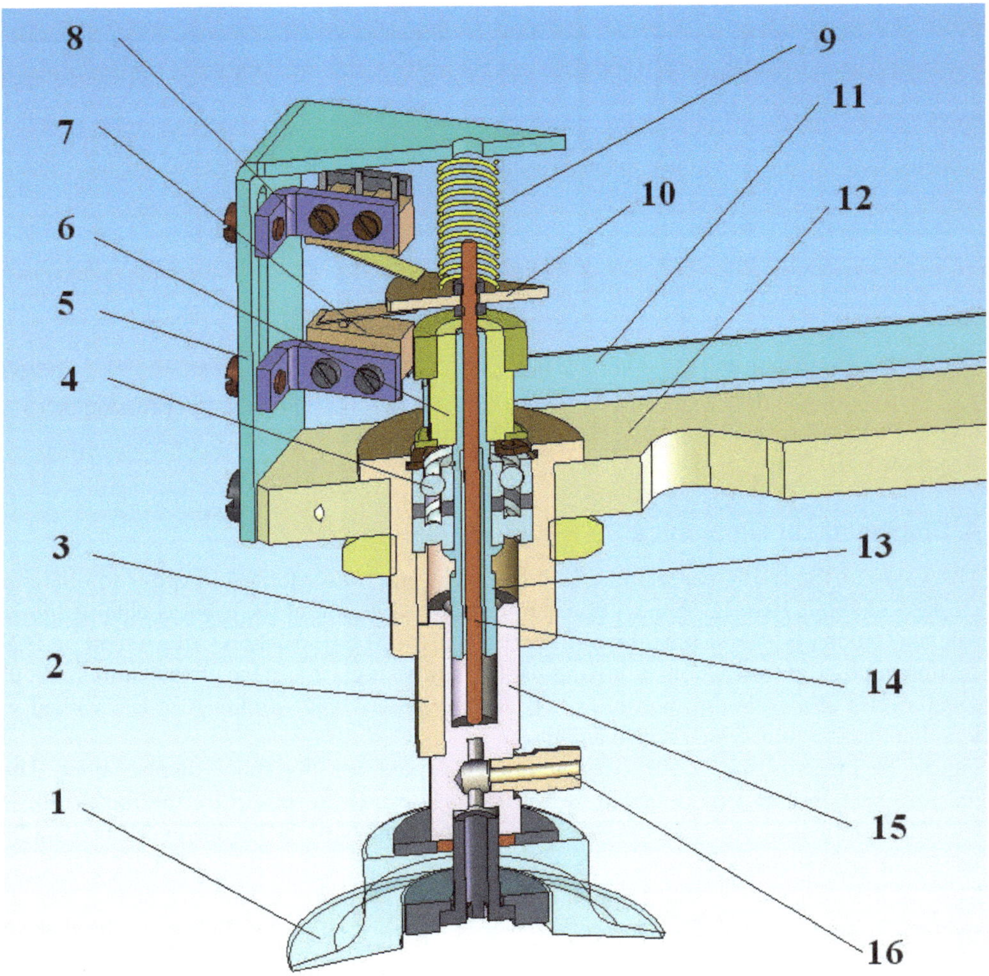

Fig. 14. Detail regarding the actuation of the displacement relative to the suction cups.

a

b

c

d

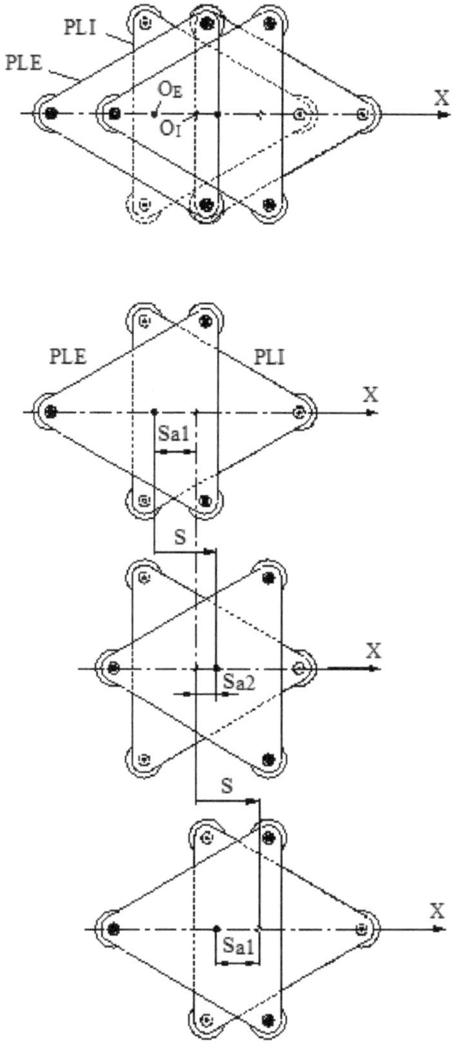

Fig. 15. One-step translation. a. Positions of the two platforms: PLE and PLI; b. Initial position; c. Position after PLE displacement; d. End position (after PLI displacement).

One-step translation of the robot involves the following sequence of movements, as shown in Figure 15. Notations from Figure 2 are used:

- PLE raising (movement $m_{2,up}$);
- PLE displacement (movement m_4);
- PLE lowering (movement $m_{2,down}$);
- PLI raising (movement $m_{1,up}$);
- PLI displacement (movement m_4);
- PLI lowering (movement $m_{1,down}$).

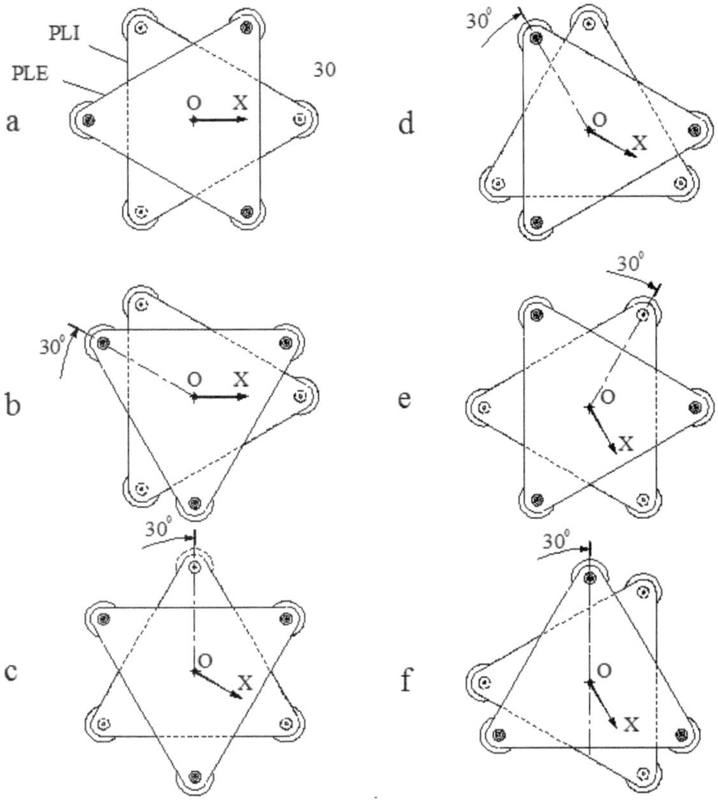

Fig. 16. Phases of a 90° clockwise rotation. a. initial state; b, d, f. Rotation FWD of PLE with 30°; c, e, g. Rotation RW of PLI with 30°.

Robot rotation is achieved by the following sequence of movements, as shown in Figure 16:

- PLE raising (movement $m_{2,up}$);
- PLE clockwise rotation of angle α (movement m_3);
- PLE lowering (movement $m_{2,down}$);
- PLI raising (movement $m_{1,up}$);
- PLI clockwise rotation of angle α (movement m_3);
- PLI lowering (movement $m_{1,down}$).

This sequence leads to a rotation of angle $\alpha = 30°$, which is the recommended maximum angle of relative rotation between platforms. The sequence is repeated till the achievement of the desired angle rotation.

The counterclockwise rotation needs only reversing the direction of the partial rotations ($FWD \leftrightarrow RW$). Axis X indicates the direction of the sliding axis of the robot. Initial centring of platforms is needed.

The combination among a translation and a number of 30° rotations leads to a pseudo-circular movement outlining a 12-sided polygon, as presented in figure 17.

The translation has to start (point A of Figure 17,a) and to stop (point C of Figure 17,a) in a centred state (the centres of the platforms coincide).

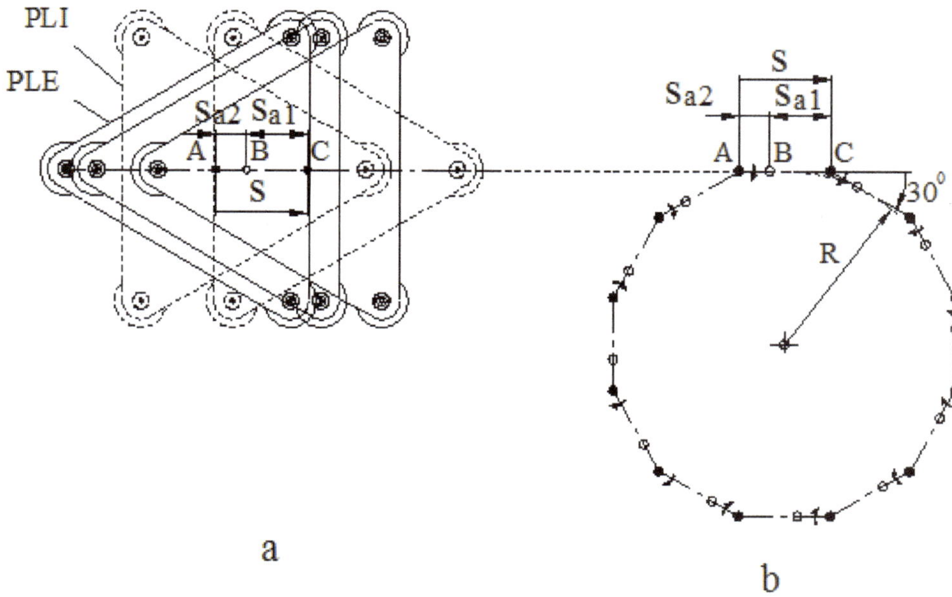

Fig. 17. Pseudo circular movement. a. the positions of the platforms PLE and PLI during translation; b. The polygon of the trajectory.

The radius of the circle inscribed in the travelled polygon is given by (1):

$$R = \frac{S}{2 \cdot tg15^{\circ}} \approx 1.867 \cdot S \tag{1}$$

5. Modelling and simulation of the robot displacement

The displacement of the robot was modelled and simulated using Cosmos Motion software (Alexandrescu, 2010a), (Apostolescu, 2010).

In order to simulate the robot translation, for the relative movement between platforms the interior platform was considered to be fixed. A parabolic variation was imposed for the acceleration. The numeric values used for simulation are: displacement $\Delta S = 100\ mm$, maximum acceleration $a_{max} = 500\ mm/s^2$ and computed maximum speed $v_{max} = 60\ mm/s$. Figures 18, 19 and 20 present the simulation results.

The simulations allowed the computation of the value of the maximum instantaneous power: $P_{tr} = 0.69$ W.

The needed power at the exit of the driving motor resulted equal to 1.42W.

The orienting rotation of the robot was simulated for rotation cycle of 30°. Figures 21, 22 and 23 present the simulation results.

In the transitory areas, it can be noticed that the variation of the angular acceleration presents deviations relatively to its theoretical shape. This phenomenon can be explained by the variation of the static charges during platform rotation.

The computed maximum value of the couple was equal to 0.204 Nm. The computed maximum power at the level of the platform was equal to Prot = 0.13 W.

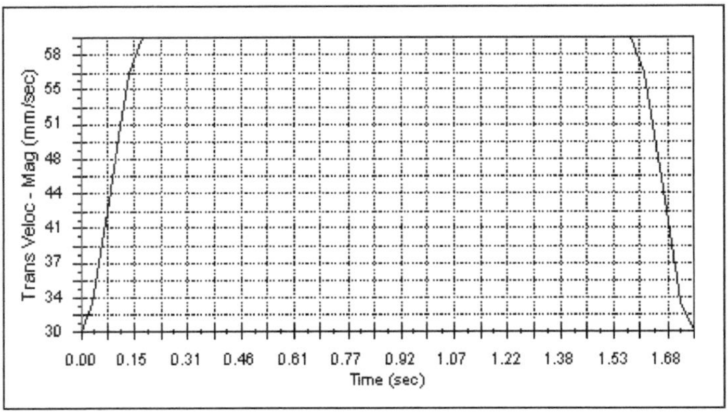

Fig. 18. Variation of translation speed.

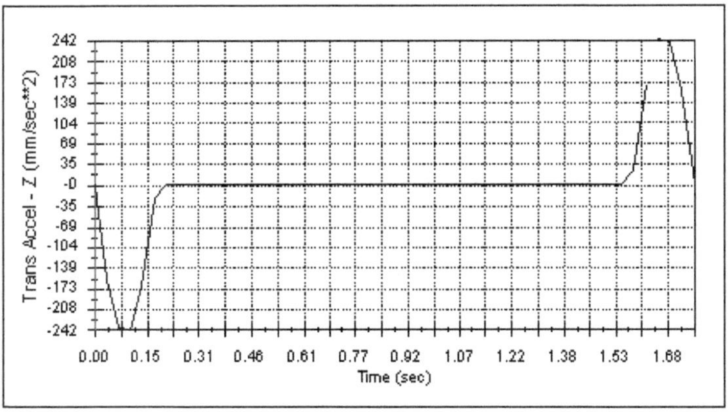

Fig. 19. Variation of translation acceleration.

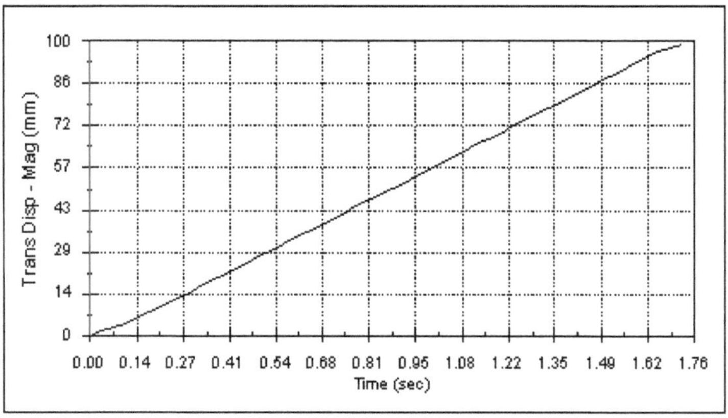

Fig. 20. Displacement during translation.

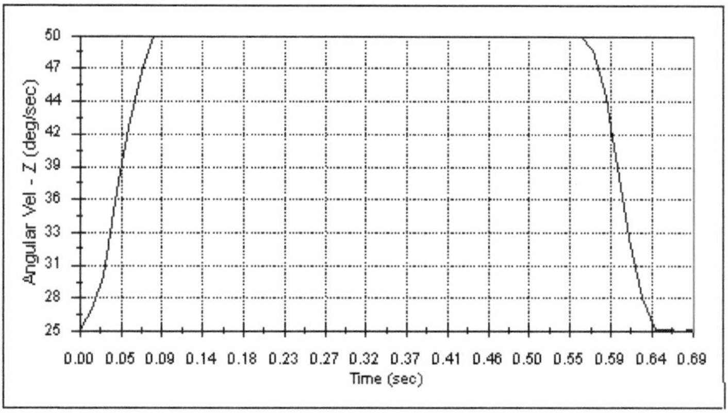

Fig. 21. Variation of angular speed during platform rotation.

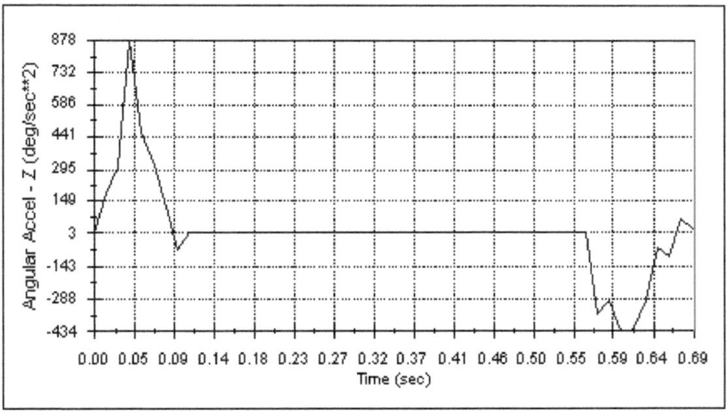

Fig. 22. Variation of angular acceleration during platform rotation.

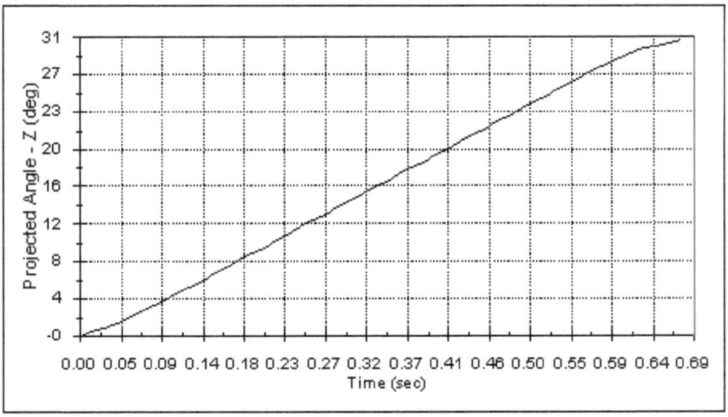

Fig. 23. Angle variation during rotation.

6. Robot control

The robot can be controlled with a data acquisition board 7344 National Instruments and LabVIEW programming or with microcontrollers. The microcontroller BS2 (Parallax) is used, easy to program but with a number of limitations concerning the control of motor speeds.

Using a data acquisition board allows introducing home switches for each of four servo axes in order to find the reference position. For axis 1 (robot translation) and axis 4 (orienting rotation), home switches are mounted between the limiting micro switches. For axes 2 and 3, representing cup translations for PLE and PLI, respectively, micro switches are used only as stroke limit. The reference position is found with the help of a photoelectric system, as shown in Figure 24. The system consists of a light stop 6 fixed on the mobile plate 4 whose displacement *s* gives the position of the cups. The light stop moves between the sides of the photoelectric sensor 2 (of type OPB 916).

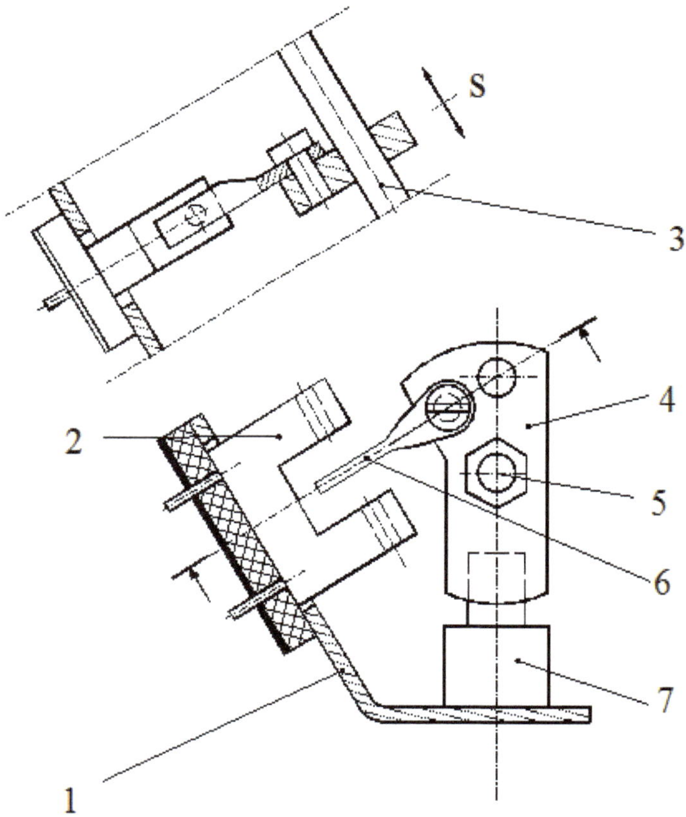

Fig. 24. Photoelectric system used in order to establish the reference position of axes 2 and 3. 1 - corner support; 2 - photoelectric sensor; 3 - rod for movement obstruction; 4 - plate attached to the mobile rod; 5 - mobile rod; 6 - light stop; 8 - microswitch.

Figure 25 presents the scheme of the photoelectric system used to determine the reference position. When the light stop reaches the optical axis of the sensor, the state of its output changes. The emitter diode is supplied through the resistor R_a for current limitation. A power amplifier OPB916 is connected at the circuit output. The suppressor diode D_s protects the transistor during its disconnection. Connection to the data acquisition board is made through the NO contact of the relay Rel.

The LabVIEW software program that allows founding the home switch is shown in Figure 26. The activation of limit switches is also needed during the search. After the reference is found, the position counter is reset. The program is applied for each of four axes of the acquisition board.

The programs consists of two sequences introduced by the cycle 10. The first one searches the reference position and the second resets the position counter (subVI 9).

The subVI 1 loads the maximum search speed and performs axis selection. The subVIs 2 and 3 load the maximum acceleration and deceleration. The subVI 4 defines the movement kinematics (S curve of the speed). A *while* type cycle is introduced. The subVI 7 reads the state of the search. The subVI 12 seizes various interruption cases. The subVI 13 stops the cycle.

In order to clean glass surfaces, the robot must cover the whole window area, paying especial attention to corners. The main control program of the robot controls the travel on the vitrified surface by horizontal and vertical movements, as well as by rotations that allow changing the direction. An ultrasonic PING sensor (Parallax) was introduced as decision element for changing the direction and stopping. The sensor is mounted on the PLI platform using the corner 3 and the jointed holder 2, as shown in Figure 27.

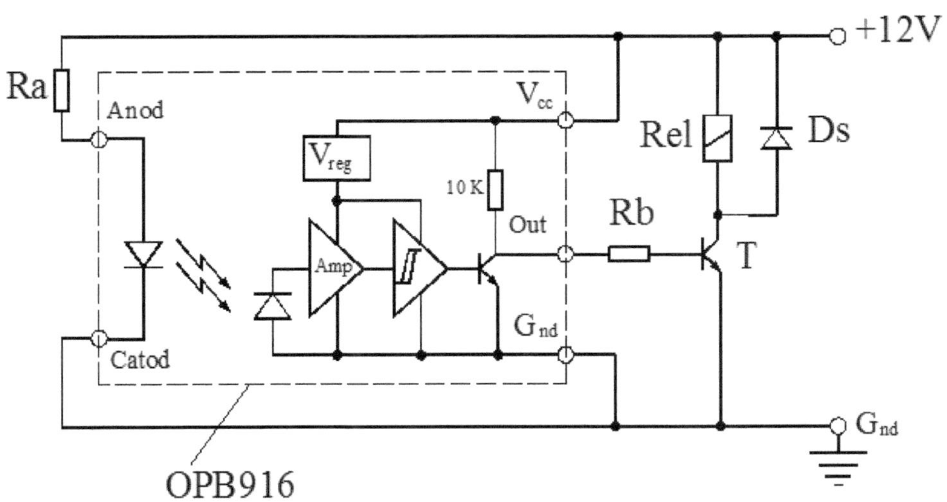

Fig. 25. Scheme of the photoelectric circuit.

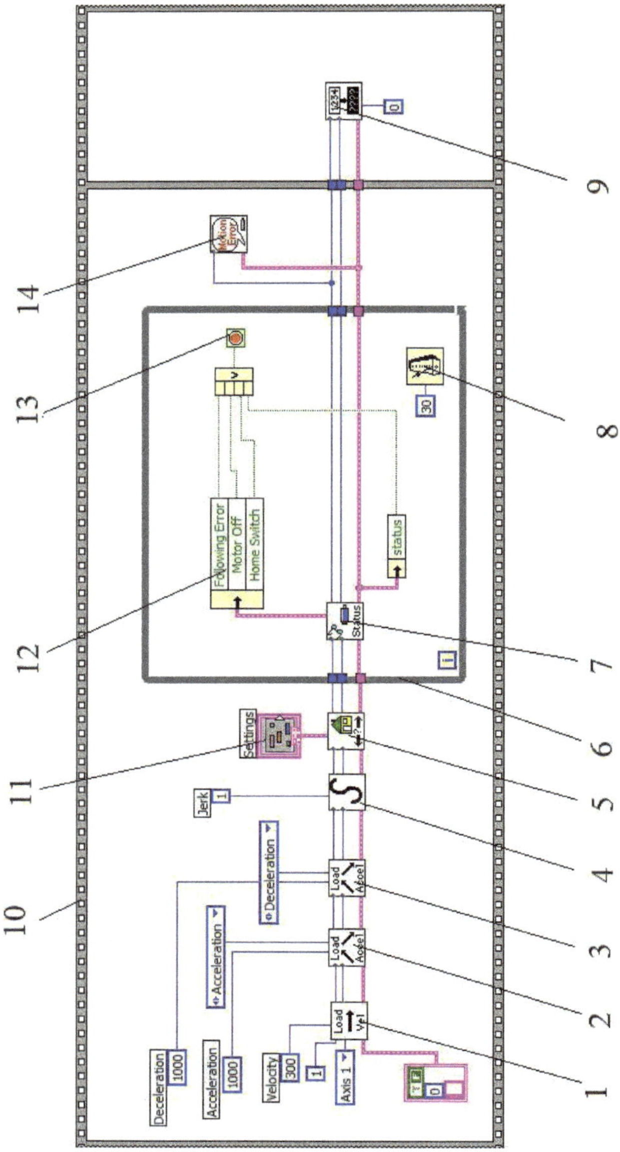

Fig. 26. LabVIEW program for founding the home switch: 1 - maximum speed load; 2 - acceleration load; 3 - deceleration load; 4 - elements of curve S (kinematics without jerk); 5 - home switch use; 6 -*while* type cycle; 7 - reading of search state; 8 - delay producing; 9 - position counter reset; 10 - sequential cycle with two sequences; 11 - search settings; 12 - reading of different interrupt situations; 13 -end cycle condition; possible errors indication of possible errors.

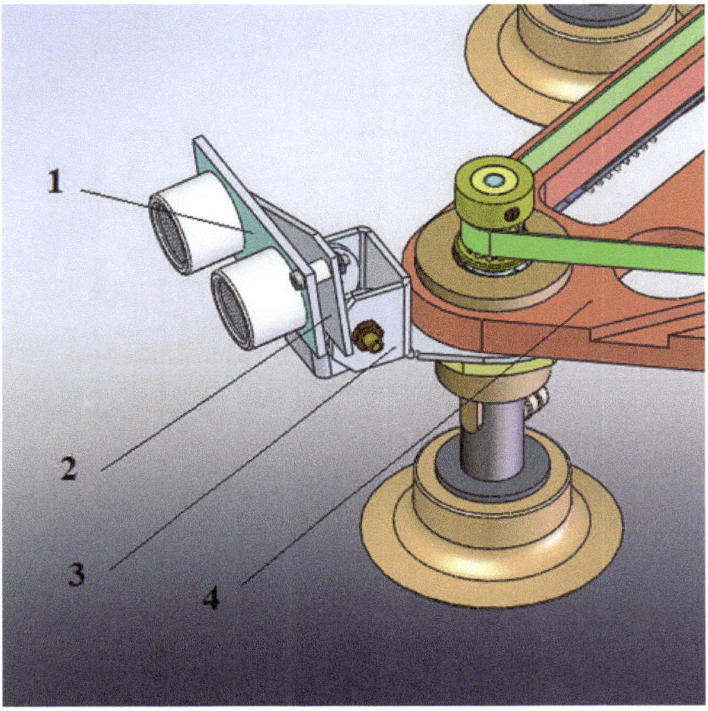

Fig. 27. The ultrasonic sensor mounted on the interior platform: 1 - sensor; 2 - sensor holder; 3 - corner; 4 - interior platform PLI of the robot.

Figure 28 presents a sequential cycle of travel. The cycle consists of the following sequences: sequential translation from left to right (this sequence ends when the proximity of the right side rim is sensed); 90° clockwise rotation; lowering with a step; 90° clockwise rotation; sequential translation from right to left (this sequence ends when the proximity of the left side rim is sensed); 90° counterclockwise rotation; lowering with a step; 90° counterclockwise rotation.

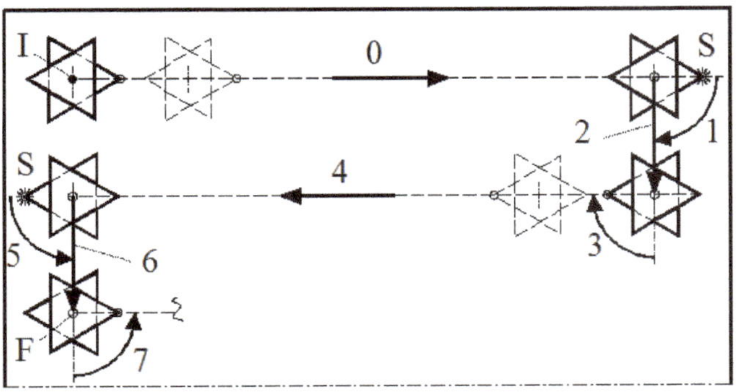

Fig. 28. Travel cycle of the robot.

The robot covers the whole window area by repeating the travel cycle. The robot stops if the sensor *S* sends the signal of proximity of the bottom rim of the vitrified surface.
The block diagram of the program is shown in Figure 29.

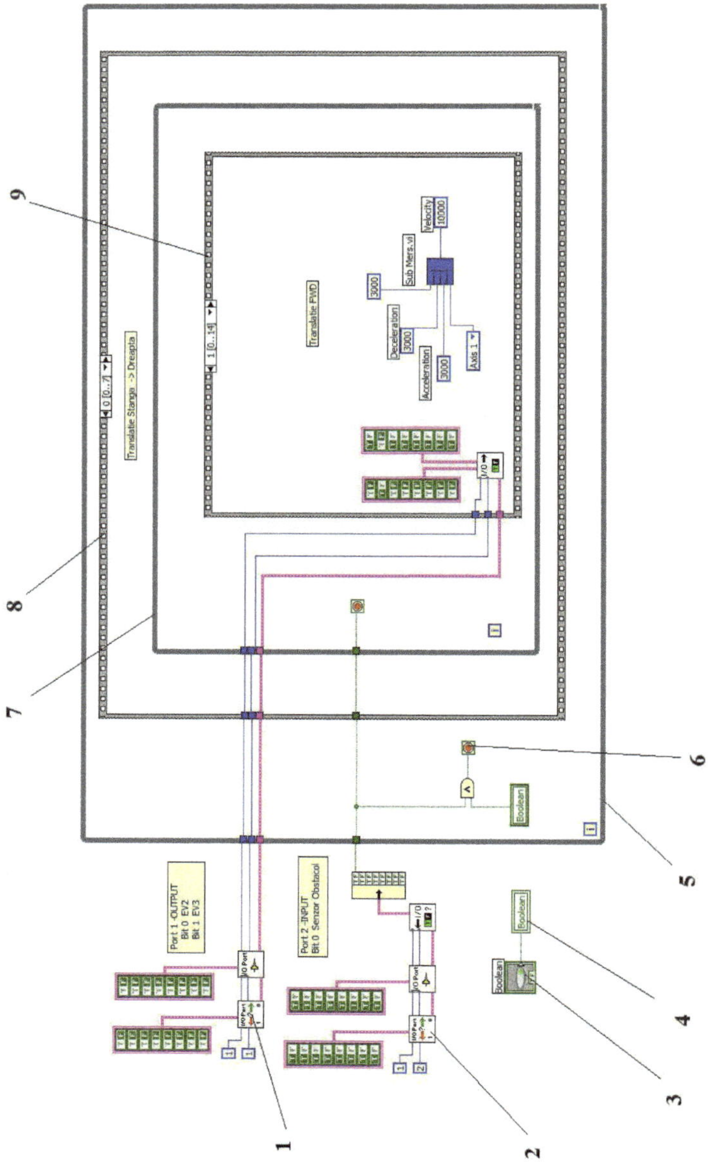

Fig. 29. Block diagram of main control program of the robot: 1 - setting port 1 as output; 2 - setting port 2 as input; 3 - initialization of local variable; 4 - boolean local variable; 5 – *while* cycle of the travel program; 6 – travel stop; 7 – first order *while* cycles; 8 – sequences of the first order cycles; 9 – sequences of the first order cycles.

The program uses the ports 1 and 2 of the acquisition board. The port 1 is used as program output, sending the commands towards the electro valves. The port 2 is used as input, receiving the signal from the sensor *S*. The control 3 initializes the boolean local variable as "False". The variable changes its state to "True" during vertical displacement. The signal from the sensor *S* is used also for stopping the horizontal translation sequences.

7. Conclusion

The chapter reports a number of very important results regarding the design and control of a prototype of climbing autonomous robot with vacuum attachment cups.
The robot construction is able to perform its intended function: the efficient cleaning of glass surfaces. The vacuum attachment system ensures good contact with the support surface, is simple and reliable. The modelling and simulation of the robot functioning, developed for platform translation as well as for relative rotation of the platforms, certifies that its performances are comparable to similar solutions conceived worldwide.
The overall size of the robot, 350mm x 350mm x 220mm, proves an optimal degree of robot miniaturization.

8. References

Alexandrescu, N.; Apostolescu, T.C.; Udrea, C.; Duminică, D.; Cartal, L.A. (2010). Autonomous mobile robots with displacements in a vertical plane and applications in cleaning services. *Proc. 2010 IEEE International Conference on Automation, Quality and Testing, Robotics*, Cluj-Napoca, Romania, 28-30 May 2010, Tome I, IEEE Catalog Number CFP10AQT-PRT, ISBN 978-1-4244-6722-8, pp. 265-270

Alexandrescu, N., Udrea, C., Duminică, D., & Apostolescu, T.C. (2010), Research of the Vacuum System of a Cleaning Robot with Vertical Displacement, *Proc. 2010 International Conference on Mechanical Engineering, Robotics and Aerospace ICMERA 2010*, Bucharest, Romania,2-4 December 2010, IEEE Catalog Number CFP1057L-ART, ISBN 978-1-4244-8867-4, pp. 279-283

Apostolescu, T.C. (2010) Autonomous robot with vertical displacement and vacuummetric attachment system (I Romanian), Ph.D. Thesis, POLITEHNICA University of Bucharest, 2010

Belforte G., Mattiazzo G., & Grassi R. (2005). Innovative solution for climbing and cleaning on smooth surfaces. *Proceedings of the 6th JFPS International Symposium on Fluid Power*, pp. 251-255,Tsukuba, Japan.

Cepolina, F.; Michelini, R.; Razzoli, R.; Zoppi, M. (2003). Gecko, a climbing robot for wall cleaning. *1 st Int. Workshop on Advances in service Robotics ASER03*, March 13-15, Bardolino, Italia, 2003, Available from http://www.dimec.unige.it/PMAR/

Miyake, T.; Ishihara, H.; Yoshimura, M. (2007). Basic studies on wet adhesion system for wall climbing robots. *Proc. 2007 IEEE/RSJ International Conference on Intelligent Robots and Systems*, San Diego, CA, USA, Oct.29-Nov.2, 2007, pp. 1920-1925

Novotny F.; Horak, M. (2009). Computer Modelling of Suction Cups Used for Window Cleaning Robot and Automatic Handling of Glass Sheets. In: *MM Science Journal*, June, 2009, pp. 113-116

Sun D., Zhu J., & Tso S. K. (2007), A climbing robot for cleaning glass surface with motion planning and visual sensing, In: *Climbing & walking robots: towards new applications*, pp.219-234, Hao Xiang Zhang (ed.), InTech, Retrieved from <http://www.intechopen.com/books/show/title/climbing_and_walking_robots_towards_new_applications>

11

A Kinematical and Dynamical Analysis of a Quadruped Robot

Alain Segundo Potts and José Jaime da Cruz
University of São Paulo
Brazil

1. Introduction

In general, legged locomotion requires higher degrees of freedom and therefore greater mechanical complexity than wheeled locomotion. Wheeled robots are simple in general, and more efficient than legged locomotion on flat surfaces. Yet as the surface turns softer, wheeled locomotion becomes inefficient due to rolling friction. Furthermore, in some cases, wheeled robots are unable to overcome small obstacles. On the other hand, legged robots are more easily adaptable to different kinds of terrains due to the fact that only a set of point contacts is required; thus, the quality of the ground between those points does not matter as long as the robot can maintain appropriate ground clearance.

Legged robots appear as the sole means of providing locomotion in highly unstructured environments. However, they cannot traverse every type of uneven terrain because they are of limited dimensions. Hence, if there are terrain irregularities such as a crevasse wider than the maximum horizontal leg reach or a cliff of depth greater than the maximum vertical leg reach, then the machine is prevented from making any progress. This limitation, however, can be overcome by providing the machine with the capability of attaching its feet to the terrain. Moreover, machine functionality is limited not only by the topography of the terrain, but also by the terrain constitution. Whereas hard rock poses no serious problem to legged robots, muddy terrain can hamper its operation to the point of jamming the machine. Still, under such adverse conditions, legged robots offer a better maneuverability than other vehicles (Angeles, 2007; Siegwart & Nourbakhsh, 2004).

The main disadvantages of legged locomotion include power and mechanical complexity. The leg, which may include several degrees of freedom, must be capable of sustaining part of the robotŠs total weight and, in many robots, must be capable of lifting and lowering the robot. Additionally, high maneuverability will only be achieved if the legs have a sufficient number of degrees of freedom to impart forces in a number of different directions.

In the last few years, this feature has given rise to a number of research activities on the subject. Despite all these efforts, the performance of legged robots is still far from what could be expected from them. This is true particularly because the robots performance depends on several factors, including the mechanical design, which sometimes may not be changed by the control designer (Estremera & Waldron, 2008).

Legged robots present some problems that are not usual in wheeled robots. For example, problems such as trajectory planning and stability analysis need a good kinematics and dynamics model of the system.

Fig. 1. Kamambaré I robot

Herein will be presented a kinematical and dynamical analysis of a quadruped robot named Kamambaré I (Bernardi & Da Cruz, 2007).
Like all the mobile robots with legs the topology of Kamambaré is time variant. Deu to his own gait, we have two different problems to solve. First when there is at least one closed kinematic chain between the support surface and the platform, the robot's behavior will be similar to a parallel robot. On the other hand, when a leg of the robot is in the air looking for a new point of grasping, the model that best describes it is an open kinematic chain model, similar to the models of a serial industrial manipulator. Through this work we will refer to these two topological model as the platform for the parallel case of modeling and model of the leg for the second case reviewed like in (Potts & Da Cruz, 2010).
The analysis above, is important for bringing the platform or the gripper to some desired position in the space, but in our case it is not sufficient. To move the platform or the gripper along some desired path with a prescribed speed, the motion of the joints must be carefully coordinated. There are two types of velocity coordination problems, direct and inverse. In the first case, the velocity of the joints is given and the objective is to find the velocity of the end effector (platform or leg); in the other case, the velocity of the end effector is given and the input joint rates required to produce the desired velocity are to be found.

2. Kinematics model

Kamambaré I is a symmetrical quadruped robot. It was developed for climbing vertical objects such as trees, power poles, bridges, etc. Each of its legs with four revolution joints. See Fig. 1. At the end of each leg, there is a gripper. All joints are powered by DC motors.
The basic gait of the robot simulates the walking trot of a quadruped mammal. In this type of gait, the diagonals legs move in tandem. While a pair of legs is fixed to the supporting surface and pushes the robot forward the other pair is on the air, seeking a new foothold, see Figure 2. According to that description, there are two basics stages for the legs, which will be named: "leg on the air" to represent the leg seeking for the new foothold, and "pushing stage" when the leg is fixed and pushing the body to a given direction.

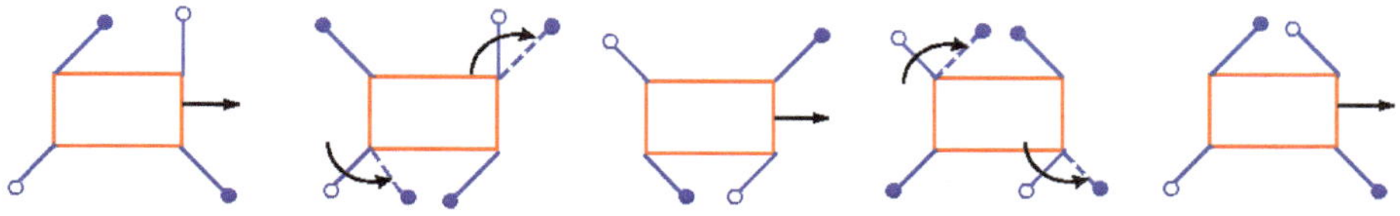

Fig. 2. Gait graphs for the trot of the Kamambaré robot. Leg on the air ∘, leg attached to the surface •.

For a robot to move to a specific position, the location of the center of its body relative to the base should be established first. This is called by some authors *position analysis problem* (Tsai, 1999). There are two types of position analysis problems: direct kinematics and inverse kinematics problems. In the first one, the joint variables are given and the problem is to find the location of the body of the robot; for the inverse kinematics, the location of the body is given and the problem is to find the joint variables that correspond to it (Kolter et al., 2008). Two approaches will be taken herein for the complete modeling of the robot in accordance with its topology. Firstly, for the robot in the pushing stage the model will be like a parallel robot with a closed chain between the two legs that are supporting the platform. Then when the leg is "on the air" the model is of a serial manipulator attached at one of the corners of the platform.

2.1 Direct kinematics problem of the platform

In this section, the direct kinematics problem of the platform will be solved. The system is modeled as a parallel robot and the legs are stuck between the supporting surface and the platform. The analysis is performed using the Denavit-Hartenberg (D-H) parametrization, starting at the surface and advancing towards the platform.

i	α_{i-1}	a_{i-1}	d_i	θ_{i_l}
4	0	0	L_4	θ_{4_l}
3	$\frac{\pi}{2}$	0	0	θ_{3_l}
2	0	L_3	0	θ_{2_l}
1	$-\frac{\pi}{2}$	L_2	0	θ_{1_l}
	0	L_1	0	0

Table 1. Denavit-Hartenberg parameters for leg l in the pushing stage

Table 1 shows the D-H parameters for the "pushing stage". Frames $\{B_l\}, \{C_l\}, \{D_l\}$ and $\{E_l\}$ are attached to links 4, 3, 2 and 1, respectively, as shown by Figure 3. Frame $\{O\}$ is attached at a point of the climbing surface, $\{A_l\}$ is attached to the gripping point and $\{P\}$ is attached to the robotic platform. The lengths of the links are L_5, L_4, L_3, L_2 and L_1, respectively, starting at the point ${}^{O}A_l$, origin of the frame $\{A_l\}$. Index l, $(l = 1, \ldots, 4)$ is used to indicate the leg of the robot, while index i, $(i = 1, \ldots, 4)$ is used to indicate the *i-th* joint of the *l-th* leg. In this paper, vector ${}^{O}\vec{A_l B_l}$ relative to frame $\{O\}$ is assumed to be orthogonal to the climbing surface, (Bernardi et al., 2009).
Denoting by ${}^{Y_l}T_{X_l}$ the homogeneous transformation from the coordinate systems $\{X_l\}$ to coordinate system $\{Y_l\}$ of the *l-th* leg, ${}^{O}T_P$ can be expressed as:

$$ {}^{O}T_P = {}^{O}T_{A_l} \cdot {}^{A_l}T_{B_l} \cdot {}^{B_l}T_{C_l} \cdot {}^{C_l}T_{D_l} \cdot {}^{D_l}T_{E_l} \cdot {}^{E_l}T_P \tag{1} $$

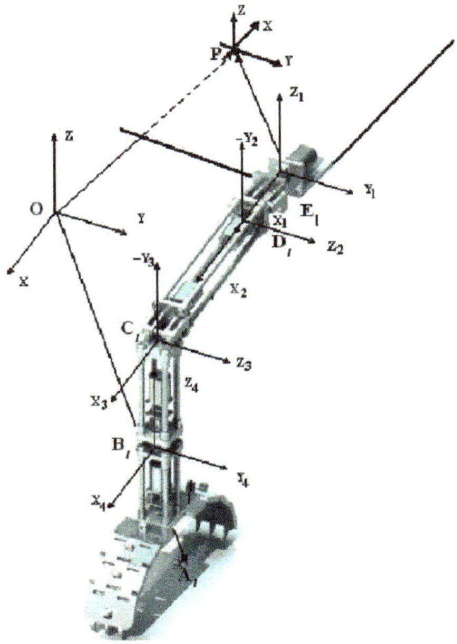

Fig. 3. Scheme of *l-th* leg.

Recalling that the structure of ${}^{O}T_{A_l}$, ${}^{A_l}T_{B_l}$ and ${}^{B_l}T_P$ are:

$$ {}^{O}T_{A_l} = \begin{bmatrix} {}^{O}R_{A_l} & | & {}^{O}\vec{OA_l} \\ -- & + & -- \\ \mathbf{0} & | & 1 \end{bmatrix}, \tag{2} $$

$$ {}^{A_l}T_{B_l} = \begin{bmatrix} {}^{A_l}R_{B_l} & | & {}^{O}\vec{A_lB_l} \\ -- & + & -- \\ \mathbf{0} & | & 1 \end{bmatrix} \tag{3} $$

and

$$ {}^{B_l}T_P = \begin{bmatrix} {}^{B_l}R_P & | & {}^{O}\vec{B_lP} \\ -- & + & -- \\ \mathbf{0} & | & 1 \end{bmatrix}, \tag{4} $$

Assume that matrixes ${}^{O}R_{A_l}$ and ${}^{A_l}R_{B_l}$ are equal to identity matriz I and deu to a sequence of straightforward computations the rotation matrix ${}^{B_l}R_P$ is equal to:

$$ {}^{B_l}R_P = \begin{bmatrix} c\theta_{4_l}c\theta_{2_l3_l}c\theta_{1_l} + s\theta_{4_l}s\theta_{1_l} & -c\theta_{4_l}c\theta_{2_l3_l}s\theta_{1_l} + s\theta_{4_l}c\theta_{1_l} & c\theta_{4_l}s\theta_{2_l3_l} \\ s\theta_{4_l}c\theta_{2_l3_l}c\theta_{1_l} - c\theta_{4_l}s\theta_{1_l} & -s\theta_{4_l}c\theta_{2_l3_l}s\theta_{1_l} - c\theta_{4_l}c\theta_{1_l} & s\theta_{4_l}s\theta_{2_l3_l} \\ s\theta_{2_l3_l}c\theta_{1_l} & -s\theta_{2_l3_l}s\theta_{1_l} & -c\theta_{2_l3_l} \end{bmatrix} \tag{5} $$

and to the position of the origin of system $\{P\}$ with respect to $\{B\}$

$$ {}^{O}\vec{B_lP} = \begin{bmatrix} (c\theta_{4_l}c\theta_{2_l3_l}c\theta_{1_l} + s\theta_{4_l}s\theta_{1_l})L_1 + c\theta_{4_l}(c\theta_{2_l3_l}L_2 + c\theta_{3_l}L_3) \\ (s\theta_{4_l}c\theta_{2_l3_l}c\theta_{1_l} - c\theta_{4_l}s\theta_{1_l})L_1 + s\theta_{4_l}(c\theta_{2_l3_l}L_2 + c\theta_{3_l}L_3) \\ s\theta_{2_l3_l}c\theta_{1_l}L_1 + s\theta_{2_l3_l}L_2 + s\theta_{3_l}L_3 + L_4 \end{bmatrix} \tag{6} $$

for $l = 1,2,3,4,$.
Then, using 6, the direct kinematics problem of the platform can be solved by the vector equation:

$$^{O}\vec{OP} = {}^{O}\vec{OA_l} + {}^{O}\vec{A_lB_l} + {}^{O}\vec{B_lP} \tag{7}$$

for a know coordinates of points $^{O}A_l$ and $^{O}B_l$ relatives to frame $\{O\}$.

2.2 Direct kinematics problem of the leg

Since any homogeneous transformation matrix $^{Y_l}T_{X_l}$ is non-singular it is possible to use that expression to solve the direct kinematics problem for the leg on the air:

$$^{E_l}T_{A_l} = \begin{bmatrix} ^{E_l}R_{A_l} & | & ^{E_l}\vec{E_lA_l} \\ -- & + & -- \\ \mathbf{0} & | & 1 \end{bmatrix} \tag{8}$$

where:

$$^{E_l}R_{A_l} = \begin{bmatrix} c\theta_{4_l}c\theta_{2_l3_l}c\theta_{1_l} + s\theta_{4_l}s\theta_{1_l} & s\theta_{4_l}c\theta_{2_l3_l}c\theta_{1_l} - c\theta_{4_l}s\theta_{1_l} & c\theta_{1_l}s\theta_{2_l3_l} \\ -c\theta_{4_l}c\theta_{2_l3_l}s\theta_{1_l} + s\theta_{4_l}c\theta_{1_l} & -s\theta_{4_l}c\theta_{2_l3_l}s\theta_{1_l} - c\theta_{4_l}c\theta_{1_l} & -s\theta_{1_l}s\theta_{2_l3_l} \\ s\theta_{2_l3_l}c\theta_{4_l} & s\theta_{2_l3_l}s\theta_{4_l} & -c\theta_{2_l3_l} \end{bmatrix} \tag{9}$$

and the position of the gripper relative to frame $\{E_l\}$ is given by:

$$^{E_l}\vec{E_lA_l} = \begin{bmatrix} (c\theta_{2_l3_l}\bar{L}_4 + c\theta_{2_l}L_3 + L_2)c\theta_{1_l} \\ (c\theta_{2_l3_l}\bar{L}_4 + c\theta_{2_l}L_3 + L_2)s\theta_{1_l} \\ s\theta_{2_l3_l}\bar{L}_4 - s\theta_{2_l}L_3 \end{bmatrix} \tag{10}$$

where $\bar{L}_4 = L_4 + L_5$
The use of 4 or 8 depends on which part of the gait is active. In other words, if the leg l of the robot is in the air, the transformations between joint frames occur based on the frame $\{E_l\}$. On the other hand, if the leg is clung to the surface, the reference coordinate system is $\{O\}$.

2.3 Inverse kinematics problem for the platform

Since each leg has only four degrees of freedom, the position and orientation of the platform must be specified in accordance with the constraints imposed by the joints.
Using equations 7 and 6, it is possible to solve the inverse kinematics problem. If both, clinging point $^{O}A_l$ and the position and orientation of the platform $[^{O}P_x, ^{O}P_y, ^{O}P_z, \psi_P]$ are given as well as the geometrical and mathematical constraints are respected, from equation 6 we have:

$$c\theta_{4_l} = \frac{y_{PAB_l}s\psi_P L_1 \pm x_{PAB_l}\sqrt{x^2_{PAB_l} + y^2_{PAB_l} - s\psi_P^2 L_1^2}}{x^2_{PAB_l} + y^2_{PAB_l}} \tag{11}$$

where: $x_{PAB_l} = {}^{O}P_x - {}^{O}A_{x_l} - {}^{O}B_{x_l}$ and $y_{PAB_l} = {}^{O}P_y - {}^{O}A_{y_l} - {}^{O}B_{y_l}$.
Equation 11 is subject to:

$$x^2_{PAB_l} + y^2_{PAB_l} \neq 0, \tag{12}$$

$$x^2_{PAB_l} + y^2_{PAB_l} \geq s\psi_P^2 L_1^2 \tag{13}$$

and

$$\left| \frac{y_{PAB_l} s\psi_P L_1 \pm x_{PAB_l} \sqrt{x_{PAB_l}^2 + y_{PAB_l}^2 - s\psi_P^2 L_1^2}}{x_{PAB_l}^2 + y_{PAB_l}^2} \right| \leq 1. \tag{14}$$

Whit respect to θ_{2_l} we have:

$$c\theta_{2_l} = \frac{x_{PAB_l}^2 + y_{PAB_l}^2 + z_{PAB_l}^2 - L_3^2 - \bar{L}_2^2 - s\psi_P^2 L_1^2}{2L_3\bar{L}_2} \tag{15}$$

where $\bar{L}_2 = c\psi_P L_1 + L_2$ and $z_{PAB_l} = {}^{O}P_z - {}^{O}A_{z_l} - {}^{O}B_{z_l} - L_4$.
As $|c\theta_{2_l}| \leq 1$, equation 15 is subject to:

$$x_{PAB_l}^2 + y_{PAB_l}^2 + z_{PAB_l}^2 \leq (L_3 + L_2)^2 + 2L_1 c\psi_P (L_3 + L_2) + L_1^2 \tag{16}$$

and

$$x_{PAB_l}^2 + y_{PAB_l}^2 + z_{PAB_l}^2 \geq (L_3 - L_2)^2 - 2L_1 c\psi_P (L_3 - L_2) + L_1^2 \tag{17}$$

finally

$$s\theta_{3_l} = \frac{(c\theta_{2_l}\bar{L}_2 + L_3) z_{PAB_l} \pm s\theta_{2_l}\bar{L}_2 \sqrt{L_3^2 + 2c\theta_{2_l} L_3 \bar{L}_2 + \bar{L}_2^2 - z_{PAB_l}^2}}{L_3^2 + 2c\theta_{2_l} L_3 \bar{L}_2 + \bar{L}_2^2} \tag{18}$$

subject to:

$$L_3^2 + 2c\theta_{2_l} L_3 \bar{L}_2 + \bar{L}_2^2 - z_{PAB_l}^2 \geq 0, \tag{19}$$

$$L_3^2 + 2c\theta_{2_l} L_3 \bar{L}_2 + \bar{L}_2^2 \neq 0, \tag{20}$$

and

$$\left| \frac{(c\theta_{2_l}\bar{L}_2 + L_3) z_{PAB_l} \pm s\theta_{2_l}\bar{L}_2 \sqrt{L_3^2 + 2c\theta_{2_l} L_3 \bar{L}_2 + \bar{L}_2^2 - z_{PAB_l}^2}}{L_3^2 + 2c\theta_{2_l} L_3 \bar{L}_2 + \bar{L}_2^2} \right| \leq 1 \tag{21}$$

the last constraints are verified when relation 13 and $x_{PAB_l}^2 + y_{PAB_l}^2 + z_{PAB_l}^2 - s\psi_P^2 L_1^2 \neq 0$ are satisfied for $l = 1, 2, 3, 4$. In addition, from 19 and 20: $z_{PAB_l} \neq 0$.
Hence, the inverse kinematics problem is solved. Now the orientation of the body has to be defined. A usual way of defining it is through the Euler angles. Denoting by ϕ_P, θ_P and ψ_P the Euler angles associated to Z-Y-Z convention, the rotation matrix with respect to system $\{O\}$, ${}^{O}\bar{R}_P$, is given by:

$${}^{O}\bar{R}_P = \begin{bmatrix} c\phi_P c\theta_P c\psi_P - s\phi_P s\psi_P & -c\phi_P c\theta_P s\psi_P - s\phi_P c\psi_P & c\phi_P s\theta_P \\ s\phi_P c\theta_P c\psi_P + c\phi_P s\psi_P & -s\phi_P c\theta_P s\psi_P + c\phi_P s\psi_P & s\phi_P s\theta_P \\ -s\theta_P c\psi_P & s\theta_P s\psi_P & c\theta_P \end{bmatrix} \tag{22}$$

Equaling 5 and 22 it follow that:

$$c\theta_P = -c\theta_{2_l 3_l}, \tag{23}$$

$$t\psi_P = t\theta_{1_l} \tag{24}$$

and

$$t\phi_P = t\theta_{4_l}, \tag{25}$$

where $\theta \neq 0, \pi$ for $l = 1,2,3,4$.
As said in the last section, angles θ_{2_l}, θ_{3_l} and θ_{4_l} are not independent. Hence, $\theta_P = f(P_x, P_y, P_z, \theta_{1_l})$ and $\phi_P = f(P_x, P_y, P_z, \theta_{1_l})$ for a given point ${}^{O}A_l$.

2.4 Inverse kinematics problem for the leg

In this section, the inverse kinematics problem for the leg on the air will be solved. The starting point for the solution of the inverse kinematic problem of the gripper is equation 10. For points ${}^{E_l}E_l$ and ${}^{E_l}A_l$ given, the solution is:

$$t\theta_{1_l} = \frac{y_{AE_l}}{x_{AE_l}} \tag{26}$$

where:

$$x_{AE_l} = {}^{E_l}A_{x_l} - {}^{E_l}E_{x_l}, y_{AE_l} = {}^{E_l}A_{y_l} - {}^{E_l}E_{y_l}, c\theta_{1_l} \neq 0$$

and

$$x_{AE_l} \neq 0$$

.

After finding θ_{1_l}, the next step is to computate θ_{3_l}:

$$c\theta_{3_l} = \frac{x^2_{AE_l} + y^2_{AE_l} \pm 2L_2\sqrt{x^2_{AE_l} + y^2_{AE_l}} + z^2_{AE_l} + L_2^2 - \bar{L}_4^2 - L_3^2}{2\bar{L}_4 L_3} \tag{27}$$

where: $x^2_{AE_l} + y^2_{AE_l} \neq 0$ and

$$(\bar{L}_4 + L_3)^2 \leq \left(\sqrt{x^2_{AE_l} + y^2_{AE_l}} - L_2\right)^2 + z^2_{AE_l} \leq (\bar{L}_4 - L_3)^2 \tag{28}$$

Finally:

$$s\theta_{2_l} = \frac{z_{AE_l}(c\theta_{3_l}L_4 + L_3) \pm s\theta_{3_l}L_4\sqrt{L_4^2 + 2c\theta_{3_l}L_4L_3 + L_3^2 - z^2_{AE_l}}}{\bar{L}_4^2 + 2c\theta_{3_l}L_3L_4 + L_3^2} \tag{29}$$

where:

$$L_4^2 + 2c\theta_{3_l}L_4L_3 + L_3^2 - z^2_{AE_l} \geq 0, \tag{30}$$

$$L_4^2 + 2c\theta_{3_l}L_4L_3 + L_3^2 \neq 0 \tag{31}$$

and

$$\left|\frac{z_{AE_l}(c\theta_{3_l}L_4 + L_3) \pm s\theta_{3_l}L_4\sqrt{L_4^2 + 2c\theta_{3_l}L_4L_3 + L_3^2 - z^2_{AE_l}}}{\bar{L}_4^2 + 2c\theta_{3_l}L_3L_4 + L_3^2}\right| \leq 1 \tag{32}$$

Inequations 30, 31 and 32 are satisfied for:

$$c\theta_{3_l} \neq -\frac{L_4^2 + L_3^2}{2L_4L_3}, \tag{33}$$

and

$$x^2_{AE_l} + y^2_{AE_l} \geq L_2^2 \tag{34}$$

Besides from 30 and 31 we have the condition $z_{AE_l} \neq 0$. Equations 26, 27 and 29 give multiples solutions for the system. The orientation of gripper is represent by φ_l and its value coincides directly with the value of θ_{4_l}

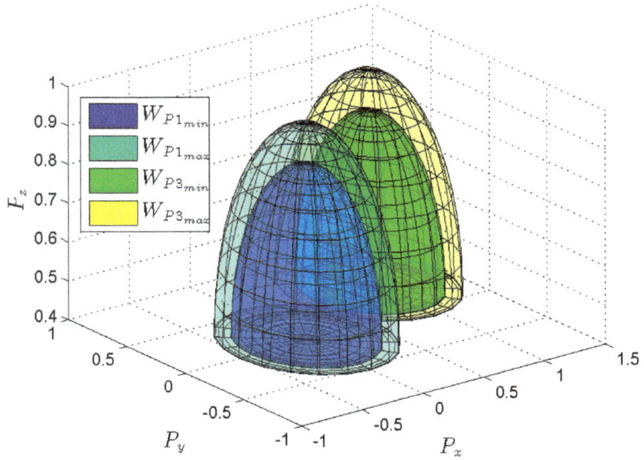

Fig. 4. Workspaces of the center of platform associated to legs 1 and 3.
$(\mathcal{W}_{P_1} = \mathcal{W}_{P_{1max}} - \mathcal{W}_{P_{1min}}$ and $\mathcal{W}_{P_3} = \mathcal{W}_{P_{3max}} - \mathcal{W}_{P_{3min}})$

2.5 Workspace

The workspace is formed by the set of points of the reachable workspace where the robot can generate velocities that span the complete tangent space at that point.

The relationships between joint space and Cartesian space coordinates are generally multiple-valued: the same position can be reached in different ways, each with a different set of joint coordinates. Hence, the reachable workspace of the robot is formed by the configurations, in which the kinematic relationships are locally one-to-one (Pieper, 1968).

2.6 Workspace of the platform

The workspace of the platform is formed by the set of points $P = (P_x, P_y, P_z)$ that satisfy equation 7 subject to constrains imposed by 13 and 17.

In a graphic form was defined by $\mathcal{W}_{P_l}$, the workspace of the center of platform relative to leg l, and if there is more than one leg support the platform the final workspace will be the intersection of all the $\mathcal{W}_{P_l}$ of the legs clung to the surface. In a general case:

$$\mathcal{W}_P = \mathcal{W}_{P_1} \cap \mathcal{W}_{P_2} \cap \ldots \mathcal{W}_{P_4} \tag{35}$$

Figure 4 show the workspace formed by the intersection of set $\mathcal{W}_{P_1}$ and $\mathcal{W}_{P_3}$, and sets $\mathcal{W}_{P_{lmin}}$ and $\mathcal{W}_{P_{lmax}}$ represents the minimum and maximum values of the workspace of each legs. The lengths of the limbs are showed in table 2.

2.7 Workspace of the leg

The workspace of leg $\mathcal{W}_{G_l}$, when it is in the air, corresponds to its reachable Cartesian space . In this case, $\mathcal{W}_{G_l}$ is formed by the admissible solutions of equations 28. The geometrical form of this workspace is shown in figure 5.

Table 2. Dimensions of the limbs

Limbs	Length(m)	Weight(kg)	Moment of Inertia($N-m^2$)
L_5	0.2	0.25	$1.25 \cdot 10^{-5}$
L_4	0.2	0.2	$17.6 \cdot 10^{-5}$
L_3	0.3	0.25	$130.83 \cdot 10^{-5}$
L_2	0.1	0.06	$1.625 \cdot 10^{-5}$
Platform: (L_1)	0.3	5.25	$2932.45 \cdot 10^{-5}$

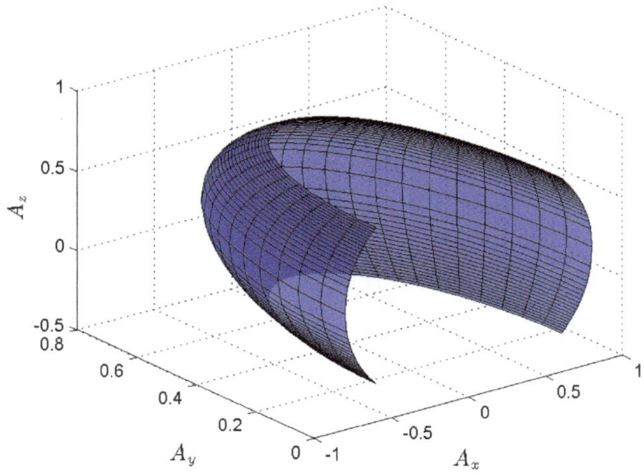

Fig. 5. Workspace for the gripper.

3. Singularity analysis

In previous sections, the problems of direct and inverse kinematics were discussed, both for the platform and for the leg. Such analysis is important for bringing the platform or the gripper to some desired position in the space, but in our case it is not sufficient. The motion of the joints must be carefully coordinated to move the platform or the gripper along some desired path with a prescribed speed, . There are two types of velocity coordination problems namely, direct and inverse. In the first case, the velocity of the joints is given and the objective is to find the velocity state of the end effector (platform or leg); in the later case, the velocity state of the end effector is given and the input joint rates required to produce the desired velocity are to be found. (Tsai, 1999)

Thus, the matrix that transforms the joint rates in the actuator space into the velocity state in the end effector space is called Jacobian matrix.

3.1 Singularity analysis for the platform

Due to the characteristics of the gait chosen for the robot, there will always be a closed-chain kinematics formed by the legs clung to the climbing surface. The closed-chain is also characterized by a set of inputs (denoted here by a vector $\mathbf{q}$), which correspond to the powered joints, and by a set of output coordinates (denoted here by a vector $\mathbf{x}$). These input and output vectors depend on the nature and purpose of the kinematics chain (Goselin & Angeles, 1990).

The orientation of the platform relative to system $\{O\}$ is given by matrix ${}^{O}R_{P}$. Then the platform angular velocity with respect to $\{O\}$, is:

$$\begin{bmatrix} {}^{O}\omega_{p_x} \\ {}^{O}\omega_{p_y} \\ {}^{O}\omega_{p_z} \end{bmatrix} = \begin{bmatrix} 0 & -s\phi_P & s\theta_P c\phi_P \\ 0 & c\phi_P & s\theta_P s\phi_P \\ 1 & 0 & c\theta_P \end{bmatrix} \begin{bmatrix} \dot{\phi}_P \\ \dot{\theta}_P \\ \dot{\psi}_P \end{bmatrix} \quad (36)$$

The linear velocity of point E_l is given by

$${}^{O}\vec{v}_{E_l} = {}^{O}\vec{v}_P + {}^{O}\vec{\omega}_p \times ({}^{O}R_P \cdot {}^{O}\vec{PE_l}) \quad (37)$$

where ${}^{O}\vec{v}_{E_l}$ and ${}^{O}\vec{v}_P$ are respectively the linear velocities of points E_l and P, with respect to $\{O\}$.
The left-hand side of 37 can be rewritten as:

$${}^{O}\vec{v}_{E_l} = J_{ql} \begin{bmatrix} \dot{\theta}_{4_l} \\ \dot{\theta}_{3_l} \\ \dot{\theta}_{2_l} \end{bmatrix} \quad (38)$$

where:

$$J_{ql} = \begin{bmatrix} -s\theta_{4_l}(c\theta_{3_l 2_l}L_2 + c\theta_{3_l}L_3) & -c\theta_{4_l}(s\theta_{3_l 2_l}L_2 + s\theta_{3_l}L_3) & -c\theta_{4_l}L_2 s\theta_{3_l 2_l} \\ c\theta_{4_l}(c\theta_{3_l 2_l}L_2 + c\theta_{3_l}L_3) & -s\theta_{4_l}(s\theta_{3_l 2_l}L_2 + s\theta_{3_l}L_3) & -s\theta_{4_l}L_2 s\theta_{3_l 2_l} \\ 0 & c\theta_{3_l 2_l}L_2 + c\theta_{3_l}L_3 & c\theta_{3_l 2_l}L_2 \end{bmatrix} \quad (39)$$

On the right-hand side of 37, the product $\vec{\omega}_p \times ({}^{O}R_P \cdot \vec{PE_l})$ can be rewritten as:

$${}^{O}\vec{\omega}_p \times ({}^{O}R_P \cdot {}^{O}\vec{PE_l}) = \Omega_{F_l} \begin{bmatrix} {}^{O}\omega_{p_x} \\ {}^{O}\omega_{p_y} \\ {}^{O}\omega_{p_z} \end{bmatrix} \quad (40)$$

where:

$$\Omega_{F_l} = \begin{bmatrix} 0 & Y_{z_l} & -Y_{y_l} \\ -Y_{z_l} & 0 & Y_{x_l} \\ Y_{y_l} & -Y_{x_l} & 0 \end{bmatrix} \quad (41)$$

and $\vec{Y}_l = {}^{O}R_P \cdot {}^{O}\vec{PE_l}$
Substituting 36 and 40 into 37 gives:

$${}^{O}\vec{v}_{E_l} = J_{x1_l} J_{x2} \begin{bmatrix} {}^{O}v_{P_x} \\ {}^{O}v_{P_y} \\ {}^{O}v_{P_z} \\ \dot{\phi} \\ \dot{\theta} \\ \dot{\psi} \end{bmatrix} \quad (42)$$

where:

$$J_{x1_l} = \begin{bmatrix} I_{3\times3} & \Omega_{F_l} \end{bmatrix} \quad (43)$$

and

$$J_{x2} = \begin{bmatrix} I_{3\times3} & 0_{3\times3} \\ 0_{3\times3} & \begin{bmatrix} 0 & -s\phi_P & s\theta_P c\phi_P \\ 0 & c\phi_P & s\theta_P s\phi_P \\ 1 & 0 & c\theta_P \end{bmatrix} \end{bmatrix} \quad (44)$$

Finally, for the four legs:

$$J_q \dot{\vec{q}} = J_x \dot{\vec{x}}_P \tag{45}$$

where:

$$J_x = J_{xP1} J_{x2}, \tag{46}$$

$$J_q = diag(J_{q1}, \ldots, J_{q4}), \tag{47}$$

$$J_{xP1} = \begin{bmatrix} I_{3\times3} & \Omega_{F_1} \\ \vdots & \vdots \\ I_{3\times3} & \Omega_{F_4} \end{bmatrix}, \tag{48}$$

$\vec{q} = [\theta_{4_1}, \theta_{3_1}, \theta_{2_1} \ldots \theta_{4_4}, \theta_{3_4}, \theta_{2_4}]^T$ and $\vec{x}_P = [x_p, y_p, z_p, \phi_p, \theta_p, \psi_p]^T$.
The elements of $\vec{q}$ correspond to the set of active joints. This set may vary with the robot gait, with the number of legs clung to the climbing surface and with the eventual use of an optimal control policy.
Vector $\vec{x}_P$ contains the position and the Euler angles that define the orientation of the platform. When the lengths of the input and output vectors are not the same, there are redundancies (Lenarcic & Roth, 2006). These are eliminated when there are only two legs holding the robot: $\vec{q} = [\theta_{4_1}, \theta_{3_1}, \theta_{2_1}, \theta_{4_3}, \theta_{3_3}, \theta_{2_3}]$. Variables $x_p, y_p, z_p, \phi_p, \theta_p$ and ψ_p are not all arbitrary, but must satisfy the constraints imposed on the kinematics equations.

3.1.1 Inverse Kinematics Singularity of the platform:
Inverse kinematics singularity occurs when:

$$det(J_q) = 0. \tag{49}$$

This kind of singularity consists of the set of points where different branches of the inverse kinematics problem meet, being the inverse kinematics problem understood here as the computation of the values of the input variables from given values of the output variables. Since the dimension of the null space of J_q is nonzero in the presence of a singularity of this kind, we can find nonzero vectors $\dot{\vec{q}}$ for which $\dot{\vec{x}}$ will be equal to zero and, therefore, some of the velocity vectors $\dot{\vec{q}}$ cannot be produced at the output (Goselin & Angeles, 1990).
From 47, it follows that

$$det(J_q) = det(J_{q_1})det(J_{q_2})det(J_{q_3})det(J_{q_4}) \tag{50}$$

where, from 39,

$$det(J_{q_l}) = -L_2 L_3 s\theta_{2_l}(c\theta_{2_l 3_l} L_2 + c\theta_{3_l} L_3) \tag{51}$$

for $l = 1, \ldots, 4$.
The singularities occur when $\theta_{2_l} = 0, \pm\pi, \ldots, \pm n\pi, \forall\, n \in \mathbb{N}$ or when:

$$c\theta_{3_l} = \pm \frac{|s\theta_{2_l}|}{\sqrt{\frac{L_3^2}{L_2^2} + 2\frac{L_3}{L_2} c\theta_{2_l} + 1}} \tag{52}$$

where:

$$c\theta_{2_l} > -\frac{\frac{L_3^2}{L_2^2} + 1}{2\frac{L_3}{L_2}} \tag{53}$$

for $l = 1, \ldots, 4$.
According to 52, for a given value of θ_{2_l} there will be two solutions for θ_{3_l}.

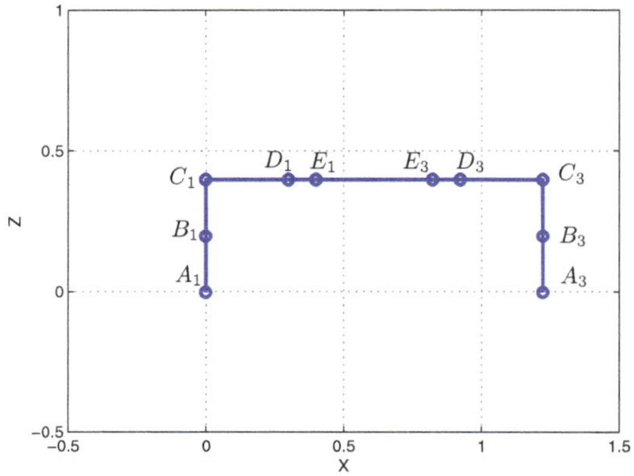

Fig. 6. Side view for the first condition of singularity.

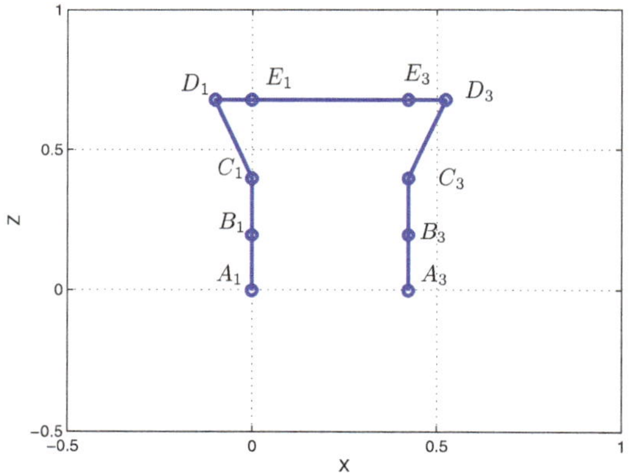

Fig. 7. Side view for the second condition of singularity.

The first condition of singularity, $\theta_{2_l} = 0, \pi, \ldots, n\pi$, means that L_2 is fully aligned with L_3. See Fig. 6.
The second condition of singularity, 52, means that joints E_l, C_l e B_l are vertically aligned in the same plane. See Fig. 7.
Provided that there are three parallel or coplanar axes, a singularity configuration will occur. (Murray et al., 1994)
In such a configuration, we say that the output link looses one or more degrees of freedom; this implies that the output link can resist to one or more components of force or moment with no torque or force applied at the powered joints. This condition can be useful if the robot needs to support heavy loads, forces or torques with little effort or low power consumption.

3.1.2 Direct kinematics singularity of the platform:

This kind of singularity occurs when

$$det(J_x) = 0. \tag{54}$$

This corresponds to configurations in which the platform is locally movable with all the actuated joints locked. The values of the output variables from given values of the input variables should be obtained. Since, in this case, the nullspace of J_x is non-empty, there exists nonzero output rate vectors $\dot{\mathbf{x}}$ which are mapped into the origin by J_x, i.e., which will correspond to null velocities of the input joints.

According to 46, $det(J_x)$ is null when $det(J_{x_1}) = 0$ or $det(J_{x_2}) = 0$.

Matrix J_{x_1} can be square, for example, when the robot is clinging to the surface with two legs, while the other two are in the air. Thus, the matrix J_{x_1} has size 6×6 and singularity occurs when $det(\Omega_{F_j} - \Omega_{F_k}) = 0$ for $j \neq k$.

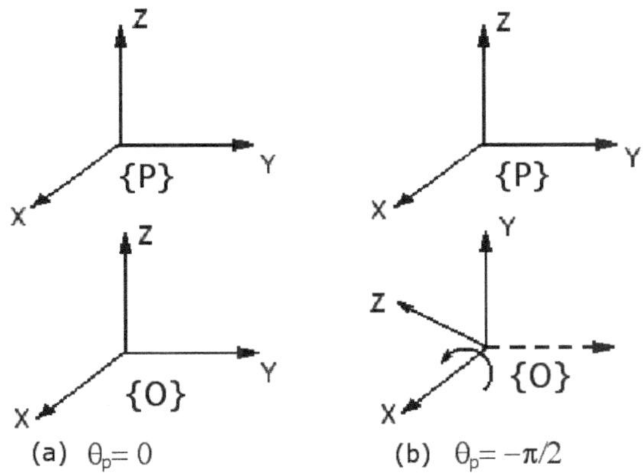

Fig. 8. Singularities for $\theta = 0$ (a) and $\theta = -\frac{\pi}{2}$(b)

On the other hand, $det(J_{x_2}) = 0$ for $\theta = 0, \pi, \ldots, n\pi$, $\forall\ n \in N$. This singularity is associated with the Euler angle convention used. For the Z-Y-Z Euler angle convention, this kind of singularity will occur for all horizontal orientations of the platform. Since this situation is not allowed in this particular application, this problem can be solved by either changing the Euler angle convention or by changing the coordinate system assigned to the climbing surface $\{O\}$ as in Fig. 8, (Harib & Srinivasan, 2003). Now the singularity will occur for $\theta = \frac{\pi}{2}, \frac{3\pi}{2} \ldots, \frac{(2n+1)\pi}{2}$, $\forall\ n \in N$, which means that the platform is completely at vertical with respect to the gripping surface (situation rather unlikely to occur in our case).

In such a configuration, we say that the output link gains one or more degrees of freedom, which implies that the output link cannot resist one or more components of force or moment even when all actuators are locked.

3.1.3 Combined singularities of the platform:

The third kind of singularity is of a slightly different nature since it requires conditions on the linkage parameters. This occurs when, for certain configurations, both $det(J_x)$ and $det(J_q)$ become simultaneously singular. If some specific conditions on the linkage parameters are satisfied, the chain can reach configurations at which the relation given by 45 degenerates.

This corresponds to configurations at which the chain can undergo finite motions when its actuators are locked or at which a finite motion at the inputs produces no motion at the outputs (Tsai, 1999).

3.2 Jacobian matrix for the leg

The study of the singularity of the leg is similar to the analysis of the serial manipulator attached to point E_l. Differently from the analysis of singularity of the platform, when the rank of the Jacobian of the serial manipulator loses its full rank (singularity condition), it may only lose degrees of freedom.

$$\vec{v}_{A_l} = J_{A_l}\dot{\vec{x}}_{A_l} \tag{55}$$

where:

$$\dot{\vec{x}}_{A_l} = \begin{bmatrix} {}^{E_l}v_{A_{x_l}} \\ {}^{E_l}v_{A_{y_l}} \\ {}^{E_l}v_{A_{z_l}} \\ \dot{\varphi}_l \end{bmatrix} \tag{56}$$

To calculate the J_{g_l} is necessary to difference equation 10 relative to time.

$$J_{A_l} = \begin{bmatrix} -s\theta_{1_l}(c\theta_{2_l3_l}\bar{L}_4 + c\theta_{2_l}L_3 + L_2) & (-s\theta_{2_l3_l}\bar{L}_4 - s\theta_{2_l}L_3)c\theta_{1_l} & -s\theta_{2_l3_l}\bar{L}_4 & 0 \\ c\theta_{1_l}(c\theta_{2_l3_l}\bar{L}_4 + c\theta_{2_l}L_3 + L_2) & (-s\theta_{2_l3_l}\bar{L}_4 - s\theta_{2_l}L_3)s\theta_{1_l} & -s\theta_{2_l3_l}\bar{L}_4 & 0 \\ 0 & c\theta_{2_l3_l}\bar{L}_4 - c\theta_{2_l}L_3 & c\theta_{2_l3_l}\bar{L}_4 & 0 \\ 0 & 0 & 0 & 1 \end{bmatrix} \tag{57}$$

In this case, the singularity of matrix J_{g_l} occurs when $det(J_{g_l}) = 0$. This condition is present for $c\theta_{2_l3_l}\bar{L}_4 + c\theta_{2_l}L_3 + L_2 = 0$ or in other words when:

$$s\theta_{2_l} = \frac{s\theta_3\bar{L}_4L_2 \pm (c\theta_{3_l}\bar{L}_4 + L_3)\sqrt{\bar{L}_4^2 + 2c\theta_{3_l}L_3\bar{L}_4 + L_3^2 - L_2^2}}{\bar{L}_4^2 + 2c\theta_{3_l}L_3\bar{L}_4 + L_3^2} \tag{58}$$

where:

$$c\theta_{3_l} \geq \frac{L_2^2 - L_3^2 - \bar{L}_4^2}{2L_3\bar{L}_4} \tag{59}$$

Another case of singularity of the leg, but, on the border of the workspace, occurs when a leg is fully extended horizontally as shown in Figure 11. This kind of singularity is deu to condition $z_{AE_l} = 0$.

4. Dynamics model

The dynamics of walking machines involves special features that render these systems more elaborate from the dynamics viewpoint, for they present a time-varying topology. What this means is that these systems include kinematic loops that open when a leg takes off and open chains that close when a leg touches the ground (Angeles, 2007). This fact implies in a degree of freedom time-varying. (Pfeiffer et al., 1995).

There are some techniques to analyze the dynamics of robots. In this section, two different methods will be used. Firstly, for the analysis of the dynamics of the platform, the Principle of Virtual Works is used and, for the analysis of the dynamics of the leg, the Newton-Euler formulation is chosen. In both cases, the notations used in (Tsai, 1999) are employed.

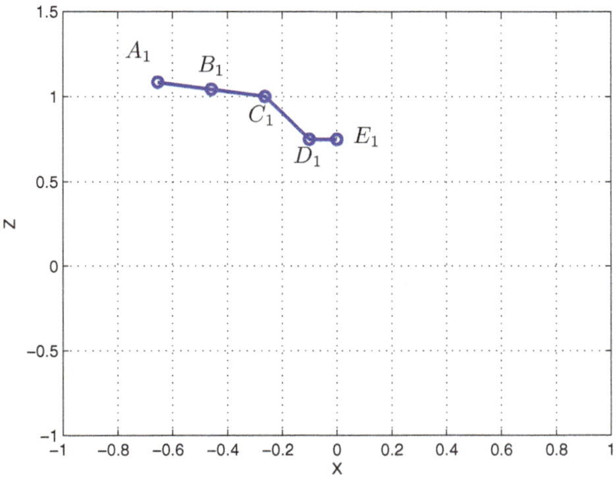

Fig. 9. Singular configuration for the leg in the ar and $\theta_{3_1} = -\frac{\pi}{4}$. First solution.

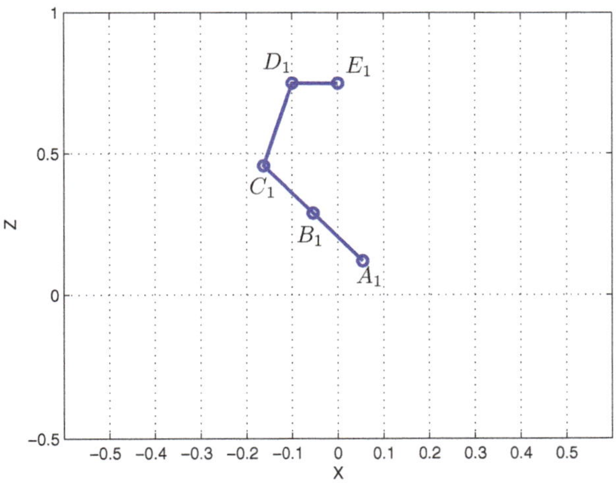

Fig. 10. Singular configuration for the leg in the ar and $\theta_{3_1} = -\frac{\pi}{4}$. Second solution.

- f_{i_l}: resulting force (excluding the actuator force) exerted at the center of mass of link i of leg l.
- $f^*_{i_l}$: inertia force exerted at the center of mass of link i of leg l, $f^*_{i_l} = -m_{i_l}\dot{\vec{v}}_{i_l}$
- $\hat{f}_{i_l} = f_{i_l} + f^*_{i_l}$
- f_p: resulting force exerted at the center of mass of the moving platform.
- $f^*_{i_l}$: inertia force exerted at the center of mass of the moving platform, $f^*_p = -m_p\dot{\vec{v}}_p$
- $\hat{f}_p = f_p + f^*_p$

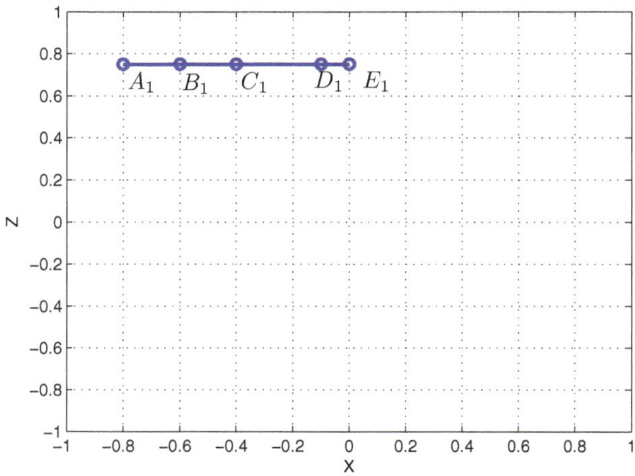

Fig. 11. Singular configuration in the border of the workspace for the leg in the ar.

- n_{i_l}: resulting torque (excluding the actuator torque) exerted at the center of mass of link i of leg l.
- $n^*_{i_l}$: inertia torque exerted at the center of mass of link i of leg l, $n^*_{i_l} = -^{i}I_{i_l}\dot{\vec{\omega}}_{i_l} -^{i_l}\vec{\omega}_{i_l} \times (^{i_l}I_{i_l}\vec{\omega}_{i_l})$
- $\hat{n}_{i_l} = n_{i_l} + n^*_{i_l}$
- n_p: resulting torque exerted at the center of mass of the moving platform.
- n^*_p: inertia torque exerted at the center of mass of the moving platform, $n^*_p = -I_p\dot{\vec{\omega}}_p - \vec{\omega}_p \times (I_p\vec{\omega}_p)$
- $\hat{n}_p = n_p + n^*_p$
- x_{i_l} six-dimensional vector describing the position and orientation of link i of leg l.
- $\delta(\cdot)$: virtual displacement of $(\cdot)$.
- $\vec{\tau} = [\tau_{1_1}, \tau_{2_1}, \ldots, \tau_{n_l}]$: vector of actuator torques applied at the active joints $1 \le i \le n$ at the leg $l = 1, 2, ..., 4$

In addition, the next vectors are defined:

$$\hat{F}_{i_l} = \begin{bmatrix} \hat{f}_{i_l} \\ \hat{n}_{i_l} \end{bmatrix}$$

where $i \in A, 1 \le i \le 4$, and

$$\hat{F}_p = \begin{bmatrix} \hat{f}_p \\ \hat{n}_p \end{bmatrix}.$$

As the velocities and accelerations of the robot are low, without losing accuracy, it is possible to assume that the link has its mass lumped at its center of mass. This approach was demonstrated in Almeida & Hess-Coelho (2010) sufficiently accuracy for modeling purposes. In both cases, the methods do not take into account all the effects that act on the joints and links. They consider only the dynamics of the rigid body under the action of gravity. A very

important force that was not included in the model is the friction force. As each joint of the Kamambaré is subject to reduction gears, in these circumstances the effects of friction can represent up to 25% of torque needed to trigger a joint in typical situations (Craig, 1989).
The effects of viscous and Coulomb friction can then be modeled by a simplified equation:

$$\tilde{\tau}_{i_l} = c \cdot sgn(\dot{\theta}_{i_l}) + b \cdot \dot{\theta}_{i_l} \tag{60}$$

where b and c are constants.

4.1 Dynamics model of the platform

The Principle of Virtual Work can be written as:

$$\delta q^T \vec{\tau} + \delta \vec{x}_p^T \hat{F}_p + \sum_{l=1}^{2} \sum_{i=1}^{n} \delta \vec{x}_{i_l}^T \hat{F}_{i_l} = 0 \tag{61}$$

As usual, the virtual displacement must be compatible with both the geometrical and kinematical constraints of the system. It is thus necessary to express the displacement as a function of a set of independent generalized virtual displacements. In accordance with that, it is convenient to choose the coordinates of the moving platform $\vec{x}_p$ as the generalized coordinates (Merlet, 2006; Tsai, 1999).
Denoting by J_p and J_{i_l} the jacobian matrices, respectively, of the moving platform and of the link:

$$\delta \vec{x}_{i_l} = J_{i_l} \delta \vec{x}_p \tag{62}$$

and

$$\delta \vec{q} = J_p \delta \vec{x}_p, \tag{63}$$

then equation 61 leads to

$$\vec{\tau} = -J_p^{-T} (\hat{F}_p + \sum_{l=1}^{2} \sum_{i=1}^{n} (J_{i_l}^T \hat{F}_{i_l})). \tag{64}$$

In addition to 64, a more accurate model of the leg dynamics could include various sources of flexibility, deflection of the links under load and vibrations (Bobrow et al., 2004). Nevertheless, this model is sufficiently accurate for our purposes since these effects are not significant for the leg under consideration.
Due to the fact, that the number of actuators is greater than the number of degrees of freedom of the robot, there is an infinite number of solutions for $\vec{\tau}$. Hence, a minimum norm solution can be adopted by applying the pseudo-inverse technique.
To solve equation (64), it is necessary to compute: i) the linear and angular velocities of each link, performing the inverse kinematics analysis; ii) the jacobian matrices of the links and of the moving platform; iii) the forces and torques of the links and of the moving platform.

4.2 Dynamics model of the leg

For the analysis of the dynamics model of the leg, the recursive Newton-Euler formulation was chosen. This formulation uses all the forces acting on the individual links of the robot leg. Hence, the resulting dynamical equation includes all the forces of constraint between two adjacent links.
The method consists of a forward computation of the velocities and accelerations of each link, followed by a backward computation of the forces and moments in each joint. (Tsai, 1999)

4.2.1 Forward computation

The first step is to calculate the angular velocity, angular acceleration, linear velocity and linear acceleration of each link. The best form to calculate these velocities is using a recursive algorithm starting at the first moving link and advancing to the gripper.

$$^{i_l+1}\omega_{i_l+1} = ^{i_l+1}R_{i_l} \cdot ^{i_l}\omega_{i_l} + \dot{\theta}_{i_l+1} \cdot ^{i_l+1}\hat{Z}_{i_l+1} \quad (65)$$

where $^{i_l+1}\hat{Z}_{i_l+1}$ is the versor of the joint axe expressed in the frame $\{i_l + 1\}$ and $^{i_l+1}\omega_{i_l+1}$ is the angular velocity of joint $i_l + 1$.

$$^{i_l+1}\dot{\omega}_{i_l+1} = ^{i_l+1}R_{i_l} \cdot ^{i_l}\dot{\omega}_{i_l} + \ddot{\theta}_{i_l+1} \cdot ^{i_l+1}\hat{Z}_{i_l+1} + ^{i_l+1}R_{i_l} \cdot ^{i_l}\omega_{i_l} \times \dot{\theta}_{i_l+1} \cdot ^{i_l+1}\hat{Z}_{i_l+1} \quad (66)$$

$$^{i_l+1}v_{i_l+1} = ^{i_l+1}R_{i_l}(^{i}v_{i_l} + ^{i_l}\omega_{i_l} \times ^{i_l}0_{i_l+1}) \quad (67)$$

$$^{i_l+1}\dot{v}_{i_l+1} = ^{i_l+1}R_{i_l}[^{i_l}\dot{\omega}_{i_l} \times ^{i_l}0_{i_l+1} + ^{i_l}\omega_{i_l} \times (^{i_l}\omega_{i_l} \times ^{i_l}0_{i_l+1}) + ^{i_l}\dot{v}_{i_l}] \quad (68)$$

For the calculation, it is assumed that velocities of base ω_0, $\dot{\omega}_0$, v_0 and $\dot{v}_0$ are known and are equal to the platform.
If the center of mass of each link $^{i_l+1}O_{C_{i_l+1}}$ is known, its acceleration may be calculated by equation 69.

$$^{i_l+1}\dot{v}_{C_{i_l+1}} = ^{i_l+1}\dot{\omega}_{i_l+1} \times ^{i_l+1}O_{C_{i_l+1}} + ^{i_l+1}\omega_{i_l+1} \times (^{i_l+1}\omega_{i_l+1} \times ^{i_l+1}O_{C_{i_l+1}}) + ^{i_l+1}\dot{v}_{i_l+1} \quad (69)$$

where $^{i_l+1}\dot{v}_{C_{i_l+1}}$ is the velocity of the center of mass of link $i_l + 1$.

4.2.2 Backward computation

Once the velocities and accelerations of the link are calculated, the joint forces and moments can be computed, one link at time, starting from the gripper and ending at the platform.

$$^{i_l}\hat{f}_{i_l} = ^{i_l}R_{i_l+1} \cdot ^{i_l+1}f_{i_l+1} + ^{i_l}\hat{f}^*_{i_l} \quad (70)$$

$$^{i_l}n_{i_l} = ^{i_l}n^*_{i_l} + ^{i}R_{i_l+1} \cdot ^{i_l+1}n_{i_l+1} + ^{i_l}0_{c_{i_l}} \times ^{i_l+1}\hat{f}^*_{i_l+1} + ^{i_l}0_{i_l+1} \times ^{i_l}R_{i_l+1} \cdot ^{i_l+1}f_{i_l+1} \quad (71)$$

Finally, the torques are obtained by projecting the forces or moments onto their corresponding joint axes.

$$\tau_{i_l} = ^{i_l}n^T_{i_l} \cdot ^{i_l}\hat{Z}_{i_l} \quad (72)$$

5. Illustrative example of the robot gait

This section, presents the performance of the robot in stages I and II (see figure 2). For this, we want to displacement of the center of the platform ^{O}P along the Y axis relative to frame $\{O\}$ from the point $^{O}P(0) = [0.392, 0, 0.231]$ to the point $^{O}P(I) = [0.392, 0.39, 0.231]$. At the starting point, legs $l = 1, 3$ are stuck to the surface and legs $l = 2, 4$ are in the air. Frame O is attached to point $^{O}A_1$ like in figure 12.
Table 3 show the control parameters used in this example.
Where N is the fator of discretization of the signals, t_f the time to execute the task, and $I_{max_{i_l}}$ and $V_{max_{i_l}}$ the maximus values of the current and voltage than can be apply to the joints motors.
In this case some conditions must be respected:

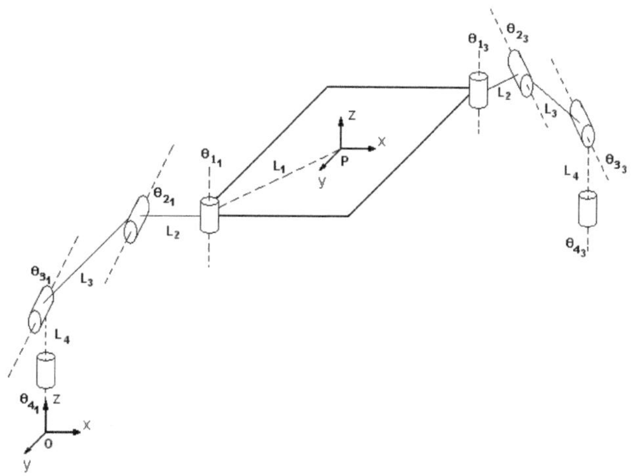

Fig. 12. Mechanical model

Parameters	Movement of the Platform	Movement of the Legs
N	100	100
t_f	$40s$	$60s$
$I_{max_{i_l}}$	3.5A	2.8A
$V_{max_{i_l}}$	12v	12v

Table 3. Control Parameters

1. Vector ${}^{O}\vec{A_l B_l}$ is always orthogonal to the surface in the pushing stage.
2. The legs in the air are locked when the platform is moving.
3. At the pushing stage, joints θ_{1_l} are passives.

The robot move was controlled by an optimal control law that minimize the loss energy in the actuator. The law of control was based in the independent joints control strategy. The objective of the simulation is to show the performance of the system in a basic cycle gait. The gait control was implemented according with the flowchart showed in figure 13.
Figures 14, 15 and 16 show the characteristics of the robot move at the "pushing stage". When the desirable position of the platform is reached, the next step is to move the legs that are in the air to the next clinging point. At this moment the stage "leg on the air" begins. Figures 17, 18 and 19 show the performance of one leg in this stage. The orientation of the gripper is the same all the time and it is $\varphi = 0$
When the four legs are clung to the surface, the basic cycle gait is over, and the robot is ready to calculate the new path to go. The total displacement of the robot was from position ${}^{O}P(0)$ to position ${}^{O}P(I)$ with an average speed of displacement in the Y axis of about ${}^{O}\bar{v}_Y = 0.00975m/s$.

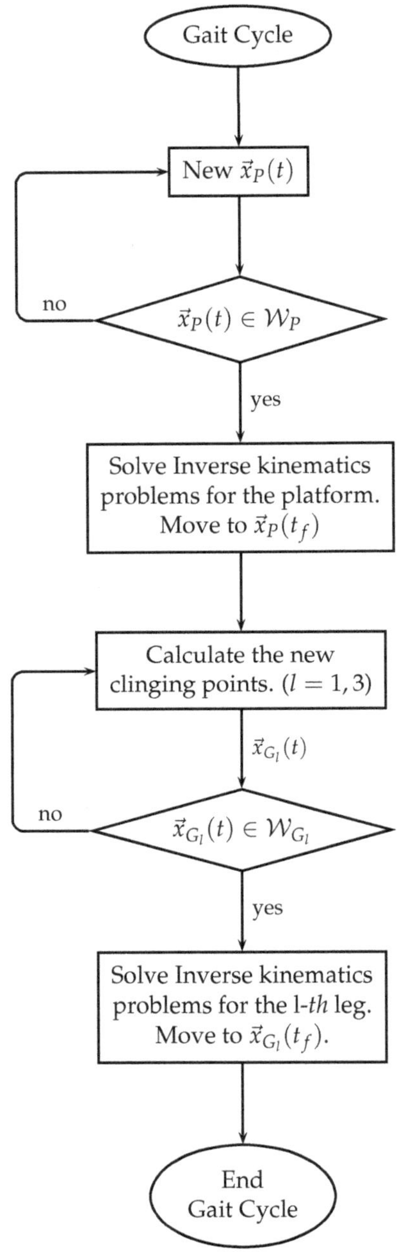

Fig. 13. Control flowchart for a cycle gait

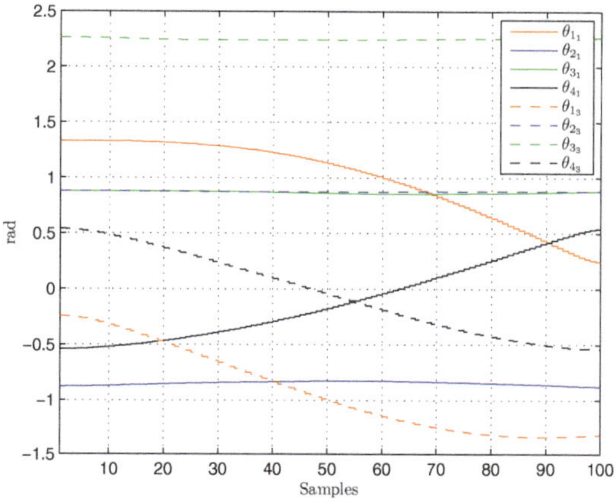

Fig. 14. Joint space in the "pushing stage".

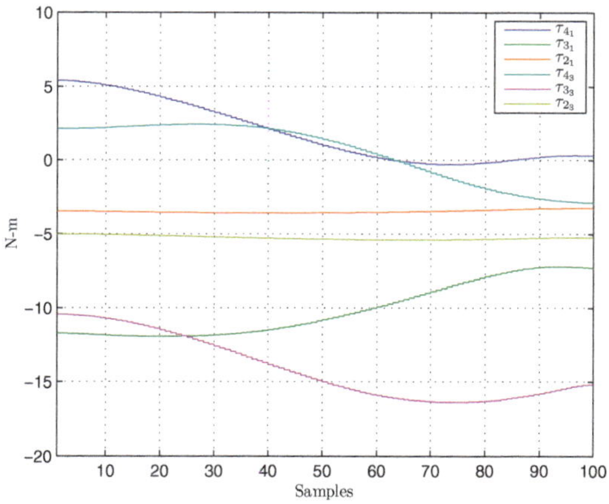

Fig. 15. Joint torques in the "pushing stage".

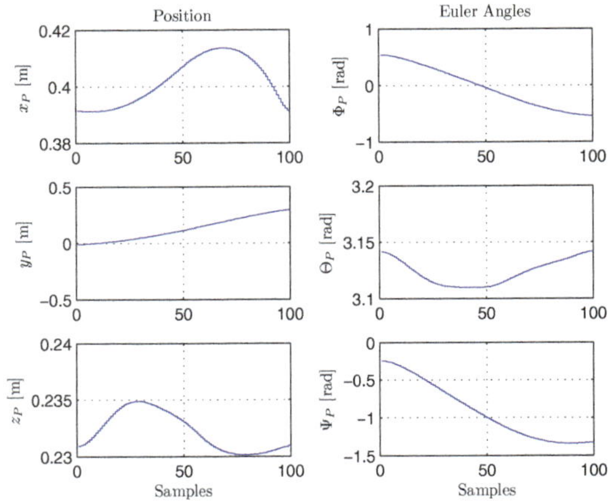

Fig. 16. Movement and orientation of the center of platform in the Cartesian space in the "pushing stage".

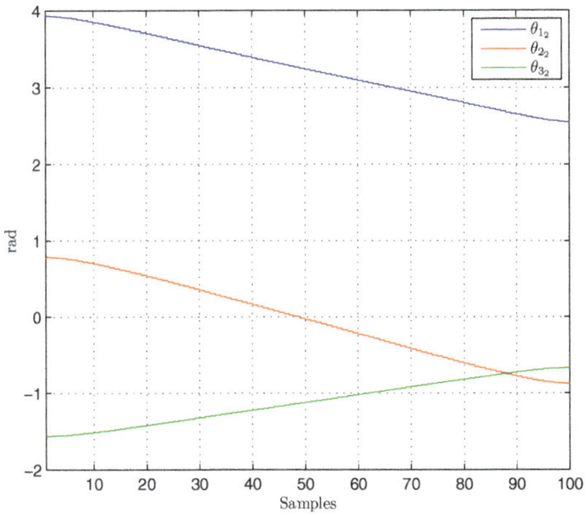

Fig. 17. Joint space for one leg 1 in the stage "leg on the air".

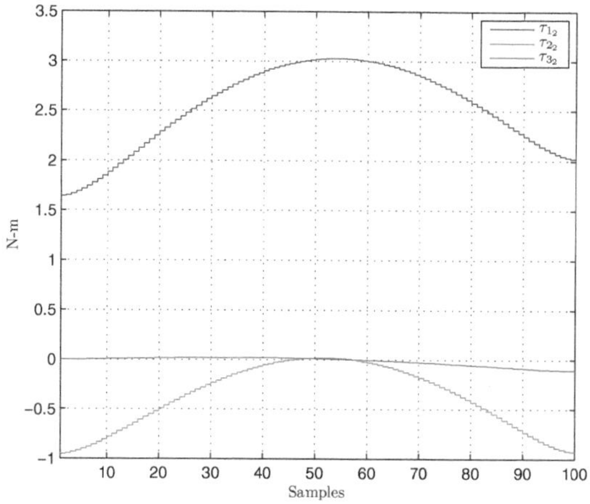

Fig. 18. Joint torques for one leg 1 in the stage "leg on the air".

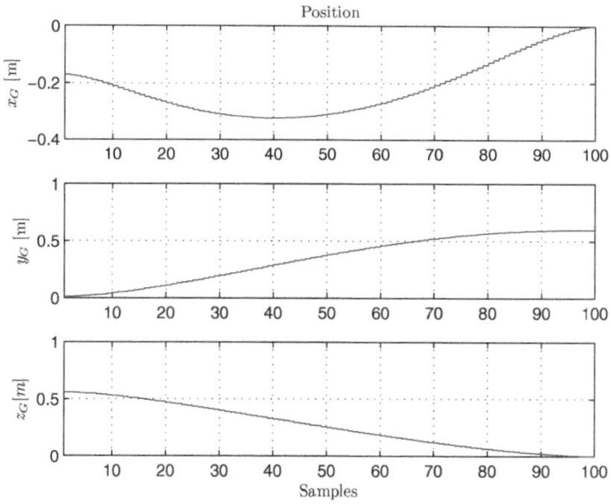

Fig. 19. Movement of the gripper of a leg 1 in the Cartesian space in the stage "leg on the air".

6. Conclusion

This paper discussed an important issue related to legged robots: the kinematics and dynamics model of the quadruped robot. The analysis done for each model was always presented in two parts, the platform and the legs, according to a time-varying topology and a time-varying degree of freedom of the system.

Several methods were used in each modeling process always trying to use those which brought to better performance in accordance with the topology modeled and that could be easily implemented in programming languages of high level. Then were used the Denavit-Hartenberg parameters for solving the direct position kinematics of the platform and leg, the Principle of Virtual Work or the d'Alembert for dynamic modeling of the platform and the Newton-Euler dynamic model for leg in the air.

Special attention was given in the section of the singularities, where the study of all the singularities in the parallel topology were presented. For that, the complete criterion of singularity for parallel robots proposed in Goselin & Angeles (1990) was used. In addition, the principals configurations of the singularities were showed through figures.
Finally, the performance of the robot in a cycle gait was presented. As a result of this example, the space joints, the torque of the joints and the cartesian space relative to this gait were displayed in figures.

7. References

Almeida, R. Z. H. & Hess-Coelho, T. A. (2010). Dynamic model of a 3-dof asymmetric parallel mechanism, *The Open Mechanical Engineering Journal* 4.

Angeles, J. (2007). *Fundamentals of Robotic Mechanical Systems. Theory, Methods, and Algorithms*, Springer.

Bernardi, R. & Da Cruz, J. J. (2007). Kamanbaré: A tree-climbing biomimetic robotic platform for environmental research., *International Conference on Informatics in Control, Automation and Robotics (ICINCO)*.

Bernardi, R., Potts, A. S. & Cruz, J. (2009). An automatic modelling approach to mobile robots, *in* F. B. Troch (ed.), *International Conference on Mathematical Modelling*, MATHMOD, Vienna, pp. 1906–1912.

Bobrow, J., Park, F. & Sideris, A. (2004). Recent advances on the algorithmic optimization of robot motion., *Technical report*, Departament of Mechanical and Aerospace Engineering, University of California.

Craig, J. (1989). *Introduction to Robotics. Mechanics and Control*, Addison Wesley Longman.

Estremera, J. & Waldron, K. J. (2008). Thrust control, stabilization and energetics of a quadruped running robot, *The International Journal of Robotics Research* 27(10): 1135–1151.

Goselin, C. & Angeles, J. (1990). Singularity analysis of closed-loop kinematic chains, *IEEE Transactions on Robotics and Automation* 6: 281–290.

Harib, K. & Srinivasan, K. (2003). Kinematic and dynamic analysis of stewart platform-based machine tool structures, *Robotica* 21: 241–254.

Kolter, J. Z., Rodgers, M. P. & Ng, A. Y. (2008). A control architecture for quadruped locomotion over rough terrain, *Technical report*, Computer Science Department, Stanford University, Stanford.

Lenarcic, J. & Roth, B. (eds) (2006). *Advances in Robots Kinematics. Mechanisms and Motion*, Springer.

Merlet, J. (2006). *Parallel Robots*, 2nd edn, Springer.

Murray, R. M., Li, Z. & Sastry, S. S. (1994). *A Mathematical Introduction to Robot Manipulation*, CRC Press.

Pfeiffer, F., Eltze, J. & Weidemann, H.-J. (1995). The tum walking machine, *Intelligent Automation and Soft Computing. An International Journal* 1: 307–323.

Pieper, D. (1968). The kinematics of manipulators under computer control., *Technical report*, Department of Mechanical Engineering, Stanford University.

Potts, A. & Da Cruz, J. (2010). Kinematics analysis of a quadruped robot, *IFAC Symposium on Mechatronics Systems*, Boston, Massachusetts.

Siegwart, R. & Nourbakhsh, I. R. (2004). *Introduction to Autonomous Mobile Robots*, The MIT Press.

Tsai, L. (1999). *Robot Analysis. The Mechanical of Serial and Paralel Manipulators*, John Wiley & Sons.

12

Development of Mobile Robot Based on I^2C Bus System

Surachai Panich
Srinakharinwirot University
Thailand

1. Introduction

Mobile robots are widely researched in many applications and almost every major university has labs on mobile robot research. Also mobile robots are found in industry, military and security environments. For entertainment they appear as consumer services. They are most generally wheeled, but legged robots are more available in many applications too. Mobile robots have ability to move around in their environment. Mobile robot researches as autonomously guided robot use some information about its current location from sensors to reach goals. The current position of mobile robot can be calculated by using sensors such motor encoders, vision, Stereopsis, lasers and global positioning systems. Engineering and computer science are core elements of mobile robot, obviously, but when questions of intelligent behavior arise, artificial intelligence, cognitive science, psychology and philosophy offer hypotheses and answers. Analysis of system components, for example through error calculations, statistical evaluations etc. are the domain of mathematics, and regarding the analysis of whole systems physics proposes explanations, for example through chaos theory.

This book chapter focuses on mobile robot system. Specifically, building a working mobile robot generally requires the knowledge of electronic, electrical, mechanical, and computer software technology. In this book chapter all aspects of mobile robot are deeply explained, such as software and hardware design and technique of data communication.

2. History of mobile robots (Wikipedia, 2011)

During World War II, the first mobile robots emerged as a result of technical advances on a number of relatively new research fields like computer science and cybernetics. W. Grey Walter constructed Elmer and Elsie, which were equipped with a light sensor. If they found a light source, they would move towards it, avoiding or moving obstacles. The Johns Hopkins University develops mobile robot named Beast. Beast used sonar to move around. Mowbot was the first automatically mobile robot to mow the lawn. The Stanford Cart for line follower was a mobile robot, which can follow a white line by using a camera. The mobile robot is developed to navigate its way through obstacle courses and make maps of its environment. The Stanford Research Institute researched on Shakey mobile robot. Shakey had a camera, a rangefinder, bump sensors and a radio link. The Soviet Union explores the surface of the moon with Lunokhod 1, a lunar rover as shown in Fig.1.

Fig. 1. A model of the Soviet Lunokhod-1 Moon rover released by the Science Photo Library

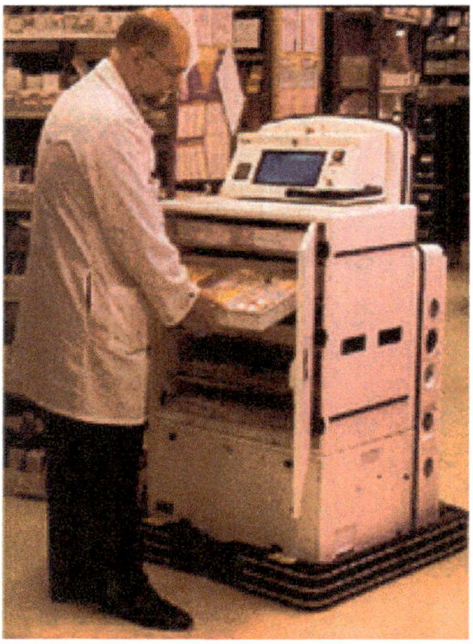

Fig. 2. Helpmate, autonomous mobile hospital robots

The team of Ernst Dickmanns at Bundeswehr University Munich built the first robot cars, driving up to 55 mph on empty streets. The Hughes Research Laboratories demonstrated the first cross-country map and sensor-based autonomous robotic vehicle. Mark Tilden invented BEAM robotics. Joseph Engelberger worked with colleagues to design the first commercially available autonomous mobile hospital robots named Helpmate as shown in Fig.2. The US Department of Defense funds the MDARS-I project for the Cyber-motion indoor security robot. Edo Franzi, André Guignard and Francesco Mondada developed Khepera, an autonomous small mobile robot. Dante I and Dante II as shown in Fig.3, walking robots used to explore live volcanoes were developed by Carnegie Mellon University.

Fig. 3. Dante II developed by Carnegie Mellon University

The twin robot vehicles VaMP and VITA-2 of Daimler-Benz and Ernst Dickmanns of UniBwM drive more than one thousand kilometers on a Paris three-lane highway in standard heavy traffic at speeds up to 130 km/h.

Fig. 4. Sojourner Rover, NASA

Semi-autonomous ALVINN steered a car coast-to-coast under computer control for all but about 50 of the 2850 miles. Throttle and brakes, however, were controlled by a human driver. The Pioneer programmable mobile robot becomes commercially available at an affordable price, enabling a widespread increase in robotics research and university study over the next decade as mobile robotics becomes a standard part of the university curriculum. NASA sends the Mars Pathfinder with its rover Sojourner to Mars as shown in Fig.4. The rover explores the surface, commanded from earth. Sojourner was equipped with a hazard avoidance system. Sony introduces Aibo as shown in Fig.5, a robotic dog capable of seeing, walking and interacting with its environment.

Fig. 5. Aibo, **Sony**

The PackBot remote-controlled military mobile robot is introduced as shown in Fig.6. PackBot is current base model using a videogame style hand controller to make easy control. PackBot is designed for improvised explosive device identification and disposal, infantry troops tasked with improvised explosive device inspection, help SWAT teams and other first responders with situational awareness.

Fig. 6. PackBot demonstrated by the French military

Swarm bots resemble insect colonies as shown in Fig.7. Typically they consist of a large number of individual simple robots that can interact with each other and together perform complex tasks.
Robosapien, a biomorphic commercially available robot designed by Mark Tilden as shown in Fig.8. The autonomous robots work together to make a map of an unknown environment and search for objects within the environment.

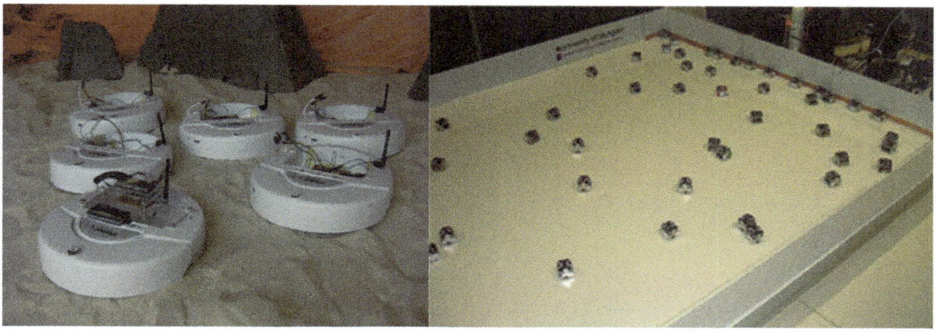

Fig. 7. Swarm bots: insect colonies behavior

Fig. 8. Robosapien designed by Mark Tilden

Sony introduced an autonomous service robot system named a lower-cost PatrolBot as shown in Fig. 9. The mobile robot becomes continue commercial product.

Fig. 9. PatrolBot introduced by Sony

The Tug as shown in Fig.10 becomes a popular means for hospitals to move large cabinets of stock from place to place for carrying blood and other patient samples from nurse stations to various labs.

Fig. 10. The Tug, Aethon's Automated Robotic Delivery System

Fig. 11. BigDog developed by Boston Dynamics

Boston Dynamics released video footage of a new generation BigDog as shown in Fig.11 able to walk on icy terrain and recover its balance when kicked from the side.

3. Related work

3.1 Development of the mobile robot system to aid the daily life for physically handicapped (Interface using internet browser)

Yoshiyuki Takahashi and et al. from Shibaura institute of technology developed the mobile robot system (Yoshiyuki Takahashi, et. al., 1998), which could bring daily using objects putting them somewhere in the room with semi-automatically control as shown in Fig.12.

Fig. 12. Mobile robot system developed by Shibaura institute of technology

It is necessary to use graphical and interactive interface to operate the robot, because the operator of this robot is physically handicapped and not always engineer and professional about the robot knowledge as shown in Fig.13.

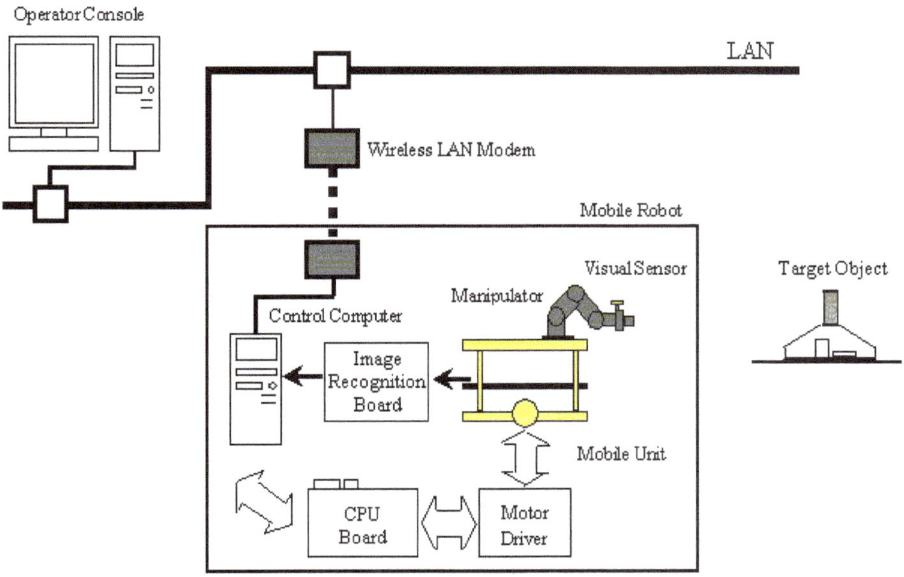

Fig. 13. System structure of mobile robot

3.2 Communication framework for sensor-actuator data in mobile robots

Fernandez proposes the architecture of the robot is designed in Fig.14, which shows several modules inter-connected with CAN bus (Fernandez J., et al., 2007). Each module performs one specific task in the distributed architecture. The actuator and the sensory modules are executed with basic control algorithms.

The communications protocol and CAN master process are implemented in figure 14 showing the different slaves are connected to the master. Control system for all modules worked on PC attached to CAN bus using a CAN-USB adapter.

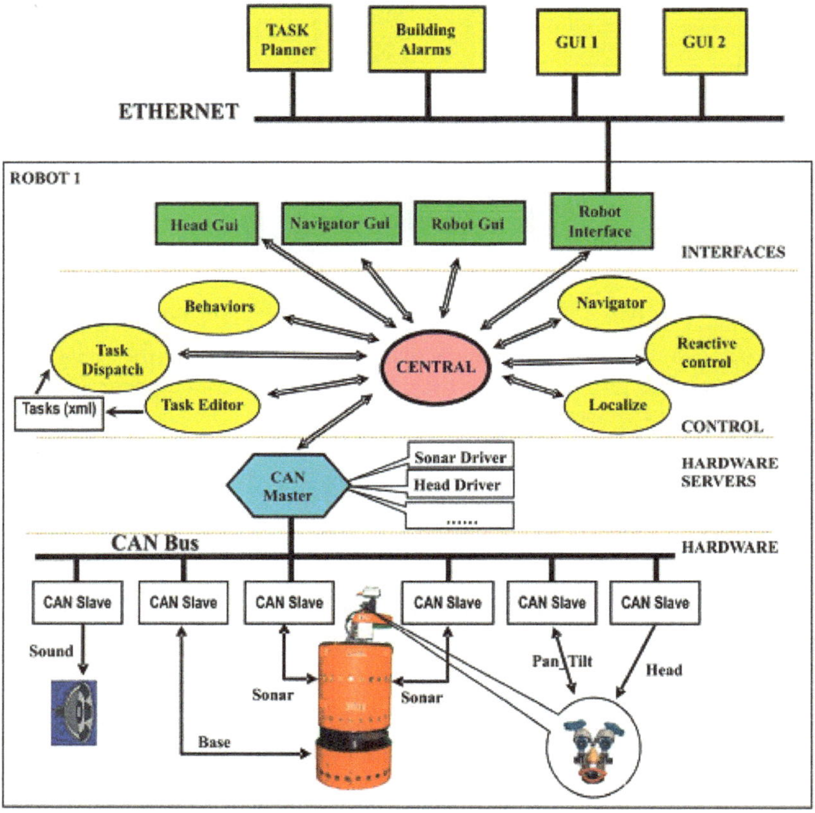

Fig. 14. Robot architecture of CAN system based on sensor and actuator and PC control module communications.

The connection and disconnection of the different slaves by loading the corresponding driver is operated by CAN server. The CAN server will first register and initialize the new module connection, for example if a module with sonar sensors is connected then it will be identified and the module will hand over the messages to the sonar module. In the case of connecting a new module, the master will send information to configure the slave and change the watchdog time. If the master does not receive the watchdog of a slave for a period longer that a timeout, it assumes the slave is disconnected or has some error and it will be notified to the control programs interested in the slave data.

3.3 SMARbot: A miniature mobile robot paradigm for ubiquitous computing

Yan Meng from Stevens Institute of Technology, Hoboken introduce SMARbot paradigm (Meng Yan, et al., 2007). The software reconfiguration of microprocessor and a core component for hardware reconfiguration is implemented in the FPGA. Multiple sensors and actuators with corresponding device drivers and signal processing modules are in the sensor or actuator layer. Each control module consists of one or more input ports, one or more output ports, and any number of other connections. The functionality of the module is implemented to provide automatic integration of the control modules. The information flow, communication and synchronization should be handled automatically by the operating system.

4. The I²C bus system overview

The standard Inter-IC (Integrated Circuit) bus named I²C is shorthand providing a good support for communication with various peripheral devices (Philips Semiconductor, 2000). It is a simple, low-bandwidth, short-distance protocol. There is no need for chip select or arbitration logic, making it cheap and simple to implement in hardware. Most I²C bus devices operate at speeds up to 400 Kbps. The I²C bus system is easy to link multiple devices together since it has a built-in addressing format. The I²C bus is a two wire serial bus as shown in Fig. 15. The two I²C signals are serial data (SDA) and serial clock (SCL).

Fig. 15. The I²C bus has only two lines in total

It is possible to support serial transmission of eight-bit bytes with seven-bit bytes device addresses plus control bits over the two wire serial bus. The device called the master starts a transaction on the I²C bus. The master normally controls the clock signal. A device controlled and addressed by the master is called a slave. The I²C bus protocol supports multiple masters, but most system designs include only one. There may be one or more slaves on the bus. Both masters and slaves can receive and transmit data bytes. The slave device with compatible hardware on I²C bus is produced with a predefined device address, which may be configurable at the board device.

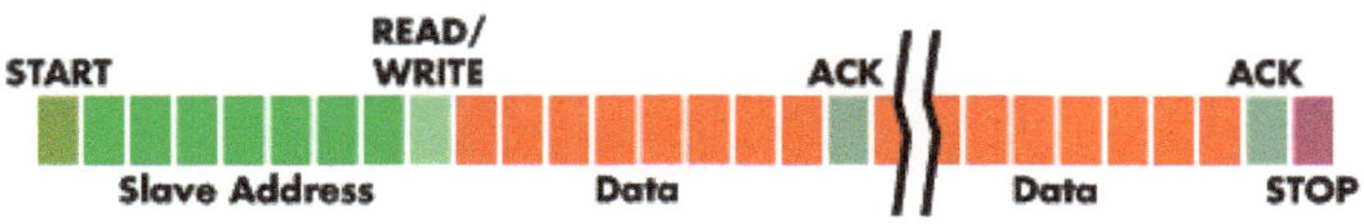

Fig. 16. The I²C bus communication

The master must send the device address of the slave at the beginning of every transaction. Each slave is responsible for monitoring the bus and responding only to its own address. As shown in Fig. 16, the master begins to communicate by issuing the start condition. The master continues by sending seven-bit slave device address with the most significant bit.

The eighth bit (read or write bit) after the start bit specifies whether the slave is now to receive or to transmit information. This is followed by an ACK bit issued by the receiver, acknowledging receipt of the previous byte. Then the transmitter (slave or master, as indicated by the bit) transmits a byte of data starting with the MSB. At the end of the byte, the receiver (whether master or slave) issues a new ACK bit. This 9-bit pattern is repeated if more bytes need to be transmitted. In a write transaction (slave receiving), when the master is done transmitting all of the data bytes it wants to send, it monitors the last ACK and then issues the stop condition. In a read transaction (slave transmitting), the master does not acknowledge the final byte it receives. This tells the slave that its transmission is done. The master then issues the stop condition.

5. Development of mobile robot based on I^{2}C bus system

In this book chapter, the system of mobile robot named AMRO (Surachai, 2010c) is deeply explained as example for understanding. This robot is developed by student team from Measurement and Mobile Robot laboratory. Its hardware is constructed and combined with the electronic components including the control program.

5.1 Hardware development for AMRO

The mobile robot is designed based on differential drive system (Byoung-Suk Choi 2009; Surachai and et al., 2009) as shown in Fig. 17. The combination of two driven wheels allows the robot to be driven straight, in a curve, or to turn on the spot. The translation between driving commands, for example a curve of a given radius and the corresponding wheel speeds are controlled by software.

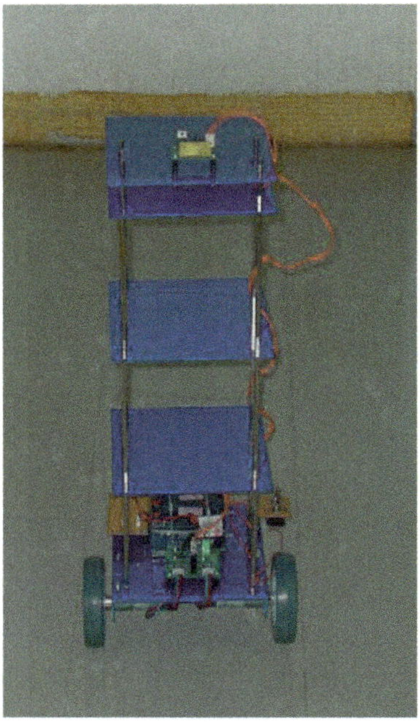

Fig. 17. The Autonomous Mobile Robot (AMRO), Measurement and Mobile robot laboratory

The AMRO is driven by two wheels and a caster powered by an MD25 Dual 5A controller. The MD25 motor driver is designed with 12v battery, which drives two motors with independent or combined control as shown in Fig. 18.

Fig. 18. The MD25 motor driver integrated on AMRO

It reads motors encoders and provides counts for determining distance traveled and direction. Motor current is readable and only 12v is required to power the module. Onboard 5v regulator can supply up to 1A peak, 300 mA continuously to external circuitry Steering feature, motors can be commanded to turn by sent value.

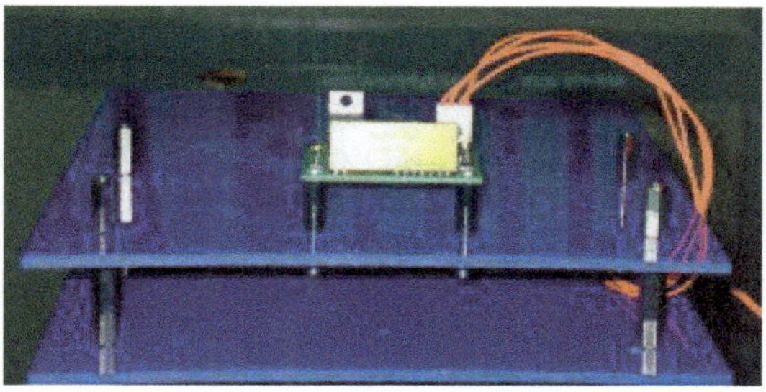

Fig. 19. The CM02 Radio communications module integrated on AMRO

The CM02 Radio communications module as shown in Fig. 19 works together with its companion RF04 module from a complete interface between PC and I^2C devices. The commands can be sent to the robot and receive telemetry data back up to the PC. The CM02 module is powered from battery, which can be anything from 6-12v. There are four I^2C connectors on the CM02, but it is not limited to four I^2C devices. The CM02 radio module provides communication with an RF04 module connected to the PC's USB port. It also provides the MD25 and I^2C devices with 5v supply from its on-board 5v regulator. The AMRO is powered by battery which goes to the CM02 module and also to the MD25 for motor power. All of the modules are connected together with a four wire I^2C loop, which are 5v, 0v, SCL and SDA lines. The PC can now control robot's motors and receive encoder information from AMRO. That means the PC now becomes the robot's brain as shown in Fig.20.

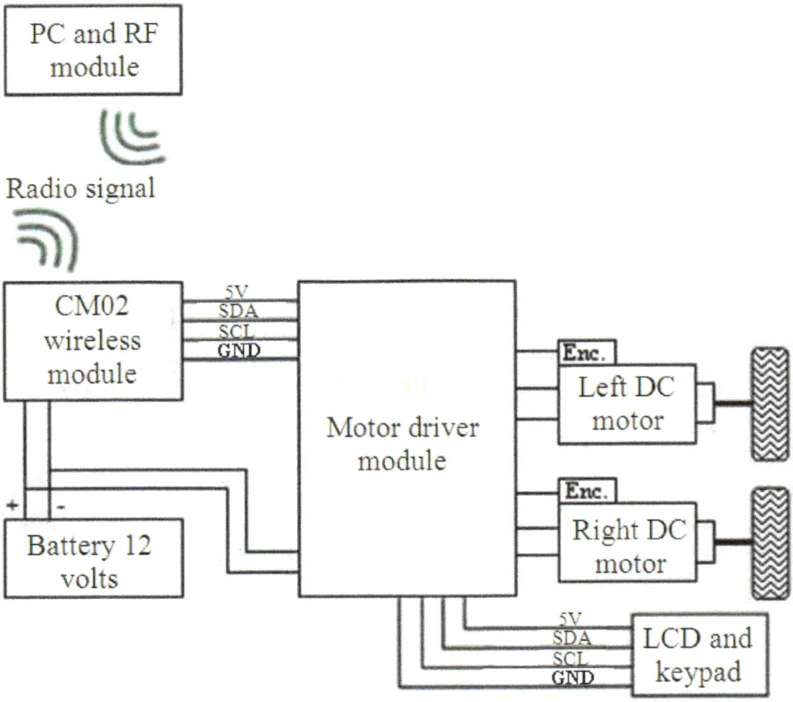

Fig. 20. The AMRO's system control

5.2 The communication between robot and I^2C bus devices (Panich, 2008)

Surachai developed the I^2C bus system to control information between the robot and sensors or additional devices.

In order that the signal line of serial port from the robot can connect to devices on I^2C bus, the electrical master module is designed to generate signal SDA and SCL as shown in Fig. 21.

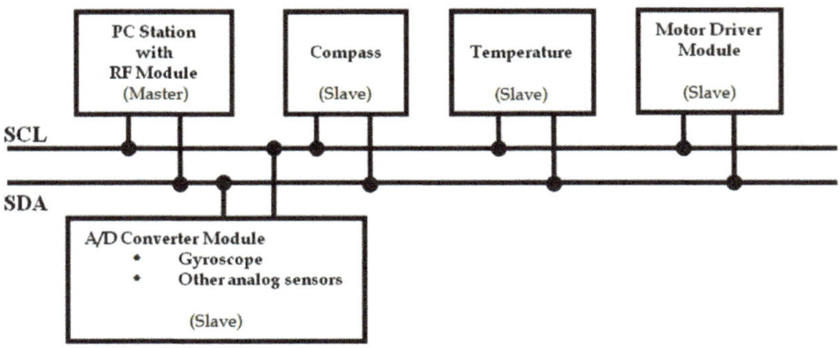

Fig. 21. Hardware communication between Mobile robot (AMRO) and I^2C devices

The PC station and RF radio module are selected to produce signal and work as master device. The system of mobile robot can directly connect to I²C devices and other devices, which cannot support I²C system, they can be connected through microcontroller with interface circuit.

5.2.1 The I²C bus devices

Standard I²C devices operate up to 100Kbps, while fast-mode devices operate at up to 400Kbps. A 1998 revision of the I²C specification (v. 2.0) added a high-speed mode running at up to 3.4Mbps. Most of the I²C devices available today support 400Kbps operation. Higher-speed operation may allow I²C to keep up with the rising demand for bandwidth in multimedia and other applications.

5.2.1.1 Compass sensor

The first I²C slave device is compass sensor module as shown in Fig. 22. This sensor can work on the I²C bus without addition circuit. This compass module has been specifically designed for use in robots as an aid to navigation. The aim was to produce a unique number to represent the direction the robot is facing. The compass uses the Philips KMZ51 magnetic field sensor, which is sensitive enough to detect the earth's magnetic field. The output from two of them mounted at right angles to each other is used to compute the direction of the horizontal component of the earth's magnetic field. The compass module requires a 5v power supply at a nominal 15mA. The pulse width varies from 1mS (0°) to 36.99mS (359.9°) - in other words 100uS/° with a +1mS offset. On I²C bus, there is an important consideration that consists of the address from manufacturer and the address from user.

Fig. 22. Compass module slave device

5.2.1.2 Gyroscope sensor

A gyroscope is a device for measuring or maintaining orientation, based on the principles of angular momentum (Komoriya, K. and Oyama, E., 1994). A mechanical gyroscope is essentially a spinning wheel or disk whose axle is free to take any orientation. This orientation changes much less in response to a given external torque than it would without the large angular momentum associated with the gyroscope's high rate of spin. Since external torque is minimized by mounting the device in gimbals, its orientation remains nearly fixed, regardless of any motion of the platform on which it is mounted. Because this gyroscope is not designed for the I²C bus system, it must be connected through microcontroller and read its information as shown in Fig. 23. In order to microcontroller can work on the I²C bus system, it must be specified in I²C format. Now gyroscope with microcontroller works as slave device. This module can read information from gyroscope and send to master device.

Fig. 23. Gyroscope with microcontroller worked on I²C bus as slave

5.2.1.3 Temperature sensor (DS1621)

The DS1621 as shown in Fig. 24 supports I²C bus and data transmission protocol. A device sends data onto the bus defined as a transmitter, and a device receiving data as a receiver. The device controls the message called a master and devices are controlled by the master called slaves. The bus must be controlled by a master device which generates the serial clock (SCL), controls the bus access, and generates the START and STOP conditions. The DS1621 operates as a slave on the 2-wire bus. Connections to the bus are made via the open-drain I/O lines SDA and SCL. A control byte is the first byte received following the START condition from the master device. The control byte consists of a 4-bit control code set as 1001 binary for read and write operations.

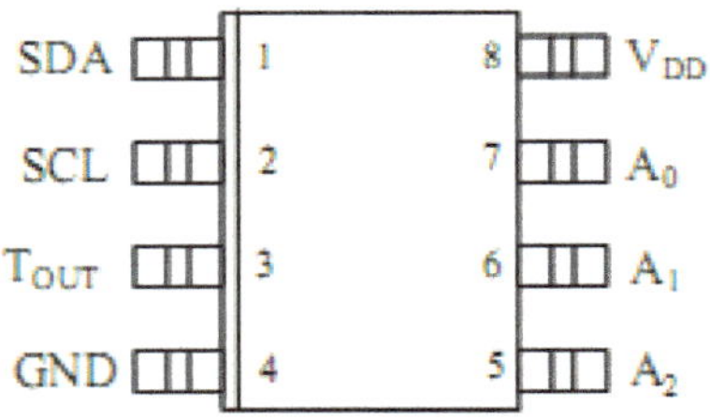

Fig. 24. DS1621 temperature sensor

The next 3 bits of the control byte are the device select bits (A2, A1, A0). They are used by the master device to select which of eight devices are to be accessed. These bits are in effect the 3 least significant bits of the slave address. The last bit of the control byte (R/ W) defines the operation to be performed. When set to a "1" a read operation is selected, when set to a "0" a write operation is selected. Following the START condition the DS1621 monitors the SDA bus checking the device type identifier being transmitted. Upon receiving the 1001 code and appropriate device select bits, the slave device outputs an acknowledge signal on the SDA line.

5.3 Software development for AMRO

A software is developed based on PC application which now manually controls AMRO. The control window contains three sections, Wireless Serial Connection, Manual Control and

Input Velocity. During connection with robot, the PC can send command to the robot and receives sensor information from robot by using I2C bus system. The START button is used to establish the connection with real robot using wireless serial connection. The STOP button is used to disconnect the robot from the PC. It terminates the communication between robot and PC. The desired velocity can be defined in mm sec^{-1} for forward and backward movement in this Input Velocity block. In this program, the robot velocity are limited up to maximum 500 mm sec^{-1} for safety reason. Left and right turning is done by heading setting, which has already defined in the program. This is the important one to navigate the robot in the environment. As shown in Fig. 25 there are total five buttons to control the robot, which are Forward, Backward, Turn Left, Turn Right and Stop. After establishing connection with the robot through wireless communication and entering all velocity values, the robot will be ready to move in the desired directions. By pressing ***F*** button robot starts to run forward with the given velocity. To stop the robot in forward moving, the **S** button must be pressed. The ***B*** button for backward moving works as same as the ***F*** button. The ***L*** and ***R*** button are used to turn the robot in left and right direction respectively. After the ***L*** or ***R*** button are pressed, the **S** button can use to stop the robot in desired position. The operation is same like for left button.

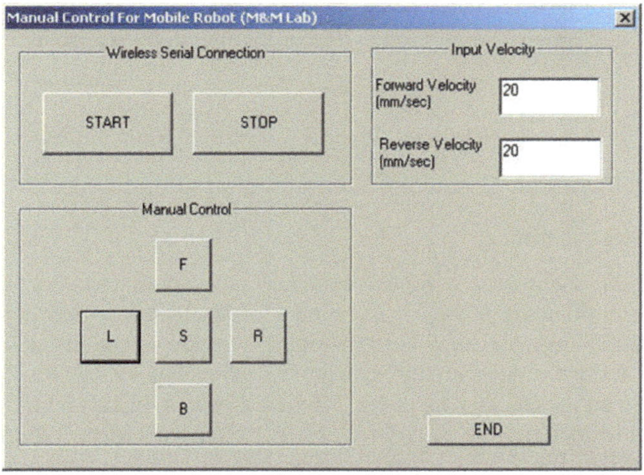

Fig. 25. Manual control window for AMRO

As above mentioned, the **S** button is very useful to stop the robot motion. In any condition of the robot, this button plays an important role to restrict the further motion of the robot without disconnecting from PC too. The software is designed to control data between the robot and sensor device, which is programmed based on Visual C^{++}. The software must be able to control SCL and SDA line. To control data line on the I2C bus, the step ordering of function based on Visual C^{++} must be carefully accurately considered and programmed, because if one step miss or does not complete, all devices on this bus will fail. The main function used for programming will be now detailed. All conditions are only generated by robot (as master device). The two main functions are I2C_START ( ) and I2C_STOP ( ). The I2C_START ( ) function produces the START condition as shown in Fig.26.

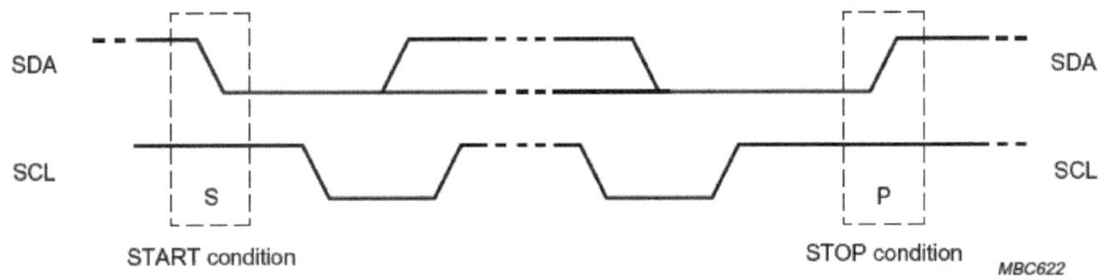

Fig. 26. Start and stop condition of two lines

This condition is a HIGH to LOW transition on the SDA line while SCL is HIGH. And the I2C_STOP ( ) will produce the STOP condition. This procedure is a LOW to HIGH transition on the SDA line while SCL is HIGH. Before the START condition will begin and after STOP condition finished, the bus is considered to be always free condition, the both lines must be HIGH. The next main function is I2C_ACK ( ). As shown in Fig.27, after the robot (master device) sent data to slave device finished, the slave device must send back the acknowledgement that the slave device received data already.

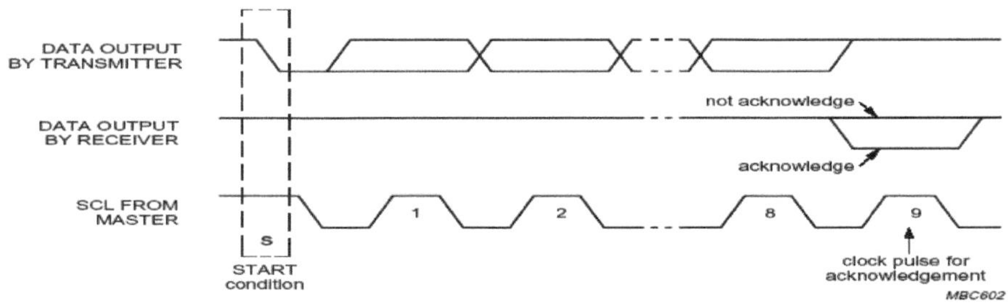

Fig. 27. Acknowledge condition

As shown in Fig.28, it shows a complete transfer cycle associated with a frame of data. Firstly, the master initiates a write by asserting logic-0 at bit-8, where a slave address is defined by the other 7-bits. A acknowledge signal then follows from the slave as specified in bit-9. The second and third bytes are the data and acknowledge signal. The 7-bits addressing allows 127 devices on the I²C bus, by using 2-bytes address, which can be extended further. The last two main functions are to control and get data from slave device. It consists of I2C_SEND ( ) and I2C_RECEIVE ( ) function. This I2C_SEND ( ) function is sent always by robot (master device) to control and set slave devices configuration and the I2C_RECEIVE ( ) function is also sent by robot to receive data from slave devices.

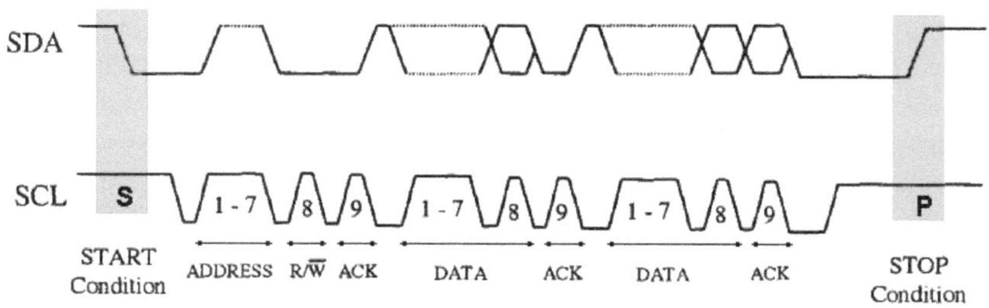

Fig. 28. Data transfer on the I²C bus

The examples of function ordering to connect a slave device through I²C bus to receive data are detailed below.

5.3.1 Function ordering for encoder information reading

The encoders are mounted at motor and their information is read by motor controller (MD25). The MD25 is designed to operate in a standard I²C bus system. To read encoder information, the function ordering can follow as below.

Step 1 : I2C_START ( ),
Step 2 : I2C_SEND ( ) - Call encoder address and write mode configuration,
Step 3 : I2C_ACK ( ),
Step 4 : I2C_SEND ( ) - Write the register number,
Step 5 : I2C_ACK ( ),
Step 6 : I2C_START ( ),
Step 7 : I2C_SEND ( ) - Call encoder address again,
Step 8 : I2C_ACK ( ),
Step 9 : I2C_RECEIVE ( ),
Step 10 : I2C_ACK ( ),
Step 11 : I2C_STOP ( ).

By using the feedback information from the two encoders on the left and right wheels of mobile robot, the position and heading angles of the mobile robot can be estimated. The distance and heading increment can be obtained as follows. (Hye Ri Park, et. Al., 2009; Surachai Panich, 2010a; Surachai Panich, 2010b; Surachai Panich and Nitin V. Afzulpurkar, 2011).

$$\Delta d_k^{encoder} = \frac{\Delta d_{R,k}^{encoder} + \Delta d_{L,k}^{encoder}}{2} \tag{1}$$

$$\Delta \psi_k^{encoder} = \frac{\Delta d_{R,k}^{encoder} - \Delta d_{L,k}^{encoder}}{B_k} \tag{2}$$

Then the position can be estimated as

$$X_{k+1} = X_k + \Delta d_{x,k}^{encoder} \tag{3}$$

$$Y_{k+1} = Y_k + \Delta d_{y,k}^{encoder} \tag{4}$$

$$\psi_{k+1} = \psi_k + \Delta \psi_k^{encoder} \tag{5}$$

where

$$\Delta d_{x,k}^{encoder} = \Delta d_k^{encoder} * \cos\psi_k$$

$$\Delta d_{y,k}^{encoder} = \Delta d_k^{encoder} * \sin\psi_k$$

The estimated position and heading can be detailed in software called PAMRO developed by Visual C^{++} as shown in Fig. 28.

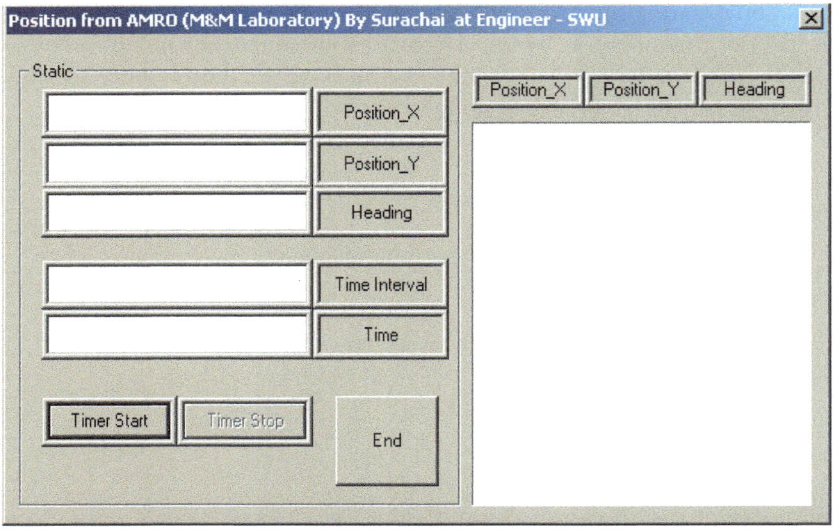

Fig. 28. The software PAMRO

5.3.2 Function ordering for compass information reading

The function ordering for compass module is detailed below.

Step 1 : I2C_START ( ),
Step 2 : I2C_SEND ( ) – Call compass address from robot and write mode configuration,
Step 3 : I2C_ACK ( ),
Step 4 : I2C_SEND ( ) – Write the register number,
Step 5 : I2C_ACK ( ),
Step 6 : I2C_START ( ),
Step 7 : I2C_SEND ( ) – Call compass address again,
Step 8 : I2C_ACK ( ),
Step 9 : I2C_RECEIVE ( ),
Step 10 : I2C_ACK ( ),
Step 11 : I2C_STOP ( ).

5.3.3 Function ordering for gyroscope information reading

To read gyroscope information, it needs device to convert analog to digital signal, because this gyroscope is not designed to work on I^2C bus system. The IC-PCF8591is selected supported the I^2C bus system to convert analog signal from gyroscope information.

Step 1 : I2C_START ( ),
Step 2 : I2C_SEND ( ) – Call slave device address from robot and write mode configuration,

Step 3 : I2C_ACK ( ),
Step 4 : I2C_SEND ( ) - Enable analog output, single mode and analog channel selection,
Step 5 : I2C_ACK ( ),
Step 6 : I2C_STOP ( ),
Step 7 : I2C_START ( ),
Step 8 : I2C_SEND ( ) - Read mode; start to read data from analog input of selected channel,
Step 9 : I2C_ACK ( ),
Step 10 : I2C_RECEIVE ( ),
Step 11 : I2C_ACK ( ),
Step 12 : I2C_STOP ( ).

5.3.4 Function ordering for temperature reading

To read temperature, it must be programmed in 3 steps; start to convert data, stop to convert data and read data.

Start to convert data:

Step 1 : I2C_START ( ),
Step 2 : I2C_SEND ( ) - Call slave device address from robot and write mode configuration,
Step 3 : I2C_ACK ( ),
Step 4 : I2C_SEND ( ) - Send command register, start to convert data,
Step 5 : I2C_ACK ( ),
Step 6 : I2C_STOP ( ),

Stop to convert data:

Step 1 : I2C_START ( ),
Step 2 : I2C_SEND ( ) - Call slave device address from robot and write mode configuration,
Step 3 : I2C_ACK ( ),
Step 4 : I2C_SEND ( ) - Send command register, stop to convert data,
Step 5 : I2C_ACK ( ),
Step 6 : I2C_STOP ( ),

Read data:

Step 1 : I2C_START ( ),
Step 2 : I2C_SEND ( ) - Call slave device address from robot and write mode configuration,
Step 3 : I2C_ACK ( ),
Step 4 : I2C_SEND ( ) - Send command register, stop to convert data,
Step 5 : I2C_ACK ( ),
Step 6 : I2C_START ( ),
Step 7 : I2C_SEND ( ) - Send command register, stop to convert data,
Step 8 : I2C_ACK ( ),
Step 9 : I2C_RECEIVE ( ) - Read MSB of temperature register,

Step 10 : I2C_ACK ( ),
Step 11 : I2C_RECEIVE ( ) - Read LSB of temperature register,
Step 12 : I2C_ACK ( ),
Step 13 : I2C_STOP ( ),

With these steps, the robot can get data from only one module for one time. If the robot wants to get data from this module once again or from another module, the robot must rerun once again from first to last step. The software is developed by Visual C++ to read information from I2C devices as shown in Fig.29.

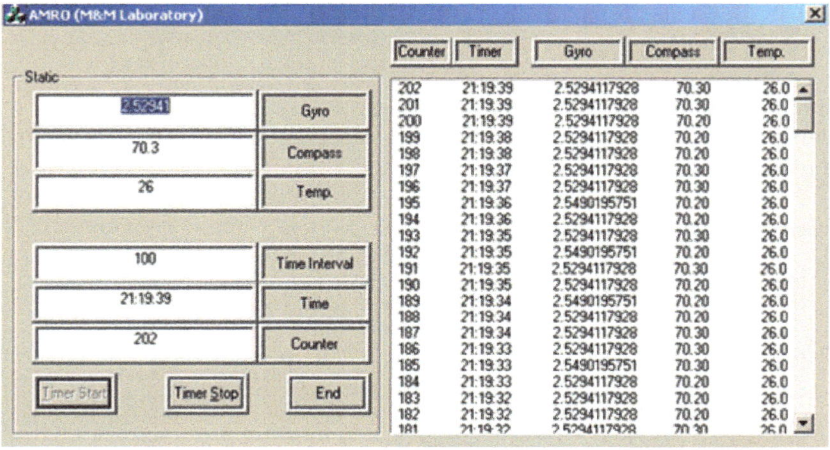

Fig. 29. Software to read data from sensors integrated on AMRO by I2C bus system

6. Conclusion

For this book chapter has mainly purpose to introduce basic structure of mobile robot and communication between mobile robot and sensors by using I2C bus system. The mobile robot named AMRO is introduced as example. Its hardware is constructed and combined with the electronics components. The real robot is tested successfully using manual controls (GUI) developed by using C++ programming. This small GUI is very useful for new user to test the various robot movements. The software is developed with manual controls for real robot. After establishing connection with the robot through wireless communication and entering all velocity values, the robot will be ready to move in the desired directions. By pressing ***F*** button robot starts to run forward with the given velocity. The I2C bus system is selected for main system of our mobile robot, because it can be conveniently developed and modified, if new sensors must be integrated or more analog sensors are used later. The I2C bus system consists of master and slave devices that the master device in this work is PC station and ADC-Converter for analog sensors (gyroscope), compass module and temperature sensor work as slave devices integrated on the mobile robot (AMRO). The I2C bus can work very well and has no problem with AMRO system. The software is developed to control the robot and to read all analog sensors based on the I2C bus format. The developed software is programmed based on Visual C++. The software devleopment must be programmed in the I2C bus format to control SDA and SCL lines.

7. Acknowledgement

This research from Measurement and Mobile Robot Laboratory (M & M LAB) was supported by Faculty of Engineering, Srinakharinwirot University under grant 180/2552.

8. References

Byoung-Suk Choi (2009). Mobile Robot Localization in Indoor Environment using RFID and Sonar Fusion System, *IEEE/RSJ International Conference on Intelligent Robots and Systems,* October 11-15, 2009 St. Louis, USA

Fernandez J., et al., 2007. Communication framework for sensor-actuator data in mobile robots, *ISIE 2007. IEEE International Symposium on Industrial Electronics, 2007,* Digital Object Identifier: 10.1109/ISIE.2007.4374825 Publication Year: 2007 , Page(s): 1502 - 1507

Hye Ri Park, et al., 2009. A Dead Reckoning Sensor System and a Tracking Algorithm for Mobile Robot. *ICROS-SICE International Joint Conference 2009,* August 18-21, Fukuoka International Congress Center, Japan.

Komoriya, K. and Oyama, E., 1994. Position estimation of a mobile robot using optical fiber gyroscope (OFG). *Proceedings of the Intelligent Robots and Systems, Advanced Robotic Systems and the Real World, IROS '94.* IEEE/RSJ/GI International Conference on 12-16 Sept. 1994.

Meng Yan, et al., 2007. SMARbot:_A_Miniature_Mobile_Robot_Paradigm_for_Ubiquitous Computing, *The 2007 International Conference on Intelligent Pervasive Computing, 2007 IPC,* Digital Object Identifier: 10.1109/IPC.2007.80 Publication Year: 2007, Page(s): 136 - 139

Panich, S. (2008). A mobile robot with an inter-integrated circuit system, *10th International Conference on Control, Automation, Robotics and Vision,* Publication Year: 2008, Page(s): 2010 - 2014, Digital Object Identifier: 10.1109/ICARCV.2008.4795839

Philips Semiconductor (2000), *I²C Bus Specification.* Version 2.1, 2000.

Surachai, P.; Thiraporn, T.; Chaiyaporn, L.; Kittichai, T.; Tayawat, S., 2009. Sensor fusion for differential encoder integrated with light intensity sensors and accelerometer. *2009 IEEE International Symposium on Computational Intelligence in Robotics and Automation (CIRA),* DOI: 10.1109/CIRA.2009.5423180, Page(s): 349 - 354

Surachai Panich (2010a). Mobile Robot Driven by Odometry System Integrated with Accelerometer. *Far East Journal of Electronics and Communications,* Volume 4, Issue 2 (June 2010) Page: 113 - 122

Surachai Panich (2010b). Dynamics, Control and Simulation of Robots with Differential Drive Integrated with Light Intensity Sensors. *Far East Journal of Dynamical Systems,* Volume 13, Issue 2 (July 2010) Page: 157 - 164

Surachai Panich (2010c). Mathematic model for mobile robot. *Far East Journal of Mathematical Sciences (FJMS),* Volume 46, Issue 1 (November 2010) Page: 23 - 32

Surachai Panich and Nitin V. Afzulpurkar (2011), Sensor Fusion Techniques in Navigation Application for Mobile Robot, in *Sensor Fusion - Foundation and Applications,* edited by Dr. Ciza Thomas, pp. 101-120, Published by InTech - Open Access Publisher, ISBN 978-953-307-446-7, Rijeka, Croatia.

Wikipedia (June 2011). Mobile robot, *Wikipedia; the free encyclopedia,* Available URL: http://en.wikipedia.org/wiki/Mobile_robot

Yoshiyuki Takahashi, et al., 1998. Development of the Mobile Robot System to aid the daily life for physically handicapped (Interface using internet browser), *The TIDE 98 Congress, Technology for Inclusive Design and Equality,* July, Helsinki

13

Epi.q Robots

Giuseppe Quaglia[1], Riccardo Oderio[1], Luca Bruzzone[2]
and Roberto Razzoli[2]
[1]*Politecnico di Torino*
[2]*University of Genova*
Italy

1. Introduction

Over the last few years there have been great developments and improvements in the mobile robotics field, oriented to replace human operators especially in dangerous tasks, such as mine-sweeping operations, rescuing after earthquakes or other catastrophic events, fire-fighting operations, working inside nuclear power stations and exploration of unknown environments.

Different locomotion systems have been developed to enable robots to move flexibly and reliably across various ground surfaces. Usually, mobile robots are wheeled, tracked and legged ones, even if there are also robots that swim, jump, slither and so on. *Wheeled robots* are robots that use wheels for moving; they can move fast with low energy consumption, have few degrees of freedom and are easy to control, but they cannot climb great obstacles (in comparison with robot dimensions) and can lose grip on uneven terrain. *Tracked robots* are robots that use tracks for moving; they are easily controllable, also on uneven terrain, but are slower than wheeled ones and have higher energy consumption. *Legged robots* are robots that use legs for moving; they possess great mobility and this makes them suitable for applications on uneven terrain; conversely, they are relatively slow, require much energy and their structure needs several actuators, with increased control complexity. Of course each robot class has advantages and drawbacks, thus scientists designed new robots, trying to comprise the advantages of different robot classes and, at the same time, to reduce the disadvantages: these robots are called *Hybrid robots*.

1.1 Background

Literature presents numerous interesting solutions for robots moving in structured and unstructured environments: some of them are here presented. The Spacecat, Whegs and MSRox can be considered smart reference prototypes for this work; the others are interesting solutions that, using different mechanisms, accomplish similar tasks.

Spacecat (Siegwart et al., 1998) is a smart rover developed at the École Polytechnique Fédérale de Lausanne (EPFL) by a team leaded by prof. Roland Siegwart, in collaboration with Mecanex S.A. and ESA. The locomotion concept is a hybrid approach called *Stepping triple wheels*, that shares features with both wheeled and legged locomotion. Two independently driven sets of three wheels are supported by two frames. The frames can rotate independently around the main body (payload frame) and allow the rover to actively lift one wheel to step climb the obstacle. Eight motors drive each wheel and frame independently. During climbing operation, the center of gravity of the rover is moved outside the contact surface formed by the four wheels. Thus the rover gets out of balance and falls with its upper wheel onto the obstacle; nevertheless no displacement of the center of gravity is required when the rover moves over a small rock; therefore, small object can be passed without any special control commands.

Whegs and Mini-Whegs (Allen et al., 2003; Quinn et al., 2003; Schroer et al., 2004) are hybrid mobile robots developed at the Center for Biologically Inspired Robotics Research at Case Western Reserve University, Cleveland, Ohio. The *Whegs* were designed using abstracted principles of cockroach locomotion. A cockroach has six legs, which support and move its body. It typically walks and runs in a tripod gait where the front and rear legs on one side of the body move in phase with the middle leg on the other side. The front legs swing head-high during normal walking so that many obstacles can be surmounted without significant gait changes. These robots are characterized by three-spoke locomotion units; they move faster than legged vehicles and climb higher barriers than wheeled ones of similar size. A single propulsion motor drives both front and rear axles and a servo actuated system controls the steering, similarly to automobile vehicle. With regard to Whegs locomotion: while the robot is walking on flat ground, three of the wheel-legs are 60° out of phase with the other three wheel-legs, which allows the robot to use an alternating tripod gait. This gait requires that the two front wheel-legs be out of phase with each other. When an obstacle is encountered, passive mechanical compliance allows the front legs to come back into phase with each other, so that they can both be used to pull the robot up and over the obstacle. After the robot has pulled itself over the obstacle, the front legs fall back into the previous pattern, thus the robot returns to an alternating tripod gait. *Whegs II*, the next generation of Whegs vehicles, incorporates a body flexion joint in addition to all of the mechanisms that were implemented in Whegs I. This actively controlled joint allows the robot to change its posture in a way similar to the cockroach, thus enabling it to climb even higher obstacles. The active body joint also allows the robot to reach its front legs down to contact the substrate during a climb and to avoid the instability of high-centering. Its aluminum frame and new leg design contributed in making Whegs II more robust than Whegs I. *Whegs VP* is a hybrid of the Whegs I and II vehicles. It is most similar in design to Whegs II, but lacks the body flexion joint. It combines the simplicity and agility of Whegs I with the durability and robustness of Whegs II. Improved legs and gait adaptation devices were implemented in its design. The *Mini-Whegs* are highly mobile, robust, and power-autonomous vehicles employ the same abstracted principles as Whegs, but on a scale more similar to the cockroach and using only four locomotion units. These robots, 90 mm long, can run at sustained speeds of over 10 body lengths per second and climb obstacles higher than the length of their legs. One version, called Jumping Mini-Whegs, has also a self-resetting jump mechanism that enables it to surmount obstacles as high as 220 mm, such as a stair.

MSRox (Dalvand & Moghadam, 2006) is an hybrid mobile robot developed by prof. Moghaddam and Dalvand at Tarbiat Modares University, Tehran, Iran. The MSRox employs an hybrid driving unit called *Star-Wheel*, designed for traversing stairs and obstacles. It is a three-legged wheel unit having three radially located wheels, mounted at the end of each spoke. Each Star-Wheel has two rotary axes: one for the rotation of the wheels, when MSRox moves on flat surfaces or passes over uphill, downhill, and slope surfaces; the other for the rotation of the Star-Wheel, when MSRox climbs or descends stairs and traverses obstacles. The four locomotion units are assembled on a central body. The robot can advance on ground, when only the wheel rotation is driven, or climb over an obstacle, when only the locomotion unit is driven. The presented version of MSRox has only two motors: one motor controls the rotation of the 12 wheels while the other controls the rotation of the Star-Wheels; the steering function is not implemented.

RHex (Saranli et al., 2001; 2004), developed first at the McGill University and University of Michigan and then at the Carnegie Mellon Robotics Institute, is characterized by compliant leg elements that provide dynamically adaptable legs and a mechanically self-stabilized gait. This hexapod robot, cockroach-inspired, uses a simple mechanical design with one actuator per leg and it is capable of doing a wide variety of tasks, such as walking, running, leaping over obstacles and climbing stairs.

Hylos (Grand et al., 2004), developed at the Université Pierre et Marie Curie, is characterized by a wheel-legged locomotion unit. Legs and wheels are independently actuated, therefore it uses wheels for propulsion and internal articulation to adapt its posture. It is a lightweight mini-robot with 16 actively actuated degrees of freedom.

VIPeR (Galileo Mobility Instruments & Elbit Systems Ltd, 2009), codeveloped by Elbit System and Galileo Mobility Instruments, is characterized by the *Galileo Wheel*, a patented system developed by Galileo Mobility Instruments ltd. The Galileo Wheel combines wheel and track in a single component, switching back and forth between the two modes within seconds. This technology enables the device to use wheels whenever possible, and tracks whenever needed.

Lego Mindstorm Artic Snow Cat (Lego Mindstorm, 2007) is characterized by four sets of triangular tracked treads that can rotate in two ways. In standard drive the treads move like a tank. When the going gets tough it can turn all four treads on the center axis, or to go through deep water it can run on the ends of its triangular treads for extra lift.

Packbot (iRobot, 2010; Mourikis et al., 2007), developed by iRobot, is a tracked vehicle with *flippers*. The flippers enable the robot to climb over obstacles, self right itself and climb stairs, enhancing ability over a simple tracked robot.

Scout II (Poulakakis et al., 2006; 2005) is characterized by a fast and stable quadrupedal locomotion. It consists of a rigid body with four compliant rigid prismatic legs. One single actuator per leg, located at the hip, allows active rotation of the leg. Each leg assembly consists of a lower and an upper part, connected via springs to form a compliant prismatic joint.

2. Mechanical architecture

Epi.q robots can be classified as hybrid robots, since their locomotion system shares features with both wheeled and legged robots. They are smart mini robots able to move in structured and unstructured environments, to climb over obstacles and to go up and down stairs. The robots do not need to actively sense obstacles for climbing them, they simply move forward and let their locomotion passively adapt to ground conditions and change accordingly without active control intervention: from rolling on wheels to stepping on rotating legs and vice-versa. Using wheels whenever possible and legs only when needed, their energy demand is really low in comparison with tracked and legged robots having similar obstacle crossing capability.

2.1 Chassis

Epi.q mechanical architecture consists of: a forecarriage, a central body and a rear axle, as shown in Figure 1. The forecarriage is composed of a frame linked to two driving units, that generate robot traction. The forecarriage frame houses motors and electronics, protecting them from dust and from potentially dangerous impacts against obstacles. The driving units are three-legged wheel units having attached thereto three wheels; they house the

transmission system and therefore they control robot locomotion. The rear axle comprises two idle wheel units, consisting of an idle three-legged wheel unit with three radially located idle wheels, mounted at the end of each spoke. The central body is a platform which connects forecarriage and rear axle, where a payload can be placed.
Two passive revolute joints, mutually perpendicular, link front and rear part of the robot,

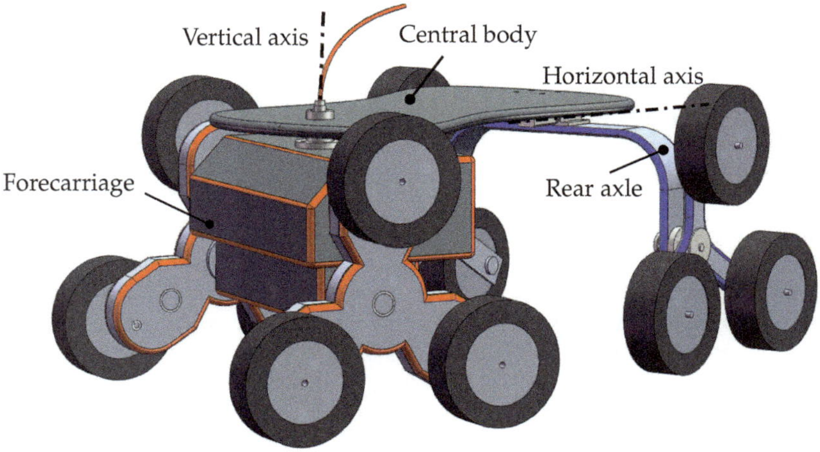

Fig. 1. Epi.q mechanical architecture

as shown in Figure 1. The vertical joint allows robot steering, while the horizontal joint guarantees a correct contact between wheels and ground, also in presence of uneven terrain.The angular excursion of the vertical and horizontal joints is limited by means of suitable mechanical stops.
Epi.q robots implement a differential steering, that provides both driving and steering functions. Differently choosing driving unit speeds, differently the instantaneous center of rotation is positioned along the common driving unit axis, so that an angle between front and rear part is generated by kinematic conditions and the robot can follow a specific path. Basically, a differential steering vehicle consists of two wheels mounted onto a device along the same axis, independently powered and controlled, and usually an idle caster wheel forms a tripod-like support structure for the body of the robot. In Epi.q robots the driven wheels are substituted by driving units and the Epi.q vertical joint accomplishes the same task of the caster wheel joint, as shown in Figure 2. If both the driving units are driven in the same direction and speed, the robot goes in a straight line. If one driving unit rotates faster than the other, the robot follows a curved path, turning inward toward the slower driving unit. If one of the driving units is stopped while the other continues to turn, the robot pivots around the stopped driving unit. If the driving units turn at equal speed but in opposite directions, both driving units traverse a circular path around a point centered half way between the two driving units, therefore the forecarriage pivots around the vertical axis.
For a classic differential steering robot, shown in Figure 2 on the left, when the velocities of the two driven wheels are chosen, the position of the instantaneous center of rotation is fixed too:

$$\frac{v_{fl}}{d+i/2} = \frac{v_{fr}}{d-i/2} \tag{1}$$

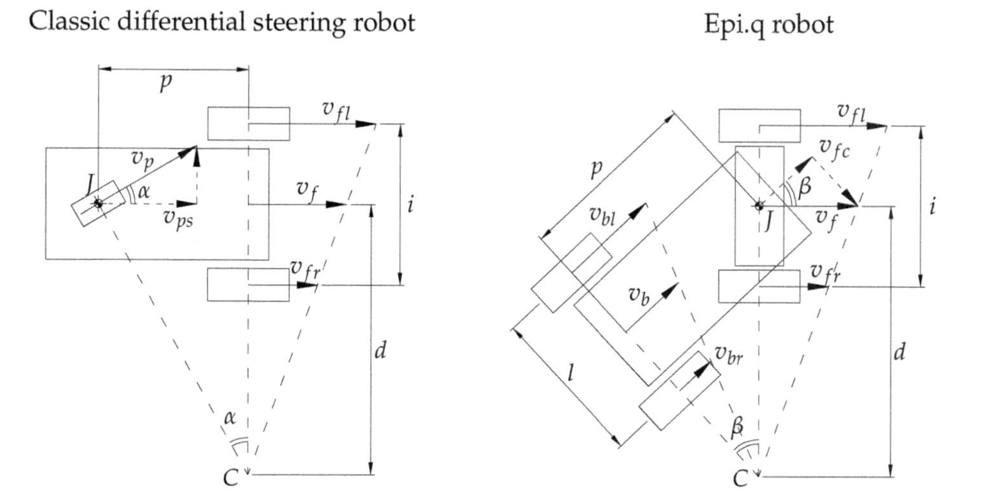

Fig. 2. Differential steering systems

$$d = \frac{v_{fl} + v_{fr}}{v_{fl} - v_{fr}} \cdot \frac{i}{2} \tag{2}$$

Consequently the velocity of a point centered half way between the two wheels is known:

$$v_f = \frac{v_{fl} + v_{fr}}{2} \tag{3}$$

and this velocity is equal to the component of the caster wheel velocity in the motion direction, otherwise there would be a deformation into robot body. During this operation the idle caster wheel is positioned by kinematic conditions and turns until it becomes orthogonal to the segment that links J and C; its velocity is a function of the driven wheel velocities:

$$v_p = \frac{v_{fl} + v_{fr}}{2\cos\alpha} \tag{4}$$

where

$$\alpha = \arctan\frac{p}{d} \tag{5}$$

For an Epi.q robot, shown in Figure 2 on the right, the mathematical treatment is quite similar. When the velocities of the two driving units are chosen, the position of the instantaneous center of rotation is fixed too:

$$\frac{v_{fl}}{d + i/2} = \frac{v_{fr}}{d - i/2} \tag{6}$$

$$d = \frac{v_{fl} + v_{fr}}{v_{fl} - v_{fr}} \cdot \frac{i}{2} \tag{7}$$

Therefore the velocity of a point centered half way between the two driving units is known:

$$v_f = \frac{v_{fl} + v_{fr}}{2} \tag{8}$$

and this point coincides with the vertical revolute joint. An angle between front and rear part of the robot is generated by kinematic conditions, that position the rear wheel unit axis in order to pass through the instantaneous center of rotation C. The component of the vertical joint velocity in rear axle direction is equal to the rear axle velocity, otherwise there would be a deformation of the robot central body:

$$v_b = \frac{v_{fl} + v_{fr}}{2} \cos \beta \tag{9}$$

where

$$\beta = \arcsin \frac{p}{d} \tag{10}$$

and consequently the velocity of the two rear idle wheel units are:

$$v_{bl} = \frac{v_{fl} + v_{fr}}{2} \cdot \frac{d \cos \beta + l/2}{d} \tag{11}$$

$$v_{br} = \frac{v_{fl} + v_{fr}}{2} \cdot \frac{d \cos \beta - l/2}{d} \tag{12}$$

2.2 Multi-leg wheel unit

A multi-leg wheel unit consists of a plurality of radially located spokes that end with a wheel. Both the forecarriage and the rear axle employ multi-leg wheel units.

A multi-leg wheel unit has a plurality of equally spaced wheels. If the number of wheels increases, the polygon defined by the wheel centers tends to become a circle and its side length decreases; thus the step overcoming capability is reduced but, on the other hand, the rotating leg motion is improved in terms of motion smoothness. Epi.q robots employ a three-legged wheel unit because it maximizes the step overcoming capability, for a given driving unit height, and the motion smoothness is guaranteed due to the fact that these robots use wheels whenever possible and legs only when needed.

Although a multi-leg wheel unit generates more friction than a single wheel during steering operations, this solution is advantageous: when the robot is moving on uneven terrain, actually its pitching is significantly reduced; when it is facing an obstacle, actually a multi-leg wheel unit can climb over higher obstacles and generally the velocity component in motion direction of the wheel unit presents smaller discontinuities.

When Epi.q robots are moving on rough ground their body vertical displacement is significantly decreased with respect to a robot that uses single wheels. Actually, as illustrated in Figure 3, if h_o is the height of an obstacle small enough to be contained between the wheels of a three-legged wheel unit, the height of the wheel unit axis can be expressed as:

$$h_a^{'} = l_l \sin 30^\circ + r_{du} \tag{13}$$

$$h_a^{''} = l_l \sin (30^\circ + \alpha) + r_{du} \tag{14}$$

The inclination α of the wheel unit can be related with the obstacle height:

$$2 l_l \cos 30^\circ \sin \alpha = h_o \tag{15}$$

therefore the vertical displacement Δh_{du} of a three-legged wheel unit follows from Equations 13, 14 and 15:

$$\begin{aligned} \Delta h_{du} = h_a{''} - h_a{'} &= l_l \sin 30^\circ \cos \alpha + l_l \cos 30^\circ \sin \alpha - l_l \sin 30^\circ = \\ &= \frac{h_o}{2} - \frac{l_l}{2} (1 - \cos \alpha) \end{aligned} \tag{16}$$

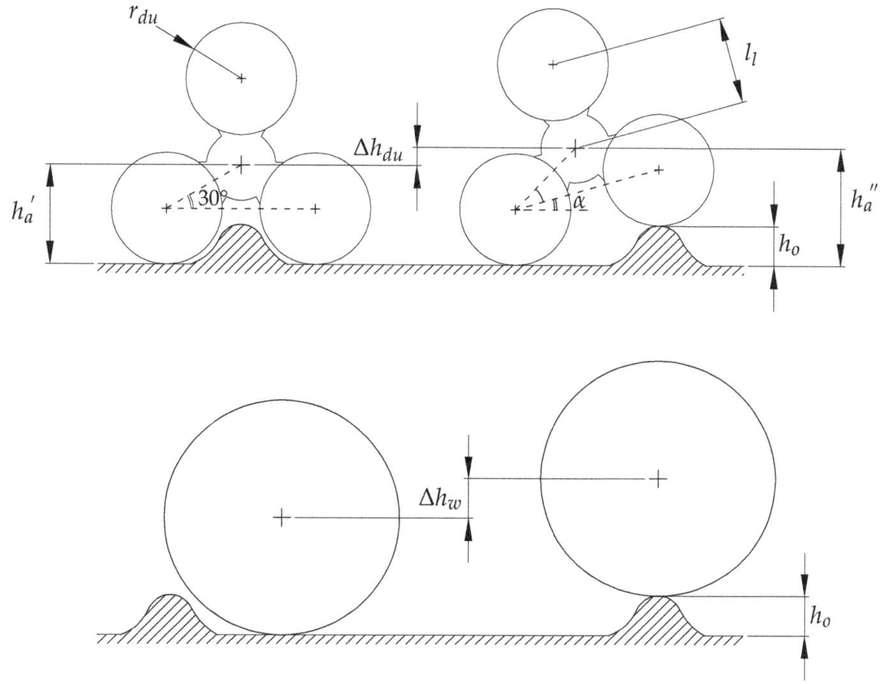

Fig. 3. Vertical displacement in presence of little unevenness, a comparative sketch between a three-legged wheel unit and a single wheel with same overall dimensions

that is always smaller or equal to half obstacle height:

$$\Delta h_{du} \leq \frac{h_o}{2} \tag{17}$$

while the vertical displacement of a single wheel, Δh_w, is always equal to obstacle height. Consequently, when the robot is moving on uneven terrain the pitching is significantly reduced with the use of a three-legged wheel unit instead of a single wheel.

As regards the ability of climbing an obstacle, a multi-leg wheel unit can climb over higher steps than a single wheel with the same overall dimensions actually, as shown in Figure 4, the maximum step that a single wheel can climb over measures a fraction of its radius while, for a multi-leg wheel unit, it is a fraction of its height: for example it was experimentally testes that the Epi.q-2 driving unit can climb over obstacles that measure till 84% of its height, see Section 4.

In case of steps that can be overcome both by a multi-leg wheel unit or by a single wheel, generally the velocity component in the motion direction presents smaller discontinuities with the multi-leg wheel unit than with a single wheel. Considering a three-legged wheel unit and a single wheel with same overall dimensions that are advancing at the same speed, shown in Figure 4, it is possible to identify a β angle:

$$\sin \beta = 1 - \frac{h_o}{r_w} \tag{18}$$

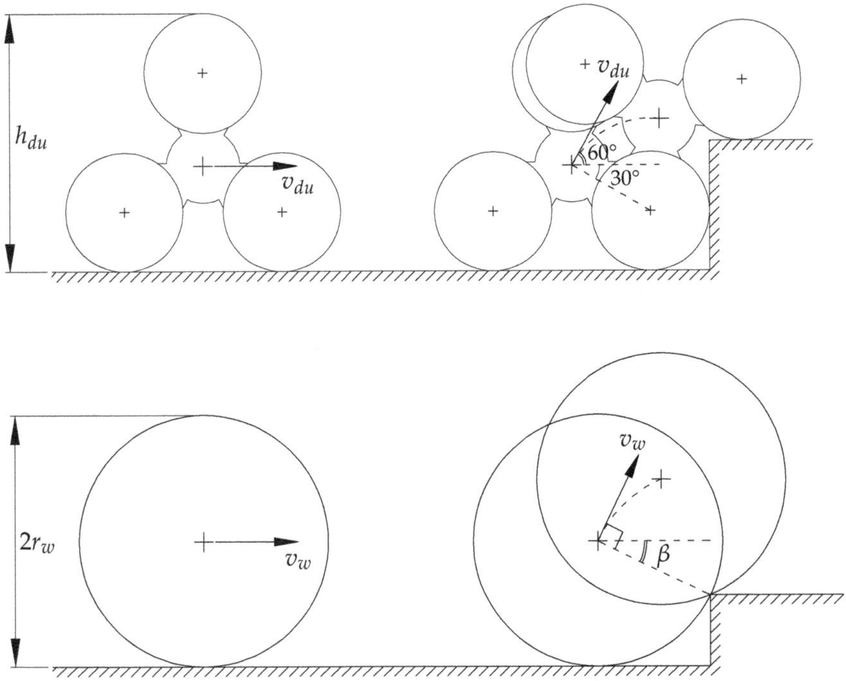

Fig. 4. Step climbing, a comparative sketch between a three-legged wheel unit and a single wheel, with same overall dimensions

The horizontal component of the wheel unit speed presents smaller discontinuities then a single wheel if the β angle is lower than 30°, that means when the obstacle is higher than a quarter of the wheel unit height or, equivalently, half wheel radius:

$$h_o > h_{du}/4 \tag{19}$$

$$h_o > r_w/2 \tag{20}$$

Moreover this discontinuity is also reduced on driving units by the fact that Epi.q robots have different velocities when they are moving on wheels or on legs, even if the gear-motors still continue to rotate at the same speed, as it will be explained in Section 3.

3. Driving unit

In this section a special focus on driving unit is discussed. The driving unit is a three-legged wheel unit having three radially located wheels, mounted at the end of each spoke. The driving units, housing the transmission system, control robot locomotion.

The driving unit concept takes place from the idea that a robot can passively modify its locomotion, from rolling on wheels to stepping on rotating legs, simply according to local friction and dynamic conditions. Actually, the driving unit is designed to have a limit torque that triggers different locomotions: if the torque required for moving on wheels exceeds the torque required for moving on legs, the robot will change its locomotion accordingly, from rolling on wheels to stepping on legs and vice versa. Thus only one motor per driving unit is required both for wheeled and legged locomotion.

3.1 Driving unit kinematic analysis

Considering the driving unit as a planar mechanism, angularly it has two degrees of freedom: the angular position of the driving unit frame and the angular position of the wheels. Actually,

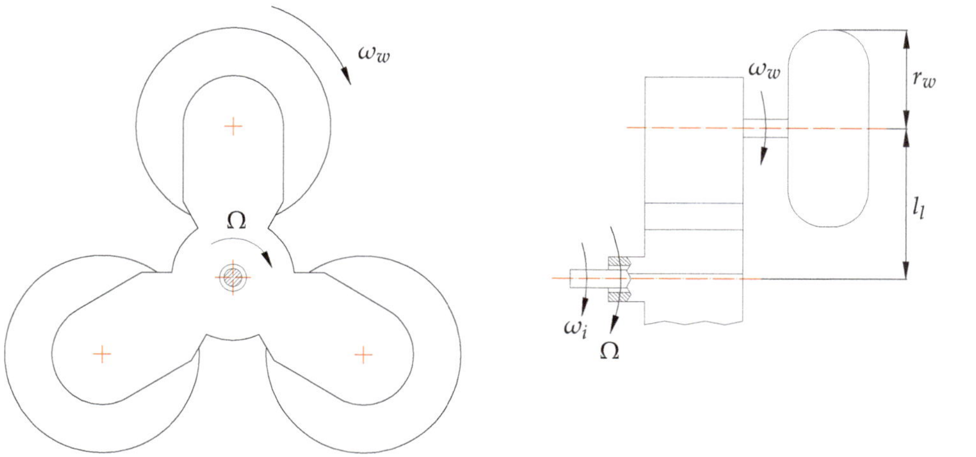

Fig. 5. Driving unit scheme

the transmission system links the input shaft angular velocity ω_i with both the angular velocity of the driving unit frame Ω, and the angular velocity of the wheels ω_w, that is the same for all the three wheels since the transmission system has the same gear ratio along each leg. Considering an observer placed on the driving unit frame, the transmission system is seen as an ordinary gearing, therefore the gear ratio (with sign) of the driving unit transmission system k_{ts} can be easily expressed as follows:

$$\frac{\omega_w - \Omega}{\omega_i - \Omega} = k_{ts} \tag{21}$$

and making ω_i explicit, it becomes:

$$\omega_i = \frac{1}{k_{ts}}\,\omega_w + \frac{k_{ts}-1}{k_{ts}}\,\Omega \tag{22}$$

When the robot is moving on wheels, *advancing mode*, the robot weight and the contact between wheels and ground constrain driving unit angular position.

If the robot is moving on a flat ground, the driving unit angular velocity is null:

$$\Omega = 0 \tag{23}$$

therefore Equations 21 and 23 lead to identify the velocity ratio i_{ad} and the driving unit linear velocity v_a, shown in Figure 6, as follows:

$$i_{ad} = \left.\frac{\omega_w}{\omega_i}\right|_{\Omega=0} = k_{ts} \tag{24}$$

$$v_{ad} = \omega_w \cdot r_w = \omega_i \cdot i_{ad} \cdot r_w = \omega_i \cdot k_{ts} \cdot r_w \tag{25}$$

where r_w is wheel radius.
When the robot bumps against an obstacle, if the local frictions between front wheel and obstacle are sufficient to stop the wheel, the driving unit starts to rotate around the stopped wheel center, allowing the robot to climb over the obstacle, *automatic climbing mode*. In this occurrence the wheel angular velocity is null:

$$\omega_w = 0 \tag{26}$$

and consequently, from Equations 21 and 26, the velocity ratio i_{ac} and the driving unit linear velocity v_{ac}, shown in Figure 6, are respectively:

Fig. 6. Driving unit linear velocity during advancing mode, on the left, and automatic climbing mode, on the right

$$i_{ac} = \left.\frac{\Omega}{\omega_i}\right|_{\omega_w=0} = \frac{k_{ts}}{k_{ts}-1} \tag{27}$$

$$v_{ac} = \Omega \cdot l_l = \omega_i \cdot i_{ac} \cdot l_l = \omega_i \cdot \frac{k_{ts}}{k_{ts}-1} \cdot l_l \tag{28}$$

where l_l is the length of the driving unit leg.
Finally, taking into account Equations 24 and 27, it is possible to rewrite Equation 22 as follows:

$$\omega_i = \frac{\omega_w}{i_{ad}} + \frac{\Omega}{i_{ac}} \tag{29}$$

3.2 Driving unit design

During the design phase it is important to establish the correct driving unit parameters, for this reason some preliminary reflections can be helpful.
The locomotion transition between wheeled and legged motion is only triggered by driving unit torque demand therefore, since driving unit motors must rotate in the same direction both for advancing mode and for automatic climbing mode, the driving unit will work properly only if the velocity ratios i_a and i_{ac} have the same sign. Equations 24 and 27 lead to identify a low limit value for the driving unit gear ratio:

$$k_{ts} > 1 \tag{30}$$

Consequently, a suitable transmission system has a gear ratio k_{ts} bigger than one and positive: if, for example, the chosen transmission system is only made of external toothed gears, this condition will lead to choose an odd number of gears with appropriate gear radii.
A second consideration regards robot motion continuity during the locomotion transition.

The ratio between driving unit linear velocity during advancing mode and automatic climbing mode, considering only the component parallel to the ground, can be identified by a coefficient β that, from Equations 25 and 28, can be expressed as:

$$\beta = \frac{v_{ac} \cdot \cos(60^\circ)}{v_{ad}} = \frac{1}{2(k_{ts} - 1)} \cdot \frac{l_l}{r_w} \tag{31}$$

Therefore the β coefficient contains information regard motion continuity: if this value is close to the unit value, motion continuity will be preserved.
A third consideration takes into account driving unit application. Considering driving units with similar overall dimensions, different capabilities can be obtained varying the r_w/l_l parameter, as shown in Figure 7: if the r_w/l_l value decreases, the robot will be more oriented towards legged locomotion and it will be able to climb over higher obstacles, otherwise the robot will be more oriented towards wheeled locomotion, with wheels that will better protect driving unit from shocks caused by the contact with obstacles. The highest limit value for the

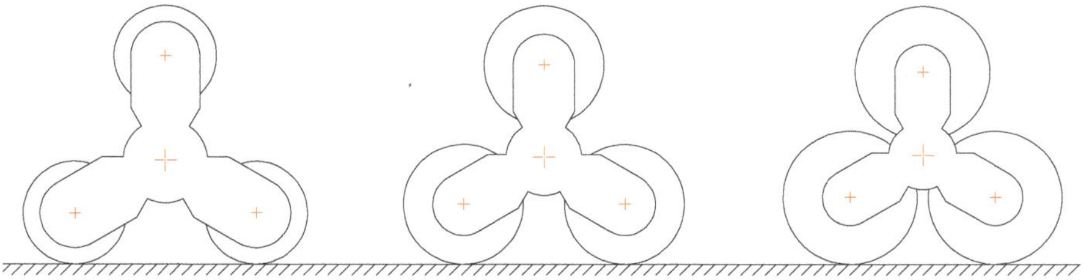

Fig. 7. Driving units with different r_w/l_l ratios, increasing value from left to right

r_w/l_l parameter corresponds to the condition in which driving unit wheels are in interference limit conditions:

$$2r_w = 2l_l \cdot \cos(30^\circ) \tag{32}$$

therefore the r_w/l_l value must be chosen accordingly to robot application and always smaller than:

$$\frac{r_w}{l_l} < \frac{\sqrt{3}}{2} \tag{33}$$

Once robot specifications are fixed and consequently the r_w/l_l and β parameters are chosen, Equation 31 identifies a first attempt value for the driving unit gear ratio:

$$k_{ts} = \frac{1}{2\beta} \cdot \frac{l_l}{r_w} + 1 \tag{34}$$

At this step it still remains to verify the predicted transition conditions between advancing mode and automatic climbing mode; if this conditions were not satisfactory, it would be necessary to relax some robot specifications.
When the gear ratio k_{ts} and the driving unit kinematic chain, as well, are chosen, lots of possible combination of mechanical components still remain to be identified: a further suggestion would be to choose the gearing that better reduce risk of interferences between driving unit frame and obstacles.
Finally, it is necessary to identify a scale factor, that will depend on the robot application field, thus the driving unit geometry is completely identified.

3.3 Epi.q-1 driving unit

The Epi.q-1 driving unit, as shown in Figures 8, is mainly composed of: an input ring gear (1) (directly linked to a gear-motor), a planet carrier (2) (rotationally free respect to robot chassis by means of bearing), three planet gears (3), a solar gear (4), a sliding solar gear (5), three leg planet gears (6), three legs (7), three planet pulleys (8), three belts (9), three wheel pulleys (10), three wheels (11), an axial device (12) linked to a mini-motor (13). The planet pulleys (8) are always rigidly connected with the planet gears (3), the wheel pulleys (10) with the wheels (11) and the leg planet gears (6) with the legs (7). An axial device controls the axial position of the

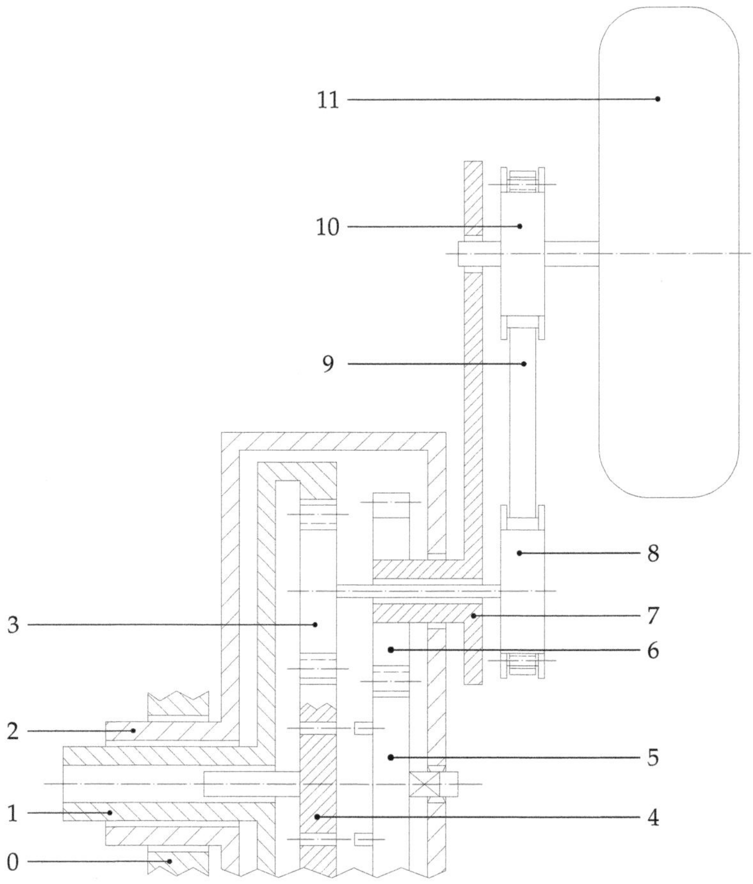

Fig. 8. Functional scheme of the Epi.q-1 driving unit, only one arm is represented

sliding solar gear (5) in order to alternatively link the leg planet gear (6) with the planet carrier (2) or with the solar gear (4), as shown in Figure 9

The operative conditions of the transmission system can be described using some kinematic equations: some of them are always valid for every operative conditions, three represent the

Nomenclature	Radius	Rot. speed	Label
Chassis			0
Input ring gear	r_r	ω_r	1
Planet carrier		Ω	2
Planet gears	r_p	ω_p	3
Solar gear	r_s	ω_s	4
Sliding solar gear	r_{ss}	ω_{ss}	5
Leg planet gears	r_{lp}	ω_l	6
Legs		ω_l	7
Planet pulleys	r_{pp}	ω_p	8
Toothed belts			9
Wheel pulleys	r_{wp}	ω_w	10
Wheels	r_w	ω_w	11

Table 1. Nomenclature of Epi.q-1 driving unit

meshing conditions in the epicyclic gearing:

$$\frac{\omega_p - \Omega}{\omega_r - \Omega} = +\frac{r_r}{r_p} = k_{e1} \tag{35}$$

$$\frac{\omega_s - \Omega}{\omega_p - \Omega} = -\frac{r_p}{r_s} = k_{e2} \tag{36}$$

$$\frac{\omega_l - \Omega}{\omega_{ss} - \Omega} = -\frac{r_{ss}}{r_{lp}} = k_{el} \tag{37}$$

while a fourth one describes the belt transmission system:

$$\frac{\omega_w - \omega_l}{\omega_p - \omega_l} = +\frac{r_{pp}}{r_{wp}} = k_b \tag{38}$$

Moreover the gear ratii k_{e1} and k_{e2} are linked by geometrical constraints:

$$r_r = r_s + 2r_p$$

$$\frac{r_r}{r_p} = \frac{r_s}{r_p} + 2$$

therefore:

$$k_{e2} = \frac{1}{2 - k_{e1}} \tag{39}$$

Other equations, describing physical constraints introduced by the sliding solar gear meshing conditions and by robot-terrain contact, are univocally determined by robot operative conditions; these equations will be introduced in the following.

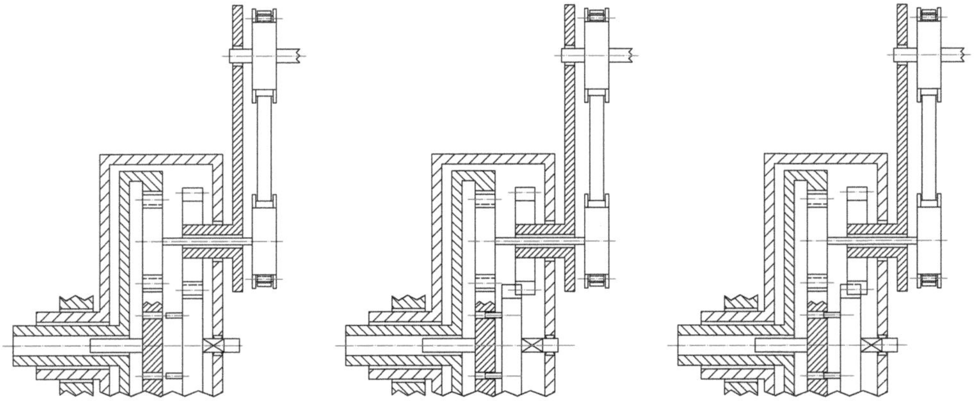

Fig. 9. Driving unit configurations for different operative modes: advancing and automatic climbing modes (on the left); changing configuration mode (in the middle); rotating leg mode (on the right)

3.3.1 Advancing & automatic climbing modes

During advancing mode the robot weight and the contact between bottom wheels and ground constrain driving unit angular position. When the robot bumps against an obstacle, if the local frictions between front wheel and obstacle are sufficient to stop the wheel, the driving unit starts to rotate around that wheel, automatic climbing mode, allowing the robot to climb over the obstacle. The robot passively changes its locomotion simply according to the torque required.

In both advancing mode and automatic climbing mode, the sliding solar gear (5) is prismatically coupled with the planet carrier (2), so that a relative rotation between them is hindered, as shown in Figure 9 on the left. The sliding solar gear (5) is always meshed with the leg planet gears (6), in order to prevent a relative rotation between legs (7) and planet carrier (2), this way legs (7) are locked to the planet carrier (2) in a prefixed position.

Advancing mode

In advancing mode the robot is moving with its legs rigidly connected to the planet carrier, actually the sliding solar gear is coupled with the planet carrier:

$$\omega_{ss} = \Omega \tag{40}$$

and it meshes with the leg planet gears, therefore:

$$\omega_l = \Omega \tag{41}$$

The planet carrier is free to rotate around its axis, but the driving unit balance and the contact between wheels and ground constrain its angular position. In the hypothesis of locomotion on flat ground, the contact between wheels and ground hinders planet carrier rotation:

$$\Omega = 0 \tag{42}$$

Equations from 35 to 42 lead to identify the advancing velocity ratio i_a and the robot speed in advancing mode v_a as a function of the gear-motor rotation speed:

$$i_{ad} = \left.\frac{\omega_w}{\omega_m}\right|_{\Omega=0} = k_{e1} \cdot k_b \tag{43}$$

$$v_a = \omega_w \cdot r_w = \omega_r \cdot i_{ad} \cdot r_w = \omega_m \cdot k_{e1} \cdot k_b \cdot r_w \tag{44}$$

where ω_m is the angular velocity of the gear-motor output shaft, directly linked to the input ring gear.

Comparing Equations 43 and 44 with the generic Equations 24 and 25, a relationship between gear ratios can be established:

$$k_{ts} = k_{e1} \cdot k_b \tag{45}$$

Summary: during the advancing mode the sliding solar gear is coupled with the planet carrier; the motion is transferred from the gear-motor to the wheels; the planet carrier is free to swing, reducing this way robot pitching.

Automatic climbing mode

In automatic climbing mode, since the sliding solar gear is still rigidly connected with the planet carrier, the robot is still moving with its legs rigidly connected to the planet carrier, as previously described for the advancing mode, therefore Equations 40 and 41 are still available:

$$\omega_{ss} = \Omega$$
$$\omega_l = \Omega$$

Certainly, also the planet carrier is still free to rotate around its axis but, in automatic climbing mode, the local friction between front wheel and obstacle is sufficient to block the wheel in contact with the obstacle, therefore:

$$\omega_w = 0 \tag{46}$$

Equations from 35 to 41, together with 46 allow to evaluate the planet carrier angular speed:

$$\Omega = \frac{k_{e1} \cdot k_b}{k_{e1} \cdot k_b - 1} \cdot \omega_m \tag{47}$$

thus the automatic climbing velocity ratio i_{ac} and the driving unit velocity in automatic climbing mode v_{ac} can be expressed as follows:

$$i_{ac} = \left.\frac{\Omega}{\omega_m}\right|_{\omega_w=0} = \frac{k_{e1} \cdot k_b}{k_{e1} \cdot k_b - 1} \tag{48}$$

$$v_{ac} = \Omega \cdot l_l = \omega_m \cdot \frac{k_{e1} \cdot k_b}{k_{e1} \cdot k_b - 1} \cdot l_l \tag{49}$$

that compared with Equations 27 and 28 lead to:

$$k_{ts} = k_{e1} \cdot k_b \tag{50}$$

as it was expected, confirming Equation 45.

Summary: during the automatic climbing mode the sliding solar gear is coupled with the planet carrier; the front wheel in contact with the obstacle is stopped by the friction forces due to the contact between wheel and obstacle themselves; the whole driving unit rotates around the stopped wheel, traversing the step.

3.3.2 Changing configuration mode

Driving unit can also modify its geometry from a closed configuration to an open one, as shown in Figure 10. The closed configuration is suitable to reach restricted spaces while the open one to get over obstacles.

In *changing configuration mode* the sliding solar gear (5) is no longer engaged in the planet carrier (2) whereas it is shifted towards the solar gear (4) and coupled with it, as shown Figure 9 in the middle, therefore:

$$\omega_{ss} = \omega_s \tag{51}$$

In order to avoid gearing lability this operation is split into two steps: first the sliding solar gear (5) couples with the solar gear (4), then the sliding solar gear (5) is rotationally disconnected from the planet carrier (2). A slow solar gear rotation and some elastic elements allow a correct axial engagement between the sliding solar gear (5) and solar gear (4) fittings. During the changing configuration mode the planet carrier angular speed is null:

$$\Omega = 0 \tag{52}$$

Therefore Equations from 35 to 38, together with 51 and 52, bring to evaluate the changing configuration velocity ratio as:

$$i_{cc} = \frac{\omega_l}{\omega_m} = k_{e1} \cdot k_{e2} \cdot k_{el} = \frac{k_{e1}\, k_b}{2 - k_{e1}} \tag{53}$$

Legs angular excursion is limited by suitable mechanical stops, placed on the planet carrier. A relative slippage between wheels and ground can happen during the changing configuration operations.

Summary: during the changing configuration mode the sliding solar gear is shifted and coupled with the solar gear; the planet carrier is blocked; the motion is transferred from the gear-motor to the legs, that rotate.

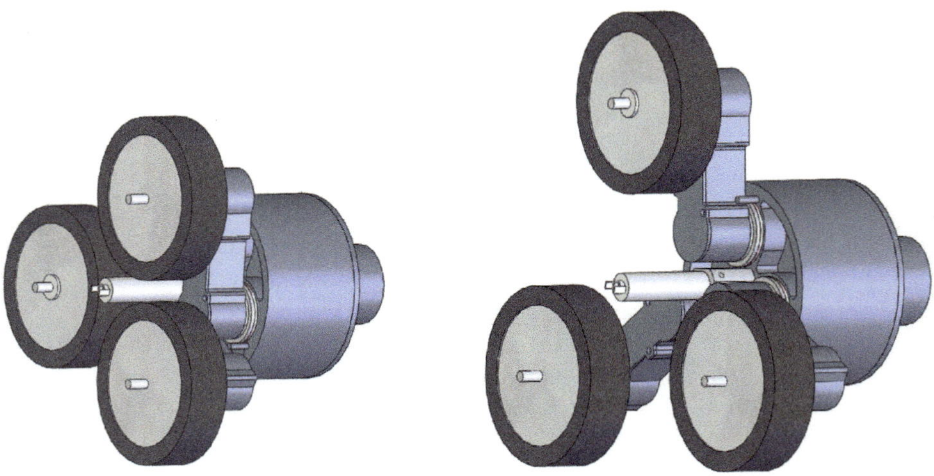

Fig. 10. Driving unit in closed and in open configuration

3.3.3 Rotating leg mode

When the robot cannot climb an obstacle in automatic climbing mode, due for example to a low friction coefficient between wheel and obstacle, it is possible to transform the driving unit into a whole rigid body which rotates around its axis, *rotating leg mode.*

Figure 9 on the right shows gearing configuration for this operative mode: the sliding solar gear (5) is shifted into an intermediate position in order to be simultaneously engaged with both the solar gear (4) and the planet carrier (2):

$$\omega_{ss} = \omega_s = \omega_l = \Omega \tag{54}$$

Equations from 35 to 38 together with 54 allow to evaluate the driving unit angular speed as follows:

$$\Omega = \omega_m \tag{55}$$

Actually, the driving unit acts as a rigid body and the motor rotation provides the rotation of the whole driving unit, as it was a single rotating body.

Wheels wear is a possible drawback of this locomotion when slipping conditions between wheel and obstacle occur.

Summary: during the rotating leg mode the sliding solar gear (5) is shifted into an intermediate position; the whole driving unit rotates as a rigid body around its central axis.

3.4 Epi.q-2 driving unit

The driving unit implemented in Epi.q-2 prototype is shown in Figure 11. With respect to Epi.q-1 version, the changing configuration ability has been removed in order to simplify the structure, to increase gearing robustness and efficiency, and to reduce overall weight.

The driving unit, as shown in Figure 11, consists of: an input solar gear (1), a planet carrier (2)

Nomenclature	Radius	Rot. speed	Label
Frame			0
Input Solar gear	r_s	ω_s	1
Planet carrier	l_l	Ω	2
First planet gear	r_{fp}	ω_{fp}	3
Second planet gear	r_{sp}	ω_w	4
Wheel	r_w	ω_w	5

Table 2. Nomenclature of Epi.q-2 driving unit

(rotational free respect to robot frame by means of bearings), three first planet gears (3), three second planet gears (4) and three wheels (5).

The transmission system can be described using some kinematic equations; the meshing conditions in the epicyclic gearing can be represented as follows:

$$\frac{\omega_{fp} - \Omega}{\omega_s - \Omega} = -\frac{r_s}{r_{fp}} = k_{e1} \tag{56}$$

$$\frac{\omega_w - \Omega}{\omega_{fp} - \Omega} = -\frac{r_{fp}}{r_{sp}} = k_{e2} \tag{57}$$

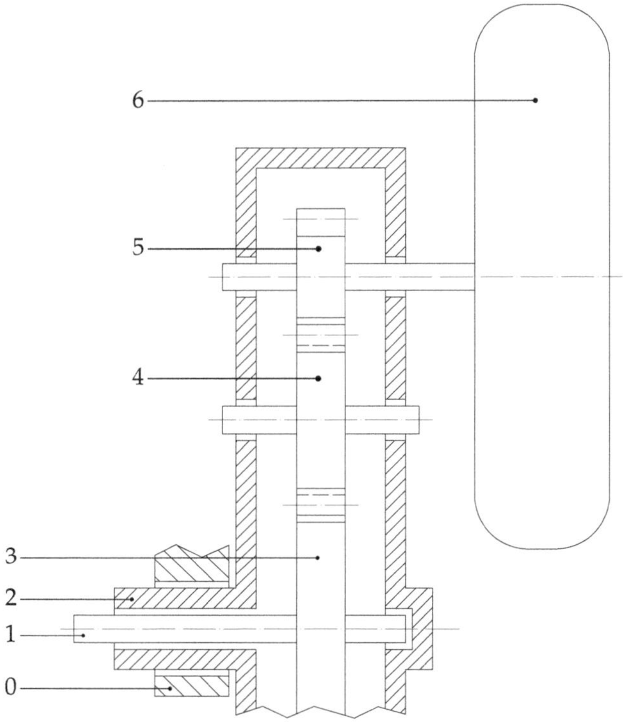

Fig. 11. Epi.q-2 driving unit

During advancing mode the robot weight and the contact between bottom wheels and ground constrain driving unit angular position.

$$\Omega = 0 \tag{58}$$

In this occurrence the velocity ratio i_a and the robot linear velocity v_a, from Equations 56, 57 and 58, are respectively:

$$i_{ad} = \left.\frac{\omega_w}{\omega_s}\right| \Omega = 0 = k_{e1} \cdot k_{e2} \tag{59}$$

$$v_{ad} = \omega_w \cdot r_w = \omega_s \cdot i_{ad} \cdot r_w = \omega_m \cdot k_r \cdot k_{ts} \cdot r_w \tag{60}$$

where k_r is the reducer gearing ratio, located between driving unit input shaft and motor, and ω_m is the angular velocity of the motor output shaft.
Comparing the Equations 59 and 60 with the generic Equations 24 and 25, it is possible to establish a relationship between gear ratii:

$$k_{ts} = k_{e1} \cdot k_{e2} \tag{61}$$

When the robot bumps against an obstacle, if the local frictions between front wheel and obstacle are sufficient to stop that wheel:

$$\omega_w = 0 \tag{62}$$

then the second planet gear rotation is hindered and consequently the driving unit rotation starts around the stopped wheel, allowing the robot to climb over the obstacle.
In this occurrence the gear ratio i_{ac} and the robot linear velocity v_{ac} are respectively:

$$i_{ac} = \frac{\Omega}{\omega_s}\bigg| \omega_w = 0 = \frac{k_{e1} \cdot k_{e2}}{k_{e1} \cdot k_{e2} - 1} \tag{63}$$

$$v_{ac} = \Omega \cdot l_l = \omega_s \cdot i_{ac} \cdot l_l = \omega_s \cdot \frac{k_{e1} \cdot k_{e2}}{k_{e1} \cdot k_{e2} - 1} \cdot l_l \tag{64}$$

that compared with Equations 27 and 28 lead to:

$$k_{ts} = k_{e1} \cdot k_{e2} \tag{65}$$

as it was expected, confirming Equation 61.

4. Tests

Epi.q robots move well on different terrains: from a structured environment, with flat surface and steps, to an unstructured one, with uneven ground and obstacles different in dimension and shape.
When the robot is moving on wheels in advancing mode, the robot weight and the contact between the bottom wheels and the ground constrains driving unit angular position, as shown in Figure 12; actually the driving units are axially joined to the forecarriage frame but

Fig. 12. Epi.q-1 in advancing mode, on uneven ground

rotationally free by means of bearings. When the robot bumps against an obstacle, if the local frictions between front wheel and obstacle are sufficient to stop that wheel, the driving unit starts to rotate around the stopped wheel center, allowing the robot to climb over the obstacle in automatic climbing mode, shown in Figure 13. The transition between wheeled and legged locomotion is passively triggered: if the torque required for moving on wheels is higher than

the torque required for moving on legs, the robot would change its locomotion from rolling on wheels to stepping on legs, and vice versa.
Experimental tests on Epi.q prototypes were conducted in order to assess their performance; tests and results are reported in the following sections.

Fig. 13. Epi.q-2 negotiating an obstacle

4.1 Step negotiating aptitude

The purpose of the test is to assess the ability of Epi.q robots to negotiate obstacles which are different in height. The robots were driven close to a step, on a flat surface. Analyzing the experimental tests, it was noticed that three different cases can occur when the rotation of the driving unit starts, as shown in Figure 14: the top wheel leans against the upper surface of the step; the top wheel leans against the front surface of the step; or an intermediate case between the two. In the first case it is always possible to overcome the step, even without

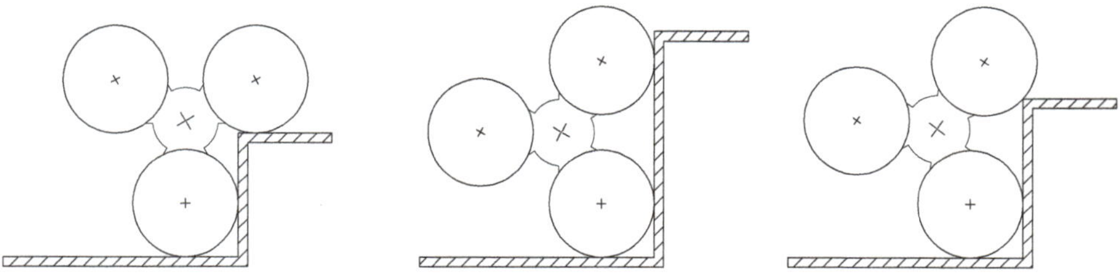

Fig. 14. Driving unit approaches three steps, different in height

initial velocity. Actually, when the rotation of the driving unit starts, the top wheel is above the step and the upper leg can lift up the robot, obviously only if the top wheel does not skid. In the second case it is never possible to overcome the step, because the rear wheel units are idle. In an intermediate case between the previous ones the possibility of overcoming a step depends on the static friction coefficient between wheels and step, on the tread pattern of the tire and on the approach speed.

4.1.1 Epi.q-1 test results

The experimental tests conducted on Epi.q-1 prototype have shown that the maximum step it can negotiate, in favourable friction conditions, is 90 mm height, equivalent to 72% of the

driving unit height.
When the driving unit was in closed configuration, Epi.q-1 crossed almost all obstacles without interference between driving unit and obstacles; actually, the driving unit was protected by the wheels.When the driving unit was in open configuration, sometimes a collision between driving unit and obstacles occurred. In this case the robot overcame the obstacles with a slightly irregular motion, combining the advancing mode and the automatic climbing mode. A step negotiating sequence is represented in Figure 15.

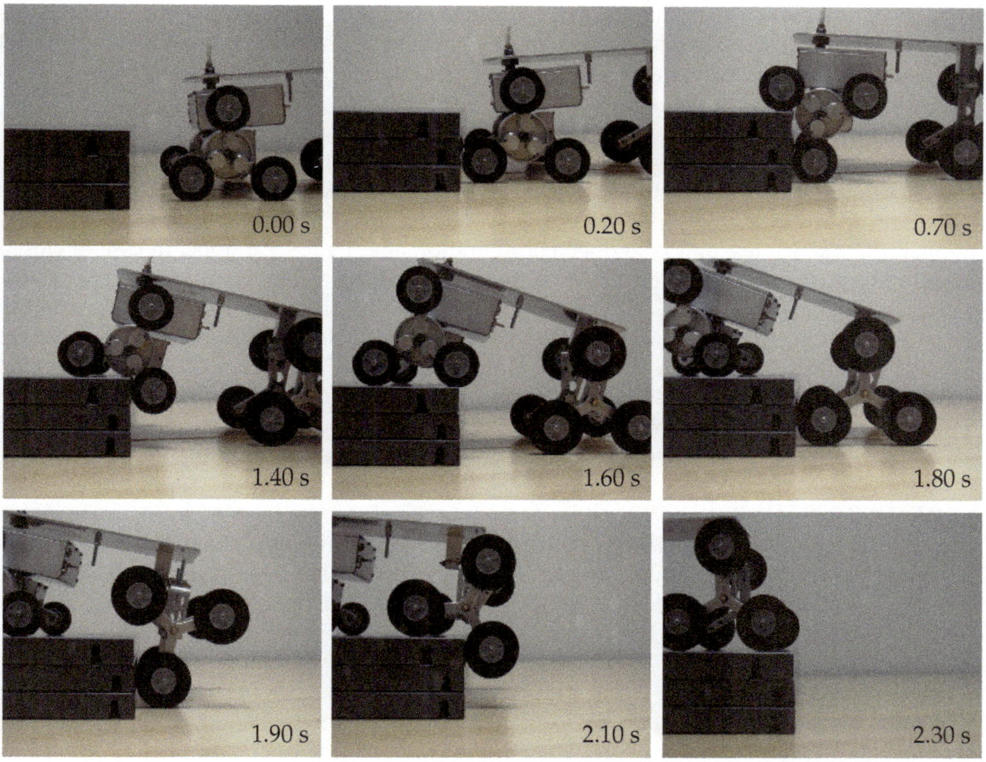

Fig. 15. Epi.q-1 negotiating a step (time stamps have been roughly estimated from a video)

4.1.2 Epi.q-2 test results

The experimental tests conducted on Epi.q-2 prototype have shown that the maximum step Epi.q-2 can climb over, in favourable friction conditions, is 110 mm height, equal to 84% of the driving unit height.
Epi.q-2 was also designed with the aim to reduce the risk of interference with obstacles; even if sometimes this condition can occur. The tests demonstrated that the robot overcomes the obstacle with a slightly irregular motion which combines the advancing mode and the automatic climbing mode.

4.2 Motion on inclined surface

The aim of the test is to assess Epi.q robot ability of moving on inclined surfaces. The robots were driven up a ramp and their behavior was observed.

The robots can drive on a slope either in advancing or in automatic climbing modes, actually there is a limit slope value that triggers the transition between wheeled and legged locomotion. Moreover there is a maximum slope value, that the robot can not overcome. This slope value is limited by motor power, by the disposition of the center of mass of the robot and by the friction coefficient between wheels and ground.

The tests highlighted a great influence of the center of mass disposition. Actually Epi.q robots does not have an active stability system and the traction is provided only by the front wheel units, which are also less loaded going uphill. Therefore, the driving units can loose traction on a slope and the wheels can start skidding.

4.2.1 Epi.q-1 test results

The experimental tests have shown that when Epi.q-1 is moving on an inclined surface the transition between advancing mode and automatic climbing mode is triggered by a 13% slope, if the the driving unit is in open configuration, and by a 9.5% slope, in case of closed configuration. The robot was tested on different surfaces with increasing friction coefficient: plexiglas, paper and plywood. When the robot is moving on plexiglas surface it can reach a slope of 29% in automatic climbing mode, if the slope is steeper the robot starts skidding while it continues to advance up to a maximum slope of 32%. When the robot moves on paper surface, the maximum slope it can reach without skidding is 40% in automatic climbing mode, but it can advance up to 43% slope. Moving on plywood the wheels never skid and the maximum slope it can reach is 45%, dur to motor torque limitations.

4.2.2 Epi.q-2 test results

The experimental tests have shown that the Epi.q-2 locomotion transition is triggered by a 31% slope. In the experimental campaign Epi.q-2 prototypes has been tested on slopes up to 33%, with friction coefficient $\mu_s = 0.83$. Obviously, when the maximum slope is limited by the friction coefficient, the traction wheels can start skidding without reaching the transition condition.

4.3 Motion on uneven and soft terrains

The purpose of the test is to assess the Epi.q robot ability of moving on different terrains.

The experimental tests have shown that, when the rotation of the wheels is hindered by the high rolling friction due to the grass or to the ground unevenness, the transition between advancing mode and automatic climbing mode occurs and the robot starts the legged locomotion, as expected.

The Epi.q robots were tested on different scenarios: on uneven terrains with grass, stones, pebbles, earth and irregular trails. In all cases they were able to advance with a motion that changed between advancing mode and automatic climbing mode: the percentage of automatic climbing mode became higher accordingly to the terrain unevenness.

4.4 Energy demand

The purpose of the test is to evaluate the energy demand of Epi.q robots. Actually, using wheels whenever possible and legs only when needed, They should require a small amount of energy.

The Epi.q-2 prototype was tested on a smooth terrain, the current demand and the speed were evaluated; the results are collected in Table 3.

Current	Speed
0.4 A	0.2 m s^{-1}
0.5 A	0.5 m s^{-1}
0.6 A	0.75 m s^{-1}

Table 3. Epi.q-2 current demand, moving on smooth terrain at different speeds

The Epi.q-2 power source, both for motor and electronics, was a removable 11 V/2200 mA h lithium-ion battery, providing more than 4 hours continuous runtime on one charge, up to 6 km of travel.

5. Technical specifications

5.1 Epi.q-1 technical specifications

Epi.q-1 weighs almost 2.6 kg and measures 160 mm × 360 mm × 280 mm (height × length × width), with a driving unit that measures 125 mm in height in open configuration and 98 mm in closed configuration.
Epi.q-1 can go up and down stairs and climb over obstacles with a maximum height of 90 mm, that is 72% of the driving unit height, as shown in Figure 16.

Fig. 16. Epi.q-1 traversing obstacle and stairs

On flat ground the maximum speed it can reach is approximately 0.5 m s^{-1}. On a slope, the theoretical maximum value it can drive (maximum gravitational stability margins) is limited to 62° when the robot is moving uphill frontwards (or downhill backwards), to 32° when it is moving downhill frontwards (or uphill backwards), and to 59° when it is driving along a cross-hill (or normal to a downhill). When the robot is driving uphill, the maximum slope value is limited to 20°, due to motor torque limitation.
Each driving unit is powered by a *Solarbotics GM17* gear-motor, declared specifications are a no load angular speed of 60 rpm and a maximum torque of almost 1 N m, when it is powered at 12 V. The axial device is powered by a *Faulhaber DC-micromotor* (series 0816) combined with a compatible planetary gearhead (series 08/01). The human operator controls the robot by means of an *Hitec - Laser 4* transmitter, and the radio signal is processed by a *Sabertooth 2X5* driver that provides the motors with the proper voltage.The power source both for motor and electronics is a removable 11 V/2200 mA h.

5.2 Epi.q-2 technical specifications

Epi.q-2 weighs almost 4 kg and measures 200 mm × 450 mm × 280 mm (height × length × width), with a driving unit that measures 130 mm in height.
Epi.q-2 can go up and down stairs and climb over obstacles with a maximum height of 110 mm, that is 84% of the driving unit height.
On flat ground the maximum speed it can reach is almost 1 m s^{-1}. On a slope, the theoretical maximum value it can drive (maximum gravitational stability margins) is limited to 70° when the robot is moving uphill frontwards (or downhill backwards), to 51° when it is moving downhill frontwards (or uphill backwards), and to 60° when it is driving along a cross-hill (or normal to a downhill). When the robot is driving uphill, the maximum slope value is limited to 15°, due to motor torque limitation.

Fig. 17. Epi.q-2 traversing an obstacle, on uneven ground

Each driving unit is powered by a gear-motor, declared specifications are a no load angular speed of about 81 rpm and a maximum torque of almost 0.5 N m, when it is powered at 12 V. The human operator controls the robot by means of a *Zebra 4* transmitter, and the radio signal is processed by a *Sabertooth 2X5* driver that provides the motors with the proper voltage. The power source both for motor and electronics is a removable 11 V/2200 mA h lithium-ion battery, providing more than 4 hours continuous runtime on one charge, up to 6 km of travel.

6. Conclusions

The paper has dealt with Epi.q robots, smart mini devices able to move in structured and unstructured environments, to climb over obstacles and to go up and down stairs. These robots do not need to actively sense obstacles for climbing them, they simply move forward and let their locomotion passively adapt to ground conditions and change accordingly: from rolling on wheels to stepping on legs and vice-versa.
Epi.q robots are mainly composed of three parts: a forecarriage, a central body and a rear axle. The forecarriage consists of a box linked to two driving units that, housing the transmission system, control robot locomotion. The rear axle comprises two idle three-legged wheel units with three idle wheels mounted at the end of each spoke. The central body is a metal platform connecting forecarriage and rear axle, where a payload can be placed. Two passive revolute joints, mutually perpendicular, link front and rear part of the robot: the vertical joint allows robot steering, while the horizontal joint guarantees a correct contact between wheels and ground, also in presence of uneven terrain. A differential steering is implemented on Epi.q robots, that provides both driving and steering functions.

Driving unit is the core of these devices. Epi.q driving unit concept takes place from the idea that a robot can passively modify its locomotion, from rolling on wheels to stepping on legs, simply according to local friction and dynamic condition. Actually, the driving unit is designed to have a limit torque that triggers different locomotion: if the torque required for moving on wheels exceeds the torque required for moving on legs, the robot changes its locomotion accordingly, from rolling on wheels to stepping on legs and vice versa. Thus only one motor per driving unit is required both for wheeled and legged locomotion. When the robot is moving on wheels, the robot weight and the contact between wheels and ground constrain driving unit angular position but, when the robot bumps against an obstacle, if the local frictions between front wheel and obstacle are sufficient to stop that wheel, the driving unit starts to rotate around the stopped wheel center, allowing the robot to climb over the obstacle.Therefore, wheels are used whenever possible and legs only when needed, consequently these robots require a small amount of energy if compared to tracked or legged robots.

Epi.q robots can be successfully employed in many fields: from monitoring and surveillance tasks to intervention in potentially dangerous environments like in presence of radiation, gas or explosive, from rescue operations after catastrophic events to exploration of unknown environments, and in many other fields as well.

7. References

Allen, T. J., Quinn, R. D., Bachmann, R. J. & Ritzman, R. E. (2003). Abstracted biological principles applied with reduced actuation improve mobility of legged vehicles, *Proceedings of the 2003 IEEE/RSJ - Int. Conference on Intelligent Robots and Systems*, Las Vegas, Nevada.

Dalvand, M. & Moghadam, M. (2006). Stair climber smart mobile robot (MSRox), *Autonomous Robots* 20(1): 3–14.

Galileo Mobility Instruments & Elbit Systems Ltd (2009). Elbit viper, Robot Magazine. `http://www.galileomobility.com/?page_id=12`.

Grand, C., Benamar, F., Plumet, F. & Bidaud, P. (2004). Stability and traction optimization of a reconfigurable wheel-legged robot, *The International Journal of Robotics Research* 23(10–11): 1041–1058.

iRobot (2010). Ground robots - 510 packbot. `http://www.irobot.com/gi/ground/510_PackBot`.

Lego Mindstorm (2007). Artic snow cat. `http://us.mindstorms.lego.com/en-us/Community/NXTLog/DisplayProject.aspx?id=c7d16dfe-780b-4b19-9aa3-3c0b22065dd5`.

Mourikis, A., Trawny, N., Roumeliotis, S., Helmick, D. & Matthies, L. (2007). Autonomous stair climbing for tracked vehicles, *The International Journal of Robotics Research* 26(7): 737–758.

Oderio, R. (2011). *Innovative concepts for stair climbing devices*, PhD thesis, Politecnico di Torino.

Poulakakis, I., Papadopoulos, E. & Buehler, M. (2006). On the stability of the passive dynamics of quadrupedal running with a bounding gait, *The International Journal of Robotics Research* 25(7): 669–687.

Poulakakis, I., Smith, J. & Buehler, M. (2005). Modeling and experiments of untethered quadrupedal running with a bound gait: the scout ii robot, *The International Journal of Robotics Research* 24(4): 239–256.

Quaglia, G., Bruzzone, L., Bozzini, G., Oderio, R. & Razzoli, R. (2011). Epi. q-tg: mobile robot for surveillance, *Industrial Robot: An International Journal* 38(3): 282–291.

Quaglia, G., Maffiodo, D., Franco, W., Appendino, S. & Oderio, R. (2010). The epi.q-1 hybrid mobile robot, *The International Journal of Robotics Research* 29(1).

Quinn, R. D., Nelson, G. M., Bachmann, R. J., Kingsley, D. A., Offi, J. T., Allen, T. J. & Ritzmann, R. E. (2003). Parallel complementary strategies for implementing biological principles into mobile robots, *The International Journal of Robotics Research* 22(3-4): 169–186.

Saranli, U., Buehler, M. & Koditschek, D. (2001). Rhex: A simple and highly mobile hexapod robot, *The International Journal of Robotics Research* 20(7): 616–631.

Saranli, U., Rizzi, A. & Koditschek, D. (2004). Model-based dynamic self-righting maneuvers for a hexapedal robot, *The International Journal of Robotics Research* 23(9): 903–918.

Schroer, R. T., Boggess, M. J., Bachmann, R. J., Quinn, R. D. & Ritzmann, R. E. (2004). Comparing cockroach and whegs robot body motions, *Proceedings of ICRA 2004 - International Conference on Robotics and Automation*, Las Vegas, Nevada.

Siegwart, R., Lauria, M., Maeusli, P. & Van Winnendael, M. (1998). Design and implementation of an innovative micro-rover, *Proc. of Robotics 98, the 3rd Conference and Exposition on Robotics in Challenging Environments*, Albuquerque, New Mexico.

Part 4

Localization and Navigation

14

Robotic Exploration: Place Recognition as a Tipicality Problem

E. Jauregi[1], I. Irigoien[1], E. Lazkano[1], B. Sierra[1] and C. Arenas[2]
[1]*Department of Computer Sciences and Artificial Intelligence, University of Basque Country*
[2]*Department of Statistics, University of Barcelona*
Spain

1. Introduction

Autonomous exploration is one of the main challenges of robotic researchers. Exploration requires navigation capabilities in unknown environments and hence, the robots should be endowed not only with safe moving algorithms but also with the ability to recognise visited places. Frequently, the aim of indoor exploration is to obtain the map of the robot's environment, i.e. the *mapping* process. Not having an automatic mapping mechanism represents a big burden for the designer of the map because the perception of robots and humans differs significantly from each other. In addition, the *loop-closing* problem must be addressed, i.e. correspondences among already visited places must be identified during the mapping process.

In this chapter, a recent method for topological map acquisition is presented. The nodes within the obtained topological map do not represent single locations but contain information about areas of the environment. Each time sensor measurements identify a set of landmarks that characterise a location, the method must decide whether or not it is the first time the robot visits that location. From a statistical point of view, the problem we are concerned with is the typicality problem, i.e. the identification of new classes in a general classification context. We addressed the problem using the so-called INCA statistic which allows one to perform a typicality test (Irigoien & Arenas, 2008). In this approach, the analysis is based on the distances between each pair of units. This approach can be complementary to the more traditional approach units $\times$ measurements – or features – and offers some advantages over it. For instance, an important advantage is that once an appropriate distance metric between units is defined, the distance- based method can be applied regardless of the type of data or the underlying probability distribution.

We describe the theoretical basis of the proposed approach and present extensive experimental results performed in both a simulated and a real robot-environment system. Behaviour Based philosophy is used to construct the whole control architecture. The developed system allows the robot not only to construct the map but also comes in useful for localisation purposes.

2. Literature review

Loop-closing has long been identified as a critical issue when building maps from local observations. Topological mapping methods isolate the problem of how loops are closed from the problem of how to determine the metrical layout of places in the map and how to deal with noisy sensors.

The loop-closing problem cannot be solved neither relying only on extereoceptive information (due to sensor aliasing) nor on propioceptive information (cumulative error). Both environmental properties and odometric information must be used to disambiguate locations and to correct robot position. Fraundorfer et al. (2007) present a highly scalable vision based localisation and mapping method that uses image collections, whereas Se et al. (2005) use vision mainly to detect the so called *loop-closing* –the place has already been visited by the robot– in robot localisation; Tardós et al. (2002) introduce a perceptual grouping process that permits the robust identification and localisation of environmental features from the sparse and noisy sonar data. On the other hand, the probabilistic Bayesian inference, along with a symbolic topological map is used by Chen & Wang (2006) to relocalise a mobile robot. More recently, Olson (2009) presents a new loop-closing approach based on data association, where places are recognised by performing a number of pose-to-pose matchings; a review of loop-closing approaches related to MONOSLAM can be found in (Williams et al., 2009). Within the field of probabilistic robotics (Thrun et al., 2005), Kalman filters, Bayesian Networks and particle filters are used to maintain probability distributions over the state space while solving mapping, localisation and planning.

But the mapping problem can also be stated from a classification perspective. In most classification problems, there is a training data available for all classes of instances that can occur at prediction time. In this case, the learning algorithm can use the training data to determine decision boundaries that discriminate among the classes. However, there are some problems that exhibit only a single class of instances at training time but are amenable to machine learning. At prediction time, new instances with unknown class labels can either belong to the target class or to a new class that was not available during training. In this scenario, two different predictions are possible: *target*, an instance that belongs to the class learnt during training, and *unknown*, where the instance does not seem to belong to the previously learnt class. Within the machine learning community, this kind of problems are known as *one-class* problems and as *typicality* problems within the statistics research.

To give some examples, in (Hempstalk et al., 2008) the probability distributions of the class variable known values are used to determine if a new case belongs to the known class values or if it should be considered as a different class member. One-class classification categorizers have a wide range of applications; in (Manevitz & Yousef, 2007) one-class classification is used to document categorisation in order to decide whether a reference is relevant in a database searching query. The same authors combine this approach with the Support Vector Machine (SVM) paradigm for document classification purposes (Manevitz & Yousef, 2002); and in (Sánchez-Yáñez et al., 2003) the same idea is applied to texture recognition in images. A thorough review of one-class classification can be found in (Tax, 2001).

Regarding the mobile robotics area, one-class classification approaches can be applied to robot mapping, i.e. to learn the structure of its environment in an automatic manner. In this way, Brooks & K. Iagnemma (2009) present a use of this approach to deal with terrain recognition, and Wang & Lopes (2005) use it to identify user actions in human-robot-interaction. However,

direct uses of this approach, with this particular name, have not been found in the robotics literature.
There are different approaches found in the literature to deal with the typicality problem (Bar-Hen, 2001; Cuadras & Fortiana, 2000; Irigoien & Arenas, 2008; McDonald et al., 1976; Rao, 1962). Some of them are only suitable for normal multivariate data, others are appropriate for any kind of data but are limited to $k = 2$, being k the number of classes. The latter case offers the most general framework to be applied. However, and in spite of the high diversity of the used methods, to the best of the author's knowledge, neither typicality nor one-class approaches appear in the mapping literature.
The approach proposed in this chapter combines the INCA statistic (Irigoien & Arenas, 2008) with the topological properties of the environmental locations considered and thus represents a new approach to tackling the robot mapping problem as a typicality case.

3. Typicality test by means of the INCA statistic

In this section the INCA statistic is introduced and the INCA test is proposed as a solution to the typicality problem.

3.1 Preliminaries

The data we consider are random vectors and we assume that distinct classes exist. Let C_1, C_2, ..., C_k be k classes represented as k independent S-valued random vectors $\mathbf{Y}_1$, $\mathbf{Y}_2$, ..., $\mathbf{Y}_k$, with probability density functions f_1, f_2, ..., f_k with respect to a suitable common measure λ. Let $\delta(\mathbf{y}, \mathbf{y}')$ be a distance (Gower, 1985) function on S. We say that δ is a Euclidean distance function if the metric space (S, δ) can be embedded in a Euclidean space, $\Psi : S \longrightarrow \mathbb{R}^p$, such that:

$$\delta^2(\mathbf{y}, \mathbf{y}') = \|\Psi(\mathbf{y}) - \Psi(\mathbf{y}')\|^2, \tag{1}$$

and we may understand $E(\Psi(\mathbf{Y}_i))$ as the δ-mean of $\mathbf{Y}_i$, $i = 1, ..., k$.
In this general framework the following concepts are considered. The geometric variability of C_i, $i = 1, ..., k$ with respect to δ is defined (Cuadras & Fortiana, 1995) as

$$V_\delta(C_i) = \frac{1}{2} \int_{S \times S} \delta^2(\mathbf{y}_{i1}, \mathbf{y}_{i2}) f(\mathbf{y}_{i1}) f(\mathbf{y}_{i2}) \lambda(d\mathbf{y}_{i1}) \lambda(d\mathbf{y}_{i2}).$$

This quantity is a variant of Rao's diversity coefficient (Rao, 1982). When δ is the Euclidean distance and $\Sigma_i = COV(\mathbf{Y}_i)$, then $V_\delta(C_i) = tr(\Sigma_i)$. For other dissimilarities $V_\delta(C_i)$ is a general measure of dispersion of $\mathbf{Y}_i$. In the context of discriminant analysis (Cuadras et al., 1997) the squared distance between C_i and C_j is defined by

$$\Delta^2(C_i, C_j) = \int_{S \times S} \delta^2(\mathbf{y}_i, \mathbf{y}_j) f(\mathbf{y}_i) g(\mathbf{y}_j) \lambda(d\mathbf{y}_i) \lambda(d\mathbf{y}_j) - V_\delta(C_i) - V_\delta(C_j) \tag{2}$$

This quantity is the Jensen difference (Rao, 1982) between the distributions of C_i and C_j. If the metric space (S, δ) can be embedded (see (1)) in a Euclidean space $\mathbb{R}^p$ and if $E(\|\Psi(\mathbf{Y}_i)\|)$ and $E(\|\Psi(\mathbf{Y}_i)\|^2)$ are finite, then $V_\delta(C_i) = E(\|\Psi(\mathbf{Y}_i)\|^2) - \|E(\Psi(\mathbf{Y}_i))\|^2$, $i = 1, ..., k$, and $\Delta^2(C_i, C_j) = \|E(\Psi(\mathbf{Y}_i)) - E(\Psi(\mathbf{Y}_j)\|^2$. If there is only one element $C_i = \{\mathbf{y}_0\}$, (3) gives the proximity function of $\mathbf{y}_0$ to C_j,

$$\phi^2(\mathbf{y}_0, \mathbf{Y}_j) = \int_S \delta^2(\mathbf{y}_0, \mathbf{y}_j) f(\mathbf{y}_j) \lambda(d\mathbf{y}_j) - V_\delta(\mathbf{C}_j). \tag{3}$$

In applied problems the distance function is typically a datum, but the probability distribution for each population is unknown. Natural estimators given samples $\mathbf{y}_1^1, ..., \mathbf{y}_{n1}^1, ..., \mathbf{y}_1^k, ..., \mathbf{y}_{nk}^k$, of sizes $n_1, ..., n_k$ coming from $C_1, ..., C_k$ are the following:

- The *geometric variability* of C_j,

$$\hat{V}_\delta(Cq_j) = \frac{1}{2n_j^2} \sum_{l,m} \delta^2(\mathbf{y}_l^j, \mathbf{y}_m^j).$$

- The *proximity function* of a new object $\mathbf{y_0}$ to C_j,

$$\hat{\phi}^2(\mathbf{y}_0, C_j) = \hat{\phi}_j^2(\mathbf{y}_0) = \frac{1}{n_j} \sum_l \delta^2(\mathbf{y}_0, \mathbf{y}_l^j) - \hat{V}_\delta(C_j).$$

- The *squared distance* between C_i and C_j,

$$\hat{\Delta}^2(C_i, C_j) = \hat{\Delta}_{ij}^2 = \frac{1}{n_i n_j} \sum_{l,m} \delta^2(\mathbf{y}_l^i, \mathbf{y}_m^j) - \hat{V}_\delta(C_i) - \hat{V}_\delta(C_j). \tag{4}$$

See (Arenas & Cuadras, 2002) and references therein for a review of these concepts, their application, different properties and proofs.

3.2 INCA statistic

Consider that n units are simply divided into k classes $C_1, ..., C_k$, of sizes $n_1, ..., n_k$. Consider a fixed unit $\mathbf{y}_0$, which may be an element of a C_j, $j = 1, ..., k$ or may belong to an unknown class, i.e. it may be an atypical unit. Consider a new class with δ-mean the linear combination $\sum_{i=1}^k \alpha_i E(\Psi(\mathbf{Y}_i))$, where $\mathbf{Y}_i$ is the random vector representing the class C_i, $i = 1, ..., k$. The INCA statistic is defined as follows:

$$W(\mathbf{y}_0) = \min_{\alpha_i} \{L(\mathbf{y}_0)\}, \qquad \sum_{i=1}^k \alpha_i = 1, \tag{5}$$

$$L(\mathbf{y}_0) = \sum_{i=1}^k \alpha_i \phi_i^2(\mathbf{y}_0) - \sum_{1 \le i < j \le k} \alpha_i \alpha_j \Delta_{ij}^2.$$

$\phi_i^2(\mathbf{y}_0)$ is the proximity function of $\mathbf{y}_0$ to C_i and Δ_{ij}^2 is the squared distance between C_i and C_j. The INCA statistic $W(\mathbf{y}_0) = \min_{\alpha_i} L(\mathbf{y}_0)$ trades off between minimising the weighted sum of proximities of $\mathbf{y}_0$ to classes (which takes into consideration the within-group variability) and maximising the weighted sum of the squared distances between classes (between-groups variability) - a common behaviour of a classing criterion. The values of $\alpha' = (\alpha_1, \ldots, \alpha_{k-1})$ together with $\alpha_k = 1 - \sum_{i=1}^{k-1} \alpha_i$, verifying (5) are $\mathbf{ff}' = M^{-1}N$, where M is the $(k-1) \times (k-1)$ matrix

$$M = \left(\Delta_{ik}^2 + \Delta_{jk}^2 - \Delta_{ij}^2\right)_{i,j=1,\ldots,k-1}$$

and N is the $(k-1) \times 1$ vector

$$N = \left(\Delta_{ik}^2 + \phi_k^2(\mathbf{y}_0) - \phi_i^2(\mathbf{y}_0)\right)_{i=1,\ldots,k-1}.$$

The statistic $W(\mathbf{y}_0)$ has a very nice geometric interpretation. It can be interpreted (see Figure 1) as the (squared) orthogonal distance or height h of $\mathbf{y}_0$ on the hyperplane generated by the δ-mean of C_i $(i = 1, ..., k)$, denoted in Figure 1 by a_i, $i = 1, ..., k$. Then, points which lie significantly far from this hyperplane are held to be outliers. This intuitive idea is used to detect outliers among existing classes.

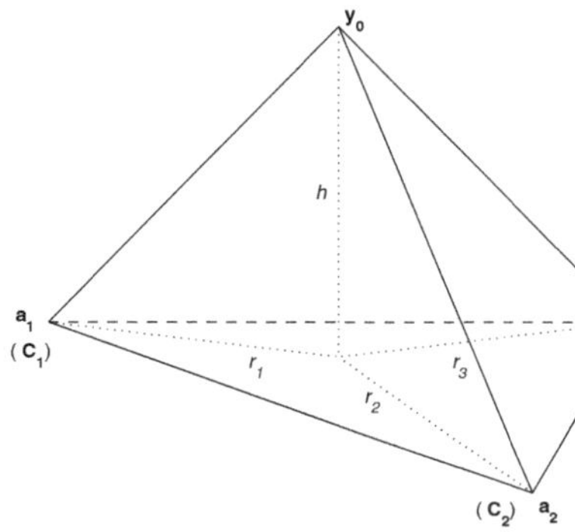

Fig. 1. For $k = 3$, new observation $\{\mathbf{y}_0\}$, centres of classes $\{\mathbf{a}_1, \mathbf{a}_2, \mathbf{a}_3\}$ and (squared) projection $\mathbf{r}_i$ of the edges $\{\mathbf{y}_0, \mathbf{a}_i\}$ on the plane $\{\mathbf{a}_1, \mathbf{a}_2, \mathbf{a}_3\}$. The (squared) height $\mathbf{h}$ is $W(\mathbf{y}_0)$

Suppose now that the data are classified in k classes. Let $\mathbf{y}_0$ be a new observation and consider the test to decide whether $\mathbf{y}_0$ belongs to one of the fixed classes C_j, $j = 1, ..., k$ or, on the contrary, it is an outlier or an atypical observation which belongs to a different and unknown class. Consider the INCA test,

$$\begin{aligned} H_0 &: \mathbf{y}_0 \text{ comes from the class with} \\ &\quad \delta\text{-mean } \sum_i \alpha_i E(\Psi(\mathbf{Y}_i)), \quad \sum_{i=1}^{k} \alpha_i = 1,\ i = 1, ..., k, \\ H_1 &: \mathbf{y}_0 \text{ comes from another unknown class,} \end{aligned}$$

and compute statistic (5). If $W(\mathbf{y}_0)$ is significant it means that $\mathbf{y}_0$ comes from a different and unknown class. Otherwise we allocate $\mathbf{y}_0$ to C_i using the rule:

$$\text{Allocate } \mathbf{y}_0 \text{ to } C_i \text{ if } U_i(\mathbf{y}_0) = \min_{j=1,\dots,k} \{U_j(\mathbf{y}_0)\}, \tag{6}$$

where $U_j(\mathbf{y}_0) = \phi_j^2(\mathbf{y}_0) - W(\mathbf{y}_0)$, $j = 1, ..., k$.

It can be observed (Irigoien & Arenas, 2008) that $U_j(\mathbf{y}_0)$ represents the (squared) projection of $\{\mathbf{y}_0, E(\Psi(\mathbf{Y}_i))\}$ on the hyper plane $\{E(\Psi(\mathbf{Y}_1)), \dots, E(\Psi(\mathbf{Y}_k))\}$. See Figure 1, where for simplicity the (squared) projection $U_j(\mathbf{y}_0)$ is denoted by r_j, $j = 1, ..., k$. Hence, criterion 6 follows the next geometric and intuitive allocation rule: Allocate $\mathbf{y}_0$ to C_i if the projection $U_i(\mathbf{y}_0)$ is the smallest.

We obtained sampling distributions of $W(\mathbf{y}_0)$ and $U_j(\mathbf{y}_0)$ $(j = 1, ..., k)$ by re-sampling methods, in particular drawing bootstrap samples as follows. Draw N units $\mathbf{y}$ with replacement from the union of $C_1, \dots, C_k$ and calculate the corresponding $W(\mathbf{y})$ and

$U_j(\mathbf{y})$ $(j = 1, ..., k)$ values. As usual, this process is repeated $10P$ times with $P \geq 1$ selected by the user. In this way, the bootstrap distributions under H_0 are obtained.

4. Behavior-Based navigation

Behavior-Based (BB) systems appeared in 1986, when R.A. Brooks proposed a bottom-up approach for robot control that imposed a new outlook for developing intelligent embodied agents capable of navigating in real environments performing complex tasks. He introduced the Subsumption Architecture (Brooks, 1986; Brooks & Connell, 1986) and developed multiple robotic creatures capable of showing different behaviours not seen before in real robots (Brooks, 1989; Connell, 1990; Matarić, 1990). Behavior-based systems are originally inspired on biological systems. Even the most simple animals show navigation capabilities with high degree of performance. For those systems, navigation consist of determining and maintaining a trajectory to the goal (Mallot & Franz, 2000). The main question to be answered for navigation is not *Where am I?* but *How do I reach the goal?* and the answer does not always require knowing the initial position. Therefore, the main abilities the agent needs in order to navigate are to move around and to identify goals.

The behavior-based approach to robot navigation relies on the idea that the control problem is better assessed by bottom-up design and incremental addition of light-weight processes, called behaviors, where each one is responsible for reading its own inputs and sensors, and deciding the adequate motor actions. There is no centralized world model and data from multiple sensors do not need to be merged to match the current system state in the stored model. The motor responses of the several behavioural modules must be somehow coordinated in order to obtain valid intelligent behavior. *Way-finding* methods rely on *local navigation strategies*. How these local strategies are coordinated is a matter of study known as *motor fusion* in BB robotics, opposed to the well known *data fusion* process needed to model data information. The aim is to match subsets of available data with motor decisions; outputs of all the active decisions somehow merge to obtain the final actions. In this case there is no semantic interpretation of the data but behavior emergence.

5. Topological places

Generally speaking, there are two typical strategies for deriving topological maps: one is to learn the topological map directly; the other is to first learn a geometric map, then to derive a topological model from it through some process of analysis (Thrun, 1999; Thrun & Bücken, 1996a;b).

As mentioned before, BB systems advocates for a functional bottom-up decomposition of the control problem in independent processes called behaviours. From this point of view, the topological "map" should be composed of tightly coupled behaviours, specific to the meaningful locations.

A topological map is formally defined as a set of nodes where each node consists of:

1. A set of inputs (from landmark identification subsystems) and outputs. These outputs should serve to reduce the distance between the current state and the goal.
2. A signature that identifies the node: $signature_i$. Each location has a signature that reflects the state of a set of specific landmarks and that is used by the robot for localisation purposes.

3. A function α_i to be executed when the node i is active and that will output the action to be performed at the node specific current state. The behaviour of the robot as well as the associated function of the nodes can be different depending on the location.
4. The location identifier that contains initial and final position of the node:

$$(x_{i0}, y_{i0}), (x_{if}, y_{if})$$

The overall "map" is then composed of sets of behaviours, each launched on a different thread. The environment is only partially unknown to the robot since it is provided with behaviour modules to properly identify certain features such as corridors, crossings or junctions and halls, each of them identifiable using distance sensors like a laser scanner. Each landmark identifier outputs a confidence level (cl) as a measure of the confidence of the identification process. These values are filtered through node signatures, giving at each time step the node activation level according to the sensor readings.

- Corridors: the robot is considered to be in a corridor if the place is between 1.6 and 2.4 m wide. To that aim, left and right side shortest readings are summed and stored in a FIFO buffer. The mean of the buffer is used in a Gaussian function that gives the confidence level of being in a corridor.
- Halls: as opposed to corridors, halls are wide areas. Therefore, the confidence level of being in a hall is defined as 1 minus the probability of being in a corridor.
- Crossings or junctions: these locations are areas where two or more alternative ways are possible. It is mandatory for the robot to identify junctions in order to choose the right way when looking for goals. Depending on the destination, the robot must select one way or another. Crossing areas usually come at the end of a corridor or hall and lead to a new area. Hence, left and right minimum distances are looked for and these minimum values are used as reference for searching continuous interval of readings that exceed the minimum values. The orientations of the possible alternative ways at the junctions are registered according to the robot heading provided by the compass sensor and the indexes of the laser scan that define the different intervals, the orientation of the possible alternative ways at the junctions are registered.

The goal of the mapping process is to fill in the nodes with the information that they must contain. More precisely, the contents of the signature and the location identifier. For this aim, during the learning process and depending on the state of the landmark identification subsystems, i.e. the confidence level of the corridor/hall/junction (cl_{corr}, cl_{hall} and cl_{cross}), the following information is given to the INCA test:

- Initial and mean heading values: θ_0, θ_{mean}.
- Initial and final pose obtained by the odometric subsystem. These poses correspond to the position values of the robot when the node signature activates/deactivates: $(x_0, y_0), (x_f, y_f)$.
- Length (previously named as *duration*) of the area calculated using the initial and final pose information: d.
- Number of alternative ways and their associated orientation: num_ways and $\theta_{w_1}, \cdots, \theta_{w_{num\_ways}}$.

These measurements will constitute the observations of the random vectors **Y** considered in the INCA statistic, as represented in Equation 7.

$$
\begin{array}{l}
\text{Corridors, Halls:} \\
\mathbf{Y} = (\sin(\theta_0), \cos(\theta_0), \sin(\theta_{mean}), \cos(\theta_{mean}), (x_0, y_0), (x_f, y_f), d) \\
\\
\text{Junctions:} \\
\mathbf{Y} = (\sin(\theta_0), \cos(\theta_0), \sin(\theta_{mean}), \cos(\theta_{mean}), \theta_{w_1}, \cdots, \theta_{w_{num\_ways}}, (x_0, y_0))
\end{array}
\tag{7}
$$

Note that there are two types of measurements: variables type coordinates in meters and variables type orientation in degrees.
The corridors/halls/crosses can differ in their orientation (mean compass value that the robot maintains when going through them in its canonical path). This is why each physical place will correspond to two or more different nodes in the topological map.

6. Proposed approach

The locations the robot must identify are not only single points but areas surrounding these points. Therefore, we propose firstly, a data generation approach to characterise the areas; and secondly, the application of the INCA test.
Let us assume that the robot has recorded the geometric information (see section 5) of k different places $C_1, \ldots, C_k$, all of them of the same type. There is only one $\mathbf{y}_i$ measurement for each place C_i ($i = 1, \ldots, k$). However, the place we want to identify topologically is not just a spot but an area or neighbourhood of the recorded measurement $\mathbf{y}_i$. In order to do so we generate $n_i - 1$ new observations for each place i which will make up the observations corresponding to the place C_i. These new observations are generated as $\mathbf{y}_i^l = \mathbf{y}_i + U(-u, u)$, $l = 2, \ldots, n_1$, where $U(-u, u)$ stands for the uniform distribution with parameters $-u$ and u ($u > 0$). Taking into account that the robot records two kinds of variables, metres and degrees, we consider two kinds of values for the parameter of the uniform distribution, let us call them, u_M and u_{DEG}, respectively.
Once the data corresponding to the k classes -places- are generated, and given $\mathbf{y}_0$, the information the robot has recorded when he arrives at a new place, the INCA test can be applied and consequently it is possible to decide whether or not $\mathbf{y}_0$ corresponds to a new place. In case it is decided $\mathbf{y}_0$ is not a new place, the conclusion is that $\mathbf{y}_0$ is one of the places $C_1, \ldots, C_k$ according to rule (6).
Pearson distance has been used for the calculus of the interdistances $\delta(\mathbf{y}, \mathbf{y}')$ between $\mathbf{y}$ and $\mathbf{y}'$. The parameter values used during the experimental phase were $n_i = 10$, $u_M = 2$ and $u_{DEG} = 30$. These values were chosen experimentaly as explained in (Jauregi et al., 2011).

7. Exploration behaviour

As stated earlier, the mapping process requires an exploration strategy to guide the robot for the terrain inspection. The strategy used in this proposal, the exploration behaviour is a coordination of the local navigation strategies and landmark identification subsystems the robot is endowed with. The proper combination of these behaviours, allow the safe exploration of the environment.

- Two local navigation strategies that are combined in a cooperative manner (weighted sum): balance the free space at both sides of the robot and follow a desired compass orientation (θ_d).
- Landmark identification subsystems that allow the robot to recognise corridors, left/right walls, halls, junctions and dead-ends. These landmarks are used to change robot's desired orientation. To show an example, Figure 2 shows the coordination of the modules for the case where a dead-end is recognised.

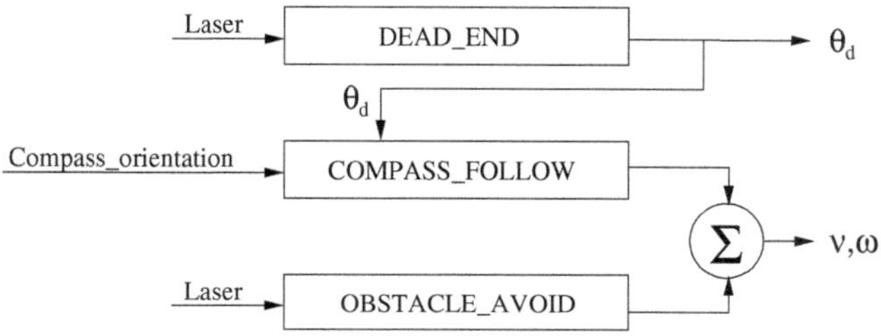

Fig. 2. Diagram of behaviours modules (v: translational velocity, w: angular velocity)

Although the robot can be positioned at any starting location, initially and until the robot reaches a dead-end the map remains empty. Hence, the map construction starts after a dead-end has been identified. This gives the correct measurement of the length of the locations (nodes). Afterwards, the first corridor, the first crossing and the first hall are always identified as new nodes since there is not any instance of the same type already stored in the map.
Once the map building process starts, each time the robot identifies a location – a corridor, a hall or a crossing – the geometric information of the identified location is recorded (the **Y** vector, Equation 7), and then the INCA test is applied to evaluate if they are locations already visited or new ones. When the location corresponds to a crossing, i.e. a junction, the orientations of the alternative ways the robot can choose are recorded. If the location has been visited before, one of the non-explored paths is randomly selected. In this way, the robot has the chance to cover all the environment. The robot finishes the exploration process when all the alternatives of the crossing nodes have been tried.

8. Simulated experiments

Experiments were carried out in the third floor of the Faculty of Computer Science. This environment is a semi-structured office-like common environment, with regular geometry as can be seen in Figure 3.
The parameter selection obtained in the previous experimental phase was applied to the more general problem of identifying the whole set of environmental locations during an exploration phase performed in simulation. To this purpose the *Stage* simulation tool was used together with the *Player* robot server.
In order to have a wider view of the mapping process, we let the robot move in the environment for a long time (more than 6500 seconds). On the left of the Figure 4 shows the robot's path starting from the dead-end at the bottom left corner and on the right the complete path followed during the exploration of the environment.

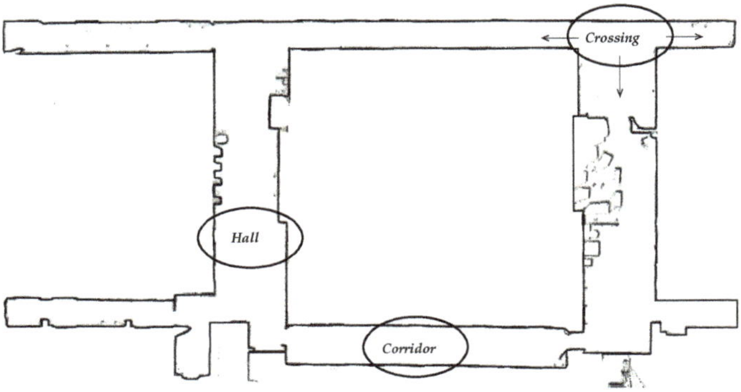

Fig. 3. Third floor of the Faculty of Computer Science. Approx. 60 × 22 meters

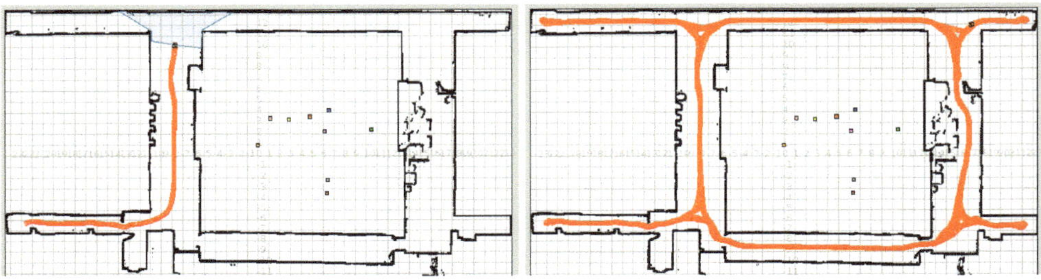

Fig. 4. The simulation path resulting from the exploration process

Related to the number of nodes, the map converged to 38 nodes: 17 corridors, 8 halls and 13 crosses (Figure 5). Table 1 shows the number of nodes that have been traversed in the path followed by the robot.

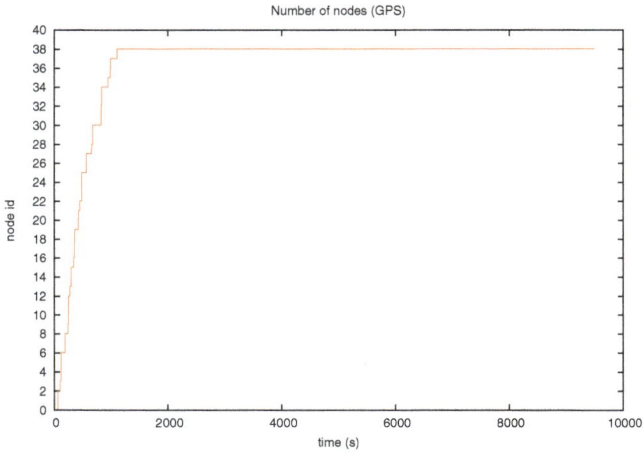

Fig. 5. Evolving number of nodes

As it can be seen, all the nodes are correctly classified:

		Corr	Hall	Cross
New	Expected	17	8	13
	Found	100%	100%	100%
Known	Traversed	47	23	38
	Classified	100%	100%	100%

Table 1. Experimental results

- Each of the 17 existing corridors were properly labelled as new places the first time the robot went along them; the same happened with the new traversed halls and crossing nodes.
- The nodes visited more than once by the robot in this long journey were also properly classified with their corresponding label; a total number of 47 corridors, 23 halls and 38 crossing nodes were visited in the robot path.

Figure 6 shows the distribution of the locations (plotted according to their central poses) and the evolution of the number of nodes over time.

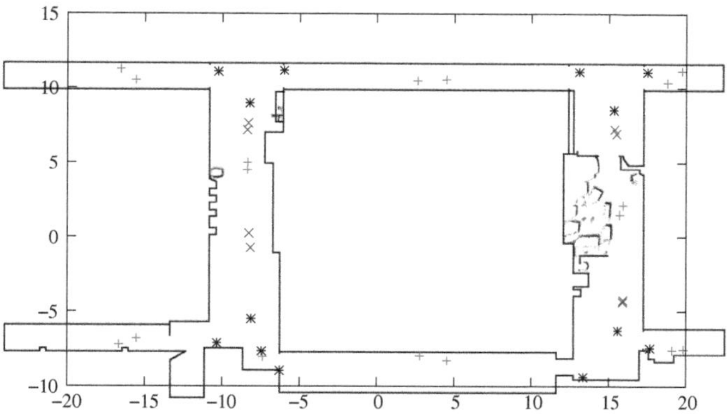

Fig. 6. Location distribution over the map. Corridors: +; Halls: x; Crossings: *

In spite of the degree of symmetry of the environment, the spatial configuration of the obtained locations does not show the same degree of symmetry. This is due to the fact that robot's and humans' perception differ from each other, and since the robot navigates according to a desired compass heading, depending on its orientation it makes the same physical place correspond to several nodes in the topological representation.

9. Experiments in the real robot/environment system

The simulation experiments showed that the proposed approach can solve the stated problem. To test the robustness of the approach experiments were extended to the real robot-environment system. The robot Tartalo–a *PeopleBot* robot form *MobileRobots* equipped with a Canon VCC5 monocular PTZ vision system, a Sicl LMS laser, a TCM2 compass and several sonars and bumpers– has been used for the empirical evaluation of the mapping system developed. But instead of relying on raw odometry information, two odometry correction methods were tested to smooth the positioning error:

- *Laser stabilised odometry* (by means of the LODO driver provided by *Player*). Laser data is used to correct the raw odometry estimate that once corrected exhibits a drift rate that is an order of magnitude less than the rate observed using pure odometry (Howard, 2005).
- Compass based odometry (CODO), where compass heading is used to correct raw odometric poses.

Experiments were performed in the third floor of the Faculty of Computer Sciences. On the left of Figure 7 shows the path completed by the robot (according to compass based odometry) and on the right shows the evolution of the number of nodes over time (s) for the different positioning methods. Clearly, the compass odometry obtained with the proposed approach offers the most precise position information.

On the other hand, Figure 8 shows the distribution of locations of the different nodes obtained from the run performed by the robot using CODO (Figure 7). As mentioned in the previous section, the difference in perception explains the fact that the number of nodes acquired by robot and humans differ from each other.

And, as expected, the number of nodes is higher when the mapping is performed by the robot because of its perception of the environment and its positioning error. However, although the number of junction nodes identified is higher in the real run, this is mainly due to the people and furniture the robot comes across, which produce nodes that lead to any number of alternative paths. However, after an exploration of about an hour and a half (more than 500 meters), the robot was able to close the loop and to recognise several times the final location as the starting one, thus confirming the suitability of the proposed approach.

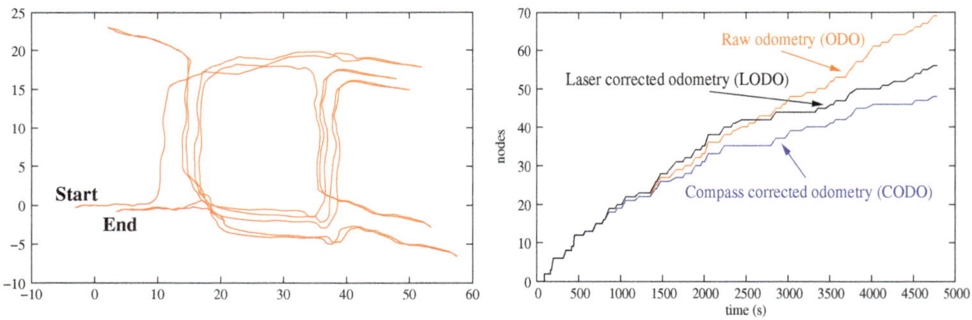

Fig. 7. Left: Robot's path (CODO). Right: Evolving number of nodes

As mentioned earlier, the experiments performed in simulation cannot be directly compared with the experiments with the real robot; the simulated sensor readings produce nodes with different characteristics specially when junction nodes are identified. Hence, the path produced by the exploration strategy in simulation differs from the path executed by the real robot. However, it is interesting to compare the evolution of the learning process using exact odometry with the evolving number of nodes when the odometry is corrected using the compass sensor. The map obtained simulating ideal odometry converged to 38 nodes and the map obtained by the robot after 4500 seconds contained 48 nodes (see Figure 9).

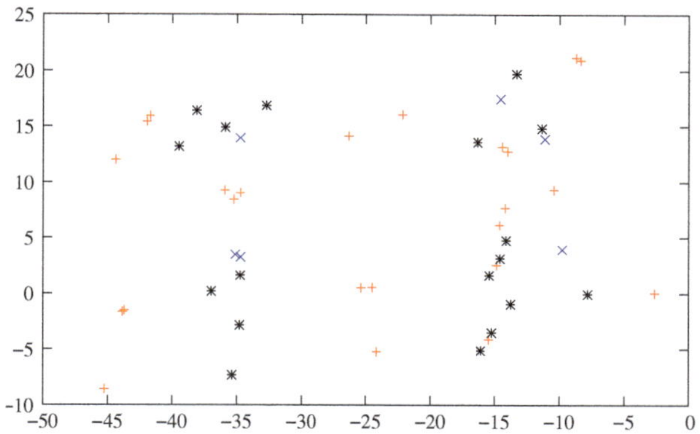

Fig. 8. Location distribution over the map. Corridors: +; Halls: x; Crossings: *

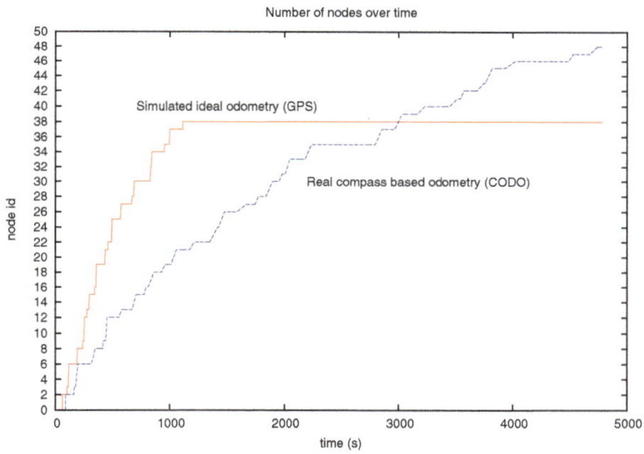

Fig. 9. Comparison with ideal odometry

10. INCA for localisation

During the previous experiments the learning process was not stopped once the loop was closed. This methodological criterion was chosen to asses the appropriateness of the approach, and as a result, there was a slow increase in the number of nodes over time mainly due to odometry error. However, in practical terms the map learning process can be stopped and then use the learnt map for localisation purposes.

The experiments described in this section were carried out to measure the usefulness of the acquired map for localisation. In this occasion, instead of a non-stop learning process, a criterion was set so that the generation of the map would stop once a certain number of nodes was included. Once the procedure reaches this value, no more nodes are allowed to be created and hence, classification rule 6 (Section 3.2) gives the closest node according to the available data. In this manner, after the map is completed the robot continues moving according to its exploration strategy while the mentioned rule gives its localisation. It is worth to mention that classification rule 6 is equivalent to the distance based classifier introduced in (Cuadras, 1992).

Experiments were conducted both in simulation and in the real robot/environment system.

10.1 Simulated experiments

Once more the *Stage* simulator was used to simulate the robot and its environment. The criterion to stop the learning process was established in 38 nodes, which was the number of nodes the map converged to when simulating the mapping process with ideal odometry. Two experiments were carried out in the simulator:

- Ideal odometry (GPS). Figure 10 shows the journey together with the spatial node configuration learnt whereas Figure 11 shows the results of the mapping and localisation process, and thus the identified set of nodes over time. The mapping process lasted about 1100 seconds and the fact that no error occurred during the localisation phase (seconds 1100-12000) confirmed that INCA is a valid approach also for localisation. Once the map has been generated, the trajectory of the robot is randomly decided at run-time. The resulting unpredictability means that instead of following a static route, the robot will randomly select the orientation at each junction. As a consequence, the robot does not produce repeatable sequences of nodes in the path, but the probability that it will revisit the whole set of nodes increases.
- Laser corrected odometry (LODO). A new experiment was conducted applying the default odometry error value defined in *Player/Stage* and applying the LODO driver to correct the odometry. Figure 12 shows the journey together with the spatial node configuration learnt whereas Figure 13 shows the results of the mapping and localisation process, and thus the identified set of nodes over time.

  Table 2 shows the path patterns extracted from plot in Figure 13(a). Again, their associated node sequence, the time interval and the label used in the plot to represent each pattern is included.

  The identified patterns concentrate in the first part of the plot (seconds 2000 to 12000). As times goes by the extracted paths are shorter due to localisation failures and the task becomes extremely difficult from second 12000 and there on. Although an odometry correction method is applied, the accumulating error severely affects the localisation of the robot. The type of error remaining after the LODO correction procedure produces a rotation on the robot's trajectory (see for instance Figure 12(a)) and thus, a misclassification of nodes with different orientations assigned. This effect was detected in the sequences labelled as $c0 - c1$ in Table 2. Chain P17, B34, P38 should have been P31, B34, P38. Node P31 with assigned orientation SN was misclassified as node P17 with assigned orientation WE.

Note that both procedures produced the same configuration of nodes, 17 corridors, 13 crosses and 8 halls, although their positional information differed due to odometry values.

10.2 Experiments in the real robot/environment system

A second set of experiments were carried out with the real robot. This time the node threshold was established in 44 nodes.

Figure 14 shows the robot's path and the obtained node distribution using laser corrected odometry values (LODO).

Figure 15 shows the robot's path and the obtained node distribution using compass corrected odometry values (CODO). And Figure 16 shows the mapping process, and the localisation over time for both, LODO and CODO.

The results were disappointing but confirmed what the simulated experiments showed for the LODO case. Although the robot localises properly for about 2000 seconds, afterwards the

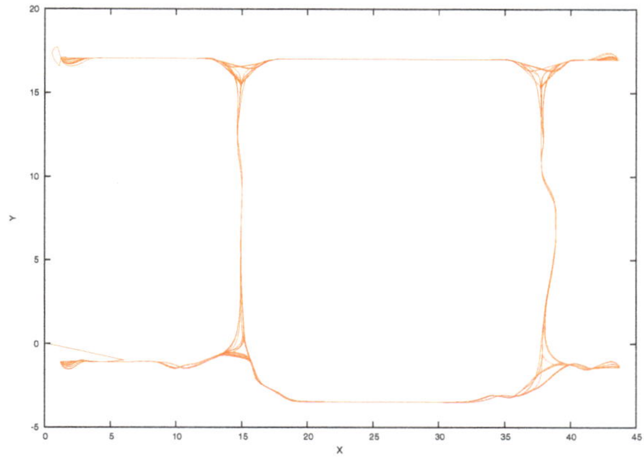

(a) Path corresponding to perfect odometry

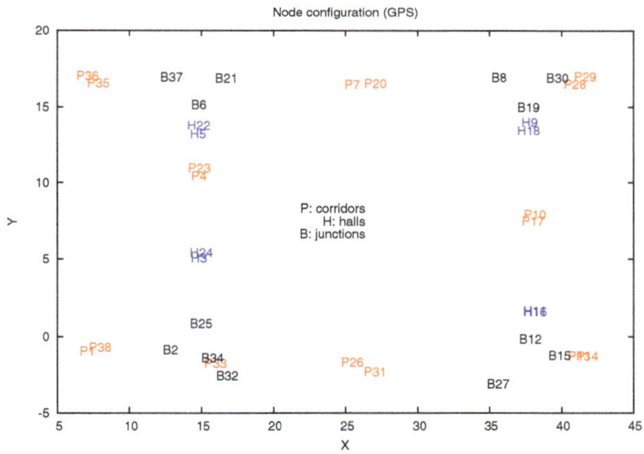

(b) Obtained map

Fig. 10. *Stage* (GPS): robot's path and the obtained map

localisation starts to degrade. It is not possible to extract valid path patterns from the plots in Figure 16. Both the LODO and CODO methods are insufficient for long term localisation. Looking at the robot's paths drawn in Figures 14(a) and 15(a), it can be stated that:

- When using the LODO correction method, the error accumulates more slowly but the error occurs in x, y and θ coordinates. According to LODO odometry, the path rotates over time. While the error is maintained within a certain range the rotation angle is small and the localisation process works correctly. Afterwards, and due to the high dependency the approach has in nominal orientations the system starts to fail and no correspondences are found.
- When using the CODO correction method, only x and y values are affected. θ value is obtained from an absolute reference value and hence, the error is not accumulative. This produces a diagonal shift on the drawn path over time. This shift led to the misclassification of the lower corridors as if they were the upper ones. Oddly, the upper corridors were always well identified.

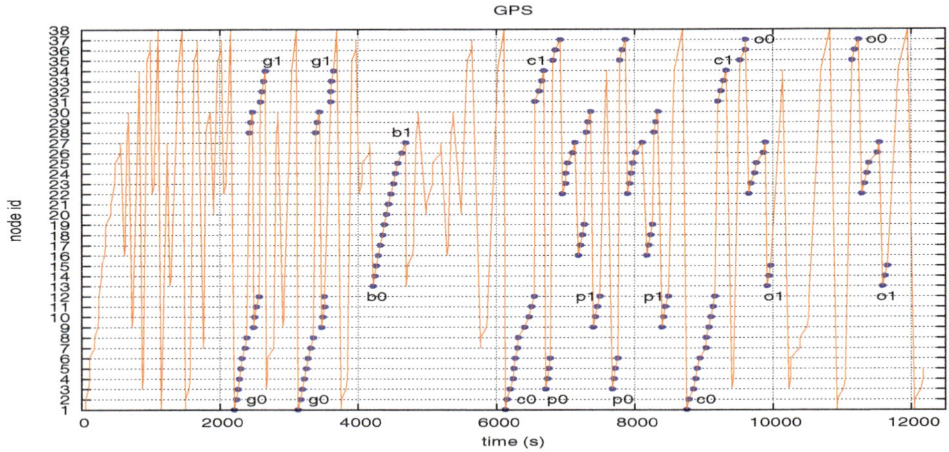

Fig. 11. *Stage* (GPS): node identification over time

Robot's path	node sequence	time stamp	label
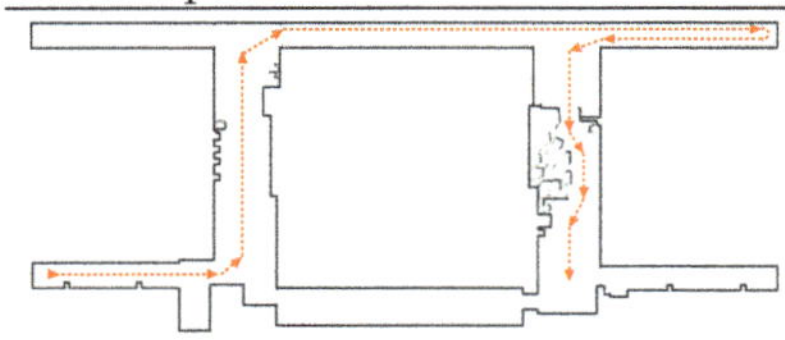	P1, B2, H3, P4, H5, B6, P7, B8, P28, P29, B30, H9, P10, H11, B12	2128-2391, 2955-3227	g0-g1
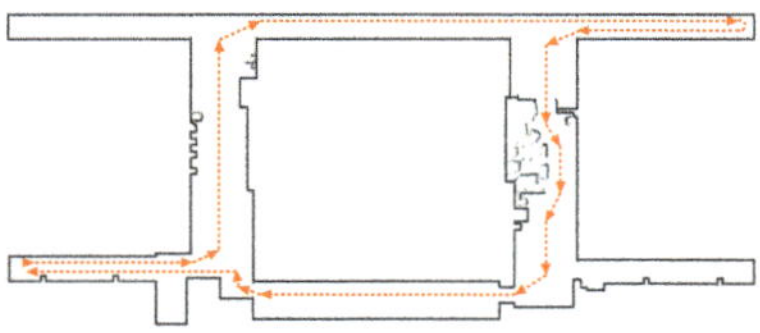	P1, B2, H3, P4, H5, B6, P7, B8, P28, P29, B30, H11, P10, H11, B12, P17*, B34, P38	5767-6217, 6301-6916, 6981-7606	c0-c1
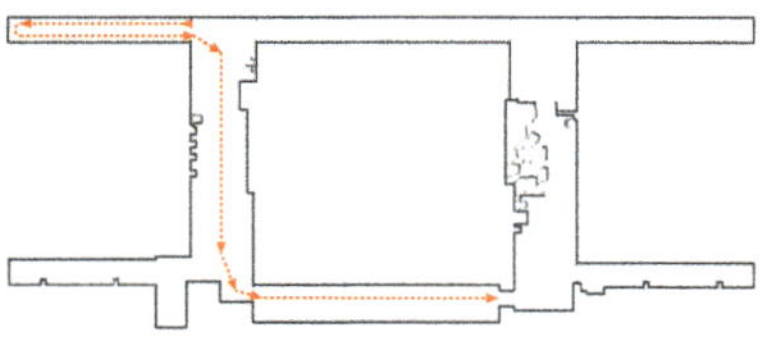	P35, P36, B37, H22, P23, H24, B25, P26, B27	3565-3781, 9130-9413, 11322-11627	m0-m1
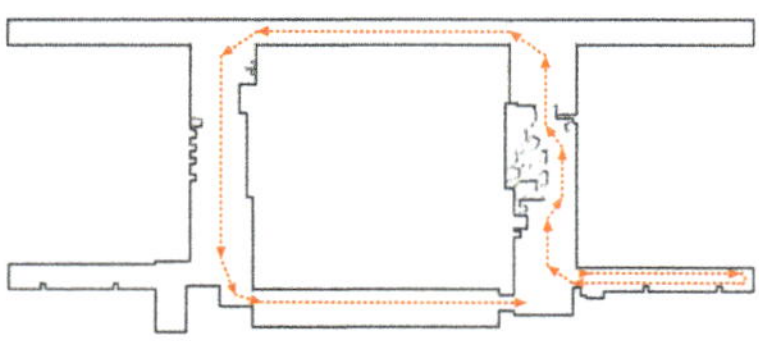	P13, P14, B15, H16, P17, H18, B19, P20, B21, H22, P23, H24, B25, P26, B27	3821-4169	b0-b1

Table 2. LODO: extracted path patterns (*: localisation error)

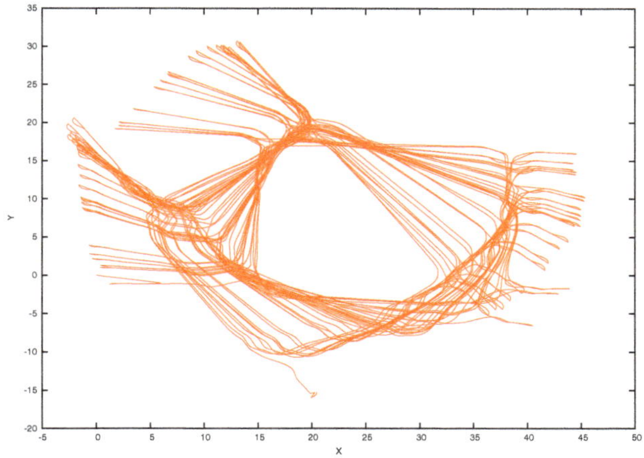

(a) Path corresponding to LODO

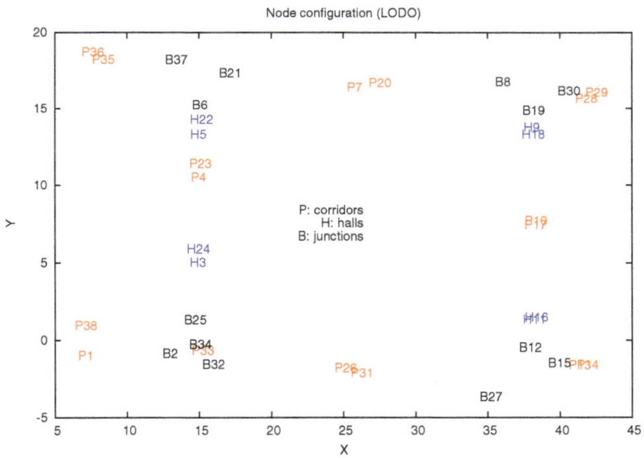

(b) Obtained map

Fig. 12. *Stage* (LODO): robot's path and the obtained map

An intuitive way of coping with this problem is to modify the positional values of the nodes each time they are revisited. Instead of keeping the acquired node information unaltered, during the localisation phase the contents of the nodes can be updated when a positive match occurs.

A last experiment was performed with the robot using CODO to correct the odometry to measure the effect of updating the contents of the node. This choice was made because of the lack of accumulated error in orientation values. Figure 17 shows the acquired map after reaching the maximum number of nodes (established in 39 nodes). The different scales of these two maps reflect the magnitude of the accumulated error in the x and y coordinates over time. Figure 18 shows the evolution of the localisation system over time.

Table 3 shows the path patterns extracted from plot in Figure 18. Again, their associated node sequence, the time interval and the label used in the plot to represent each pattern is included.

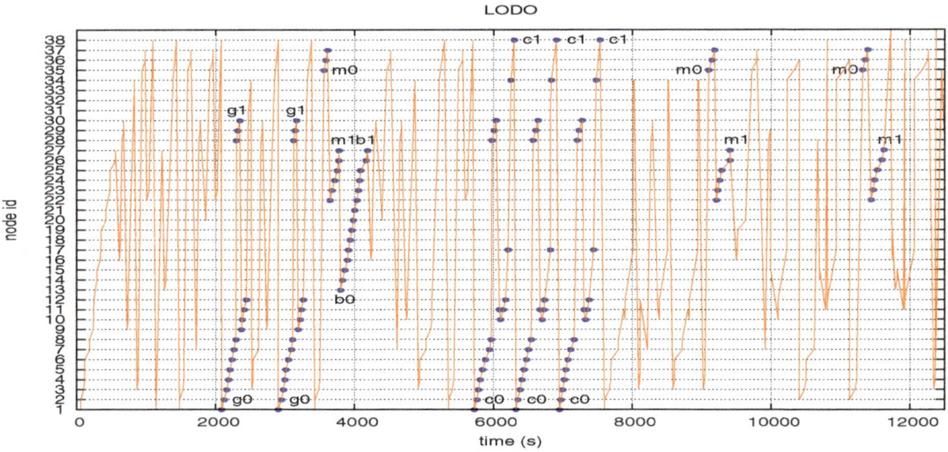

(a) Time stamp 0 to 12000

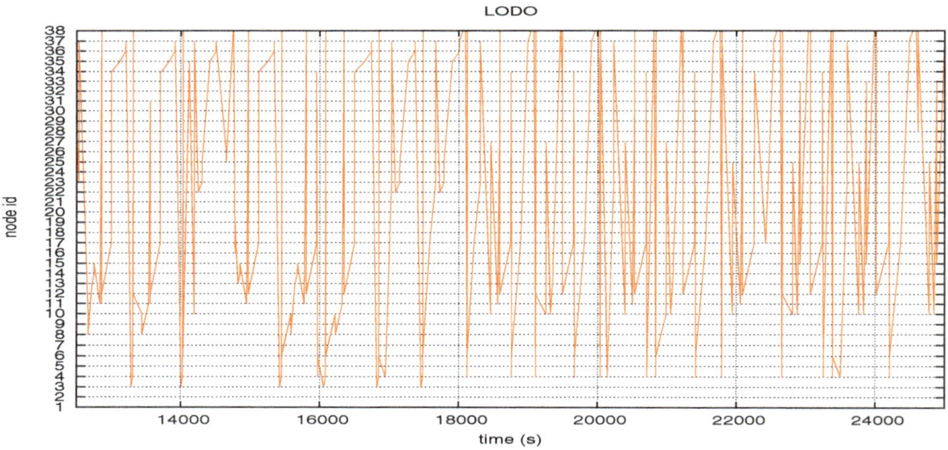

(b) Time stamp 12000 to 25000

Fig. 13. *Stage* (LODO): node identification over time

The localisation process last until the robot run out of batteries and only one location was misclassified. As mentioned in Section 6, some parameters need to be adjusted for INCA to function properly. The value u_M deeply influences the acceptable deviations from nodes' (x, y) locations. A small u_M value produces failures on loop-closings because of the odometry error. On the contrary, setting u_M to a high value produces that close areas with the same signatures remain indistinguishable. This effect was detected once during the last localisation experiment carried out. Node H30 was wrongly identified as node H32 and thus, the sequence H32, P31, H32 of time stamp 1700 should have been H30, P31, H32. Notice that nodes H30 and H32 are separated by a short corridor labelled as P31. Summarising, upgrading node information made the developed system valuable for localisation in spite of odometry error.

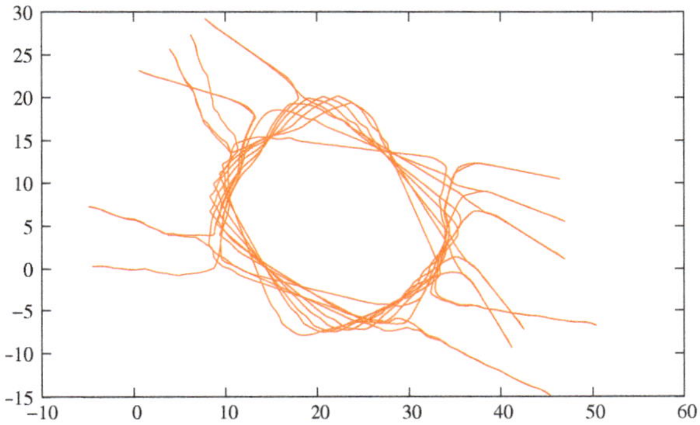

(a) Path corresponding to LODO

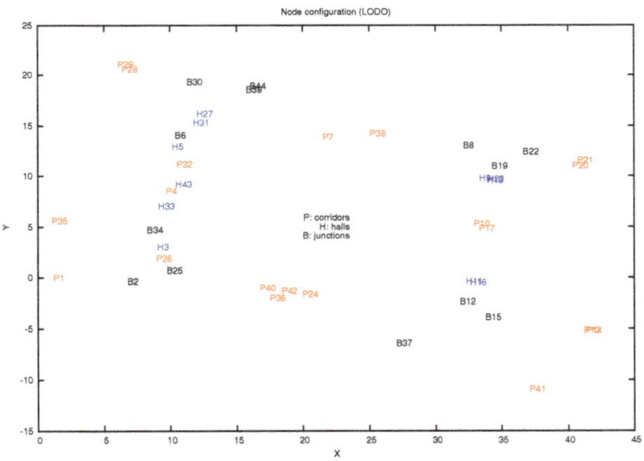

(b) Obtained map

Fig. 14. Tartalo (LODO): robot's path and the obtained map

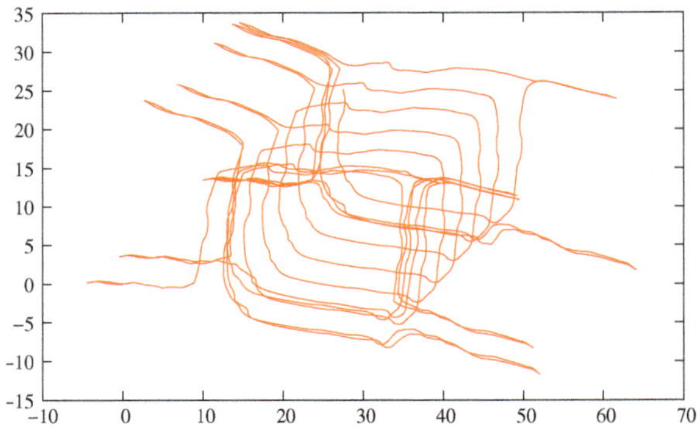

(a) Path corresponding to compass odometry

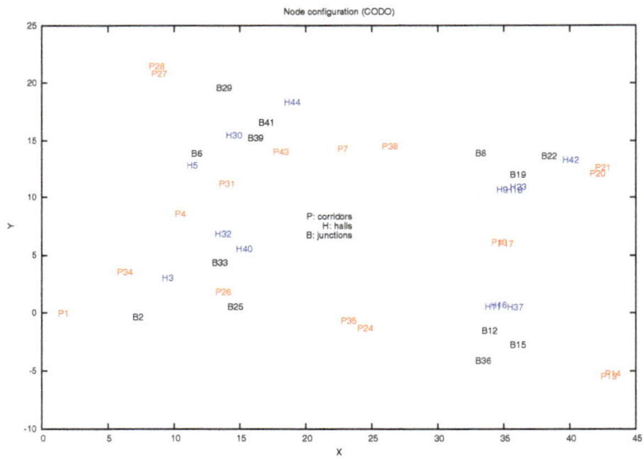

(b) Obtained map

Fig. 15. Tartalo (CODO): robot's path and the obtained map

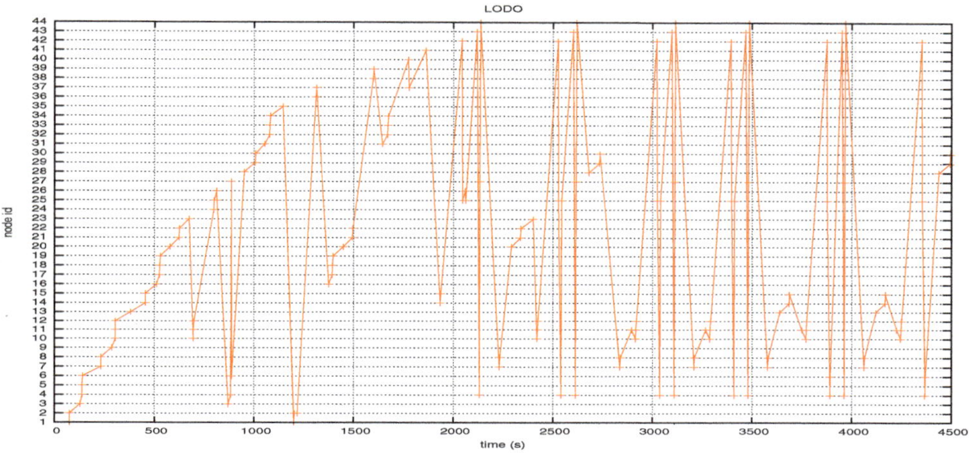

(a) Laser corrected odometry

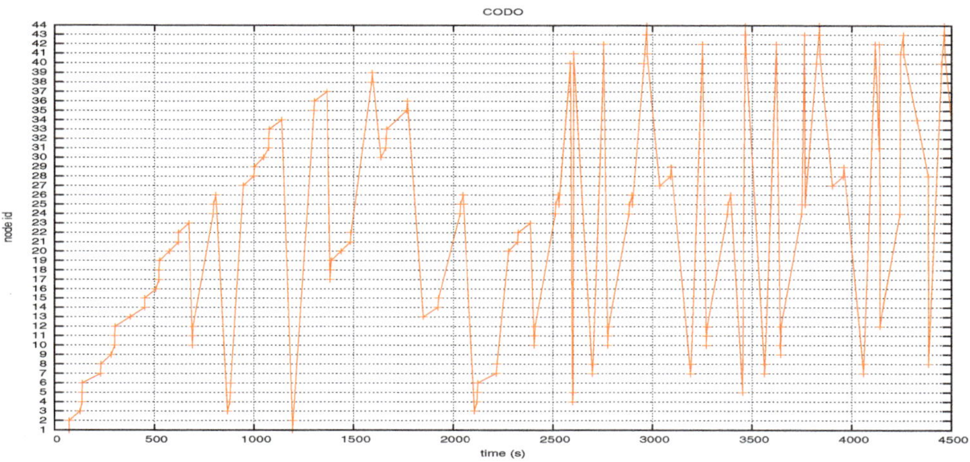

(b) Compass corrected odometry

Fig. 16. Tartalo(LODO and CODO): node identification over time. Labels are set to show extracted path patterns

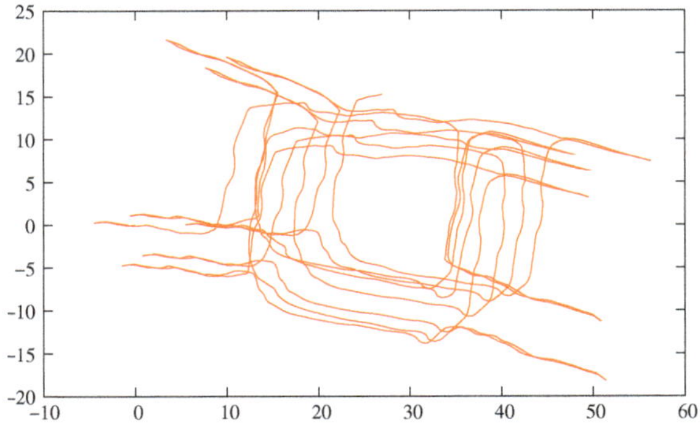

(a) Robot's path

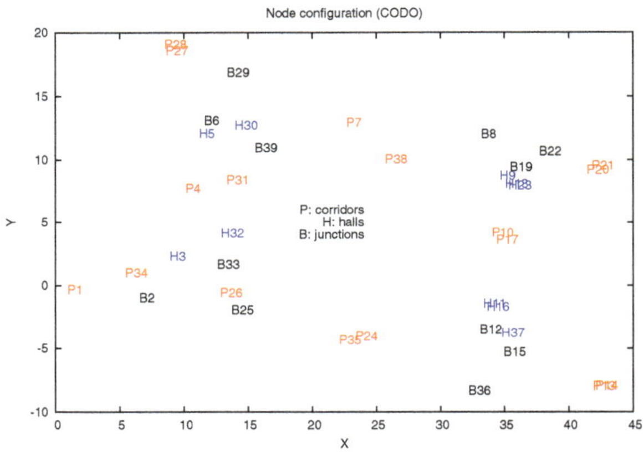

(b) Original map

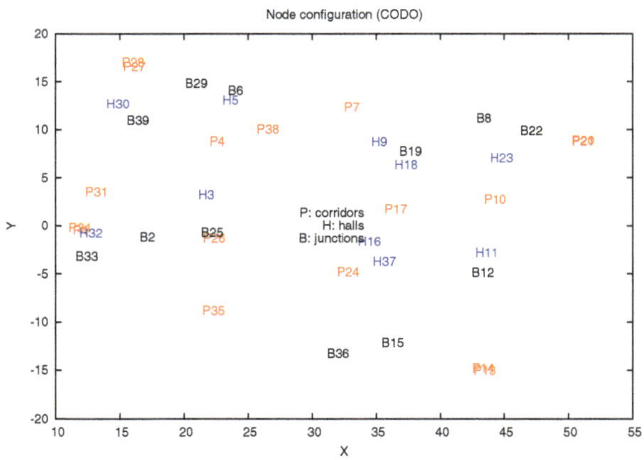

(c) Upgraded map

Fig. 17. Tartalo (CODO with adaptive node location): robot's path and the obtained map

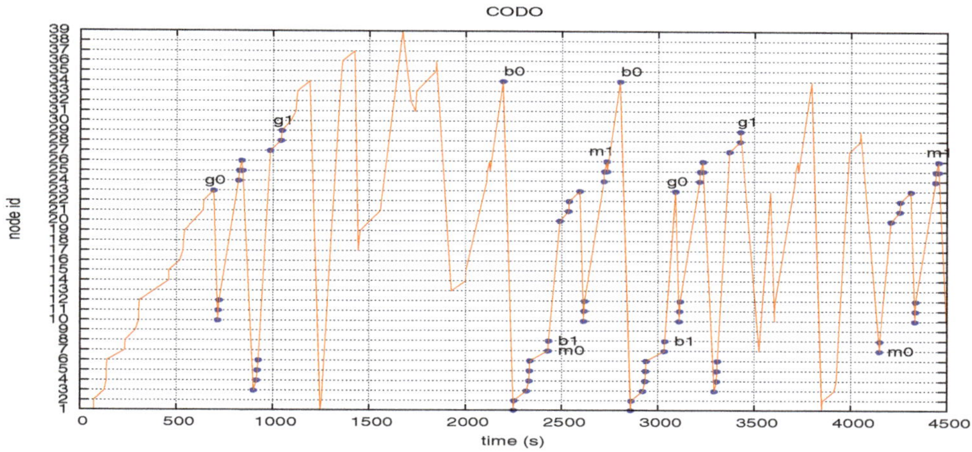

Fig. 18. Tartalo (CODO with adaptive node location): node identification over time

Robot's path	node sequence	time stamp	label
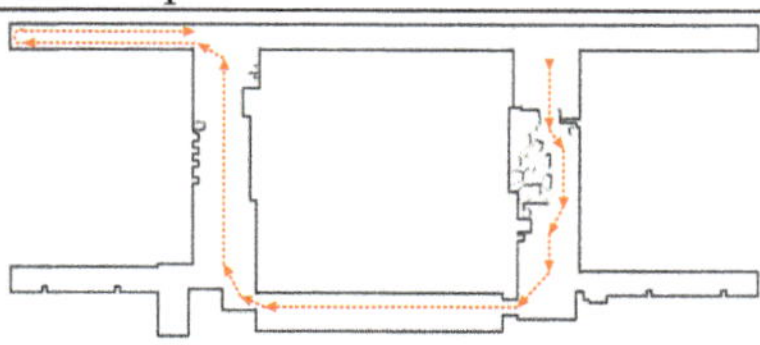	H23, P10, H11, B12, P24, B25, P26, B25, H3, P4, H5, B6, P27, P28, B29	643-1048, 3092-3428	g0-g1
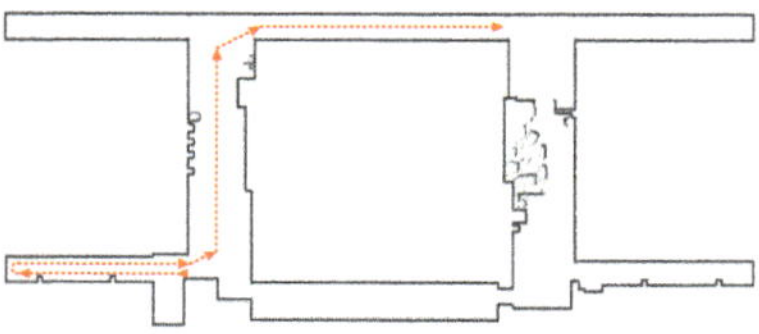	P34, P1, B2, H3, P4, H5, B6, P7	2195-2430, 2804-3034	b0-b1
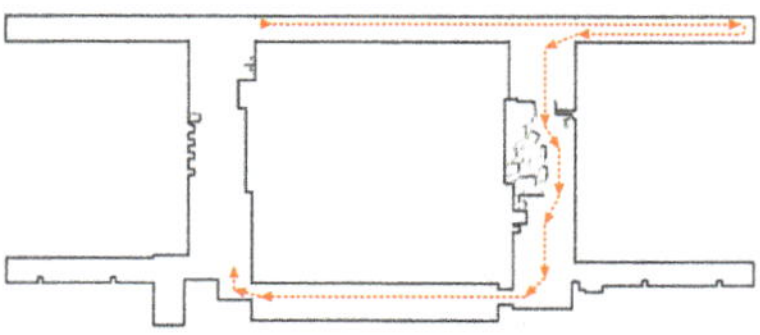	P7, B8, P20, P21, B22, H23, P10, H11, B12, P24, B25, P26, B25	2430-2737, 4150-4459	m0-m1

Table 3. CODO: extracted path patterns

11. Conclusions

In this chapter a new approach for incremental topological map construction was presented. A statistical test called INCA was used to this end, combined with a data sampling approach which decided if a topological node found by the robot had already been visited by it. The method was integrated in a behaviour-based control architecture and tested also for localisation purposes.
To measure the adequateness of the approach the map acquisition was performed non-stop until the robot run out of batteries. Afterwards, the experiments were repeated but once the number of nodes in the map reached a given threshold, the learning step was finished and the acquired map was used for localisation purposes.
However, INCA also suffers from odometry error. Of the two error correction methods used in the present work, LODO and CODO, compass corrected odometry was better suited for the developed navigation approach. A last experiment was carried out using CODO and modifying the contents of the acquired nodes each time a location was revisited. The type of error remaining after CODO facilitated the upgrade of the nodes' locations and improved drastically the localisation process.
The experiments conducted confirmed INCA based mapping and localisation as a valid approach and that BB systems can be provided with automatic map acquisition mechanisms.
To improve the efficiency of the automatic map acquisition system, when looking for correspondences the use of their associated probability value should be studied.
The criteria for stopping the learning process, i.e. the maximum number of nodes should be revised. Given that it is not possible to know a priori the number of nodes, the map should be closed when no more alternative ways remain unvisited in the junction nodes.
Some aspects of the implementation of INCA should be improved and more experiments should be conducted in a systematic manner in order to better identify the advantages and drawbacks of the test.
New local navigation strategies and several landmark identification modules need to be incorporated to increase the granularity of the environment in order to reach more interesting goals than halls and corridors, such as offices and laboratories. Adding more topological nodes would allow the generalisation of the experiments to different environments, and the comparison with other approaches. The first step should be to integrate the door identification and door crossing modules already developed, and to enrich the behaviour associated to several nodes with door crossing abilities, and a wall following behaviour. These two modules would help to cover the perimeter of small rooms and improve the exploration strategy.
Nothing has been said about planning. Up to now, the proposed modifications were tested using an exploration strategy. The overall map should be used for commanding the robot to fulfil a concrete goal and thus, to reach concrete locations.

12. References

Arenas, C. & Cuadras, C. M. (2002). Some recent statistical methods based on distances, *Contributions to Science* 2: 183–191.

Bar-Hen, A. (2001). Preliminary tests in linear discriminat analysis, *Statistica* 4: 585–593.

Brooks, C. & K. Iagnemma, K. (2009). Visual detection of novel terrain via two-class classification, *Proceedings of the ACM symposium on Applied Computing (SAC)*, ACM, New York, NY, USA, pp. 1145–1150.

Brooks, R. A. (1986). A robust layered control system for a mobile robot, *IEEE Journal of robotics and automation* RA–26: 14–23.

Brooks, R. A. (1989). A robot that walks: emergent behaviors from a carefully evolved network, *Technical Report AI MEMO 1091*, MIT.

Brooks, R. A. & Connell, J. H. (1986). Asynchronous distributed control system for a mobile robot, *Proceedings of the SPIE's Cambridge Symposyum on Optical and Optoelectronic Engineering*, pp. 77–84.

Chen, C. & Wang, H. (2006). Appearance-based topological bayesian inference for loop-closing detection in a cross-country environment, *International Journal of Robotic Research* 25(10): 953–983.

Connell, J. H. (1990). *Minimalist mobile robotics. A colony-style architecture for an artificial creature*, Academic Press, inc.

Cuadras, C. M. (1992). Some examples of distance based discrimination, *Biometrical Letters* pp. 3–20.

Cuadras, C. M. & Fortiana, J. (1995). A continuous metric scaling solution for a random variable, *Journal of Multivariate Analysis* 32: 1–14.

Cuadras, C. M. & Fortiana, J. (2000). The importance of geometry in multivariate analysis and some applications, *Statistics for the 21st Century*, Marcel Dekker, New York, pp. 93–108.

Cuadras, C. M., Fortiana, J. & Oliva, F. (1997). The proximity of an individual to a population with applications in discriminant analysis, *Journal of Classification* 14: 117–136.

Fraundorfer, F., Engels, C. & Nistér, D. (2007). Topological mapping, localization and navigation using image collections, *Intelligent Robots and Systems (IROS)*, pp. 3872–3877.

Gower, J. C. (1985). *Encyclopedia of Statistical Sciences*, Vol. 5, John Wiley & Sons, New York, chapter Measures of similarity, dissimilarity and distance, pp. 397–405.

Hempstalk, K., Frank, E. & Witten, I. H. (2008). One-class classification by combining density and class probability estimation, *ECML PKDD '08: Proceedings of the 2008 European Conference on Machine Learning and Knowledge Discovery in Databases - Part I*, Springer-Verlag, Berlin, Heidelberg, pp. 505–519.

Howard, A. (2005). Multi-robot simultaneous localization and mapping using particle filters, *In Robotics: Science and Systems*, pp. 201–208.

Irigoien, I. & Arenas, C. (2008). INCA: New statistic for estimating the number of clusters and identifying atypical units, *Statistics in Medicine* 27(15): 2948–2973.

Jauregi, E., Irigoien, I., Sierra, B., Lazkano, E. & Arenas, C. (2011). Lopp-closing: a tipicality approach, *Robotics and Autonomous Systems* 59: 218–227.

Mallot, H. A. & Franz, M. A. (2000). Biomimetic robot navigation, *Robotics and Autonomous System* 30: 133–153.

Manevitz, L. M. & Yousef, M. (2002). One-class SVMs for document classification, *J. Mach. Learn. Res.* 2: 139–154.

Manevitz, L. & Yousef, M. (2007). One-class document classification via neural networks, *Neurocomput.* 70(7-9): 1466–1481.

Matarić, M. (1990). *A distributed model for mobile robot environment-learning and navigation*, Master's thesis, MIT Artificial Intelligence Laboratory.

McDonald, L. L., Lowe, V. W., Smidt, R. K. & Meister, K. A. (1976). A preliminary test for discriminant analysis based on small samples, *Biometrics* 32: 417–422.

Olson, E. (2009). Recognizing places using spectrally clustered local matches, *Robotics and Autonomous System* 57: 1157–1172.

Rao, C. R. (1962). Use of discriminant and allied functions in multivariate analysis, *Sankhya-Serie A* 24: 149–154.

Rao, C. R. (1982). Diversity: its measurement, decomposition, apportionment and analysis, *Sankhyā. The Indian Journal of Statistics, Series A* 44: 1–22.

Sánchez-Yáñez, R. E., Kurmyshev, E. V. & Fernández, A. (2003). One-class texture classifier in the CCR feature space, *Pattern Recognition Letters* 24(9-10): 1503–1511.

Se, S., Lowe, D. G. & Little, J. J. (2005). Vision-based global localization and mapping for mobile robots, *IEEE Transactions on Robotics* 21: 364–375.

Tardós, J. D., Neira, J., Newman, P. M. & Leonard, J. J. (2002). Robust mapping and localization in indoor environments using sonar data, *The International Journal of Robotics Research* 21(4): 311–330.

Tax, D. (2001). *One-class classification; Concept-learning in the absence of counter-examples*, PhD thesis, Delft University of Technology.

Thrun, S. (1999). Learning metric-topological maps for indoor mobile robot navigation, *Artificial Intelligence* pp. 21–71.

Thrun, S. & Bücken, A. (1996a). Integrating grid-based and topological maps for mobile robot navigation, *Proceedings of the Thirteenth National Conference on Artificial Intelligence*, pp. 944–950.

Thrun, S. & Bücken, A. (1996b). Learning maps for indoor mobile robot navigation, *Technical report*, Carnegie Mellon University.

Thrun, S., Burgard, W. & Fox, D. (2005). *Probabilistic Robotics*, MIT Press.

Wang, Q. & Lopes, L. S. (2005). *Emerging Solutions for Future Manufacturing Systems*, Vol. 159, Springer Boston, chapter One-Class Learning for Human-Robot Interaction, pp. 489–498.

Williams, B., Cummins, M., Neira, J., Newman, P., Reid, I. & Tardós, J. (2009). A comparison of loop closing techniques in monocular SLAM, *Robotics and Autonomous System* 57: 1157–1172.

15

The Development of the Omnidirectional Mobile Home Care Robot

Jie-Tong Zou
Department of Aeronautical Engineering, National Formosa University
Taiwan, R.O.C.

1. Introduction

In the last few years, intelligent robots were successfully fielded in hospitals (King, S., and Weiman, C, 1990), museums (Burgard, W. et al., 1999), and office buildings/department stores (Endres, H. et al. , 1998), where they perform cleaning services, deliver, educate, or entertain (Schraft, R., and Schmierer, G. , 2005). Robots have also been developed for guiding blind people, as well as robotic aids for the elderly.
Today, the number of elderly in need of care is increasing dramatically. As the baby-boomer generation approaches the retirement age, this number will increase significantly. Current living conditions for the majority of elderly people are already unsatisfactory, and situation will worsen in the future. (Nicholas Roy et al. , 2000)
Rapid progress of standard of living and health care resulted in the increase of aging population. More and more elderly people do not receive good care from their family or caregivers. Maybe the intelligent service robots can assist people in their daily living activities. Robotics aids for the elderly have been developed, but many of these robotics aids are mechanical aids. (Song, W.-K. et al., 1998) (Dario, P. et al., 1999) (Takahashi, Y. et al., 1999). The intelligent service robot can assist elderly people with many tasks, such as remembering to take medicine or measure blood pressure on time.
The main objective of this Chapter is to develop an omnidirectional mobile home care robot. This service mobile robot is equipped with "Indoor positioning system". The indoor positioning system is used for rapid and precise positioning and guidance of the mobile robot. Five reflective infrared sensors are placed around the robot for obstacle avoidance.
The wireless IP camera is placed on the top layer of this robot. Through the internet remote control system, the live image of the IP camera on the robot can be transferred to the remote client user. With this internet remote control system, the remote client user can monitor the elderly people or the home security condition. On the aid of this system, remote family member can control the robot and talk to the elderly. This intelligent robot also can deliver the medicine or remind to measure the blood pressure or blood sugar on time. We hope this intelligent robot can be a housekeeper or family guard to protect our elderly people or our family.
The functions of the proposed robot are illustrated as follows:

- deliver medicine or food on time
- remind to measure and record the blood pressure or blood sugar of the elderly on time

- remind the elderly to do something important
- assist the elderly to stand or walk
- send a short message automatically under emergency condition
- With the remote control system, remote family member can control the robot and talk to the elderly.

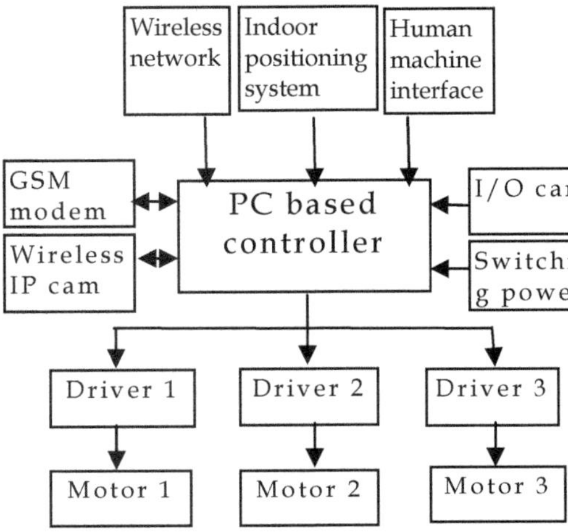

Fig. 1. Hardware structure of the omnidirectional mobile home care robot.

Hardware structure of the omnidirectional mobile home care robot is shown in Fig. 1. A PC based controller was used to control three DC servo motors. The indoor positioning system was used for rapid and precise positioning and guidance of the mobile robot. Five reflective infrared sensors are connected to an I/O card for sensor data acquisition. The GSM modem can send a short message automatically under emergency condition. The live image of the wireless IP camera on the robot can be transferred to the remote client user. The subsystems of this robot are explained in the following sections.

2. The robot mechanism

The proposed omnidirectional mobile home care robot is shown in Fig. 2 and Fig. 3. The main body of this robot is consisted with five layers of hexagonal aluminum alloy board. Many wheeled mobile robots are equipped with two differential driving wheels. Since these robots possess 2 degrees-of-freedom (DOFs), they can rotate about any point, but cannot perform holonomic motion including sideways motion (Jae-Bok Song and Kyung-Seok Byun , 2006). To increase the mobility of this service robot, three omni-directional wheels driven by three DC servo motors are assembled on the robot platform (see Fig. 4). The omni-directional mobile robots can move in an arbitrary direction without changing the direction of the wheels.
The three-wheeled omni-directional mobile robots are capable of achieving 3 DOF motions by driving 3 independent actuators (Carlisle, B., 1983) (Pin, F. & Killough, S., 1999), but they may have stability problem due to the triangular contact area with the ground, especially when traveling on a ramp with the high center of gravity owing to the payload they carry.

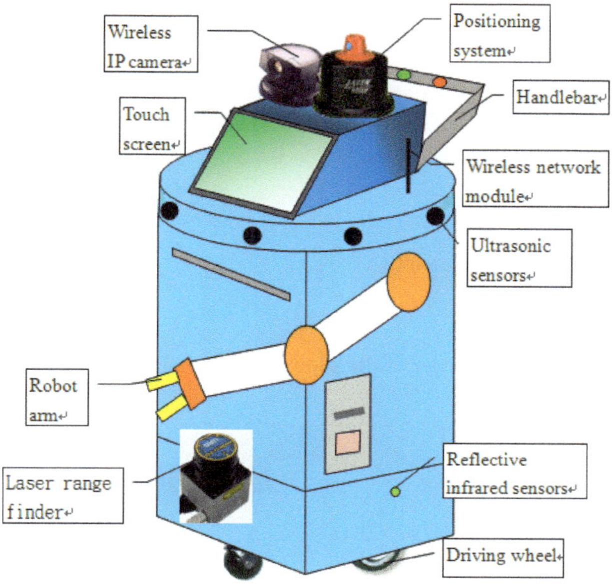

Fig. 2. Structure of the omnidirectional mobile home care robot.

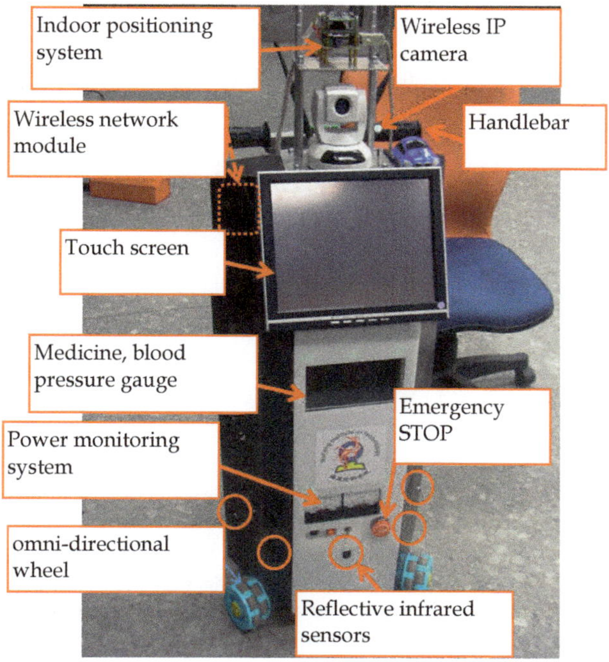

Fig. 3. Photo of the omnidirectional mobile home care robot.

Fig. 5(a) is structures of the omni-directional wheel, Fig. 5(b) is the motor layout of the robot platform. The relationship of motor speed and robot moving speed is shown as:

$$
\begin{aligned}
V_1 &= \omega_1 r = V_x + \omega_p R \\
V_2 &= \omega_2 r = -0.5V_x + 0.867V_y + \omega_p R \\
V_3 &= \omega_3 r = -0.5V_x - 0.867V_y + \omega_p R
\end{aligned} \tag{1}
$$

Where:

V_i=Velocity of wheel i

ω_i=rotation speed of motor i

ω_P= rotation speed of robot

r=radius of wheel

R=distance from wheel to center of the platform

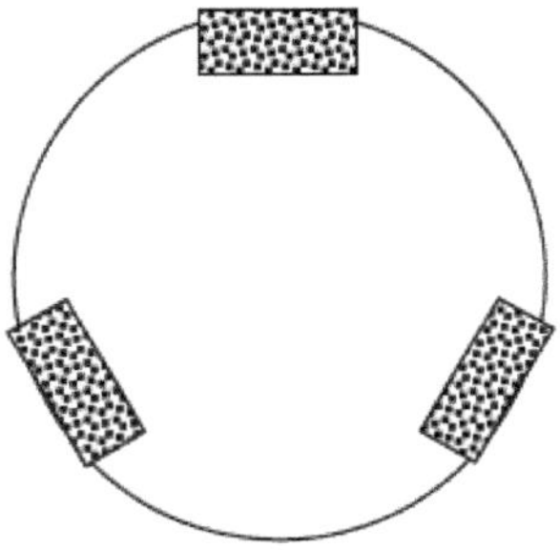

Fig. 4. Robot platform with three Omnidirectional wheels.

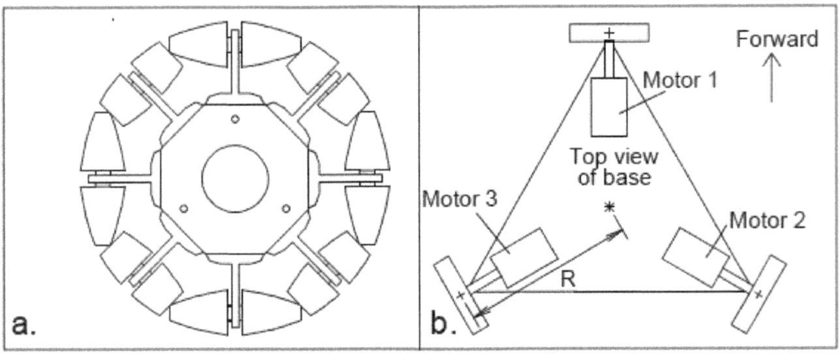

Fig. 5. (a) Structure of Omni-directional wheel; (b)Motor layout of Robot platform

As shown in Fig. 6, 3D CAD software (SolidWork) was used to design the robot platform. In Fig. 7, an omni-directional wheel composed of passive rollers or balls was driven by a DC servo motor. Fig. 8 is the photo of the robot platform with three omni-directional wheels. The omni-directional robot platform can move in an arbitrary direction and can rotate about any point. The omni-directional robot platform can enhance the mobility of mobile robots.

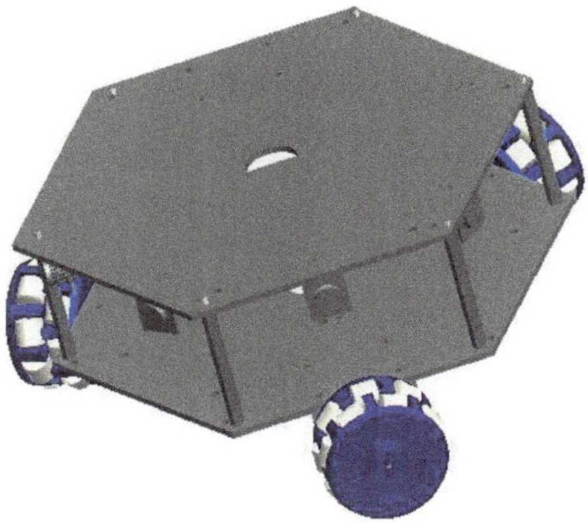

Fig. 6. 3D CAD software was used to design the robot platform

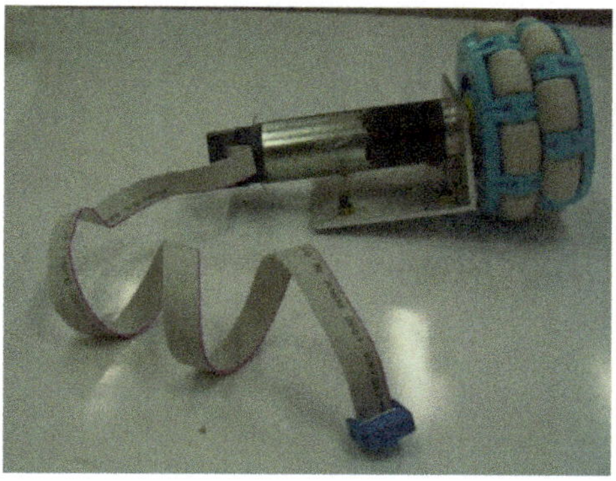

Fig. 7. An omni-directional wheel was driven by DC servo motor

Fig. 8. Photo of the robot platform with three omni-directional wheels

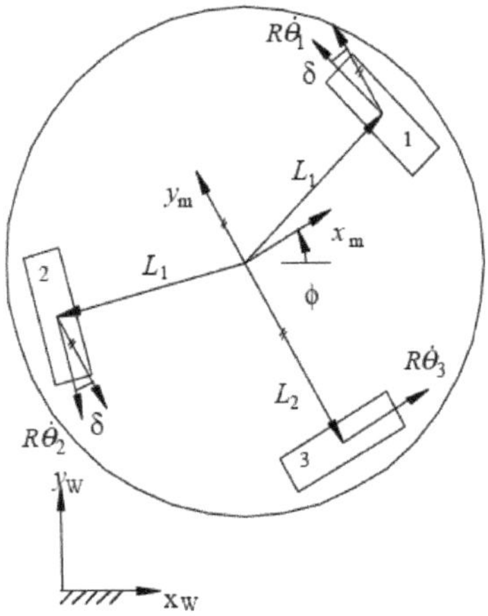

Fig. 9. Kinematic Diagram of the robot platform with three omni-directional wheels

Fig. 9 is the kinematic diagram of the robot platform with three omni-directional wheels. The inverse kinematic equations of the robot platform with three omni-directional wheels are shown as follow:

$$
\begin{aligned}
R\dot{\theta}_1 &= -\sin(\delta+\phi)\dot{x}_w + \cos(\delta+\phi)\dot{y}_w + L_1\dot{\phi} \\
R\dot{\theta}_2 &= -\sin(\delta-\phi)\dot{x}_w - \cos(\delta-\phi)\dot{y}_w + L_1\dot{\phi} \\
R\dot{\theta}_3 &= \cos(\phi)\dot{x}_w + \sin(\phi)\dot{y}_w + L_2\dot{\phi}
\end{aligned}
\tag{2}
$$

Where:

L_i=distance from center of the platform to omni-directional wheel

ψ=Orientation angle according to world coordinate $[x_w, y_w]$

θ_i=rotation angle of omni-directional wheel i

The inverse Jacobian matrix can be derived from the above equations:

$$
\begin{Bmatrix} \dot{\theta}_1 \\ \dot{\theta}_2 \\ \dot{\theta}_3 \end{Bmatrix} = \frac{1}{R}\begin{bmatrix} -\sin(\delta+\phi) & \cos(\delta+\phi) & L_1 \\ -\sin(\delta-\phi) & -\cos(\delta-\phi) & L_1 \\ \cos(\phi) & \sin(\phi) & L_2 \end{bmatrix}\begin{Bmatrix} \dot{x}_w \\ \dot{y}_w \\ \dot{\phi} \end{Bmatrix}
\tag{3}
$$

3. Indoor localization system

As shown in Fig. 10, Indoor localization system (http://www.hagisonic.com/), which used IR passive landmark technology, was used in the proposed service mobile robot. The localization sensor module (see Fig. 11) can analyze infrared ray image reflected from a passive landmark (see Fig. 12) with characteristic ID. The output of position and heading angle of a mobile robot is given with very precise resolution and high speed. The position repetition accuracy is less than 2cm; the heading angle accuracy is 1 degree.

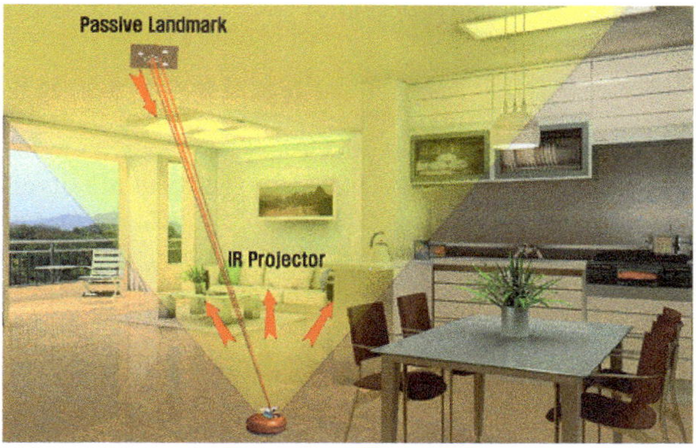

Fig. 10. Indoor localization system (Hagisonic co.)

Fig. 11. Localization sensor module (Hagisonic co.)

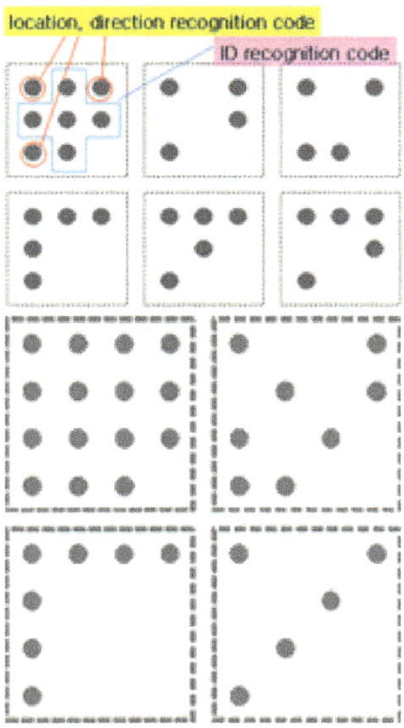

Fig. 12. IR passive landmark (Hagisonic co.)

4. Robot control system

As shown in Fig. 13, a PC based controller was used to control the mobile robot. Through RS232 interface, PC based controller can control three motor drivers to drive three DC servo motors. The PC based controller and Solid State Disks (SSD) are shown in Fig. 14. Solid State Disks have no moving parts. Consequently, SSDs deliver a level of reliability in data storage that hard drives cannot approach. In this mobile robot application that is exposed to shock or vibration, the reliability offered by SSDs is vitally important.

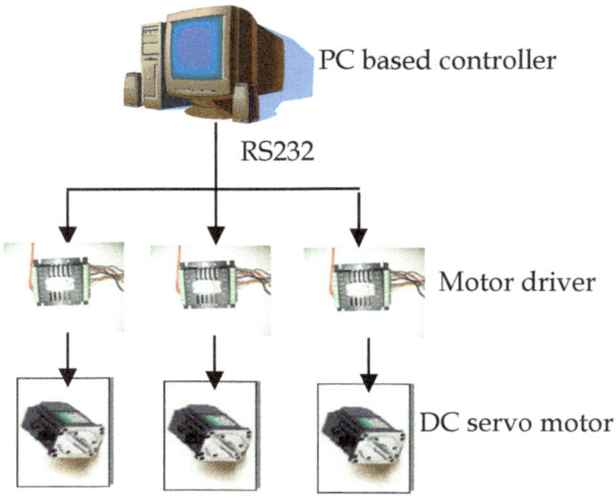

Fig. 13. PC based controller and motor drivers.

Fig. 14. PC based controller and Solid State Disk (SSD)

5. Obstacle avoidance system

Obstacle avoidance is a robotic discipline with the objective of moving vehicles on the basis of the sensorial information. As shown in Fig. 15, five reflective infrared sensors (see Fig. 16) are placed around the robot for obstacle avoidance. Five infrared sensors are numbered from 1 to 5 in a counterclockwise direction. If the obstacle is in front of the robot or on the left hand side, it will turn right. If the obstacle is on the right hand side, it will turn left.

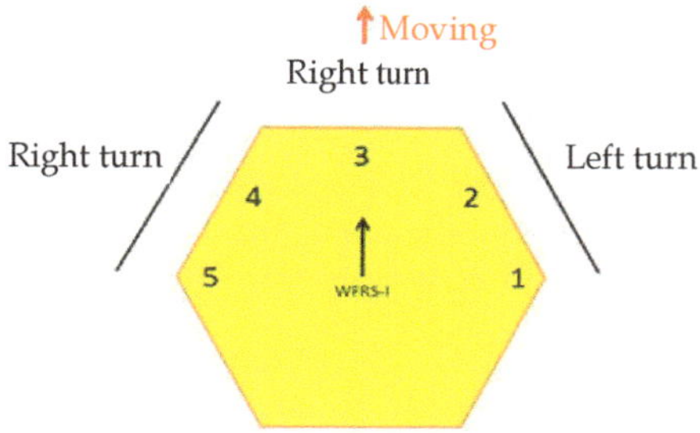

Fig. 15. Five reflective infrared sensors are placed around the robot on the bottom layer

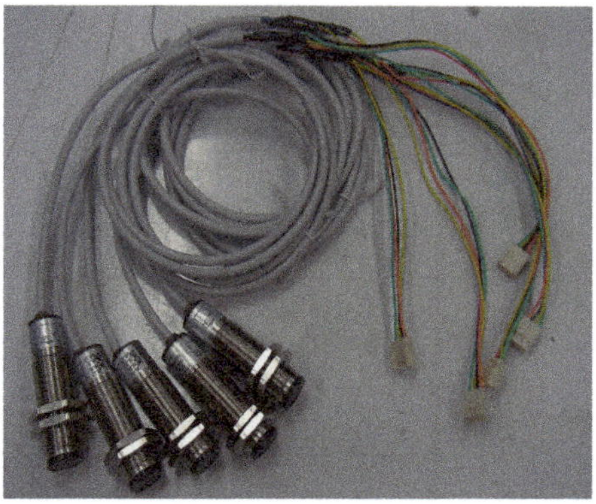

Fig. 16. Five reflective infrared sensors

6. Human-machine interface (HMI)

The human-machine interface (HMI) includes touch screen, speaker, and appliances voice control system. Touch screen can be regarded as input and display interface. Speaker can

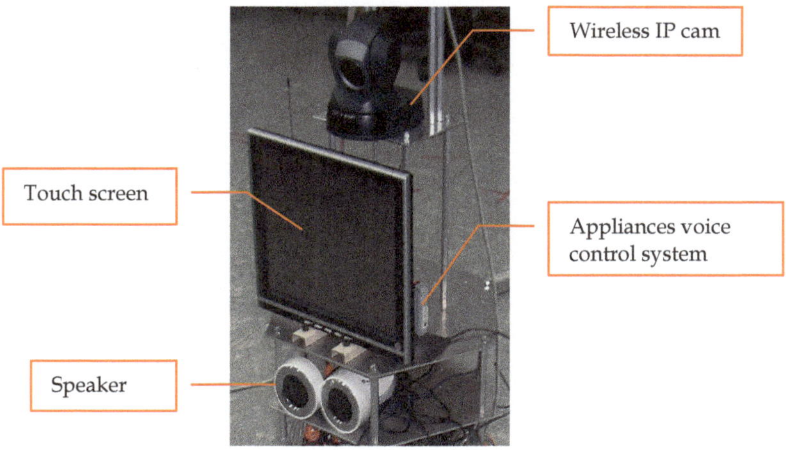

Fig. 17. Human-machine interface (HMI)

produce the voice of robot. Appliances voice control system can let users or the elderly to remote control the appliances by voice command.

7. Software interface

The software interface of the proposed robot is developed by Visual BASIC program. As shown in Fig. 18, the main interface of the proposed service robot can be divided into the following six regions:

1. **Home map region**: The home map and the targets position are displayed in this region. With the information from the indoor positioning system, the position and heading angle of the mobile robot also can be shown in this region.
2. **Robot targets setting region**: First, as a teaching stage, a user controls a robot by joystick or other interface and teaches the targets to the robot. The position and heading angle of the mobile robot on the target place can be recorded into a file. Next, as a playback stage, the robot runs autonomously on the path instructed during the teaching stage.
3. **Positioning system information region**: With the aid of the indoor positioning system, the mobile robot position (X,Y) and heading angle also can be shown in this region.
4. **Infrared sensors information**: Five reflective infrared sensors are placed around the robot for obstacle avoidance. Five reflective infrared sensors are connected to an I/O card for sensor data acquisition. Obstacles in front of the mobile robot can be displayed in this region.
5. **Robot control interface**: In this region, users can control the mobile robot to move in an arbitrary direction or rotate about any point
6. **Remote control information**: With the internet remote control system, the remote client user can monitor the elderly people or the home security condition. On the aid of this system, remote family member can control the robot and talk to the elderly. The remote user IP and the remote control command also can be shown in this region.

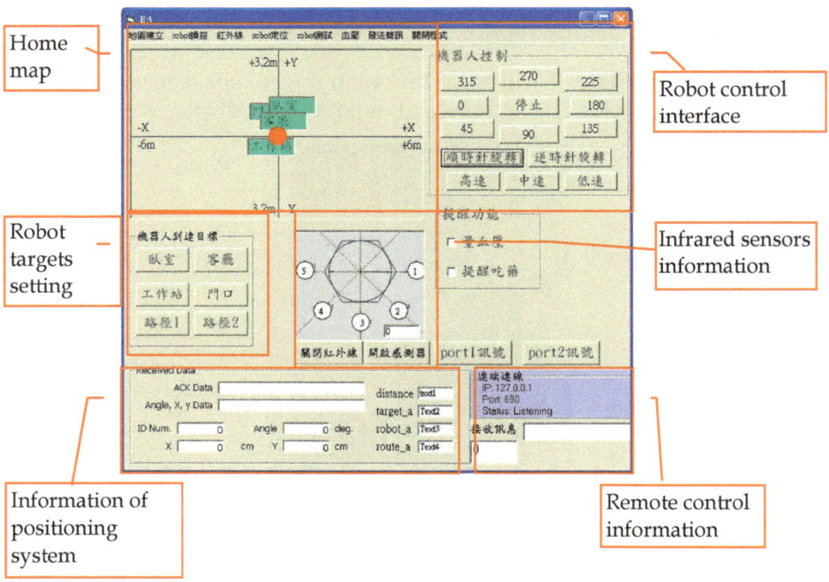

Fig. 18. Main interface of the proposed robot

The proposed service robot can remind the elderly to measure and record the blood pressure or blood sugar on time. As shown in Fig. 19, the blood pressure or blood sugar data can be displayed and recorded in this interface. If blood pressure or blood sugar data is too high, the GSM modem will send a short message automatically to the remote families.

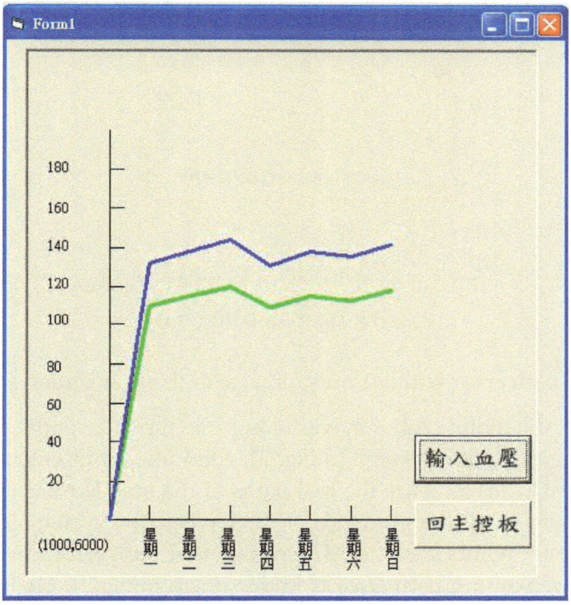

Fig. 19. Interface for blood pressure measurement

8. Experimental results

In order to understand the stability of three wheeled omni-directional mobile robot, an experiment for the straight line path error had been discussed (Jie-Tong Zou, et al., 2010). From these experimental results, when the robot moves faster or farther, the straight line error will increase. We make some experiments to measure several different paths error of the proposed mobile robot in this research.

8.1 Rectangular path error test for the omni-directional robot platform

In this experiment, the proposed mobile robot will move along a rectangular path with or without the guidance of the indoor localization system. As shown in Fig. 20, the mobile robot moves along a rectangular path (a→b→c→d→a) without the guidance of the localization system. The localization system is only used to record the real path in this experiment.

In Fig. 20, solid line represents the ideal rectangular path, dot lines (■:Test1, ▲:Test2) are the real paths of the mobile robot without the guidance of the localization system. The vertical paths (path b→c and d→a) have larger path error. Finally, the mobile robot cannot return to the starting point "a".

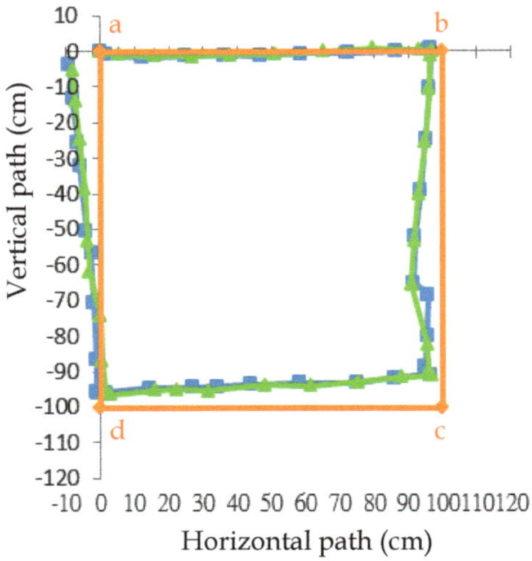

Fig. 20. Rectangular path error without the guidance of the localization system.

As shown in Fig. 21, the mobile robot moves along a rectangular path (a→b→c→d→a) with the guidance of the localization system. In Fig. 21, solid line represents the ideal rectangular path, dot lines (■:Test1, ▲:Test2) are the real paths of the mobile robot with the guidance of the localization system. With the guidance of the localization system, the mobile robot can pass through the corner points a, b, c, d. The rectangular path error in Fig.21 is smaller than that in Fig. 20. The maximum path error is under 10 cm in Fig.21. Finally, the mobile robot can return to the starting point "a". The rectangular path is closed at point "a".

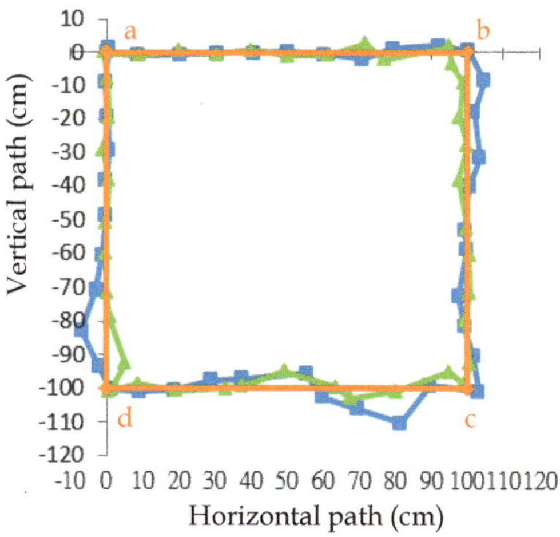

Fig. 21. Rectangular path error with the guidance of the localization system.

8.2 Circular path error test for the omni-directional robot platform

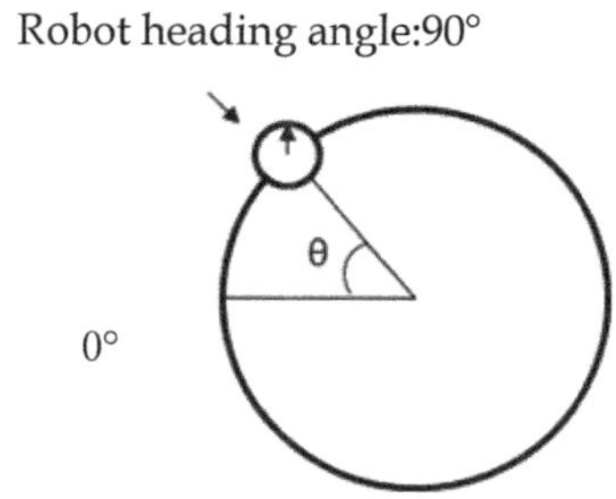

Fig. 22. Circular angle (θ) of the robot

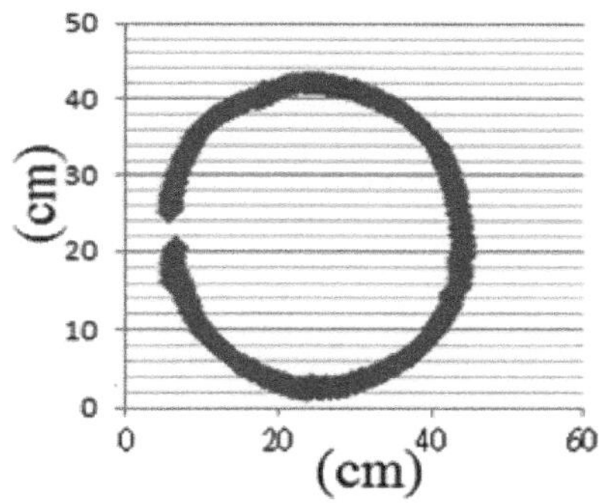

Fig. 23. Circular path without the guidance of the localization system.

The omni-directional mobile robot can move in an arbitrary direction without changing the direction of the wheels. In this experiment, the proposed mobile robot will move along a circular path with or without the guidance of the indoor localization system. As shown in Fig. 22, the mobile robot moves along a circular path without the guidance of the localization system. The robot heading angle is 90°(upwards) during this test. The localization system is only used to record the real path in this experiment.

The circular path without the guidance of the localization system is shown in Fig. 23. The shape of the real path is similar to a circle, but the starting point and the end point cannot overlap. The heading angle error with different circular angle (θ) of the robot is shown in Fig. 24. The maximum heading angle error is about 8°.

The circular path with the guidance of the localization system is shown in Fig. 25. The shape of this path is more similar to a circle; the starting point and the end point are overlapped. The heading angle error with different circular angle (θ) of the robot is shown in Fig. 26. The maximum heading angle error is about ±1°. From this experiment result, the localization system can successfully maintain the robot heading angle along a circular path.

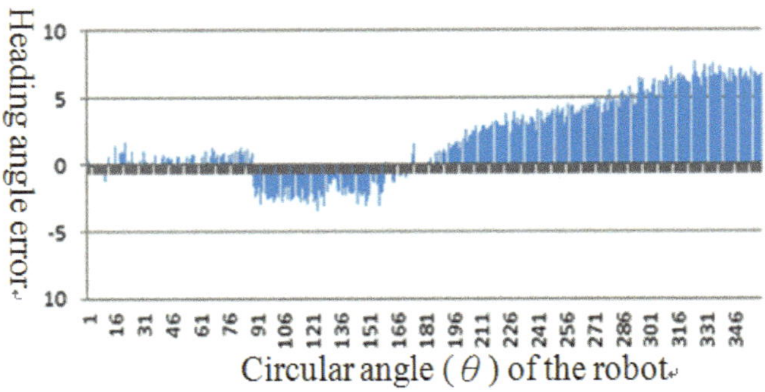

Fig. 24. Heading angle error with different circular angle (θ) of the robot

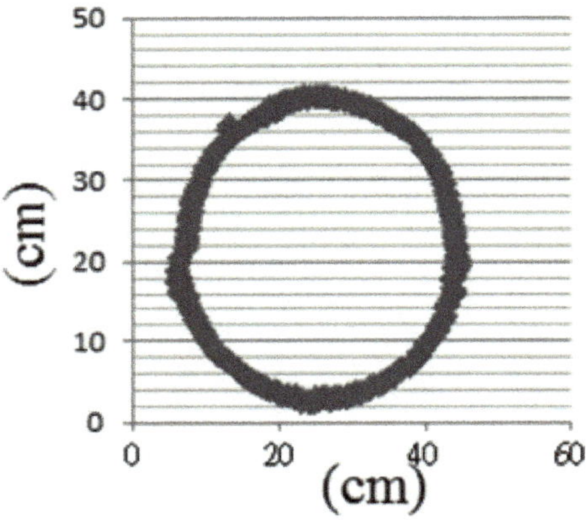

Fig. 25. Circular path with the guidance of the localization system.

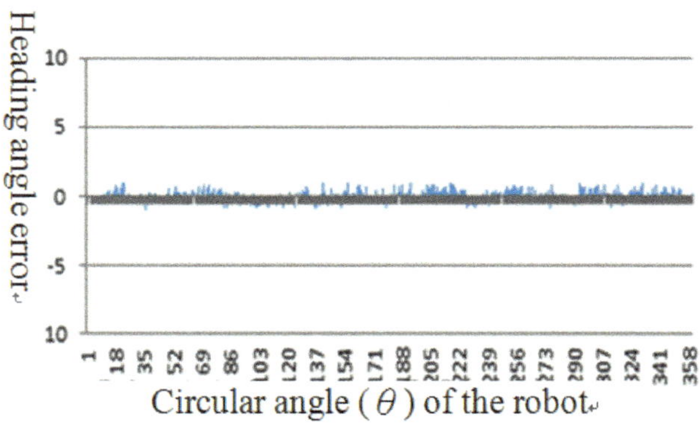

Fig. 26. Heading angle error with different circular angle (θ) of the robot

8.3 Functions test for robot taking care of the elderly

8.3.1 Delivering medicine or food on time

The elderly people usually forget to take medicine or measure blood pressure on time. It is harmful for the elderly people's health. The proposed robot can deliver medicine or food on the preset time. The robot also can remind the elderly to take medicine on time.

8.3.2 Assist the elderly to stand or walk

As shown in Fig. 27, a handlebar is placed on the rear side of the robot. With the assistance of the robot, the elderly can hold the handlebar to stand up or walk. The elderly can set the target place on the touch screen and hold on the handle bar, the mobile robot will help the elderly to the target place. There are four buttons (Start, Stop, Speed up, Slow down) on the handlebar, the elderly can control the robot speed to fit his walk speed.

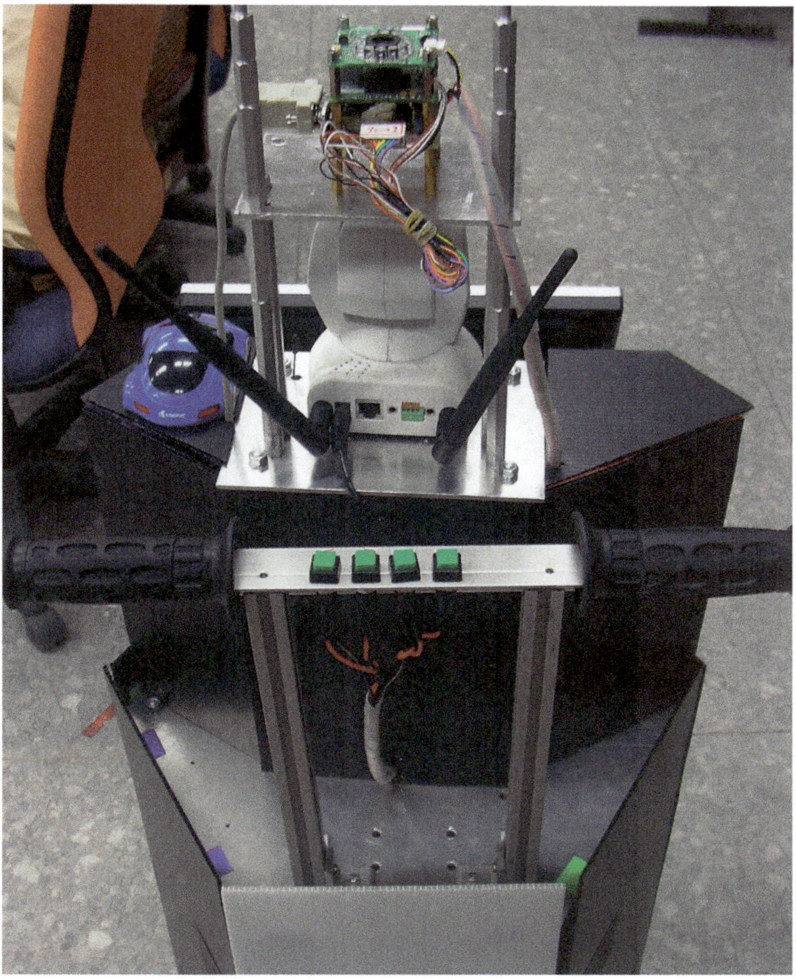

Fig. 27. Robot can assist the elderly to stand or walk

8.3.3 Send a short message automatically under emergency condition

The blood pressure measurement and short message sending interface is shown in Fig. 28. The robot will remind the elderly to take blood pressure or blood sugar on time. When the elderly finished taking blood pressure (the blood pressure gauge is shown in Fig. 29), the blood pressure data will be recorded in the robot's computer. If the blood pressure is too high, the robot will send a short message to the remote families automatically. If the elderly has emergency condition, for example, the elderly falls down, he can press the "Emergency" button, the robot also can send a short message to the families to deal with this emergency condition.

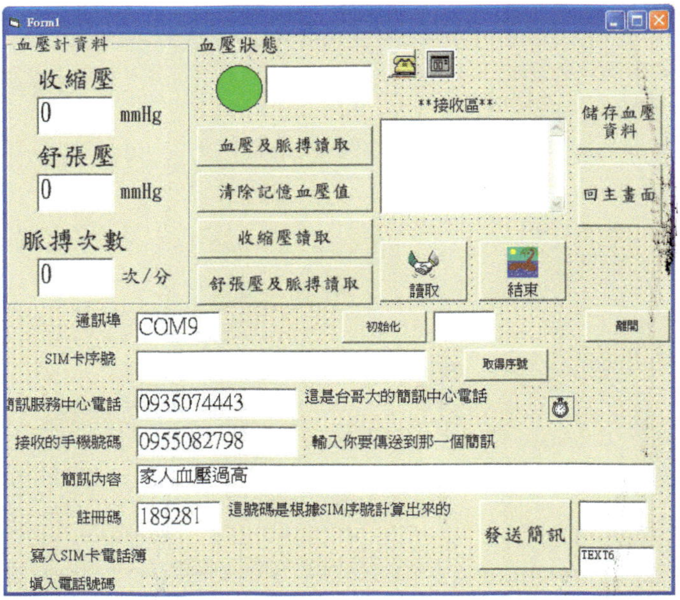

Fig. 28. Blood pressure measurement and short message sending interface

Fig. 29. Blood pressure gauge

8.3.4 Remote control system

The wireless IP camera is placed on the top layer of this robot. Through the internet remote control system, the live image of the IP camera on the robot can be transferred to the remote client user. With this internet remote control system, the remote client user can monitor the elderly people or the home security condition. On the aid of this system, remote family member can control the robot and talk to the elderly.

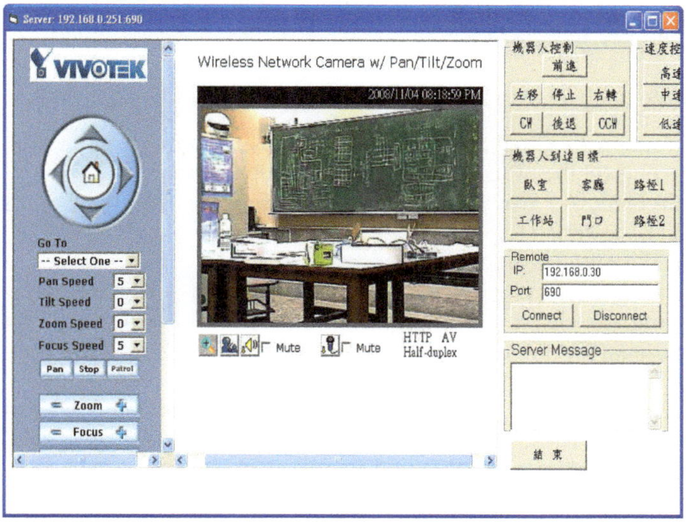

Fig. 30. Remote control interface

9. Conclusion

Today, the number of elderly in need of care is increasing dramatically. More and more elderly people do not receive good care from their family or caregivers. Maybe the intelligent service robots can assist people in their daily living activities.

The main objective of this Chapter is to present an omnidirectional mobile home care robot. This service mobile robot is equipped with "Indoor positioning system". The indoor positioning system is used for rapid and precise positioning and guidance of the mobile robot. Five reflective infrared sensors are placed around the robot for obstacle avoidance.

In order to present the stability of three wheeled omni-directional mobile robot, the ahthors make some experiments to measure the rectangular and circular path error of the proposed mobile robot in this research.

Firstly, the mobile robot moves along a rectangular path without the guidance of the localization system. The experimental paths have larger path error. Finally, the mobile robot cannot return to the starting point. To overcome this problem, the indoor localization system was used to compensate the path error. With the guidance of the localization system, the maximum path error is under 10 cm. Finally, the mobile robot can pass through the corner points and return to the starting point. The rectangular path is closed at the starting point.

Secondly, the proposed mobile robot can move along a circular path with or without the guidance of the indoor localization system. The circular path without the guidance of the localization system cannot be closed. The maximum heading angle error is about 8°.

The circular path with the guidance of the localization system is more similar to a circle; the starting point and the end point are overlapped. The maximum heading angle error is about ±1°. From this experiment result, the localization system can successfully maintain the robot heading angle along a circular path.

On the aid of the remote control system, remote family member can control the robot and talk to the elderly. This intelligent robot also can deliver the medicine or remind to measure the blood pressure or blood sugar on time. We hope this intelligent robot can be a housekeeper or family guard to protect our elderly people or our family.

10. References

King, S., and Weiman, C (1990), "Helpmate autonomous mobile robot navigation system", Proceedings of the SPIE Conference on Mobile Robots, Vol. 2352, pp.190–198.

Burgard, W.; Cremers, A.; Fox, D.; H¨ahnel, D.; Lakemeyer, G.;Schulz, D.; Steiner,W.; and Thrun, S. (1999), "Experiences with an interactive museum tour-guide robot", Artificial Intelligence.

Endres, H.; Feiten, W.; and Lawitzky, G. (1998), "Field test of a navigation system: Autonomous cleaning in supermarkets.", Proceedings of the 1998 IEEE International Conference on Robotics & Automation (ICRA 98).

Schraft, R., and Schmierer, G. (1998), "Serviceroboter", Springer verlag. In German.

Nicholas Roy, Gregory Baltus, Dieter Fox, Francine Gemperle, Jennifer Goetz, Tad Hirsch Dimitris Margaritis, Michael Montemerlo, Joelle Pineau, Jamie Schulte, Sebastian Thrun (2000), "Towards Personal Service Robots for the Elderly", Workshop on Interactive Robots and Entertainment (WIRE 2000).

Song, W.-K.; Lee, H.-Y.; Kim, J.-S.; Yoon, Y.-S.; and Bien, Z. Kares (1998), "Intelligent rehabilitation robotic system for the disabled and the elderly. Proceedings of the 20th Inter. Conf. of the IEEE Engineering in Medicine and Biology Society, Vol. 5, pp. 2682–2685.

Dario, P.; Laschi, C.; and Guglielmelli, E. (1999), "Design and experiments on a personal robotic assistant.", Advanced Robotics 13(2) ,pp.153–69.

Takahashi, Y.; Kikuchi, Y.; T.Ibaraki; and Ogawa, S. (1999), "Robotic food feeder.", Proceedings of the *38th* International Conference of *SICE*, pp. 979–982.

Jae-Bok Song and Kyung-Seok Byun (2006), "Design and Control of an OmnidirectionalMobile Robot with SteerableOmnidirectional Wheels", Mobile Robots, Moving Intelligence, pp. 576.

Carlisle, B. (1983) "An Omnidirectional Mobile Robot", Development in Robotics, Kempston, pp.79-87.

Pin, F. & Killough, S. (1999), "A New Family of Omnidirectional and Holonomic Wheeled Platforms for Mobile Robot", IEEE Transactions on Robotics and Automation, Vol. 15,No. 6, pp. 978-989. http://www.hagisonic.com/

Jie-Tong Zou , Kuo L. Su and Feng-Chun Chiang (2010), "The development of the Omnidirectional Home Care Mobile Robot" , The Fifteenth International Symposium on Artificial Life and Robotics (AROB 15th '10),B-Con Plaza, Beppu, Oita, Japan, Feb. 4- 6.

16

Dynamic Modeling and Power Modeling of Robotic Skid-Steered Wheeled Vehicles

Wei Yu, Emmanuel Collins and Oscar Chuy
Florida State University
U.S.A

1. Introduction

Dynamic models and power models of autonomous ground vehicles are needed to enable realistic motion planning Howard & Kelly (2007); Yu et al. (2010) in unstructured, outdoor environments that have substantial changes in elevation, consist of a variety of terrain surfaces, and/or require frequent accelerations and decelerations.

At least 4 different motion planning tasks can be accomplished using appropriate dynamic and power models:

1. *Time optimal motion planning.*
2. *Energy efficient motion planning.*
3. *Reduction in the frequency of replanning.*
4. *Planning in the presence of a fault, such as flat tire or faulty motor.*

For the purpose of motion planning this chapter focuses on developing dynamic and power models of a skid-steered wheeled vehicle to help the above motion planning tasks. The dynamic models are the foundation to derive the power models of skid-steered wheeled vehicles. The target research platform is a skid-steered vehicle. A skid-steered vehicle can be either tracked or wheeled . Fig. 1 shows examples of a skid-steered wheeled vehicle and a skid-steered tracked vehicle.

This chapter is organized into five sections. Section 1 is the introduction. Section 2 presents the kinematic models of a skid-steered wheeled vehicle, which is the preliminary knowledge to the proposed dynamic model and power model. Section 3 develops analytical dynamic models of a skid-steered wheeled vehicle for general 2D motion. The developed models are characterized by the coefficient of rolling resistance, the coefficient of friction, and the shear deformation modulus, which have terrain-dependent values. Section 4 develops analytical power models of a skid-steered vehicle and its inner and outer motors in general 2D curvilinear motion. The developed power model builds upon a previously developed dynamic model in Section 3. Section 5 experimentally verifies the proposed dynamic models and power models of a robotic skid-steered wheeled vehicle.

Ackerman steering, differential steering, and skid steering are the most widely used steering mechanisms for wheeled and tracked vehicles. Ackerman steering has the advantages of good lateral stability when turning at high speeds, good controllability Siegwart & Nourbakhsh (2005) and lower power consumption Shamah et al. (2001), but has the disadvantages of low maneuverability and need of an explicit mechanical steering subsystem Mandow et al.

Fig. 1. Examples of skid-steered vehicles: (Left) Skid-steered wheeled vehicle, (Right) Skid-steered tracked vehicle

(2007); Shamah et al. (2001); Siegwart & Nourbakhsh (2005). Differential steering is popular because it provides high maneuverability with a zero turning radius and has a simple steering configuration Siegwart & Nourbakhsh (2005); Zhang et al. (1998). However, it does not have strong traction and mobility over rough and loose terrain, and hence is seldom used for outdoor terrains. Like differential steering, skid steering leads to high maneuverability Caracciolo et al. (1999); Economou et al. (2002); Siegwart & Nourbakhsh (2005), faster response Martinez et al. (2005), and also has a simple Mandow et al. (2007); Petrov et al. (2000); Shamah et al. (2001) and robust mechanical structure Kozlowski & Pazderski (2004); Mandow et al. (2007); Yi, Zhang, Song & Jayasuriya (2007). In contrast, it also leads to strong traction and high mobilityPetrov et al. (2000), which makes it suitable for all-terrain traversal.
Many of the difficulties associated with modeling and operating both classes of skid-steered vehicles arise from the complex wheel (or track) and terrain interaction Mandow et al. (2007); Yi, Song, Zhang & Goodwin (2007). For Ackerman-steered or differential-steered vehicles, the wheel motions may often be accurately modeled by pure rolling, while for skid-steered vehicles in general curvilinear motion, the wheels (or tracks) roll and slide at the same time Mandow et al. (2007); O. Chuy et al. (2009); Yi, Song, Zhang & Goodwin (2007); Yi, Zhang, Song & Jayasuriya (2007). This makes it difficult to develop kinematic and dynamic models that accurately describe the motion. Other disadvantages are that the motion tends to be energy inefficient, difficult to control Kozlowski & Pazderski (2004); Martinez et al. (2005), and for wheeled vehicles, the tires tend to wear out faster Golconda (2005).
A kinematic model of a skid-steered wheeled vehicle maps the wheel velocities to the vehicle velocities and is an important component in the development of a dynamic model. In contrast to the kinematic models for Ackerman-steered and differential-steered vehicles, the kinematic model of a skid-steered vehicle is dependent on more than the physical dimensions of the vehicle since it must take into account vehicle sliding and is hence terrain-dependent Mandow et al. (2007); Wong (2001). In Mandow et al. (2007); Martinez et al. (2005) a kinematic model of a skid-steered vehicle was developed by assuming a certain equivalence with a kinematic model of a differential-steered vehicle. This was accomplished by experimentally determining the instantaneous centers of rotation (ICRs) of the sliding velocities of the left

and right wheels. An alternative kinematic model that is based on the slip ratios of the wheels has been presented in Song et al. (2006); Wong (2001). This model takes into account the longitudinal slip ratios of the left and right wheels. The difficulty in using this model is the actual detection of slip, which cannot be computed analytically. Hence, developing practical methods to experimentally determine the slip ratios is an active research area Endo et al. (2007); Moosavian & Kalantari (2008); Nagatani et al. (2007); Song et al. (2008).

To date, there is very little published research on the experimentally verified dynamic models for general motion of skid-steered vehicles, especially wheeled vehicles. The main reason is that it is hard to model the tire (or track) and terrain interaction when slipping and skidding occur. (For each vehicle wheel, if the wheel linear velocity computed using the angular velocity of the wheel is larger than the actual linear velocity of the wheel, slipping occurs, while if the computed wheel velocity is smaller than the actual linear velocity, skidding occurs.) The research of Caracciolo et al. (1999) developed a dynamic model for planar motion by considering longitudinal rolling resistance, lateral friction, moment of resistance for the vehicle, and also the nonholonomic constraint for lateral skidding. In addition, a model-based nonlinear controller was designed for trajectory tracking. However, this model uses Coulomb friction to describe the lateral sliding friction and moment of resistance, which contradicts the experimental results Wong (2001); Wong & Chiang (2001). In addition, it does not consider any of the motor properties. Furthermore, the results of Caracciolo et al. (1999) are limited to simulation without experimental verification.

The research of Kozlowski & Pazderski (2004) developed a planar dynamic model of a skid-steered vehicle, which is essentially that of Caracciolo et al. (1999), using a different velocity vector (consisting of the longitudinal and angular velocities of the vehicle instead of the longitudinal and lateral velocities). In addition, the dynamics of the motors, though not the power limitations, were added to the model. Kinematic, dynamic and motor level control laws were explored for trajectory tracking. However, as in Caracciolo et al. (1999), Coulomb friction was used to describe the lateral friction and moment of resistance, and the results are limited to simulation. In Yi, Song, Zhang & Goodwin (2007) a functional relationship between the coefficient of friction and longitudinal slip is used to capture the interaction between the wheels and ground, and further to develop a dynamic model of skid-steered wheeled vehicle. Also, an adaptive controller is designed to enable the robot to follow a desired trajectory. The inputs of the dynamic model are the longitudinal slip ratios of the four wheels. However, the longitudinal slip ratios are difficult to measure in practice and depend on the terrain surface, instantaneous radius of curvature, and vehicle velocity. In addition, no experiment is conducted to verify the reliability of the torque prediction from the dynamic model and motor saturation and power limitations are not considered. In Wang et al. (2009) the dynamic model from Yi, Song, Zhang & Goodwin (2007) is used to explore the motion stability of the vehicle, which is controlled to move with constant linear velocity and angular velocity for each half of a lemniscate to estimate wheel slip. As in Yi, Song, Zhang & Goodwin (2007), no experiment is carried out to verify the fidelity of the dynamic model.

The most thorough dynamic analysis of a skid-steered vehicle is found in Wong (2001); Wong & Chiang (2001), which consider steady-state (i.e., constant linear and angular velocities) dynamic models for circular motion of *tracked* vehicles. A primary contribution of this research is that it proposes and then provides experimental evidence that in the track-terrain interaction the shear stress is a particular function of the shear displacement. This model differs from the Coulomb model of friction, adopted in Caracciolo et al. (1999); Kozlowski & Pazderski (2004), which essentially assumes that the maximum shear stress is obtained as soon as there is any relative movement between the track and the ground. This research also provides detailed analysis of the longitudinal and lateral forces that act on a tracked vehicle. But their results had not been extended to skid-steered *wheeled* vehicles. In addition, they do not consider vehicle acceleration, terrain elevation, actuator limitations, or the vehicle control system.

In the existing literature there are very few publications that consider power modeling of skid-steered vehicles. The research of Kim & Kim (2007) provides an energy model of a skid-steered wheeled vehicle in linear motion. This model is essentially the time integration of a power model and is derived from the dynamic model of a motor, including the energy loss due to the armature resistance and viscous friction as well as the kinetic energy of the vehicle. This research also uses the energy model to find the velocity trajecotry that minimizes the energy consumption. However, the energy model only considers the dynamics of the motor, but does not include the mechanical dynamics of the vehicle and hence ignores the substantial energy consumption due to sliding friction. Because longitudinal friction and moment of resistance lead to substantial power loss when a skid steered vehicle is in general curvilinear motion, the results of Kim & Kim (2007) cannot be readily extended to motion that is not linear.

The most thorough exploration of power modeling of a skid-steered (tracked) vehicle is presented in Morales et al. (2009) and Morales et al. (2006). This research develops an experimental power model of a skid-steered tracked vehicle from terrain's perspective. The power model includes the power loss drawn by the terrain due to sliding frictions, and also the power losses due to the traction resistance and the motor drivers. Based on another conceptual model, this research considers the case in which the inner track has the same velocity sign as the outer track and *qualitatively* describes the negative sliding friction of the inner track, which leads the corresponding motor to work as a generator. Experiments to apply the power model for navigation are also described. However, this research has two limitations that the current research seeks to overcome. First, as in Caracciolo et al. (1999); Kozlowski & Pazderski (2004), discussed above in the context of dynamic modeling of skid-steered vehicles, Coulomb's law is adopted to describe the sliding friction component in the power modeling, which can lead to incorrect predictions for larger turning radii. Second, since the power model is derived from the perspective of the terrain drawing power from the tracks, it does not appear possible to quantify the power consumption of the left and right side motors. This is important since the motion of the vehicles can be dependent upon the power limitations of the motors.

Building upon the research in Wong (2001); Wong & Chiang (2001), this chapter will develop dynamic models of a skid-steered wheeled vehicle for general curvilinear planar (2D) motion. As in Wong (2001); Wong & Chiang (2001) the modeling is based upon the functional relationship of shear stress to shear displacement. Practically, this means that for a vehicle tire the shear stress varies with the turning radius. This chapter also includes models of the saturation and power limitations of the actuators as part of the overall vehicle model.

Using the developed dynamic model for 2D general curvilinear motion, this chapter will also develop power models of a skid-steered wheeled vehicle based on separate power models for left and right motors. The power model consists of two parts: (1) the mechanical power consumption, including the mechanical loss due to sliding friction and moment of resistance, and the power used to accelerate vehicle; and (2) the electrical power consumption, which is the electrical loss due to the motor electrical resistance. The mechanical power consumption is derived completely from the dynamic model, while the electrical power consumption is derived using the electric current predicted from this dynamic model along with circuit

theory. This chapter also discusses the interesting phenomenon that while the outer motor always consumes power, even though the velocity of inner wheel is always positive, as the turning radius decreases from infinity (corresponding to linear motion), the inner motor first consumes power, then generates power, and finally consumes power again.
In summary, we expect this chapter to make the following two fundamental contributions to dynamic modeling and power modeling of skid-steered wheeled vehicles:

1. **A paradigm for deriving dynamic models of skid-steered wheeled vehicles.** The modeling methodology will result in terrain-dependent models that describe general general planar (2D) motion.
2. **A paradigm for deriving power models of skid-steered wheeled vehicles based on dynamic models.** The power model of a skid-steered vehicle will be derived from vehicle dynamic models. The power model will be described from the perspective of the motors and includes both the mechanical power consumption and electrical power consumption. It can predict when a given trajectory is unachievable because the power limitation of one of the motors is violated.

2. Kinematics of a skid-steered wheeled vehicle

In this section, the kinematic model of a skid-steered wheeled vehicle is described and discussed. It is an important component in the development of the overall dynamic models and power models of a skid-steered wheeled vehicle.
To mathematically describe the kinematic models that have been developed for skid-steered vehicles, consider a wheeled vehicle moving at constant velocity about an instantaneous center of rotation as shown in Fig. 2.
The global and local coordinate frames are denoted respectively by X-Y and x-y. The variables v, $\dot{\varphi}$ and R are respectively the translational velocity, angular velocity and turning radius of vehicle. The instantaneous centers of rotation for the left wheel and right wheel are given respectively by ICR_l and ICR_r. Note that ICR_l and ICR_r are the centers for left and right wheel treads (the parts of the tires that contact and slide on the terrain) Wong & Chiang (2001); Yi, Zhang, Song & Jayasuriya (2007), i.e., they are the centers for the sliding velocities of these contacting treads, but not the centers for the actual velocities of each wheel. It has been shown that the three ICRs are in the same line, which is parallel to the x-axis of the local frame Mandow et al. (2007); Yi, Zhang, Song & Jayasuriya (2007).
In the x-y frame, the coordinates of ICR, ICR_l and ICR_r are described as (x_{ICR}, y_{ICR}), (x_{ICRl}, y_{ICRl}) and (x_{ICRr}, y_{ICRr}). The vehicle velocity is denoted as $u = [v_x \ v_y \ \dot{\varphi}]^T$, where v_x and v_y are the components of v along the x and y axes. The angular velocities of the left and right wheels are denoted respectively by ω_l and ω_r. (Note that for both the left and right side of the vehicle the velocities of the front and rear wheels are the same since they are driven by the same belt, and hence, there is only one velocity associated with each side.) The parameters b, B and r are respectively the wheel width, the vehicle width, and the wheel radius.
An experimental kinematic model of a skid-steered wheeled vehicle that is developed in Mandow et al. (2007) is given by

$$\begin{bmatrix} v_x \\ v_y \\ \dot{\varphi} \end{bmatrix} = \frac{r}{x_{ICRr} - x_{ICRl}} \begin{bmatrix} -y_{ICR} & y_{ICR} \\ x_{ICRr} & -x_{ICRl} \\ -1 & 1 \end{bmatrix} \begin{bmatrix} \omega_l \\ \omega_r \end{bmatrix} \tag{1}$$

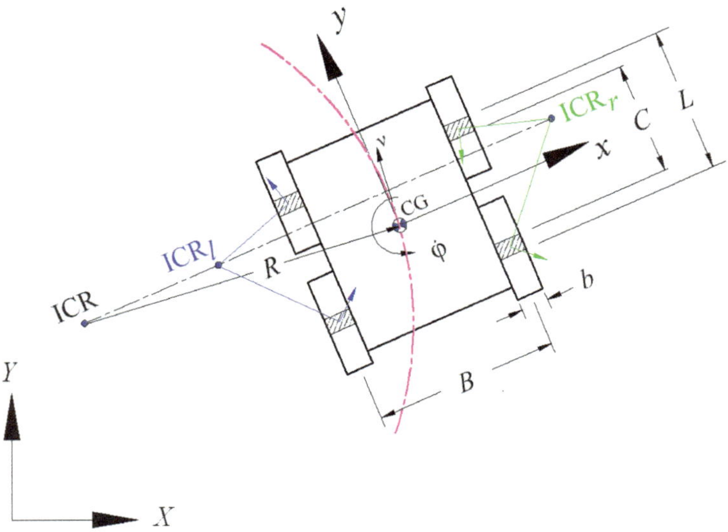

Fig. 2. The kinematics of a skid-steered wheeled vehicle and the corresponding instantaneous centers of rotation (ICRs)

If the skid-steered wheeled vehicle is symmetric about the x and y axes, then $y_{ICRl} = y_{ICRr} = 0$ and $x_{ICRl} = -x_{ICRr}$. Define the expansion factor α as the ratio of the longitudinal distance between the left and right wheels over the vehicle width, i.e.,

$$\alpha \triangleq \frac{x_{ICRr} - x_{ICRl}}{B}. \tag{2}$$

Then, for a symmetric vehicle the kinematic model (1) can be expressed as

$$\begin{bmatrix} v_y \\ \dot{\varphi} \end{bmatrix} = \frac{r}{\alpha B} \begin{bmatrix} \frac{\alpha B}{2} & \frac{\alpha B}{2} \\ -1 & 1 \end{bmatrix} \begin{bmatrix} \omega_l \\ \omega_r \end{bmatrix}. \tag{3}$$

(Note that $v_x = 0$.)

The expansion factor α varies with the terrain. Experimental results show that the larger the rolling resistance, the larger the expansion factor. For a Pioneer 3-AT, $\alpha = 1.5$ for a vinyl lab surface and $\alpha > 2$ for a concrete surface. Equation (3) shows that the kinematic model of a skid-steered wheeled vehicle of width B is equivalent to the kinematic model of a differential-steered wheeled vehicle of width αB. Note that when $\alpha = 1$, (3) becomes the kinematic model for a differential-steered wheeled vehicle.

A more rigorously derived kinematic model for a skid-steered vehicle is presented in Moosavian & Kalantari (2008); Song et al. (2006); Wong (2001). This model takes into account the longitudinal slip ratios i_l and i_r of the left and right wheels and for symmetric vehicles is given by

$$\begin{bmatrix} v_y \\ \dot{\varphi} \end{bmatrix} = \frac{r}{B} \begin{bmatrix} \frac{(1-i_l)B}{2} & \frac{(1-i_r)B}{2} \\ -(1-i_l) & (1-i_r) \end{bmatrix} \begin{bmatrix} \omega_l \\ \omega_r \end{bmatrix}, \tag{4}$$

where $i_l \triangleq (r\omega_l - v_{l\_a})/(r\omega_l)$, $i_r \triangleq (r\omega_r - v_{r\_a})/(r\omega_r)$ and v_{l_a} and v_{r_a} are the actual velocities of the left and right wheels. We have found that when

$$\frac{i_l}{i_r} = -\frac{\omega_r}{\omega_l} \quad and \quad \alpha = \frac{1}{1 - \frac{2 i_l i_r}{i_l + i_r}}, \tag{5}$$

(3) and (4) are identical. Currently, to our knowledge no analysis or experiments have been performed to verify the left hand equation in (5) and analyze its physical significance. However, for a limited range of turning radii experimentally derived expressions for i_l/i_r, essentially in terms of ω_l and ω_r, are given in Endo et al. (2007); Nagatani et al. (2007).

3. Dynamic modeling of a skid-steered wheeled vehicle

This section develops dynamic models of a skid-steered wheeled vehicle for the cases of general 2D motion. In contrast to dynamic models described in terms of the velocity vector of the vehicle Caracciolo et al. (1999); Kozlowski & Pazderski (2004), the dynamic models here are described in terms of the angular velocity vector of the wheels. This is because the wheel (specifically, the motor) velocities are actually commanded by the control system, so this model form is particularly beneficial for control and planning.

Following Kozlowski & Pazderski (2004), the dynamic model considering the nonholonomic constraint is given by

$$M\ddot{q} + C(q,\dot{q}) + G(q) = \tau, \tag{6}$$

where $q = [\theta_l \ \theta_r]^T$ is the angular displacement of the left and right wheels, $\dot{q} = [\omega_l \ \omega_r]^T$ is the angular velocity of the left and right wheels, $\tau = [\tau_l \ \tau_r]^T$ is the torque of the left and right motors, M is the mass matrix, $C(q,\dot{q})$ is the resistance term, and $G(q)$ is the gravitational term. The primary focus of the following subsection is the derivation of $C(q,\dot{q})$ to properly model the ground and wheel interaction. In the following content, it is assumed that the vehicle is symmetric and the center of gravity (CG) is at the geometric center.

When the vehicle is moving on a 2D surface, it follows from the model given in Kozlowski & Pazderski (2004), which is expressed in the local x-y coordinates, and the kinematic model (3) that M in (6) is given by

$$M = \begin{bmatrix} \frac{mr^2}{4} + \frac{r^2 I}{\alpha B^2} & \frac{mr^2}{4} - \frac{r^2 I}{\alpha B^2} \\ \frac{mr^2}{4} - \frac{r^2 I}{\alpha B^2} & \frac{mr^2}{4} + \frac{r^2 I}{\alpha B^2} \end{bmatrix}, \tag{7}$$

where m and I are respectively the mass and moment of inertia of the vehicle. Since we are considering planar motion, $G(q) = 0$. $C(q,\dot{q})$ represents the resistance resulting from the interaction of the wheels and terrain, including the rolling resistance, sliding frictions, and the moment of resistance, the latter two of which are modeled using Coulomb friction in Caracciolo et al. (1999); Kozlowski & Pazderski (2004). Assume that $\dot{q} = [\omega_l \ \omega_r]^T$ is a known constant, then $\ddot{q} = 0$ and (6) becomes

$$C(q,\dot{q}) = \tau. \tag{8}$$

Previous research Caracciolo et al. (1999); Kozlowski & Pazderski (2004) assumed that the shear stress takes on its maximum magnitude as soon as a small relative movement occurs between the contact surface of the wheel and terrain. Instead of using this theory for tracked vehicle, Wong (2001) and Wong & Chiang (2001) present experimental evidence to show that the shear stress of the tread is function of the shear displacement. The maximum shear stress is practically achieved only when the shear displacement exceeds a particular threshold. In this section, this theory will be applied to a skid-steered wheeled vehicle.

Based on the theory in Wong (2001); Wong & Chiang (2001), the shear stress τ_{ss} and shear displacement j relationship can be described as,

$$\tau_{ss} = p\mu(1 - e^{-j/K}), \tag{9}$$

where p is the normal pressure, μ is the coefficient of friction and K is the shear deformation modulus. K is a terrain-dependent parameter, like the rolling resistance and coefficient of friction Wong (2001).

Fig. 3 depicts a skid-steered wheeled vehicle moving counterclockwise (CCW) at constant linear velocity v and angular velocity $\dot{\varphi}$ in a circle centered at O from position 1 to position 2. X–Y denotes the global frame and the body-fixed frames for the right and left wheels are given respectively by the x_r–y_r and x_l–y_l. The four contact patches of the wheels with the ground are shadowed in Fig. 3 and L and C are the patch-related distances shown in Fig. 3. It is assumed that the vehicle is symmetric and the center of gravity (CG) is at the geometric center. Note that because ω_l and ω_r are known, v_y and $\dot{\varphi}$ can be computed using the vehicle kinematic model (3), which enables the determination of the radius of curvature R since $v_y = R\dot{\varphi}$.

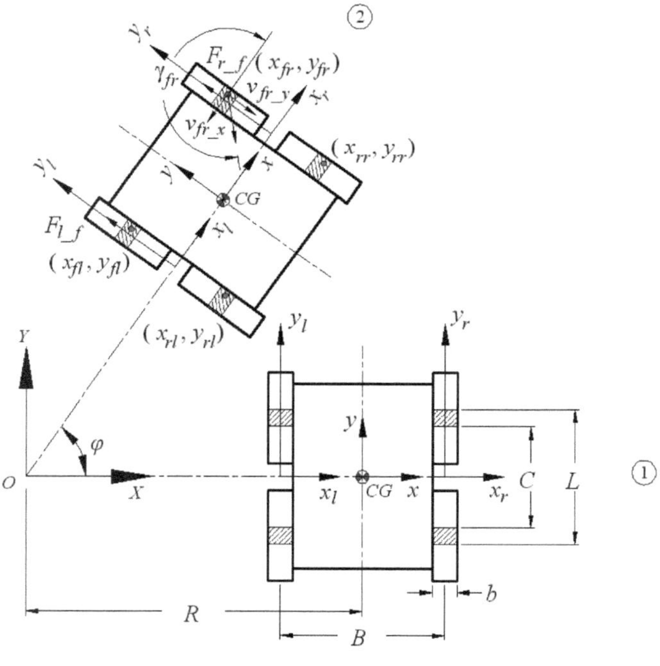

Fig. 3. Circular motion of a skid-steered wheeled vehicle

In the x_r–y_r frame consider an arbitrary point on the contact patch of the front right wheel with coordinates (x_{fr}, y_{fr}). This contact patch is *not* fixed on the tire, but is the part of the tire that contacts the ground. The time interval t for this point to travel from an initial contact point $(x_{fr}, L/2)$ to (x_{fr}, y_{fr}) is,

$$t = \int_{y_{fr}}^{L/2} \frac{1}{r\omega_r} dy_r = \frac{L/2 - y_{fr}}{r\omega_r}. \tag{10}$$

During the same time, the vehicle has moved from position 1 to position 2 with an angular displacement of φ. The sliding velocities of point (x_{fr}, y_{fr}) in the x_r and y_r directions are denoted by v_{fr_x} and v_{fr_y}. Therefore,

$$v_{fr\_x} = -y_{fr}\dot{\varphi}, \; v_{fr\_y} = (R + B/2 + x_{fr})\dot{\varphi} - r\omega_r. \tag{11}$$

The resultant sliding velocity v_{fr} and its angle γ_{fr} in the x_r-y_r frame are

$$v_{fr} = \sqrt{v_{fr\_x}^2 + v_{fr\_y}^2}, \; \gamma_{fr} = \pi + \arctan\left(\frac{v_{fr\_y}}{v_{fr\_x}}\right). \tag{12}$$

Note that when the wheel is sliding, the direction of friction is opposite to the sliding velocity, and if the vehicle is in pure rolling, v_{fr_x} and v_{fr_y} are zero.
In order to calculate the shear displacement of this reference point, the sliding velocities need to be expressed in the global X–Y frame. Let v_{fr_X} and v_{fr_Y} denote the sliding velocities in the X and Y directions. Then, the transformation between the local and global sliding velocities is given by,

$$\begin{bmatrix} v_{fr\_X} \\ v_{fr\_Y} \end{bmatrix} = \begin{bmatrix} \cos\varphi & -\sin\varphi \\ \sin\varphi & \cos\varphi \end{bmatrix} \begin{bmatrix} v_{fr\_x} \\ v_{fr\_y} \end{bmatrix}. \tag{13}$$

The shear displacements j_{fr_X} and j_{fr_Y} in the X and Y directions can be expressed as

$$\begin{aligned} j_{fr\_X} &= \int_0^t v_{fr\_X} dt = \int_{y_{fr}}^{L/2} (v_{fr\_x}\cos\varphi - v_{fr\_y}\sin\varphi)\frac{1}{r\omega_r} dy_r \\ &= (R + B/2 + x_{fr}) \cdot \{\cos\left[\frac{(L/2 - y_{fr})\dot{\varphi}}{r\omega_r}\right] - 1\} - y_{fr}\sin\left[\frac{(L/2 - y_{fr})\dot{\varphi}}{r\omega_r}\right], \end{aligned} \tag{14}$$

$$\begin{aligned} j_{fr\_Y} &= \int_0^t v_{fr\_Y} dt = \int_{y_{fr}}^{L/2} (v_{fr\_x}\sin\varphi + v_{fr\_y}\cos\varphi)\frac{1}{r\omega_r} dy_r \\ &= (R + B/2 + x_{fr}) \cdot \sin\left[\frac{(L/2 - y_{fr})\dot{\varphi}}{r\omega_r}\right] - L/2 + y_{fr}\cos\left[\frac{(L/2 - y_{fr})\dot{\varphi}}{r\omega_r}\right]. \end{aligned} \tag{15}$$

The resultant shear displacement j_{fr} in the X–Y frame is given by, $j_{fr} = \sqrt{j_{fr\_X}^2 + j_{fr\_Y}^2}$. Similarly, it can be shown that for the reference point (x_{rr}, y_{rr}) in the rear right wheel the angle of the sliding velocity γ_{rr} in the x_r-y_r frame is

$$\gamma_{rr} = \arctan\left[\frac{(R + B/2 + x_{rr})\dot{\varphi} - r\omega_r}{-y_{rr}\dot{\varphi}}\right], \tag{16}$$

and the shear displacements j_{rr_X} and j_{rr_Y} are given by

$$j_{rr\_X} = (R + B/2 + x_{rr}) \cdot \{\cos\left[\frac{(-C/2 - y_{rr})\dot{\varphi}}{r\omega_r}\right] - 1\} - y_{rr}\sin\left[\frac{(-C/2 - y_{rr})\dot{\varphi}}{r\omega_r}\right], \tag{17}$$

$$j_{rr\_Y} = (R + B/2 + x_{rr}) \cdot \sin\left[\frac{(-C/2 - y_{rr})\dot{\varphi}}{r\omega_r}\right] + C/2 + y_{rr}\cos\left[\frac{(-C/2 - y_{rr})\dot{\varphi}}{r\omega_r}\right]. \tag{18}$$

and the magnitude of the resultant shear displacement j_{rr} is $j_{rr} = \sqrt{j_{rr\_X}^2 + j_{rr\_Y}^2}$.
The friction force points in the opposite direction of the sliding velocity. Using j_{fr} and j_{rr}, derived above, with (9) and integrating along the contact patches yields that the longitudinal sliding friction of the right wheels F_{r_f} can be expressed as

$$F_{r\_f} = \int_{C/2}^{L/2} \int_{-b/2}^{b/2} p_r \mu_r (1 - e^{-j_{fr}/K_r}) \sin(\pi + \gamma_{fr}) dx_r dy_r + \int_{-L/2}^{-C/2} \int_{-b/2}^{b/2} p_r \mu_r (1 - e^{-j_{rr}/K_r}) \sin(\pi + \gamma_{rr}) dx_r dy_r, \quad (19)$$

where p_r, μ_r and K_r are respectively the normal pressure, coefficient of friction, and shear deformation modulus of the right wheels. While most of the parameters in (19) can be directly measured, as discussed further below, the parameters μ_r and K_r must be estimated.
Let f_{r_r} denote the rolling resistance of the right wheels, including the internal locomotion resistance such as resistance from belts, motor windings and gearboxes Morales et al. (2006). The complete resistance torque τ_{r_Res} from the ground to the right wheel is given by

$$\tau_{r\_Res} = r(F_{r\_f} + f_{r\_r}). \quad (20)$$

Since ω_r is constant, the input torque τ_r from right motor will compensate for the resistance torque, such that

$$\tau_r = \tau_{r\_Res}. \quad (21)$$

The above discussion is for the right wheels. Exploiting the same derivation process, one can obtain analytical expressions for the shear displacements j_{fl} and j_{rl} of the front and rear left wheels, and the angles of the sliding velocity γ_{fl} and γ_{rl}. The longitudinal sliding friction of the left wheels F_{l_f} is then given by

$$F_{l\_f} = \int_{C/2}^{L/2} \int_{-b/2}^{b/2} p_l \mu_l (1 - e^{-j_{fl}/K_l}) \sin(\pi + \gamma_{fl}) dx_l dy_l + \int_{-L/2}^{-C/2} \int_{-b/2}^{b/2} p_l \mu_l (1 - e^{-j_{rl}/K_l}) \sin(\pi + \gamma_{rl}) dx_l dy_l, \quad (22)$$

where p_l, μ_l and K_l are respectively the normal pressure, coefficient of friction, and shear deformation modulus of the left wheels. Denote the rolling resistance of the left wheels as f_{l_r}. The input torque τ_l of the left motor equals the resistance torque of the left wheel τ_{l_Res}, such that

$$\tau_l = \tau_{l\_Res} = r(F_{l\_f} + f_{l\_r}). \quad (23)$$

Fig. 4 compares the resistance torque prediction of τ_{l_Res} and τ_{r_Res} using shear stress and shear displacement function (9) and Coulomb's law when the skid-steered wheeled vehicle of Fig. 13 is in steady state rotation.

$$\tau_{ss} = p\mu \ \text{(Coulomb's Law)} \quad (24)$$

It is seen that Coulomb's law leads to a resistance torque that has the same constant value for all turning radii, which contradicts the experimental results shown in Wong (2001); Wong & Chiang (2001) for tracked vehicles and below in Fig. 15 for wheeled vehicles.
Using (21) and the left equation of (23) with (8) yields

$$C(q, \dot{q}) = [\tau_{l\_Res} \ \tau_{r\_Res}]^T. \quad (25)$$

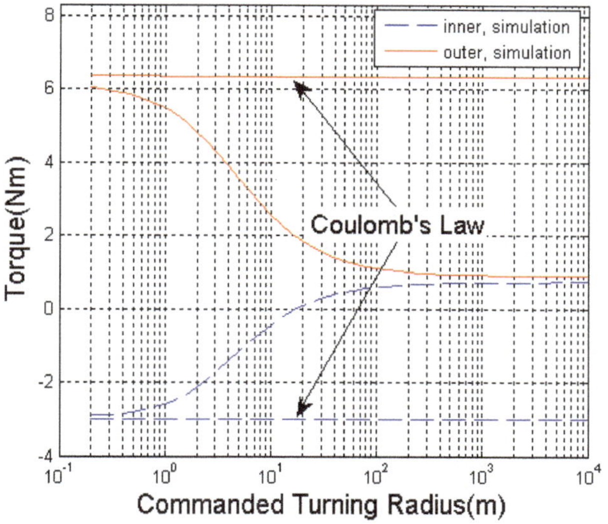

Fig. 4. Inner and outer motor resistance torque prediction using function (9) and Coulomb's law when vehicle is in steady state rotation.

Substituting (7), (25) and $G(q) = 0$ into (6) yields a dynamic model that can be used to predict 2D movement for the skid-steered vehicle:

$$\begin{bmatrix} \frac{mr^2}{4} + \frac{r^2 I}{\alpha B^2} & \frac{mr^2}{4} - \frac{r^2 I}{\alpha B^2} \\ \frac{mr^2}{4} - \frac{r^2 I}{\alpha B^2} & \frac{mr^2}{4} + \frac{r^2 I}{\alpha B^2} \end{bmatrix} \ddot{q} + \begin{bmatrix} \tau_{l\_Res} \\ \tau_{r\_Res} \end{bmatrix} = \begin{bmatrix} \tau_l \\ \tau_r \end{bmatrix}. \tag{26}$$

In summary, in order to obtain (25), the shear displacement calculation of (14), (15), (17) and (18) is the first step. The inputs to these equations are the left and right wheel angular velocities ω_l and ω_r. The shear displacements are employed in (19) and (22) to obtain the right and left sliding friction forces, F_{r_f} and F_{l_f}. Next, the sliding friction forces and rolling resistances are substituted into (20) and (23) to calculate the right and left resistance torques, which determine $C(q, \dot{q})$ using (25).

4. Power modeling of a skid-steered wheeled vehicle

This section derives power models for a skid-steered wheeled vehicle, moving as in Fig. 3. The foundation for modeling the power consumption model is the dynamic model of Section 3. The power consumption for each side of the vehicle includes the mechanical power consumption due to the motion of the wheels and the electrical power consumption due to the electrical resistance of the motors. The total power consumption of the vehicle is the sum of the power consumption of of the left and right sides.

Assume that a skid-steered wheeled vehicle moves CCW about an instantaneous center of rotation (see Fig. 3). The circuit diagram for each side of the vehicle is shown in Fig. 5. Each circuit includes a battery, motor controller, motor M and the motor electrical resistance R_e. In Fig. 5 ω_l and ω_r are the angular velocities of the left and right wheels, U_l and U_r are the output voltages of the left and right motor controllers, and i_l and i_r are the currents of the left

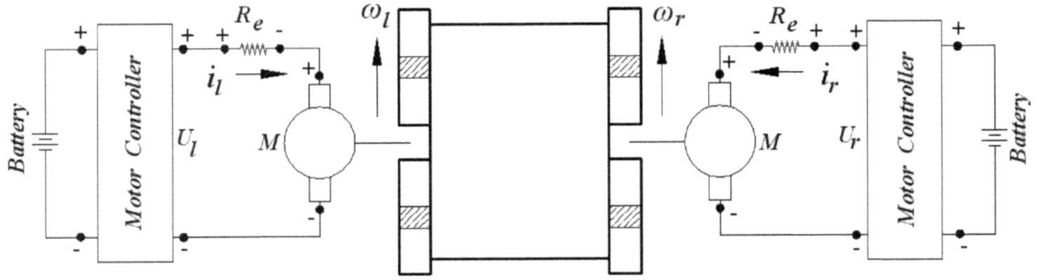

Fig. 5. The circuit layout for the left and right side of a skid-steered wheeled vehicle.

and right circuits. For the experimental vehicle used in this research (the modified Pioneer 3-AT shown in Fig. 13), ω_l and ω_r are always positive[1].
The electric model of a DC motor at steady state is given by Rizzoni (2000),

$$V_a = E_b + R_a I_a, \tag{27}$$

where V_a is the supply voltage to the motor, R_a is the motor armature resistance, I_a is the motor current, and E_b is the back EMF. The power consumption P_a of a DC motor is given by $P_a = V_a I_a$. Hence, multiplying (27) by I_a yields

$$P_a = V_a I_a = P_e + R_a I_a^2, \tag{28}$$

where P_e is the portion of the electric power converted to mechanical power by the motor and is given by

$$P_e = E_b I_a. \tag{29}$$

The mechanical power P_m of a DC motor is given by

$$P_m = \omega_m \tau, \tag{30}$$

where ω_m and τ are respectively the angular velocity and applied torque of the motor. For ideal energy-conversion case Rizzoni (2000),

$$P_m = P_e. \tag{31}$$

Substituting (29), (30) and (31) into (28) yields the power model of a DC motor used in the analysis of this research,

$$P_a = \omega_m \tau + R_a I_a^2. \tag{32}$$

In (32), the first term is the mechanical power consumption, which includes the power to compensate the left and right sliding frictions and the moment of resistance along with the power to accelerate the motor, and the second term is the electrical power consumption due to the motor electric resistance, which is dissipated as heat.
Using (32), the power consumed by the right motor P_r can be expressed as,

$$P_r = U_r i_r = P_{r,m} + P_{r,e}, \tag{33}$$

[1] Due to the torque limitations of the motors in the experimental vehicle, the minimum achievable turning radius is larger than half the width of the vehicle. This implies that the instantaneous radius of curvature is located outside of the vehicle body so that ω_l and ω_r are always positive.

where $P_{r,m}$ and $P_{r,e}$ are the mechanical power consumption and the electrical power consumption for the right side motor. In non ideal case, $P_{r,m}$ and $P_{r,e}$ in (33) are,

$$P_{r,m} = \frac{\tau_r \omega_r}{\eta}, \tag{34}$$

$$P_{r,e} = i_r^2 R_e, \tag{35}$$

where τ_r and ω_r are the same as in (26) in Section 3, and η is the motor efficiency.
For the right side motor, the output torque τ_r determined from the dynamic model (26) is given by

$$\tau_r = K_T i_r g_r \eta, \tag{36}$$

where K_T is the torque constant and g_r is the gear ratio. So the required current in right side of motor is,

$$i_r = \frac{\tau_r}{K_T g_r \eta}. \tag{37}$$

Plugging (37) into (35) yields,

$$P_{r,e} = (\frac{\tau_r}{K_T g_r \eta})^2 R_e. \tag{38}$$

Substituting (34) and (38) into (33), the power model for the right (outer) motor is,

$$P_r = \frac{\tau_r \omega_r}{\eta} + (\frac{\tau_r}{K_T g_r \eta})^2 R_e. \tag{39}$$

Notice that the only variables in (39) is the applied torques τ_r along with the angular velocities ω_r, which are available from the dynamic model in Chapter 3.
Similarly, for the left (inner) part of vehicle,

$$P_l = U_l i_l = P_{l,m} + P_{l,e}, \tag{40}$$

$$P_{l,m} = \frac{\tau_l \omega_l}{\eta}, \tag{41}$$

$$P_{l,e} = i_l^2 R_e, \tag{42}$$

$$\tau_l = K_T i_l, \tag{43}$$

$$P_{l,e} = (\frac{\tau_l}{K_T g_r \eta})^2 R_e, \tag{44}$$

$$P_l = \frac{\tau_l \omega_l}{\eta} + (\frac{\tau_l}{K_T g_r \eta})^2 R_e. \tag{45}$$

Let P denote the power that must be supplied by the motor drivers to the motors to enable the motion of a of a skid-steered wheeled vehicle and define the operator $\sigma : \mathbb{R} \rightarrow \mathbb{R}$ such that

$$\sigma(Q) = \begin{cases} Q : & Q \geq 0 \\ 0 : & Q < 0. \end{cases} \tag{46}$$

Then the entire power model of a skid-steered wheeled vehicle is,

$$P = \sigma(P_r) + \sigma(P_l). \tag{47}$$

Typically, one might expect to write $P = P_r + P_l$. However, since it turns out that P_l can be negative and that this generated power does not charge the battery in our research vehicle,

the more general form (47) is used. To enable the battery to be charged requires modifications of the motor controller, which was beyond the scope of this project.
In summary, given the input of the left wheel and right wheel angular velocities ω_l and ω_r, the first step in computing (45) and (39), the power models of the left motors and right motors, is the calculation of the left wheel and right wheel sliding frictions F_{l_f} and F_{r_f} using (22) and (19). The sliding frictions along with experimentally determined values of the rolling resistances f_{l_r} and f_{r_r} are then substituted into (23) and (20) to obtain the resistance torques τ_{l_Res} and τ_{r_Res}, which are in turn substituted into the vehicle dynamic model (26) to obtain the left wheel and right wheel torques τ_l and τ_r. Next, the left and right wheel torques are substituted into (45) and (39) to calculate the power consumption of the left and right wheels. The entire power consumption of the vehicle may then computed by substituting (45) and (39) into (47). While (45) and (39) are general equations for all DC motor driven vehicles, the τ_l and τ_r in (45) and (39) have to be calculated specifically from the skid-steered dynamic model (26).

4.1 Power models analysis

To better analyze the vehicle power model, consider the results from a series of simulations for CCW steady state rotation of the modified Pioneer 3-AT, the research platform shown in Fig. 13. For each simulation, the vehicle was commanded to have a linear velocity of 0.2 m/s. The commanded turning radius r_c is defined as the turning radius resulting from applying the wheel speeds, ω_l and ω_r, to a differential-steered kinematic model assuming no slip. The simulations corresponded to varying the commanded turning radius from $10^{-0.7}$ m to 10^4 m with the exponent increasing in increments of 0.1.
Fig. 6 and Fig. 7 were developed by using a simulation of the dynamic model (26) and respectively represent the applied torques and angular velocities for the left and right wheels. For both the right and left motors Fig. 8 compares the power consumption prediction obtained using the exponential friction model (9) with that obtained using Coulomb's law (24). (Both predictions were based on using the dynamic model (26) in conjunction with the power models (39) and (45).) It is seen that when Coulomb friction is assumed, as the turning radius increases, the power consumption of the outer motor does not converge to a small value (~3W) as is predicted using exponential friction. Instead it converges to a much larger value (~35W), which contradicts the experimental result in Fig. 16, which is almost identical to the exponential friction prediction. Fig. 9 further shows that the two models can yield dramatic differences in their predictions of the power consumption of the entire vehicle.
For the right motor Fig. 10 shows the total power consumption resulting from (39) along with its mechanical power component from (34) and its electrical power component from (35). Fig. 11 displays the same power information for the left motor using (45), (41), and (42).Although these curves correspond to a specific commanded linear velocity (0.2m/s), the shapes of these curves are typical of all velocities that have been simulated. Note that in these figures and the following discussion, $\underline{r} \approx 10^0$ m and $\bar{r} \approx 10^{1.1}$ m.
From (26), it is seen that when a vehicle is in steady state rotation,

$$\tau_{l\_Res} = \tau_l, \quad \tau_{r\_Res} = \tau_r, \tag{48}$$

which implies that for vehicle steady state rotation τ_l and τ_r only need to compensate for the resistance torques. So τ_l, τ_r in Fig. 6 also represent the resistance torques τ_{l_Res}, τ_{r_Res} for vehicle steady state rotation.

Below, Fig. 6, Fig. 7 and Fig. 8 are used to analyze the current, voltage, and power consumption of each motor in greater detail. Particular attention is given to the inner (left) motor since it sometimes generates power.

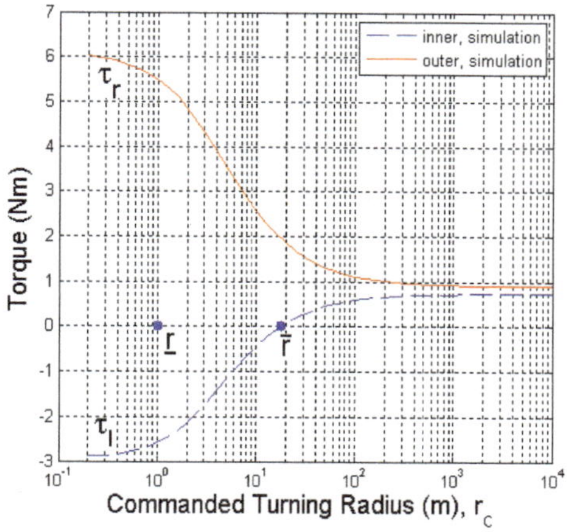

Fig. 6. Steady-state, inner and outer wheel torques vs. commanded turning radius, obtained via simulation of the dynamic model, for a commanded linear velocity of 0.2 m/s on the lab vinyl surface.

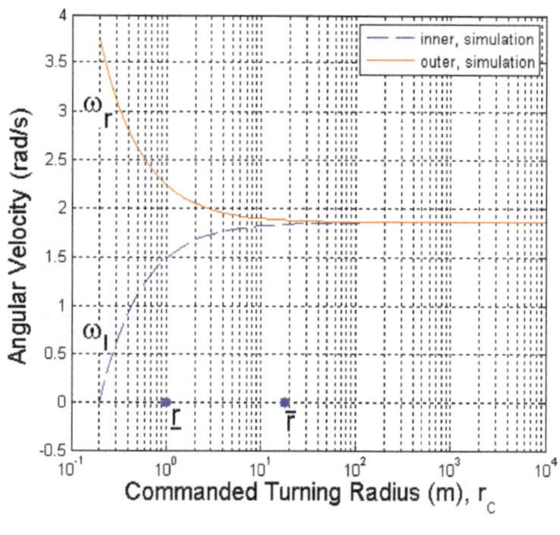

Fig. 7. Steady-state, inner and outer wheel angular velocities vs. commanded turning radius, obtained via simulation of the dynamic model for the conditions of Fig. 6.

Analysis of the Vehicle's Outer (Right) Side

Fig. 6, Fig. 7 and Fig. 8 show that τ_r, ω_r and P_r are always positive. From (36) and (33), it follows that i_r and U_r are also positive. Therefore, for the right side of the vehicle,

$$i_r > 0,\ U_r > 0,\ P_r > 0, \tag{49}$$

which implies that the outer motor always consumes power. The direction of current flow, and the voltage of the motor controller, motor resistance and motor have the same signs as in Fig. 5.

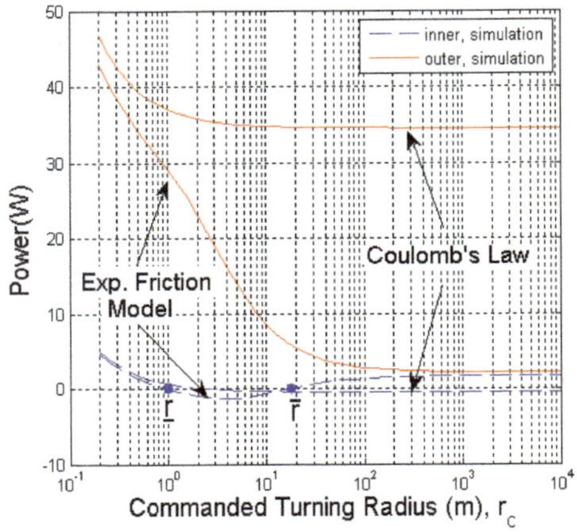

Fig. 8. Power prediction for the inner and outer motors vs. commanded turning radius using exponential friction model (9) and Coulomb's law.

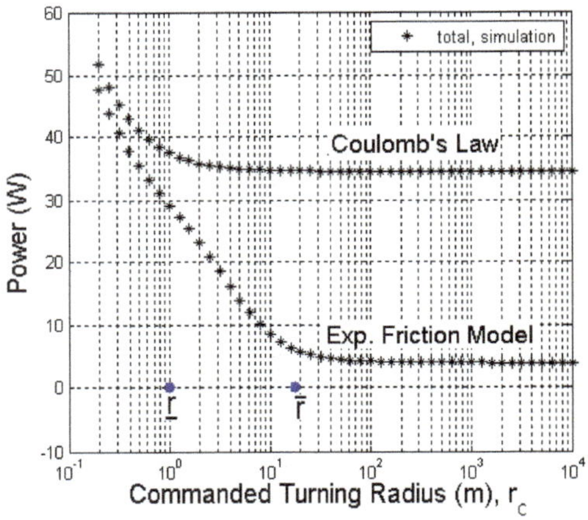

Fig. 9. Power prediction for the whole vehicle corresponding to Fig. 8 using exponential friction model (9) and Coulomb's law.

For the outer side of vehicle Fig. 10 shows that the mechanical power consumption and electrical power consumption are nearly equal for each turning radius[2]. (As discussed after (32), only the mechanical power is converted into motion while the electrical power consumption is dissipated (or wasted) as heat.) In this case, the power source is always the battery operating through the motor controller,[3] while the motor shaft motion consumes mechanical power and the motor electrical resistance consumes electrical power. Referring to (47),

$$\sigma(P_r) = P_r, \tag{50}$$

[2] As the motor electrical resistance decreases, the electrical power consumption will be smaller than the mechanical power consumption.

[3] Some of the battery's power is dissipated as heat in the motor controller's resistance. See Fig. 5

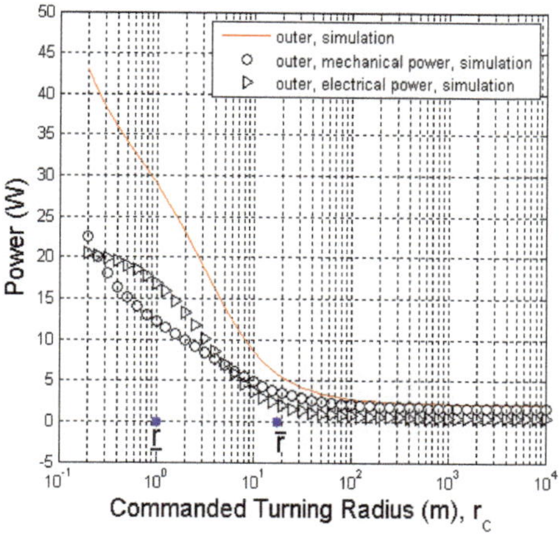

Fig. 10. Outer motor power comparison: outer total power consumption, outer mechanical power consumption and outer electrical power consumption, obtained via simulation of the dynamic model for the conditions of Fig. 6.

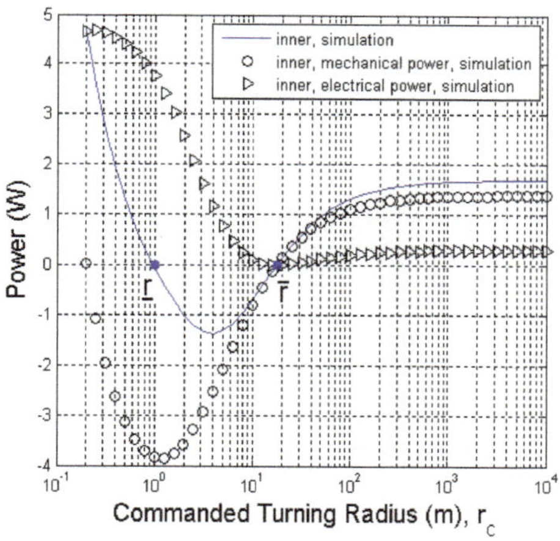

Fig. 11. Inner motor power comparison: inner total power consumption, inner mechanical power consumption and inner electrical power consumption, obtained via simulation of the dynamic model for the conditions of Fig. 6.

where P_r is given by (39).

Analysis of the Vehicle's Inner (Left) Side

Fig. 6 shows that τ_l can be either positive or negative, Fig. 7 shows that ω_l is positive, while Fig. 8 shows that P_l can be either positive or negative. The signs of τ_l and P_l depend on whether the commanded turning radius r_c is in one of three regions: 1) $r_c \geq \bar{r}$, 2) $\underline{r} < r_c < \bar{r}$, and 3) $r_c \leq \underline{r}$. These three cases are now analyzed.

Case 1 ($r_c \geq \bar{r}, P_l > 0$): Fig. 6, Fig. 7 and Fig. 8 show that $\tau_l > 0$, $\omega_l > 0$ and $P_l > 0$. From (43) and (40), it follows $i_l > 0$, $U_l > 0$. Therefore, for the left side of the vehicle in this case,

$$i_l > 0,\ U_l > 0, P_l > 0, \tag{51}$$

which implies the left motor consumes power. The direction of the motor current flow and voltage are as shown in Fig. 5.
Fig. 11 shows that for each commanded turning radius r_c satisfying $r_c \geq \bar{r}$ the total motor power consumption is dominated by the mechanical power consumption although there is a small amount of electrical power consumption. In this case, the power source is the motor controller system, while the motor shaft motion consumes mechanical power and the motor electrical resistance consumes electrical power. Referring to (47),

$$\sigma(P_l) = P_l, \tag{52}$$

where P_l is given by (45).
Case 2 ($\underline{r} < r_c < \bar{r}$, $P_l < 0$): Fig. 6, Fig. 7 and Fig. 8 show that $\tau_l < 0, \omega_l > 0$ and $P_l < 0$. From (43) and (40), it follows $i_l < 0, U_l > 0$. Therefore, for the left side of the vehicle in this case,

$$i_l < 0,\ U_l > 0, P_l < 0, \tag{53}$$

which implies that the left motor generates power. In Fig. 5 the direction of i_l and the voltage drop across R_e are reversed, while the motor controller voltage U_l and that of the motor remain as shown.
Fig. 11 shows that for each commanded turning radius r_c satisfying $\underline{r} < r_c < \bar{r}$ the mechanical power consumption is negative, and hence the motor shaft motion does not consume power but on the contrary generates power from the terrain. This is because when the vehicle rotates, the outer wheel drags the inner wheel through the vehicle body Morales et al. (2009), which leads to the backward sliding friction for the inner wheel and the generation of power for the inner motor from terrain. Since the mechanically generated power is larger than the electrical power consumption, there is a net power generation that is consumed by the motor controller system. In this case, the power source is the motor shaft, while the motor electrical resistance and the motor controller system consume power. Referring to (47),

$$\sigma(P_l) = 0. \tag{54}$$

Case 3 ($r_c \leq \underline{r}$, $P_l > 0$): Fig. 6, Fig. 7 and Fig. 8 show that $\tau_l < 0, \omega_l > 0$ and $P_l > 0$. From (43) and (40), it follows $i_l < 0, U_l < 0$. Therefore, for the left side of the vehicle in this case,

$$i_l < 0,\ U_l < 0, P_l > 0, \tag{55}$$

which implies that the left motor consumes power. In Fig. 5 the direction of i_l, the voltage drop across the R_e and the motor controller voltage U_l are reversed, while the voltage sign of the motor remains the same.
Fig. 11 shows for each commanded turning radius r_c satisfying $r_c \leq \underline{r}$ the mechanical power consumption is negative, which means, as in Case 2, the motor shaft motion does not consume power but generates power from terrain. However, unlike Case 2, the generated mechanical power is smaller than the electrical power consumption. Hence, there is a net power consumption and the motor controller system still has to supply power. In this case, the power sources are the motor shaft and the motor controller system, while the motor electrical resistance consumes power. Referring to (47),

$$\sigma(P_l) = P_l, \tag{56}$$

where P_l is given by (45).

Analysis of the Power Consumption of the Entire Vehicle

The overall power consumption of the vehicle is due to the power consumption of both the inner and outer motors and is shown in Fig. 8. Fig. 12 shows the percentage of the overall power consumption of the vehicle due to mechanical power consumption and electrical power consumption. It is of interest to note that when $r_c \geq \bar{r}$, the mechanical power consumption is dominant, while as r_c decreases in value from $\bar{r}$ the electrical heat dissipation eventually dominates. This indicates that in motion planning, as might be expected, it is more energy efficient to plan for trajectories with large turning radii.

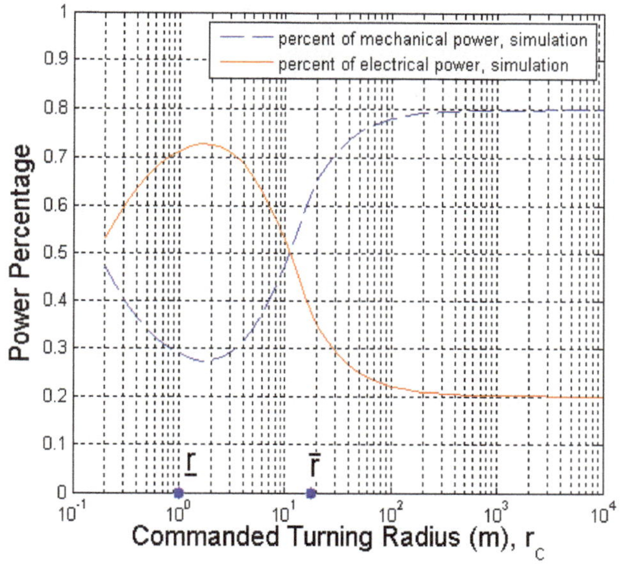

Fig. 12. Mechanical power and electrical power consumption percentages vs. commanded turning radius, obtained via simulation of the dynamic model, for the conditions of Fig. 6.

In summary, analysis of the power models of the right and left motors reveal the interesting phenomenon that for vehicle steady state rotation while the outer motor always consumes power, as the vehicle turning radius decreases, the inner motor first consumes, then generates and finally consumes power again. Since Fig. 8 is generated using the the power model (45), the model enables prediction of the two transition turning radii $\underline{r}$ and $\bar{r}$.

5. Experimental results

This section presents the experiment and simulation results to verify the proposed dynamic models and power models in this chapter.

The experimental platform is the modified Pioneer 3-AT shown in Fig. 13. The original, nontransparent, speed controller from the company was replaced by a PID controller and motor controller. PC104 boards replaced the original control system boards that came with the vehicle. Two current sensors were mounted on each side of the vehicle to provide real time measurement of the motors' currents. It was modified to run on the QNX realtime operating system with a control sampling rate of 1KHz. The mobile robot can be commanded with a linear velocity and turning radius.

Fig. 13. Modified Pioneer 3-AT

5.1 Steady state rotation for different turning radii

In this subsection, 2D steady state rotation results for different turning radii are presented. For a vehicle commanded linear velocity of 0.2m/s and turning radius changing from $10^{-0.7}$ m to 10^{4} m, Fig. 14 shows the experimental and simulation wheel angular velocity vs. commanded radius, Fig. 15 shows the experimental and simulation applied torques vs. commanded radius and Fig. 16 shows the experimental and simulated power consumption vs. commanded radius. In all figures there is good correspondence between the experimental and simulated results. If shear stress is not a function of shear displacement, but instead takes on a maximum value when there is a small relative movement between wheel and terrain, the left and right motor torques should be constant for different commanded turning radii, a phenomenon not seen in Fig. 15. Instead this figure shows the magnitudes of both the left and right torques reduce as the commanded turning radius increases. The same trend is found in Wong (2001); Wong & Chiang (2001). Note that the three cases of inner motor power consumption are observed in the experimental results of Fig. 16.

5.2 Circular movement with motor power saturation

In terms of motion planning one advantage of using the separate motor power models (39) and (45) instead of relying completely on the entire vehicle power model (47) is that the separate models enable more accurate predictions of vehicle velocity when an individual motor experiences power saturation. For a given trajectory it is possible for the entire vehicle power consumption required to achieve that trajectory to be below the total power that can be provided by the motor drive systems, but the power consumption required for one of the motors to be above the power limitations of that motor's drive system. In this case the desired trajectory cannot be achieved.

For the modified Pioneer 3-AT of Fig. 13, the maximum linear velocity is 0.93 m/s. The power limitation for each side of the motor drive system is 51 w, and total power limitation is 102 w. Fig. 17 was generated using (39) and (45) and shows the power requirements for the inner and outer motors vs. commanded turning radius when the vehicle has a linear velocity of 0.7 m/s. It also shows a line corresponding to the 51 w power limitation of each of the motor drive systems.

For a vehicle commanded linear velocity of 0.7 m/s and turning radius of 1.2 m, the predicted inner and outer motor power requirements for a commanded turning radius of 1.2 m are marked using square symbols in Fig. 17. It is seen that the outer motor power consumption is

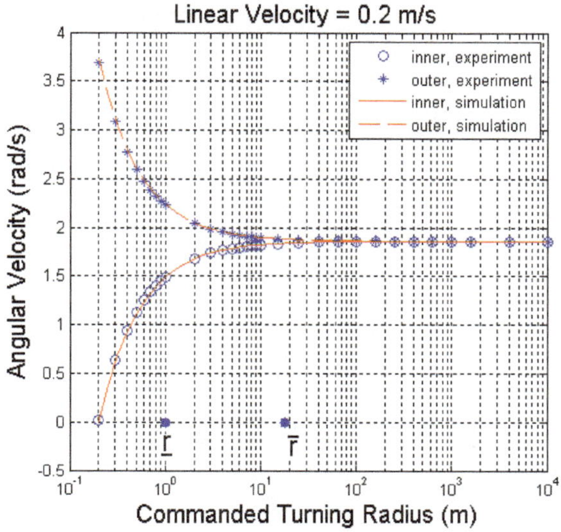

Fig. 14. Vehicle inner and outer wheel angular velocity comparison during steady-state CCW rotation for different commanded turning radii on the lab vinyl surface when the commanded linear velocity is 0.2 m/s.

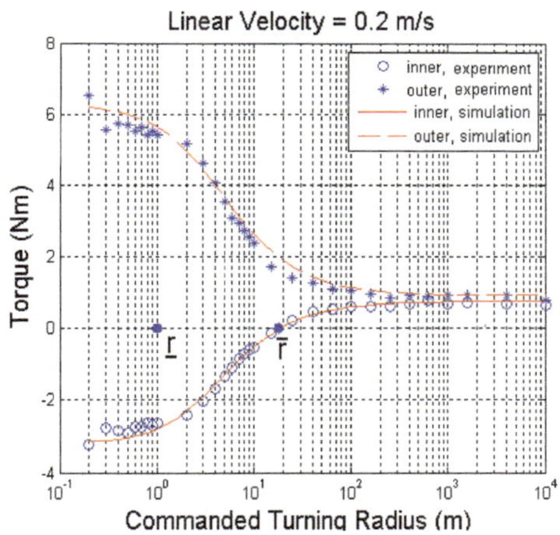

Fig. 15. Vehicle inner and outer wheel applied torque comparison corresponding to Fig. 14

approximately 58 w, which is above the 51 w power limitation, which means this velocity and radius combination cannot be achieved. However, the total power requirement in this case is also equal to 58 w. This power requirement is well below the 102 w limitation for the entire vehicle, showing the importance of estimating the power consumption of each individual motor. It is seen that as predicted by the power model for the outer wheel that this wheel was unable to achieve the desired vehicle due to the power limitation of the motor drive system. Fig. 18 compares the experimental, simulation and commanded angular velocity for the inner and outer wheels. Fig. 19 shows the experimental and simulation applied torques vs.

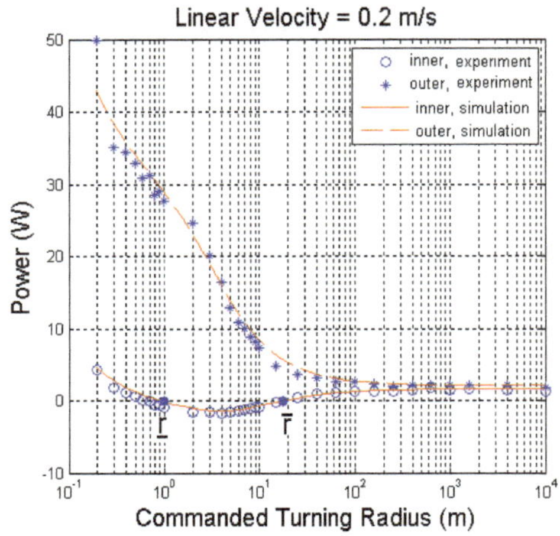

Fig. 16. Vehicle inner and outer wheel power comparison comparison corresponding to Fig. 14.

commanded radius and Fig. 20 shows the experimental and simulated power consumption vs. commanded radius.

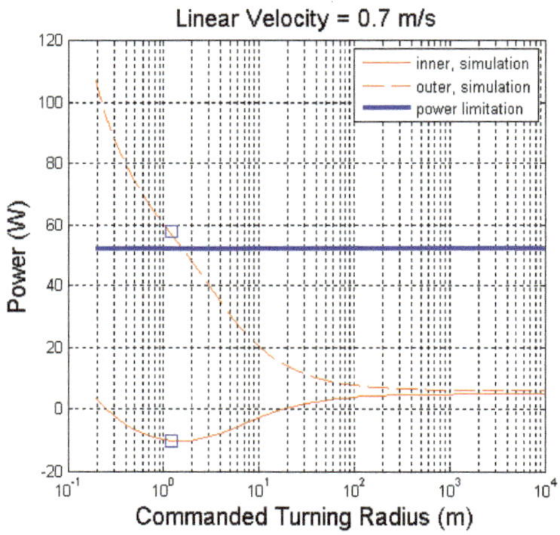

Fig. 17. Power limitation for each side of vehicle, and vehicle inner and outer wheel power prediction during steady-state CCW rotation for different commanded turning radii on the lab vinyl surface when the commanded linear velocity is 0.7 m/s

5.3 Curvilinear movement

In this subsection, the model is to be tested for general 2D curvilinear motion. The vehicle was commanded to move in lemniscate trajectory, which required it to have modest linear and angular accelerations. Fig. 21 shows one complete cycle lemniscate trajectory along with the partial lemniscate trajectory used in the experiment.

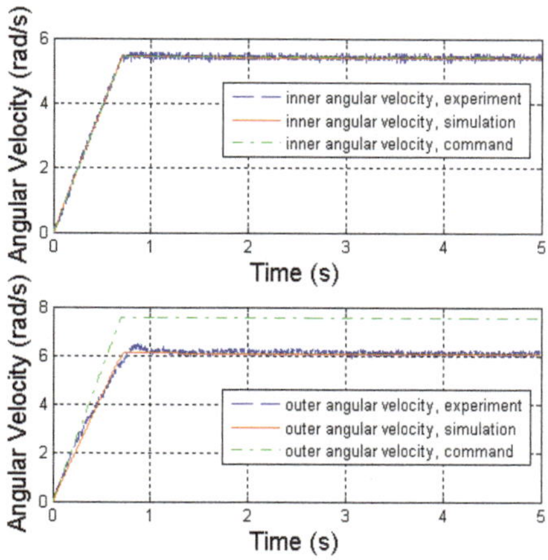

Fig. 18. Experiment, simulation and commanded angular velocity comparison for inner and outer wheels for 2D circular movement on the lab vinyl surface when the commanded radius is 1.2 m and the commanded linear velocity is 0.7 m/s.

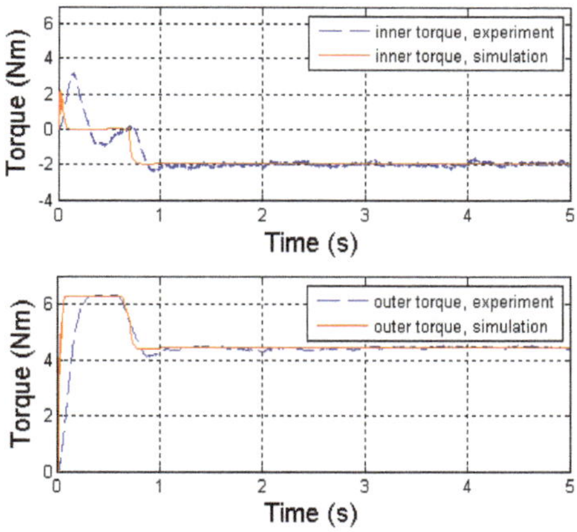

Fig. 19. Vehicle inner and outer wheels torque comparison corresponding to Fig. 18.

In Fig. 21, the lemniscate trajectory is governed by,

$$X(t) = \frac{9\cos(\frac{t}{50})}{1+\sin^2(\frac{t}{50})}, \tag{57}$$

$$Y(t) = \frac{27\sin(\frac{t}{50})\cos(\frac{t}{50})}{1+\sin^2(\frac{t}{50})}, \tag{58}$$

where t is the time, and $(X(t), Y(t))$ is the position in global coordinates. The vehicle was first commanded to go straight with an acceleration of 1 m/s^2 to achieve the initial entering velocity 0.54 m/s of the lemniscate trajectory in Fig. 21 within 3 seconds. Then it was commanded to follow the desired trajectory in Fig. 21 with changing linear velocity and

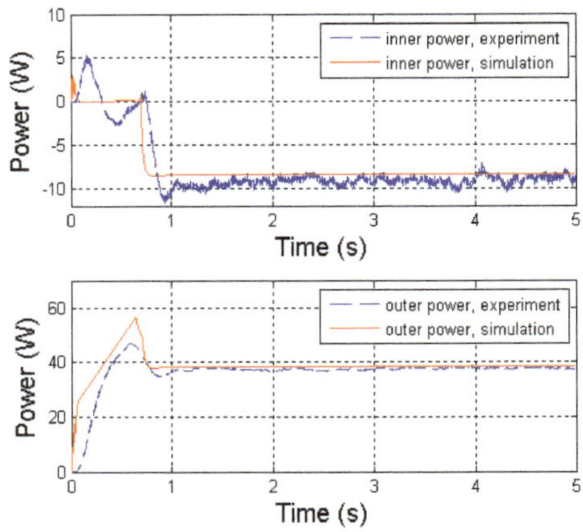

Fig. 20. Vehicle inner and outer wheels power comparison corresponding to Fig. 18.

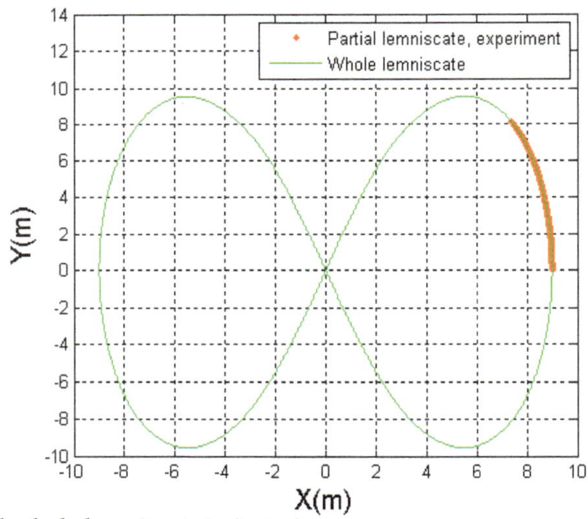

Fig. 21. Partial and whole lemniscate trajectories

turning radius for another 18 seconds. The linear acceleration changes in the range [0 0.02] m/s^2.

Fig. 22, Fig. 23 and Fig. 24 show the experimental and simulation comparisons for the vehicle linear and angular velocity, the inner and outer wheel angular velocity, and the inner and outer wheel applied torque. Fig. 25 shows the corresponding power comparison for the inner and outer part of the vehicle. These results reveal that if the vehicle turns with continually changing linear and angular accelerations of limited magnitude, the dynamic models and power models are still capable of providing high fidelity predictions for both the inner and outer part of the vehicle. Fig. 25 also shows during the lemniscate traversal the inner motor gradually changes from consuming power to generating power while the outer motor always consumes power. The transition time for the inner motor can also be predicted from the motor power model (45).

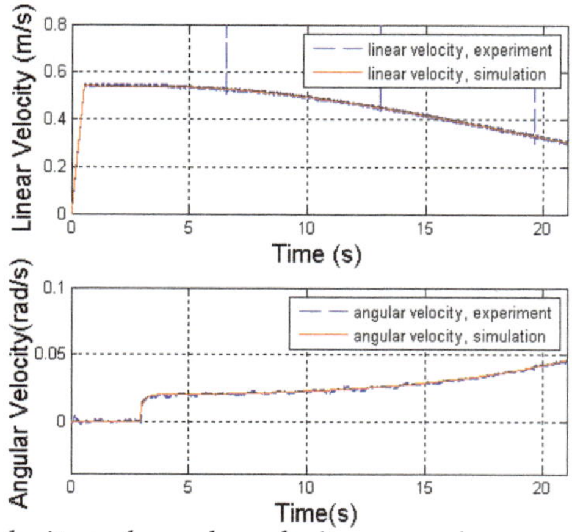

Fig. 22. Vehicle linear velocity and angular velocity comparison corresponding to lemniscate movement in Fig. 21

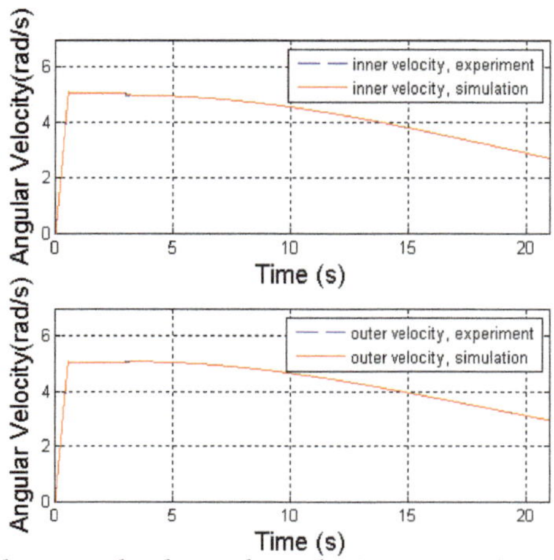

Fig. 23. Vehicle inner and outer wheel angular velocity comparison corresponding to lemniscate movement in Fig. 21

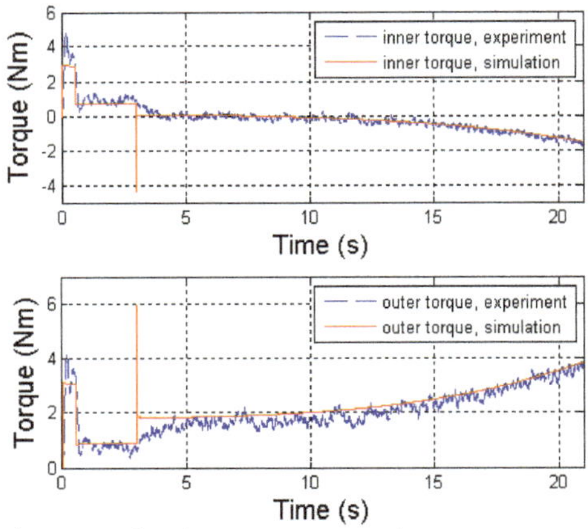

Fig. 24. Vehicle inner and outer wheel torque comparison corresponding to lemniscate movement in Fig. 21

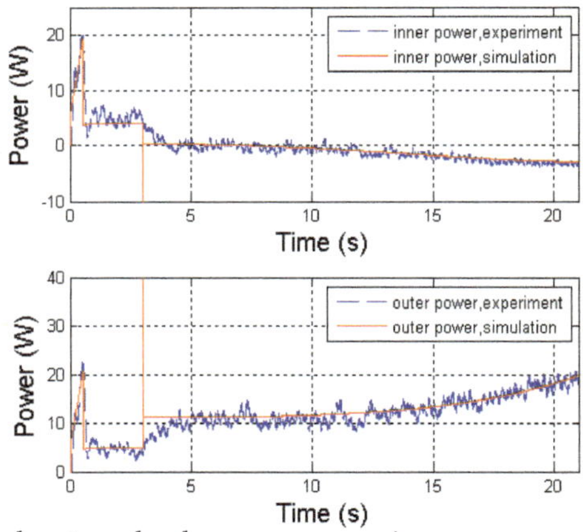

Fig. 25. Vehicle inner and outer wheel power comparison corresponding to lemniscate movement in Fig. 21

6. References

Caracciolo, L., Luca, A. D. & Iannitti, S. (1999). Trajectory tracking control of a four-wheel differentially driven mobile robot, *Proceedings of the IEEE International Conference on Robotics and Automation*, Detroit, MI, pp. 2632–2638.

Economou, J., Colyer, R., Tsourdos, A. & White, B. (2002). Fuzzy logic approaches for wheeled skid-steer vehicle, *Vehicular Technology Conference*, pp. 990–994.

Endo, D., Okada, Y., Nagatani, K. & Yoshida, K. (2007). Path following control for tracked vehicles based slip-compensation odometry, *Proceedings of the IEEE International Conference on Intelligent Robots and Systems*, pp. 2871–2876.

Golconda, S. (2005). *Steering control for a skid-steered autonomous ground vehicle at varying speed*, Master's thesis, University of Louisiana at Lafayette.

Howard, T. M. & Kelly, A. (2007). Optimal rough terrain trajectory generation for wheeled mobile robots, *International Journal of Robotics Research* pp. 141–166.

Kim, C. & Kim, B. K. (2007). Minimum-energy translational trajectory generation for differential-driven wheeled mobile robots, *Journal of Intelligent Robot System* pp. 367–383.

Kozlowski, K. & Pazderski, D. (2004). Modeling and control of a 4-wheel skid-steering mobile robot, *International Journal of Mathematics and Computer Science* pp. 477–496.

Mandow, A., Martĺınez, J. L., Morales, J., Blanco, J.-L., Garcĺıa-Cerezo, A. & Gonzalez, J. (2007). Experimental kinematics for wheeled skid-steer mobile robots, *Proceedings of the International Conference on Intelligent Robots and Systems*, San Diego, CA, pp. 1222–1227.

Martinez, J., Mandow, A., Morales, J., Pedraza, S. & Garcia-Cerezo, A. (2005). Approximating kinematics for tracked mobile robots, *International Journal of Robotics Research* pp. 867–878.

Moosavian, S. A. A. & Kalantari, A. (2008). Experimental slip estimation for exact kinematics modeling and control of a tracked mobile robot, *Proceedings of the International Conference on Intelligent Robots and Systems*, Nice, France, pp. 95–100.

Morales, J., Martinez, J. L., Mandow, A., Garcia-Cerezo, A., Gomez-Gabriel, J. & Pedraza, S. (2006). Power anlysis for a skid-steered tracked mobile robot, *Proceedings of the IEEE International Conference on Mechatronics*, pp. 420–425.

Morales, J., Martinez, J. L., Mandow, A., Garcia-Cerezo, A. J. & Pedraza, S. (2009). Power consumption modeling of skid-steer tracked mobile robots on rigid terrain, *IEEE Transactions on Robotics* pp. 1098–1108.

Nagatani, K., Endo, D. & Yoshida, K. (2007). Improvement of the odometry accuracy of a crawler vehicle with consideration of slippage, *Proceedings of the International Conference on Robotics and Automation*, Rome, Italy, pp. 2752–2757.

O. Chuy, J., E. Collins, J., Yu, W. & Ordonez, C. (2009). Power modeling of a skid steered wheeled robotic ground vehicle, *Proceedings of the IEEE International Conference on Robotics and Automation*, Kobe, Japan.

Petrov, P., de Lafontaine, J., Bigras, P. & Tetreault, M. (2000). Lateral control of a skid-steering mining vehicle, *Proceedings of the International Conference on Intelligent Robots and Systems*, Takamatsu, Japan, pp. 1804–1809.

Rizzoni, G. (2000). *Principles and Applications of Electrical Engineering*, McGraw-Hill.

Shamah, B., Wagner, M. D., Moorehead, S., Teza, J., Wettergreen, D. & Whittaker, W. R. L. (2001). Steering and control of a passively articulated robot, *SPIE, Sensor Fusion and Decentralized Control in Robotic Systems IV*, Vol. 4571.

Siegwart, R. & Nourbakhsh, I. R. (2005). *Introduction to Mobile Robotics*, MIT Press, Cambridge, MA.

Song, X., Song, Z., Senevirante, L. & Althoefer, K. (2008). Optical flow-based slip and velocity estimation technique for unmanned skid-steered vehicles, pp. 101–106.

Song, Z., Zweiri, Y. H. & Seneviratne, L. D. (2006). Non-linear observer for slip estimation of skid-steering vehicles, *Proceedings of the IEEE International Conference on Robotics and Automation*, Orlando, Fl, pp. 1499–1504.

Wang, H., Zhang, J., Yi, J., Song, D., Jayasuriya, S. & Liu, J. (2009). Modeling and motion stability analysis of skid-steered mobile robots, *Proceedings of the International Conference on Robotics and Automation*, Kobe, Japan, pp. 4112– 4117.

Wong, J. Y. (2001). *Theory of Ground Vehicle*, 3rd edn, John Wiley & Sons, Inc.

Wong, J. Y. & Chiang, C. F. (2001). A general theory for skid steering of tracked vehicles on firm ground, *Proceedings of the Institution of Mechanical Engineers, Part D, Journal of Automotive Engineerings* pp. 343–355.

Yi, J., Song, D., Zhang, J. & Goodwin, Z. (2007). Adaptive trajectory tracking control of skid-steered mobile robots, *Proceedings of the IEEE International Conference on Robotics and Automation*, Roma, Italy, pp. 2605–2610.

Yi, J., Zhang, J., Song, D. & Jayasuriya, S. (2007). IMU-based localization and slip estimation for skid-steered mobile robot, *Proceedings of the IEEE International Conference on Intelligent Robots and Systems*, San Diego, CA, pp. 2845–2849.

Yu, W., Chuy, O., Collins, E. G. & Hollis, P. (2010). Analysis and experimental verification for dynamic modeling of a skid-steered wheeled vehicle, *IEEE Transactions on Robotics* pp. 440–453.

Zhang, Y., Hong, D., Chung, J. H. & Velinsky, S. A. (1998). Dynamic model based robust tracking control of a differentially steered wheeled mobile robot, *Proceedings of The American Control Conference*, Philadelphia, PA, pp. 850–855.

17

Advances in Simulation of Planetary Wheeled Mobile Robots

Liang Ding, Haibo Gao, Zongquan Deng and Weihua Li
State Key Laboratory of Robotics and System,
Harbin Institute of Technology
P. R. China

1. Introduction

Ever since the Sojourner rover of the United States landed on Mars in 1997 (Jet Propulsion Laboratory [JPL], a), there has been an upsurge in the exploration of planets using wheeled mobile robots (WMRs or rovers). The twin rovers that followed, Spirit and Opportunity, have endured many years of activity on Mars and have made many significant discoveries (JPL, b). Several other new missions are in progress to explore Mars (Volpe, 2005; Van et al., 2008) and the Moon (Neal, 2009) using planetary rovers that are expected to traverse more challenging terrain with scientific objectives such as searching for evidence of life and investigating the origin of the solar system.

The present planet exploration rovers are advanced WMRs that show excellent performance and have integrated the cutting-edge technologies of many fields, and overcoming new frontier issues that are specific to planetary rovers has promoted the development of terrestrial WMRs.

Simulation technology plays an important role in both the research and development (R&D) and exploration phases of planetary WMRs (Ding et al., 2008). During the R&D phase of a WMR, a simulation system can be used for mechanical design (e.g., performance analysis and optimization), control algorithm verification, and performance testing, and during the exploration phase, the simulation system can be used to support three-dimensional (3D) predictive displays for successive teleoperation (such as in the case of a lunar rover) or to validate command sequences for supervised teleoperation (such as in the case of a Mars rover). As compared to conventional simulation systems used for WMRs, the simulation system for planetary rovers is characterized by high fidelity, high speed, and comprehensiveness.

The recent development and cutting-edge technologies of simulation systems for planetary rovers are summarized in this article to extend their application to conventional WMRs. The significance of simulation for planetary WMRs is discussed in Section 2. An overview of the simulation technology for planetary WMRs is presented in Section 3. Section 4 introduces key theories (models of terramechanics, dynamics, and terrain geometry) for developing a simulation system for planetary rovers. In Section 5, the research results for the simulation of planetary rovers at the State Key Laboratory of Robotics and System (SKLRS) of China, including simulation methods, systems, and verification results, are presented to provide examples of different simulation methods (based on the commercial dynamics simulation software, general simulation software, and real-time simulation software) in detail for different applications.

2. Importance of simulation for planetary rovers

Virtual simulation can guarantee successful of WMRs for exploring the planets as it plays important roles in both the R&D phase and the exploration phase of the rovers, which are shown in Fig. 1. It is of great importance to planetary rovers in three aspects.

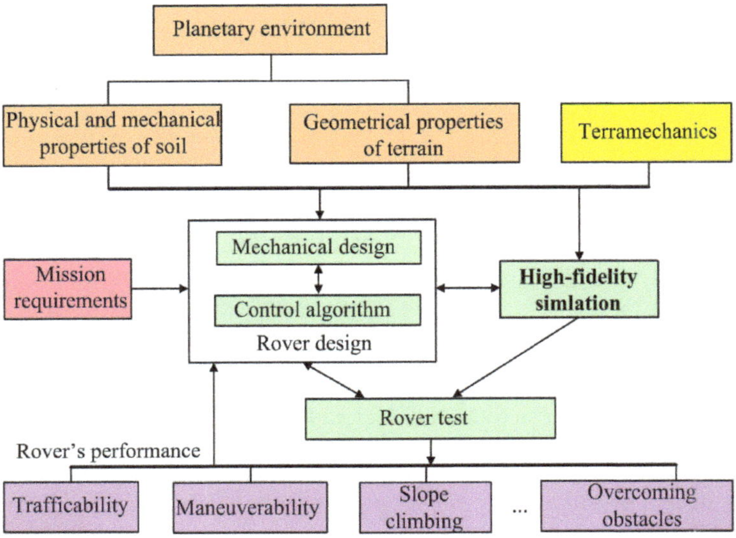

(a) R&D phase

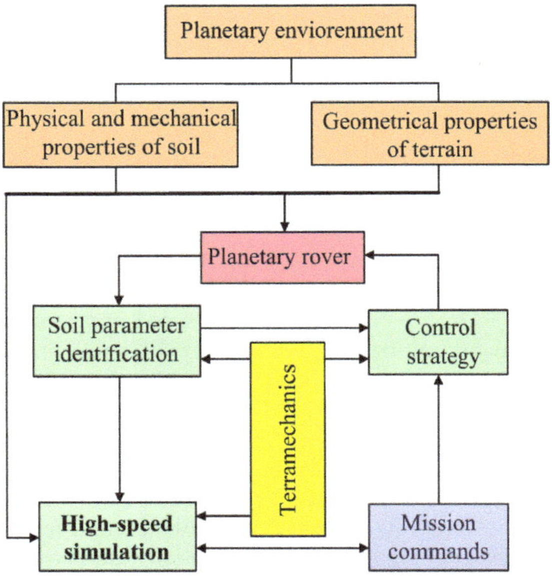

(b) Exploration phase

Fig. 1. Roles of simulation for planetary rovers

2.1 Platform for rover design, evaluation, testing, and control

During the R&D phase of a rover, virtual simulation can be used for configuration design, evaluation and optimization, mobility performance analysis, and control strategy research before a rover is manufactured. For example, before the Mars Pathfinder probe carrying the Sojourner rover was launched, with the help of virtual simulation technology, researchers at JPL predicted that it would turn over while landing because of the interaction between the breaking rocket and Martian wind. The technical program was revised to solve the problem and ensured successful soft landing on Mars (Chen et al., 2005). To support the study on Mars rovers, the United States developed a comprehensive simulation system ROAMS. Europe is committed to the development of the simulation tools RCET, RPET, and RCAST for the evaluation and optimization of Mars rovers. Ye, the general director and general designer of the Chinese Chang'e-1 satellite, advocated numerical simulation technology to study the mobility performance of lunar rovers and simulate floating lunar dust (Ye & Xiao, 2006). Researchers from the Beijing Institute of Control Engineering discussed the technical scheme for researching lunar rovers and pointed out that a virtual simulation system having the abilities of modeling, dynamics simulation, and control simulation should be developed to test the configuration and dynamics parameters of a rover, as well as the control algorithms (Liang et al., 2005).

2.2 Supporting 3D predictive display for successive teleoperation

Exploration activities and the moving path of a rover are subject to the great uncertainty in our knowledge of a planet's surface environment. Therefore, even the most advanced rovers at present, Spirit and Opportunity, do not employ a completely autonomous control strategy but are teleoperated by scientists on the Earth. The time delay in transmission between the Earth and Moon is about several seconds owing to the long distance and limited bandwidth, which makes continuous closed-loop control of a rover unstable. This problem can be solved effectively by successive remote operation using a 3D predictive display. This has been studied by researchers at Japan's Meiji University, China's Jilin University, and elsewhere (Lei et al., 2004; Kuroda et al., 2003; Kunii et al., 2001). Ground computers construct a virtual simulation environment according to the rover status and imagery information of the terrain, through which the rover will move after a time delay. Operators control the rover in a virtual environment, and the same commands are sent to the real rover after error compensation. The real rover repeats the motion of the virtual rover after a while. If the timing of the sent commands is regulated well, the rover can move successively without stopping.

2.3 Validating command sequence for supervised teleoperation

As the time delay in transmission from the Earth to Mars is tens of minutes, supervised teleoperation is adopted instead of nonfeasible continuous teleoperation to control Mars rovers. Researchers at JPL learned an important lesson from the Mars Pathfinder mission that a fast, accurate, and powerful tool for driving the rover is necessary (Maxwell et al., 2005). Thus, the rover sequencing and visualization program (RSVP) has been developed to drive rovers (Wright et al., 2006). It is a suite consisting of two main components: RoSE (rover sequence editor) and Hyperdrive. Hyperdrive is an immersive 3D simulation of the rover, and its environment enables operators to construct detailed rover motions and verify their safety. It uses the state information of a rover to analyze and review the current state, identify any anomalous issues, review previously commanded activities, and verify that the commanded activities have been completed. Scientists define science activities and plan a rover's path with 3D terrain models built according to images obtained by stereo cameras. Then, command sequences are constructed. Sequence validation is done by simulation, and the verified sequence is transmitted to the real rover for the exploration of a solar day. RSVP is used to generate the final command sequences for the mission and plan and validate all mobility and manipulation activities.

The supervised teleoperation method has its disadvantages. For instance, the rover must stop to wait for command sequences and it should be very intelligent. However, it also has obvious advantages. The science activities are defined systematically, and the operators can easily control the rover. Therefore, the two teleoperation modes are used according to the terrain and mission. No matter which is used, virtual simulation is indispensable.

3. Overview of current simulation technology for planetary WMRs

3.1 General simulation tools used in research of planetary WMRs

Various general simulation tools are used by researchers to investigate the kinematics, dynamics, control, and visualization of rovers. They are summarized in Table 1 (Ding et al., 2008). General simulation tools are used by researchers to realize different research purposes, but they do not meet the requirements of rover-based exploration missions well because of their limitations.

3.2 Virtual simulation system developed for planetary WMRs

To support the R&D of WMRs for related planetary exploration missions, customized simulation tools or tools with different simulation functions have been developed.

3.2.1 ROAMS, WITS, and RSVP developed by JPL, United States

The Mars Technology Program, in conjunction with the Mars Science Laboratory Mission, has funded three complementary infrastructure elements: ROAMS, the web interface for tele-science (WITS), and CLARAty (Volpe, 2003).

ROAMS is a comprehensive physics-based simulator for planetary-exploration roving vehicles. It includes a mechanical subsystem, electrical subsystem, internal and external sensors, on-board resources, on-board control software, a terrain environment, and terrain/vehicle interactions. ROAMS can be used for stand-alone simulation, closed-loop simulation with on-board software, and operator-in-the-loop simulation (Yen et al., 1999). It includes rover kinematics, dynamics, navigation, locomotion, and visualization. ROAMS provides simulation services for off-line analysis, and acts as a virtual rover platform for CLARAty control software (Fig. 2), which is reusable rover software architecture being developed in collaboration by JPL, NASA Ames, Carnegie Mellon University and other institutions. A goal of the CLARAty development is to provide open architecture for component algorithm developers to develop and integrate their capabilities into.

RSVP, which is capable of kinematics simulation, was developed for rover teleoperation as mentioned above. WITS provides collaborative downlink data visualization and uplink activity planning for multiple Mars lander and rover missions. For the 2003 Mars Exploration Rover mission, WITS was the primary science operations tool for downlink data visualization and uplink science activity planning. It can build scripted sequences and execute them on a 3D simulated rover, which can be controlled interactively (Backes et al., 2004).

Simulation contents	Explanation	Purpose	Examples of simulation tools
Forward kinematics	Calculates the position and yaw angle of a rover according to feedback information of pitch and roll angles and the positions of motors and joints	Determines the position and orientation of a rover (dead reckoning method)	Matlab, VC++
Inverse kinematics	Calculates the altitude, pitch, and rock angles of the rover, as well as the positions of motors and joints given the yaw angle, horizontal position of a rover, and the path on known terrain	Analyzes the stability and traversability of a rover while following a certain path	
Dynamics	Used in research on wheel-soil interaction mechanics, multi-body dynamics, and methods for solving differential equations with high efficiency	Analyzes the dynamics performance such as vibratility and the ability to overcome obstacles; the basis for designing control strategy	ADAMS, DADS, Vortex
Control strategy	Used in research on control strategies of a rover's locomotion, path planning, path following, and intelligent navigation	Develops a program for rover control	Matlab
Visualization	Builds a visual virtual simulation environment according to topography and surface features of the planet and the rover's configuration	Provides a visualization platform for virtual simulation	Vega, 3D Max, MultiGen Creator

Table 1. Rover simulation and tools

3.2.2 RCET, RPET, and RCAST developed in Europe

Mars rover chassis evaluation tools (RCET) have been developed in Europe to support the design of planetary rovers (Michaud et al., 2006). RCET, which was developed jointly by Contraves Space, Ecole Polytechnique Fédérale de Lausanne, the German Aerospace Center (DLR) and others, accurately predicts the rover performances of the locomotion subsystem. It consists of a two-dimensional (2D) rover simulator that uses a tractive prediction module to compute the wheel/ground interaction. 3D simulation can also be performed with the help of RoverGen.

The rover performance evaluation tool (RPET) consists of the rover mobility performance evaluation tool (RMPET) and mobility synthesis (MobSyn). RMPET computes mobility performance parameters such as drawbar pull, motion resistances, soil thrust, slippage, and sinkage for a mobility system selected by the user for the evaluation of particular terrain (Patel et al., 2004).

RCAST was developed to characterize and optimize the ExoMars rover mobility in support of the evaluation of locomotion subsystem designs before RCET was available (Fig. 3). It uses the AESCO soft soil tire model (AS²TM) software package for terramechanics (Bauer et al., 2005).

At present, DLR is responsible for modeling, simulating and testing the entire mobility behavior of the rover within the ExoMars mission preparation phases. The commercial software tool Simpack, which includes contact modeling based on polygonal contact modeling, is used for simulation (Schäfer et al., 2010). The terramechanics for wheel-soil contact dynamics modeling and simulation and its experimental validation for the ExoMars rover has been introduced (Schäfer et al., 2010).

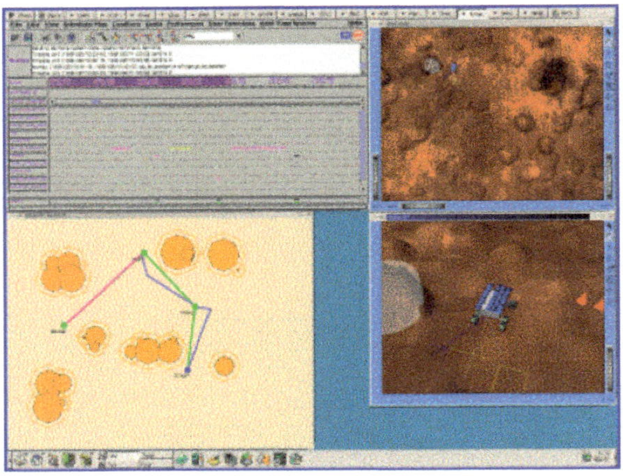

Fig. 2. Simulation with ROAMS and CLARAty

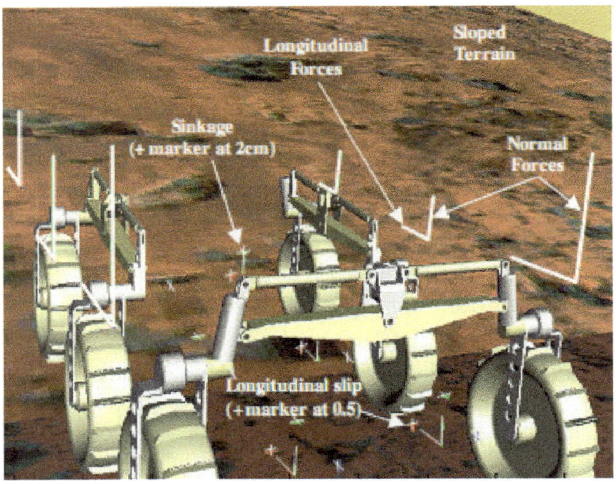

Fig. 3. ExoMars simulation with RCAST

3.2.3 SpaceDyn developed by Tohoku University, Japan

Space Robotics Laboratory (SRL) of Tohoku University developed the dynamics simulation toolbox SpaceDyn with Matlab. SpaceDyn has been successfully used for ETS-VII robot-arm simulation and touchdown of the Hayabusa spacecraft on Itokawa. SpaceDyn is also used to simulate the motion dynamics of a rover with a slip-based traction model (Yoshida & Hamano, 2002) and steering characteristics of a rover on loose soil based on terramechanics (Ishigami & Yoshida, 2005). A path-planning method taking into account wheel-slip dynamics of a planetary exploration rover was developed to generate several paths for a rover moving over rough terrain. Dynamics simulation was carried out by controlling the rover to follow the candidate paths for evaluation, as shown in Fig. 4 (Ishigami et al., 2007).

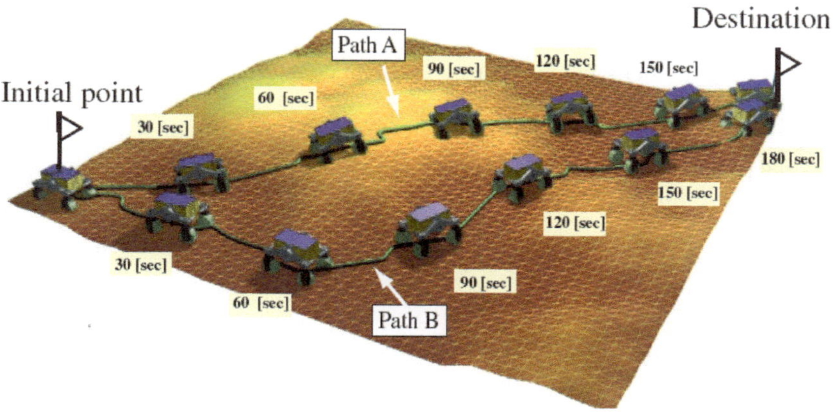

Fig. 4. Simulation of path planning and evaluation

3.2.4 Simulation platform for China's lunar rover

Researchers at the State Key Laboratory of Robotics and System (SKLRS) of the Harbin Institute of Technology developed a simulation platform entitled rover simulation based on terramechanics and dynamics (RoSTDyn) (Li et al., 2012) under a contract with the Chinese Academy of Space Technology to evaluate the performance of the Chang'e lunar rover and assist with its teleoperation.

Yang et al. of Shanghai Jiaotong University presented the framework and key technologies of a virtual simulation environment for a lunar rover (Yang et al., 2008). A fractional Brownian motion technique and statistical properties were used to generate the lunar surface. The multi-body dynamics and complex interactions with soft ground were integrated in the environment.

Researchers at Tsinghua University investigated a test and simulation platform for lunar rovers (Luo & Sun, 2002). The platform provides modules for creating the topography of the terrain and an environmental components editor. The virtual lunar environment can be constructed with the terrain modules built in advance. COM technology was used to support distributed simulation.

4. Key theories for development of simulation system for planetary WMRs

Planetary exploration missions require comprehensive simulation systems that have the abilities of modeling, kinematics, dynamics, control, and visualization, and have the characteristics of high speed and high fidelity.

Figure 5 shows the architecture of a comprehensive virtual simulation system for the high-fidelity/high-speed simulation of rovers. It comprises the RoSTDyn and interactive virtual planetary rover environment (IVPRE) systems. RoSTDyn is a comprehensive simulation system similar to ROAMS. Control commands of three levels received from itself, IVPRE, or other control software can be accepted by RoSTDyn; i.e. the goals, paths, and motor's position. Users can control the virtual rover interactively with IVPRE, which constructs a virtual lunar environment with terrain components or images from the real rover. Digital evaluation model (DEM) terrain data are then generated for RoSTDyn. It can also calculate the mechanics parameters of the soil for RoSTDyn. This system can be further developed for successive lunar rover teleoperation based on 3D predictive display. Key technologies of the simulation system include generalized dynamics modeling, wheel-soil interaction terramechanics models, and deformable rough-terrain geometry modeling.

4.1 Generalized recursive dynamics modeling

4.1.1 Recursive kinematics and Jacobian matrices

If $\mathbf{a}=[a_1 \quad a_2 \quad a_3]^{\mathrm{T}}$, $\mathbf{b}=[b_1 \quad b_2 \quad b_3]^{\mathrm{T}}$, and we define $\tilde{\mathbf{a}}=\begin{bmatrix} 0 & -a_3 & a_2 \\ a_3 & 0 & -a_1 \\ -a_2 & a_1 & 0 \end{bmatrix}$, then $\mathbf{a}\times\mathbf{b}=\tilde{\mathbf{a}}\mathbf{b}$ and $\mathbf{b}\times\mathbf{a}=-\tilde{\mathbf{a}}\mathbf{b}=\tilde{\mathbf{a}}^{\mathrm{T}}\mathbf{b}$.

Let $\mathbf{q}=[q_1 \quad q_2 \quad \cdots \quad q_{n_v}]^{\mathrm{T}}$ denote joint variables, where n_v is the number of joints. The WMRs are articulated multi-body systems with a moving base and n_w end-points (wheels). Let $\mathbf{q}_s=[q_l \quad q_m \quad q_n \quad \cdots \quad q_s]^{\mathrm{T}}$ denote a branch from the rover body to a wheel and n_s denote the number of elements in $\mathbf{q}_s$. Replace the joint number *l*, *m*, *n*, …, *s* of the branch with 1, 2, 3,…, n_s, as shown in Fig. 1, which also shows the inertial coordinate $\{\Sigma_I\}$ and the coordinates $\{\Sigma_i\}$ related to link *i* (*i* = *l*, *m*, *n*, …, *s*) and related vectors, where $\boldsymbol{p}_i$ is the position vector of joint *i*; $\boldsymbol{r}_i$ is the position vector of the centroid of link *i*; $\boldsymbol{c}_{ij}$ is the link vector from link *i* to joint *j*; $\boldsymbol{l}_{ij} = \boldsymbol{p}_j - \boldsymbol{p}_i$ is the link vector from joint *i* to joint *j*; and $\boldsymbol{l}_{ie}$ is the vector from joint *i* to end-point *e*.

The position vector of end-point $\boldsymbol{p}_e$ is

$$\mathbf{p}_e = \mathbf{r}_0 + \mathbf{c}_{01} + \sum_{i=1}^{n_s-1} \mathbf{l}_{i(i+1)} + \mathbf{l}_{n_s e} \tag{1}$$

The derivative of Eq. (1) is

$$\begin{aligned} \mathbf{v}_e &= \mathbf{v}_0 + \boldsymbol{\omega}_0 \times (\mathbf{p}_e - \mathbf{r}_0) + \sum_{i=1}^{n_s} \mathbf{A}_i {}^i\mathbf{Z}_i \times (\mathbf{p}_e - \mathbf{p}_i)\dot{q}_i \\ &= [\mathbf{J}_{BTe} \quad \mathbf{J}_{MTe}]\left[\left(\mathbf{v}_0^{\mathrm{T}} \quad \boldsymbol{\omega}_0^{\mathrm{T}}\right) \quad \dot{\mathbf{q}}^{\mathrm{T}}\right]^{\mathrm{T}} \end{aligned} \tag{2}$$

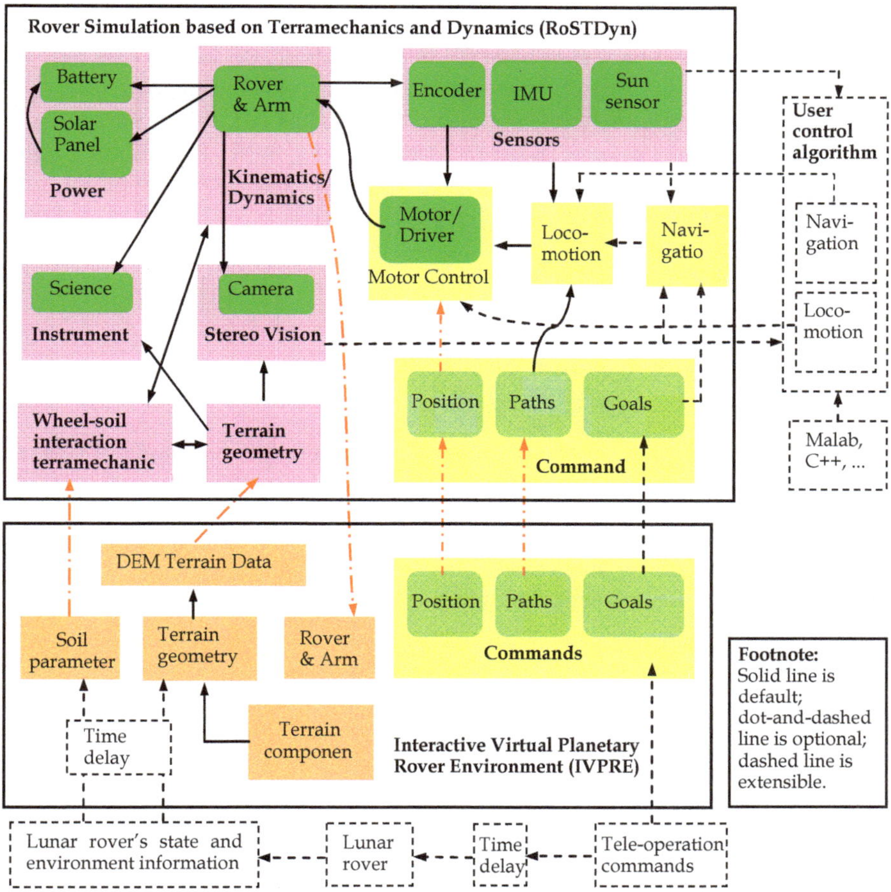

Fig. 5. Architecture of planetary rover's comprehensive simulation system

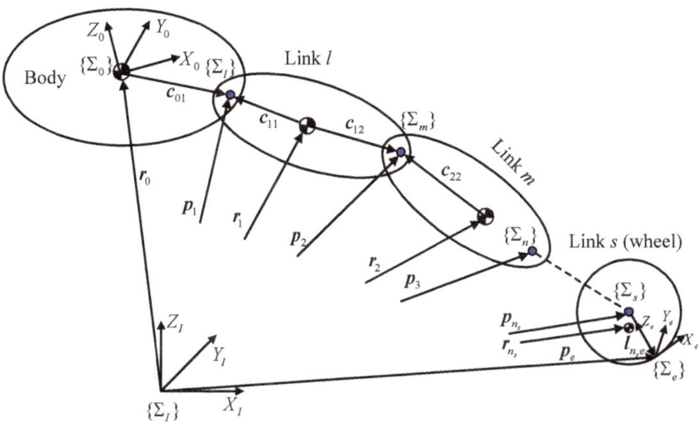

Fig. 6. Coordinates and vectors from rover body to wheel

where $\mathbf{J}_{MTe} = [\mathbf{L}_{s1}\mathbf{A}_1\,{}^1\mathbf{Z}_1 \times \mathbf{P}_{1e} \quad \mathbf{L}_{s2}\mathbf{A}_2\,{}^2\mathbf{Z}_2 \times \mathbf{P}_{2e} \quad \cdots \quad \mathbf{L}_{sn_v}\mathbf{A}_{n_v}\,{}^{n_r}\mathbf{Z}_{n_v} \times \mathbf{P}_{n_v e}]$ is a 3 × n_v matrix, $\mathbf{A}_i = {}^I\mathbf{A}_i$ is the transformation matrix from $\{\Sigma_i\}$ to $\{\Sigma_I\}$, ${}^i\mathbf{Z}_i = [0 \quad 0 \quad 1]^{\mathrm{T}}$ because the z axis is set to coincide with the joint displacement axis, $\mathbf{L}_{ij}$ is an element of matrix $\mathbf{L}_{n_v \times n_v}$ that indicates whether link j is on the access road from link 0 to link i ($\mathbf{L}_{ij}$ =1) or not ($\mathbf{L}_{ij}$ = 0), and $\mathbf{P}_{ie}$ is the vector from the origin of $\{\Sigma_i\}$ to the end-point. $\mathbf{J}_{BTe} = [\mathbf{E} \quad \tilde{\mathbf{P}}_{er0}^T]$ is a 3 × 6 matrix, where $\mathbf{P}_{er0} = \mathbf{p}_e - \mathbf{r}_0$.
The angular velocity of the end-point is

$$\boldsymbol{\omega}_e = \boldsymbol{\omega}_0 + \sum_{i=1}^{n_v} \mathbf{A}_i\,{}^i\mathbf{Z}_i\dot{q}_i = [\mathbf{J}_{BRe} \quad \mathbf{J}_{MRe}]\left[\left(\mathbf{v}_0^{\mathrm{T}} \quad \boldsymbol{\omega}_0^{\mathrm{T}}\right) \quad \dot{\mathbf{q}}^{\mathrm{T}}\right]^{\mathrm{T}}, \tag{3}$$

where $\mathbf{J}_{MRe} = [\mathbf{L}_{s1}\mathbf{A}_1\,{}^1\mathbf{Z}_1 \quad \mathbf{L}_{s2}\mathbf{A}_2\,{}^2\mathbf{Z}_2 \quad \cdots \quad \mathbf{L}_{sn_v}\mathbf{A}_{n_v}\,{}^{n_r}\mathbf{Z}_{n_v}]$ is a 3 × n_s matrix and $\mathbf{J}_{BRe} = [0 \quad \mathbf{E}]$ is a 3 × 6 matrix.

Let $\mathbf{J}_e = [\mathbf{J}_{Be} \quad \mathbf{J}_{Me}] = \begin{bmatrix} \mathbf{J}_{BTe} & \mathbf{J}_{MTe} \\ \mathbf{J}_{BRe} & \mathbf{J}_{MRe} \end{bmatrix}$ be a 6 × (6 + n_v) Jacobian matrix for mapping generalized velocities to the end-points; let $\dot{\boldsymbol{\Phi}} = \left[\left(\mathbf{v}_0^{\mathrm{T}} \quad \boldsymbol{\omega}_0^{\mathrm{T}}\right) \quad \dot{\mathbf{q}}^{\mathrm{T}}\right]^{\mathrm{T}}$ be a vector with (6 + n_v) elements, which are the linear velocities and angular velocities of the body, and joint velocities. Let $\dot{\mathbf{X}}_{ae}$ and $\mathbf{J}_{ae}$ denote the velocities of all the wheel–soil interaction points and the corresponding Jacobian matrix:

$$\dot{\mathbf{X}}_{ae} = \begin{bmatrix} \mathbf{v}_e(1) \\ \boldsymbol{\omega}_e(1) \\ \vdots \\ \mathbf{v}_e(n_w) \\ \boldsymbol{\omega}_e(n_w) \end{bmatrix}, \quad \mathbf{J}_{ae} = \begin{bmatrix} \mathbf{J}_e(1) \\ \mathbf{J}_e(2) \\ \vdots \\ \mathbf{J}_e(n_w) \end{bmatrix},$$

which are a $6n_w \times 1$ vector and $6n_w \times (n_v + 6)$ matrix, respectively. We thus obtain

$$\dot{\mathbf{X}}_{ae} = \mathbf{J}_{ae}\dot{\boldsymbol{\Phi}}. \tag{4}$$

The same method is used to deduce the Jacobian matrix by mapping the velocities from the generalized coordinates to the link centroid:

$$\dot{\mathbf{X}}_a = \mathbf{J}_a\dot{\boldsymbol{\Phi}}, \tag{5}$$

where $\dot{\mathbf{X}}_a$ $(6n_v \times 1)$ is the velocity vector of all centroids, and J_a $(6n_v \times (n_v + 6))$ is the Jacobian matrix. In Eq. (5),

$$\dot{\mathbf{X}}_a = \begin{bmatrix} \mathbf{v}_1 \\ \boldsymbol{\omega}_1 \\ \vdots \\ \mathbf{v}_{n_v} \\ \boldsymbol{\omega}_{n_v} \end{bmatrix}, \quad \mathbf{J}_a = \begin{bmatrix} \mathbf{J}_1 \\ \mathbf{J}_2 \\ \vdots \\ \mathbf{J}_{n_v} \end{bmatrix}, \quad \mathbf{J}_i = [\mathbf{J}_{Bi} \quad \mathbf{J}_{Mi}] = \begin{bmatrix} \mathbf{J}_{BTi} & \mathbf{J}_{MTi} \\ \mathbf{J}_{BRi} & \mathbf{J}_{MRi} \end{bmatrix},$$

where J_i is a $6 \times (6 + n_v)$ matrix. $\mathbf{J}_{BTi} = \begin{bmatrix} \mathbf{E} & \tilde{\mathbf{r}}_{0i}^{\mathrm{T}} \end{bmatrix}$ and $\mathbf{J}_{BRi} = \begin{bmatrix} 0 & \mathbf{E} \end{bmatrix}$ are both 3×6 matrices, and

$$\mathbf{J}_{MRi} = \begin{bmatrix} \mathbf{L}_{i1}\,{}^{1}\mathbf{Z}_1 & \mathbf{L}_{i2}\,{}^{2}\mathbf{Z}_2 & \cdots & \mathbf{L}_{in_v}\,{}^{n_v}\mathbf{Z}_{n_v} \end{bmatrix} \text{ and}$$

$$\mathbf{J}_{MTi} = \begin{bmatrix} \mathbf{L}_{i1}\,{}^{1}\mathbf{Z}_1 \times (\mathbf{r}_i - \mathbf{p}_1) & \mathbf{L}_{i2}\,{}^{2}\mathbf{Z}_2 \times (\mathbf{r}_i - \mathbf{p}_2) & \ldots & \mathbf{L}_{in_v}\,{}^{n_v}\mathbf{Z}_{n_v} \times (\mathbf{r}_i - \mathbf{p}_{n_v}) \end{bmatrix}$$

are both $3 \times n_v$ matrices.

4.1.2 Generalized dynamics model

Substituting Eq. (5) into the kinetic energy equation gives

$$T = \frac{1}{2}\sum_{i=0}^{n_v}(\boldsymbol{\omega}_i^{\mathrm{T}}\mathbf{I}_i\boldsymbol{\omega}_i + m_i\mathbf{v}_i^{\mathrm{T}}\mathbf{v}_i) = \frac{1}{2}\dot{\boldsymbol{\Phi}}^{\mathrm{T}}\mathbf{H}_{sys}\dot{\boldsymbol{\Phi}}, \tag{6}$$

where $\mathbf{H}_{sys}$ is the $(n_v+6)\times(n_v+6)$ system generalized inertia matrix (Yoshida, 2000):

$$\mathbf{H}_{sys} = \begin{bmatrix} M_a(\mathbf{E})_{3\times3} & M_a(\tilde{\mathbf{r}}_{0g}^{\mathrm{T}})_{3\times3} & (\mathbf{J}_{Tg})_{3\times n_v} \\ M_a(\tilde{\mathbf{r}}_{0g})_{3\times3} & (\mathbf{H}_{\omega})_{3\times3} & (\mathbf{H}_{\omega\phi})_{3\times n_v} \\ (\mathbf{J}_{Tg}^{\mathrm{T}})_{n_v\times3} & (\mathbf{H}_{\omega\phi}^{\mathrm{T}})_{n_v\times3} & (\mathbf{H}_{\phi})_{n_v\times n_v} \end{bmatrix}. \tag{7}$$

In Eq. (7), M_a is the overall mass of the robot,

$$\mathbf{r}_{0g} = \mathbf{r}_g - \mathbf{r}_0,\ \mathbf{H}_{\omega} = \mathbf{I}_0 + \sum_{i=1}^{n_v}(\mathbf{I}_i + m_i\tilde{\mathbf{r}}_{0i}\tilde{\mathbf{r}}_{0i}^{\mathrm{T}}) = \mathbf{I}_0 + \sum_{i=1}^{n_v}(\mathbf{I}_i + m_i\tilde{\mathbf{r}}_{0i}^{\mathrm{T}}\tilde{\mathbf{r}}_{0i}),\ \mathbf{J}_{Tg} = \sum_{i=0}^{n_v} m_i\mathbf{J}_{MTi},$$

$$\mathbf{H}_{\phi} = \sum_{i=1}^{n_v}(\mathbf{J}_{MRi}^{\mathrm{T}}\mathbf{I}_i\mathbf{J}_{MRi} + m_i\mathbf{J}_{MTi}^{\mathrm{T}}\mathbf{J}_{MTi}),\text{ and } \mathbf{H}_{\omega\phi} = \sum_{i=1}^{n_v}(\mathbf{I}_i\mathbf{J}_{MRi} + m_i\tilde{\mathbf{r}}_{oi}\mathbf{J}_{MTi}).$$

According to the Lagrange function,

$$\mathbf{F}_{sys} = \mathbf{H}_{sys}(\boldsymbol{\Phi})\ddot{\boldsymbol{\Phi}}_{sys} + \mathbf{C}(\boldsymbol{\Phi},\dot{\boldsymbol{\Phi}})\dot{\boldsymbol{\Phi}} + \mathbf{f}(\dot{\boldsymbol{\Phi}}) + \mathbf{G}(\boldsymbol{\Phi}), \tag{8}$$

where $\boldsymbol{C}$ is an $(n_v+6)\times(n_v+6)$ stiffness matrix describing the Coriolis and centripetal effects, which are proportional to $\dot{q}_i^2$ and $\dot{q}_i\dot{q}_j$, respectively; f is an $(n_v+6)\times1$ matrix that describes viscous and Coulomb friction (typically negligible in a rigid-body dynamics system); $\boldsymbol{G}$ is an $(n_v+6)\times1$ gyroscopic vector reflecting gravity loading; and $\boldsymbol{F}_{sys}$ is the vector of generalized forces:

$$\mathbf{F}_{sys} = \mathbf{N} + \mathbf{J}_{ae}{}^{\mathrm{T}}\mathbf{N}_{ae}. \tag{9}$$

In Eq. (9), $\boldsymbol{N}$ is an $(n_v+6)\times1$ matrix including the forces ($\mathbf{F}_0$) and moments ($\mathbf{M}_0$) acting on the body, and those acting on the joints ($\boldsymbol{\tau} = [\tau_1 \quad \tau_2 \quad \cdots \quad \tau_{n_v}]^{\mathrm{T}}$); $\mathbf{N}_{ae}$ is a $6n_w\times1$ vector including the external forces ($\mathbf{F}_e$) and moments ($\mathbf{M}_e$) from the soil that act on the wheel:

$$\mathbf{N}=\begin{bmatrix}\mathbf{F}_0\\ \mathbf{M}_0\\ \boldsymbol{\tau}\end{bmatrix},\ \mathbf{N}_{ae}=[\mathbf{F}_e^{\mathrm{T}}(1)\ \ \mathbf{M}_e^{\mathrm{T}}(1)\ \ \cdots\ \ \mathbf{F}_e^{\mathrm{T}}(n_w)\ \ \mathbf{M}_e^{\mathrm{T}}(n_w)]^{\mathrm{T}}.$$

The dynamics equation of a WMR including the wheel–soil interaction terramechanics is

$$\mathbf{H}_{sys}(\boldsymbol{\Phi})\ddot{\boldsymbol{\Phi}}_{sys}+\mathbf{C}(\boldsymbol{\Phi},\dot{\boldsymbol{\Phi}})\dot{\boldsymbol{\Phi}}+\mathbf{f}(\dot{\boldsymbol{\Phi}})+\mathbf{G}(\boldsymbol{\Phi})-\mathbf{N}-\mathbf{J}_{ae}{}^{\mathrm{T}}\mathbf{N}_{ae}=0. \tag{10}$$

Let $\mathbf{C}(\boldsymbol{\Phi},\dot{\boldsymbol{\Phi}})\dot{\boldsymbol{\Phi}}+\mathbf{f}(\dot{\boldsymbol{\Phi}})+\mathbf{G}(\boldsymbol{\Phi})=\mathbf{D}$. The generalized accelerations can then be calculated as

$$\ddot{\boldsymbol{\Phi}}_{sys}=\mathbf{H}_{sys}^{-1}(\mathbf{N}+\mathbf{J}_{ae}{}^{\mathrm{T}}\mathbf{N}_{ae}-\mathbf{D}). \tag{11}$$

The recursive Newton–Euler method is used to deduce an equation equivalent to Eq. (10) to calculate the unknown $\boldsymbol{D}$.
The Newton–Euler equations are

$$\begin{cases}\mathbf{F}_i=m_i\dot{\mathbf{v}}_i\\ \mathbf{N}_i=\mathbf{I}_i\dot{\boldsymbol{\omega}}_i+\boldsymbol{\omega}_i\times\mathbf{I}_i\boldsymbol{\omega}_i\end{cases}. \tag{12}$$

According to D'Alembert's principle, the effect of $\mathbf{f}_i$ and $\mathbf{m}_i$ on link i through joint i is

$$\begin{cases}\mathbf{m}_i=\mathbf{M}_i+\sum\limits_{j=i+1}^{n}\mathbf{S}_{ij}(\mathbf{l}_{ij}\times\mathbf{f}_j+\mathbf{m}_j)-\mathbf{S}_{ii}\times\\ \quad[\lambda_P(i)\mathbf{A}_i{}^i\mathbf{Z}_iq_i-\mathbf{c}_{ii}]\times(\mathbf{F}_i-m_i\mathbf{g})-\mathbf{S}_{ei}(\mathbf{l}_{ie}\times\mathbf{F}_{ei}+\mathbf{M}_{ei}),\\ \mathbf{f}_i=\mathbf{F}_i-m_i\mathbf{g}+\sum\limits_{j=i+1}^{n}\mathbf{S}_{ij}\mathbf{f}_j-\mathbf{S}_{ei}\mathbf{F}_{ei}\end{cases} \tag{13}$$

where $\lambda_P(i)$ is 1 for a prismatic joint and zero for a rotational joint, S is the incidence matrix to find the upper connection of a link, and S_{ei} indicates whether i is an end-point. The generalized force/moment of link i is

$$\tau_i=\begin{cases}\mathbf{m}_i^{\mathrm{T}}\mathbf{A}_i{}^i\mathbf{Z}_i & \text{(Rotational joint)}\\ \mathbf{f}_i^{\mathrm{T}}\mathbf{A}_i{}^i\mathbf{Z}_i & \text{(Prismatic joint)}\end{cases}. \tag{14}$$

The forces and moments that act on the body are

$$\begin{cases}\mathbf{F}_0=\sum\limits_{j=1}^{n}\mathbf{S}_{0j}\mathbf{f}_j+m_0(\dot{\mathbf{v}}_0-\mathbf{g})\\ \mathbf{M}_0=\sum\limits_{j=1}^{n}\mathbf{S}_{0j}(\mathbf{c}_{0j}\times\mathbf{f}_j+\mathbf{m}_j)+\mathbf{I}_0\dot{\boldsymbol{\omega}}_0+\boldsymbol{\omega}_0\times\mathbf{I}_0\boldsymbol{\omega}_0\end{cases}, \tag{15}$$

where S_{0j} is a flag vector that indicates whether j has a connection with the body.
Following Eq. (10), let the accelerations of all the generalized coordinates and the external forces/moments be zero; it is then possible to obtain $\boldsymbol{D}$ with Eqs. (14) and (15).

4.2 Wheel–soil interaction terramechanics models

The soil applies three forces and three moments to each wheel, as shown in Fig. 4. The normal force F_N can sustain the wheel. The cohesion and shearing of the soil can generate a resistance moment M_R and a tractive force; the resistance force is caused by the wheel sinking into the soil; the composition of the tractive and resistance forces is called the drawbar pull F_{DP}, which is the effective force of driving a wheel. As a wheel steers or when there is a slip angle, there is a side force F_S, a steering resistance moment M_S, and an overturning moment M_O acting on the wheel.

4.2.1 Driving model

Figure 7 is a diagram of the lugged wheel-soil interaction mechanics, where z is wheel sinkage; θ_1 is the entrance angle at which the wheel begins to contact the soil; θ_2 is the exit angle at which the wheel loses contact with the soil; θ_m is the the angle of maximum stress; θ_1' is the angle at which the soil starts to deform; W is the vertical load of the wheel; DP is the resistance force acting on the wheel; T is the driving torque of the motor; r is the wheel radius; h is the height of the lugs; v is the vehicle velocity; and ω is the angular velocity of the wheel. The soil interacts with the wheel in the form of continuous normal stress σ and shearing stress τ, which can be integrated to calculate the interaction mechanics. To improve the simulation speed, a simplified closed-form formula (Ding et al., 2009a) is adopted and improved considering the effect of the normal force:

$$\begin{cases} F_{DP}=[\dfrac{M_R(A^2+B^2)}{r_sAC}-\dfrac{BF_N}{A}](1+c_{P1}+c_{P2}s)(1+c_{P3}\dfrac{\overline{W}-F_N}{\overline{W}}) \\ F_N=rbA\sigma_m+r_sbB\tau_m \\ M_R=\dfrac{r_s^2CD[bc+[1+c_M(\overline{W}-F_N)/\overline{W}]F_N\tan\varphi/(rA)]}{1+r_sBD\tan\varphi/(rA)} \end{cases} . \quad (16)$$

In Eq. (16), s is the slip ratio defined by Ding et al. (2009); c_{P1} and c_{P2} are adopted to reflect the influence of the slip ratio on the drawbar pull, and θ_m can thus be simplified as half of θ_1; c_{P3} and c_M are parameters that compensate for the effect of the normal force; $\overline{W}$ is the average normal force of the wheels; and $\sigma_m=K_sr^N(\cos\theta_m-\cos\theta_1)^N$, $C=(\theta_1-\theta_2)/2$,

$$A=(\cos\theta_m-\cos\theta_2)/(\theta_m-\theta_2)+(\cos\theta_m-\cos\theta_1)/(\theta_1-\theta_m),$$

$$B=(\sin\theta_m-\sin\theta_2)/(\theta_m-\theta_2)+(\sin\theta_m-\sin\theta_1)/(\theta_1-\theta_m), \text{ and}$$

$$\begin{aligned}\tau_m &=(c+\sigma_m\tan\phi)\times \\ &(1-\exp\{-r_s[(\theta_1'-\theta_m)-(1-s)(\sin\theta_1'-\sin\theta_m)]/k\})\end{aligned}.$$

The newly introduced parameters are
$\theta_1'=\text{acos}[(r-z)/R_j]$, $K_s=k_c/b+k_\varphi$, $N=n_0+n_1s$, and $\theta_2\approx 0$.

The radius R_j is a value between r and $r+h$ that compensates for the lug effect (Ding et al., 2009b). The soil parameters in the equations are k_c, the cohesive modulus; k_φ, the frictional modulus; N, an improved soil sinkage exponent; c, the cohesion of the soil; φ, the internal friction angle; and k, the shearing deformation modulus. n_0 and n_1 are coefficients for calculating N, which are important when predicting the slip-sinkage of wheels.

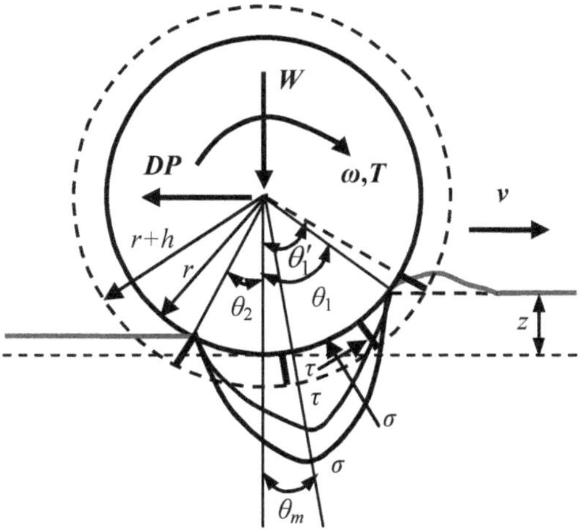

Fig. 7. Lugged wheel–soil interaction mechanics

4.2.2 Steering model

The model for calculating the side force F_S is (Ishigami & Yoshida, 2005)

$$F_S = rb\int_{\theta_2}^{\theta_1} \tau_y(\theta)d\theta + \int_{\theta_2}^{\theta_1} R_b(r - h(\theta)\cos\theta)d\theta \,, \tag{17}$$

$$\tau_y(\theta) = [c + \sigma(\theta)]\{1 - \exp[-r(1-s)(\theta_1 - \theta)\tan\beta / k_y]\} \,, \tag{18}$$

$$R_b = \cot X_c + \tan(X_c + \phi)\left\{hc + \frac{1}{2}\rho h^2\left[(\cot X_c + \frac{\cot X_c{}^2}{\cot\phi}\right]\right\}, \tag{19}$$

where $X_c = \pi / 4 - \varphi / 2$; k_y is the shearing deformation modulus in the y direction; β is the skid angle; and h is the wheel height in the soil.
The overturning moment is approximated by

$$M_O \approx F_S r \,. \tag{20}$$

The steering resistance moment M_s is considered to be zero, and the motion of steering is simulated employing the kinematics method, as the steering torque has little effect on the motion of the entire rover, and the model is still under development.

4.3 Deformable rough-terrain geometry modeling (Ding, 2009)

4.3.1 Calculation of contact area

For simplicity, the literature often assumes that wheel–soil interaction occurs at a single point, which may result in large errors when the robot moves over deformable rough terrain, and even in simulation failure because of abrupt changes in wheel sinkage and other forces. Calculating the interaction area of a wheel moving on soft soil is important for high-fidelity simulation, which is employed to predict and transform the interaction mechanics.

Figure 8 shows the interaction area of a wheel moving on rough terrain. The known parameters are the position of a wheel's center W, (x_w, y_w, z_w); the yaw angle of a wheel, φ_w; and the DEM of the terrain. The interaction area is simplified as an inclined plane determined by points P_1, P_2, and P_3, the normal vector of which is

$$\mathbf{z_e} = \begin{bmatrix} A_t \\ B_t \\ C_t \end{bmatrix} = \begin{bmatrix} x_2 - x_1 \\ y_2 - y_1 \\ z_2 - z_1 \end{bmatrix} \times \begin{bmatrix} x_3 - x_1 \\ y_3 - y_1 \\ z_3 - z_1 \end{bmatrix}. \tag{21}$$

The equation of the inclined plane $P_1P_2P_3$ is therefore

$$A_t(x - x_1) + B_t(y - y_1) + C_t(z - z_1) = 0 \tag{22}$$

P, the foot of the perpendicular line drawn from point w to plane $P_1P_2P_3$, is located on the line $(x - x_w)/A_t = (y - y_w)/B_t = (z - z_w)/C_t$. The coordinates of point E can be solved by substituting the line equation into Eq. (22). The length of wP is thus deduced:

$$\overline{wP} = \frac{|A_t(x_w - x_1) + B_t(y_w - y_1) + C_t(z_w - z_1)|}{\sqrt{A_t^2 + B_t^2 + C_t^2}}. \tag{23}$$

The wheel sinkage is then determined as

$$z = \overline{Pe} = r - \overline{wP}. \tag{24}$$

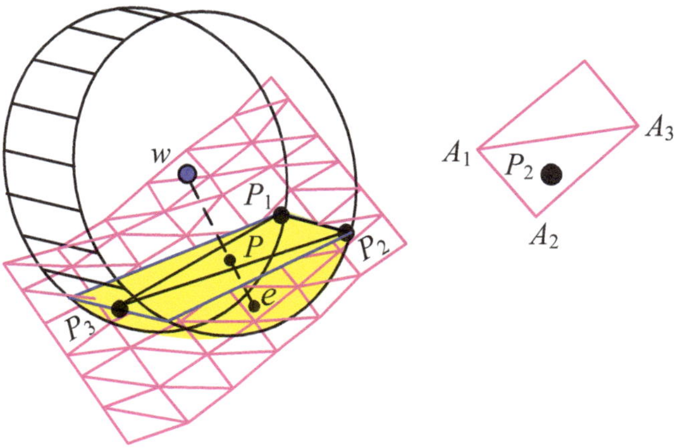

Fig. 8. Interaction area of wheel moving on deformable rough terrain

Point P_2 is used to illustrate how to obtain the coordinates of points P_1, P_2, and P_3. A wheel moving on a random plane can be decomposed into climbing up/down a slope at an angle θ_{cl} and traversing a slope with an inclination angle θ_{cr}, as shown in Fig. 4. The x and y coordinates of point P_2 are then

$$\begin{cases} x_{P2} = x_w + r\cos\theta_{cr} \\ y_{P2} = y_w + r\sin\theta_1 \cos\theta_{cl} \end{cases}. \tag{25}$$

The coordinates of points A_1, A_2, and A_3 are easy to find by referring to the DEM. z_{P2} can then be determined using the same method as that for calculating point E.

4.3.2 Terminal force transformation matrix

Figure 9 shows the forces and moments of the soil that act on a wheel. $\{\Sigma_e\}$ and $\{\Sigma_w\}$ are coordinate systems with the same orientation and different origins, at the end-point and wheel center, respectively.

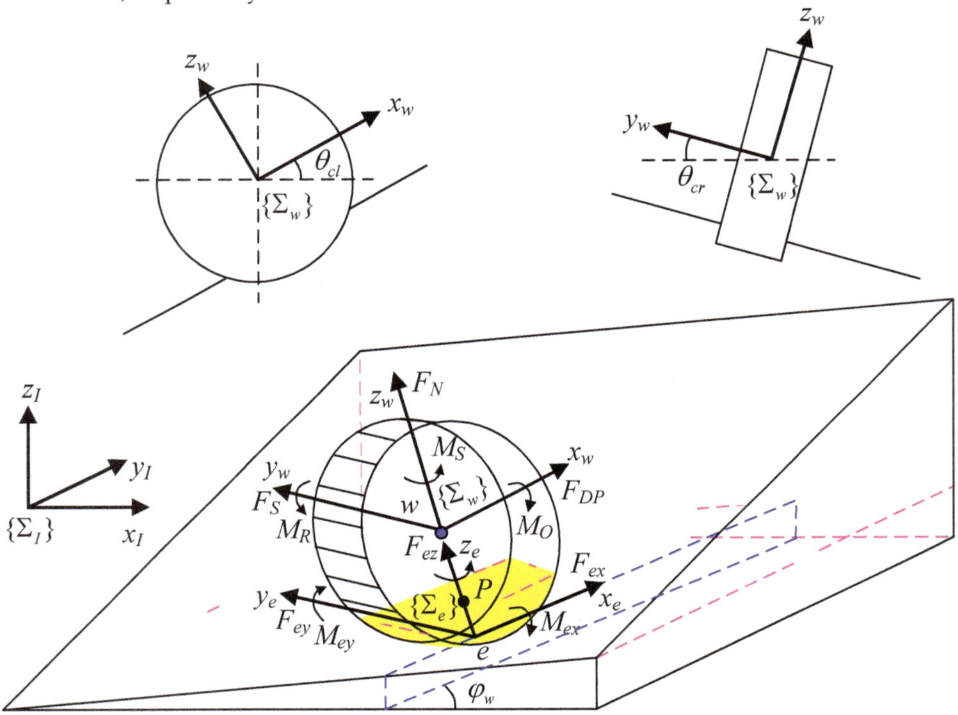

Fig. 9. Force analysis of wheel moving on random slope

x_e is the line of intersection of the wheel-soil interaction plane and the plane with an included angle of φ_w to the x axis: $x\tan\varphi_w - y + D' = 0$. It is deduced that

$$\mathbf{x_e} = \{C_t, C_t\tan\varphi_w, -A_t - B_t\tan\varphi_w\}. \tag{26}$$

The vector direction of y_e is then

$$\mathbf{y}_e = \mathbf{z}_e \times \mathbf{x}_e = \begin{bmatrix} -A_tB_t - (B_t^2 + C_t^2)\tan\varphi_w \\ C_t^2 + A_t(A_t + B_t\tan\varphi_w) \\ A_tC_t\tan\varphi_w - B_tC_t \end{bmatrix}. \tag{27}$$

θ_{cl} (θ_{cr}) is the angle between x_e (y_e) and the horizontal plane, which can be calculated as

$$\begin{cases} \theta_{cl} = \arcsin[(-A_t - B_t \tan\varphi_w) / X_1] \\ \theta_{cr} = \arcsin[C_t(A_t \tan\varphi_w - B_t) / X_2] \end{cases}, \tag{28}$$

where $X_1 = \sqrt{C_t^2(1+\tan^2\varphi_w) + (A_t + B_t\tan\varphi_w)^2}$,

$X_2 = \sqrt{X_3[A_t^2 + C_t^2 + 2A_tB_t\tan\varphi_w + (B_t^2 + C_t^2)\tan^2\varphi_w]}$, and $X_3 = \sqrt{A_t^2 + B_t^2 + C_t^2}$.

According to x_e, y_e, and z_e, the transformation matrix from $\{\Sigma_e\}$ to $\{\Sigma_I\}$ is

$$\mathbf{A}_e = \begin{bmatrix} \dfrac{C_t}{X_1} & \dfrac{-A_tB_t - (B_t^2 + C_t^2)\tan\varphi_w}{X_2} & \dfrac{A_t}{X_3} \\ \dfrac{C_t\tan\varphi_w}{X_1} & \dfrac{C_t^2 + A_t^2 + A_tB_t\tan\varphi_w}{X_2} & \dfrac{B_t}{X_3} \\ \dfrac{-A_t - B_t\tan\varphi_w}{X_1} & \dfrac{A_tC_t\tan\varphi_w - B_tC_t}{X_2} & \dfrac{C_t}{X_3} \end{bmatrix}. \tag{29}$$

The external forces and torques that act at the wheel–soil interaction point are

$$\begin{cases} {}^e\mathbf{F}_e = {}^w\mathbf{F}_e = [F_{DP} \quad F_S \quad F_N]^{\mathrm{T}} \\ {}^e\mathbf{M}_e = [M_O - rF_S \quad -M_R + rF_{DP} \quad M_S]^{\mathrm{T}} \end{cases}. \tag{30}$$

The equivalent forces and moments that act on the wheel in the inertial coordinates $\{\Sigma_I\}$ are

$$\begin{cases} \mathbf{F}_e = \mathbf{A}_e {}^e\mathbf{F}_e \\ \mathbf{M}_e = \mathbf{A}_e {}^e\mathbf{M}_e \end{cases}. \tag{31}$$

5. Research on simulation for planetary WMRs at SKLRS

Simulation technology has played an important role in the research of planetary rovers at SKLRS for more than 10 years. Different simulation methods are used for different purposes. In collaboration with Tohoku University, Japan, the Matlab toolbox entitled SpaceDyn is used to develop a simulation system for control algorithm verification, by integrating high-fidelity terramechanics models. To realize real-time simulation for hardware-in-loop testing and successive teleoperation, Vortex, a real-time simulation platform, is adopted for the development of RoSTDyn with C++ language.

5.1 ADAMS-based simulation for planetary WMRs

Commercial dynamics software ADAMS is adopted in the mechanical design of rovers for purposes such as performance analysis, optimization design, and control algorithm verification. The Contact model, Tire model, and self-developed terramechanics model are used to predict wheel–soil interaction mechanics.

5.1.1 Simulation with Contact model

Using the Contact model provided by ADAMS software is an easy way to realize dynamics simulation of a wheeled rover. The wheel-soil interaction is considered as solid-to-solid contact. Different frictional coefficients and dumping ratios can be set by the users according to their understanding of the soil properties. The structured terrain can be constructed with ADAMS or imported from computer-aided design software such as Pro/E. This approach is simple and reflects the characteristics of the mechanism, at least at the kinematics level, although the fidelity of wheel-soil interaction mechanics is poor, and it is thus used during the initial phase of rover design. For instance, Tao et al. designed a six-wheeled robot with good locomotion performance on rough terrain, and they simulated different locomotion modes with ADAMS to verify the concept of its mechanism, as shown in Fig. 10 (Tao et al., 2006).

Fig. 10. Simulation for WMR with ADAMS and Contact model

5.1.2 Comprehensive simulation using Tire model

The structure of a comprehensive virtual simulation system for lunar rovers is shown in Fig. 11 (Gao et al., 2011). The simulation system integrates the software of ADAMS, Pro/E, 3DS Max, and Matlab to realize the functions of rover vehicle modeling, terrain modeling, kinematics/dynamics analysis including the mechanics of wheel-soil interaction, and control, respectively.

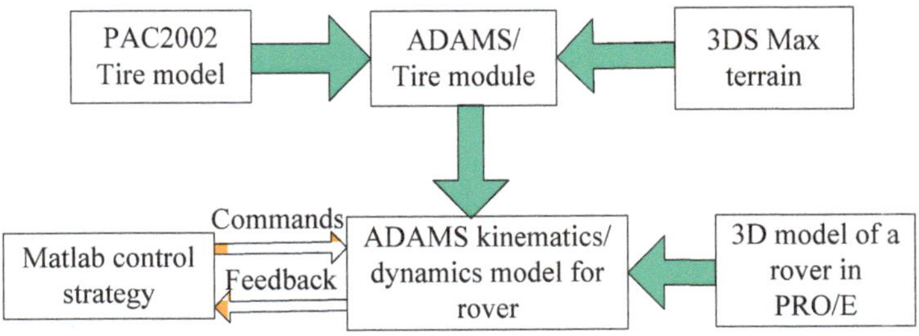

Fig. 11. Diagram of comprehensive virtual simulation system for lunar rovers

The 3D model of a lunar rover constructed using Pro/E can be imported into ADAMS through an ADAMS/Pro connection module; the rough terrain model of the lunar surface produced with 3DS Max can be imported into an ADAMS/Tire module through an interface; the PAC2002 Tire model is called by the ADAMS/Tire module for prediction of the terramechanics of the wheel-soil interaction. By configuring the simulation parameters for the rover model, the terrain, and Tire model in ADAMS, a virtual rover is constructed whose kinematics and dynamics can be solved by ADAMS/Solver. Figure 12 presents an example.

In directing a virtual rover to follow a planned path while compensating for slipping on deformable rough terrain, the required control strategy can be realized with Matlab. A virtual rover created with ADAMS software can be controlled through the interface between ADAMS/Control and Matlab/Simulink. ADAMS/Control provides an interactive environment for establishing and demonstrating an S-function "controlled object," which can be controlled with the Simulink toolbox of Matlab. In each integration step, ADAMS/Control is called as a subprogram by Simulink. The control instructions generated by Simulink are then directed to the corresponding mechanisms (such as the driving motors of the wheels) of lunar rovers in ADAMS through ADAMS/Control. The rover's motion is then calculated by ADAMS/Solver on the basis of dynamics simulation, and the related information is fed back to Simulink through ADAMS/Control. The entire transmission process is automatic and transparent.

Fig. 12. Dynamics simulation of lunar rover with ADAMS

ADAMS/Tire is a module for predicting the interactive forces and moments between wheels and roads. Several types of tire models are provided by ADAMS, including MF-Tyre, UA, Ftire, SWIFT, PAC89, and PAC2002. Each tire model has its own advantages and disadvantages. After careful analysis and comparison, the PAC2002 model was finally chosen for calculation of the mechanics of the wheel-soil interaction of a lunar rover. This model is applicable to the interaction between a wheel and 3D roads or obstacles.

The PAC2002 model uses trigonometric functions to fit the experimental data and thus predict the forces and moments that the soil exerts upon a wheel, including the drawbar pull F_x, lateral force F_y, sustaining force F_z, overturning moment M_x, resistance torque M_y, and aligning torque M_z. The general expression of the function, which is called the magic formula, is

$$Y(X) = D\cos\left(C\arctan\{BX - E[BX - \arctan(BX)]\}\right), \tag{32}$$

where $Y(X)$ is a force or moment, the independent variable X reflects the effect of the slip angle or longitudinal slip ratio of a wheel for an actual situation, parameters B, C, and D are determined by the wheel's vertical load and camber angle, while E is the curvature factor.

The unknown parameters can be determined by a data fitting method based on experimental results.

5.1.3 Simulation using self-developed terramechanics model

The simulation fidelity is improved to some extent using the Tire model instead of the Contact model. However, phenomenon such as of severe slip-sinkage and the lug effect still cannot be reflected well. ADAMS provides a method for users to embed the newly developed terramechanics models with higher precision (Jiao, 2009).

The GFORCE command is used to define the wheel-soil interaction mechanics that consists of three mutually orthogonal translational force components and three orthogonal torque components. As the force and torque expressions are lengthy and complex, the GFOSUB evaluation subroutine is used to compute wheel-soil interaction mechanics applied by a GFORCE statement with FORTRAN language. The wheel-soil interaction mechanics program is compiled and linked to generate an object file (*.obj). The Create Custom Solver command is then used to generate a dynamic link library (*.dll) file and library file (*.lib). The general force, GFORCE, which is applied to the center of the wheel, is set as shown in Fig. 13 to call the wheel-soil interaction function through the *.dll file.

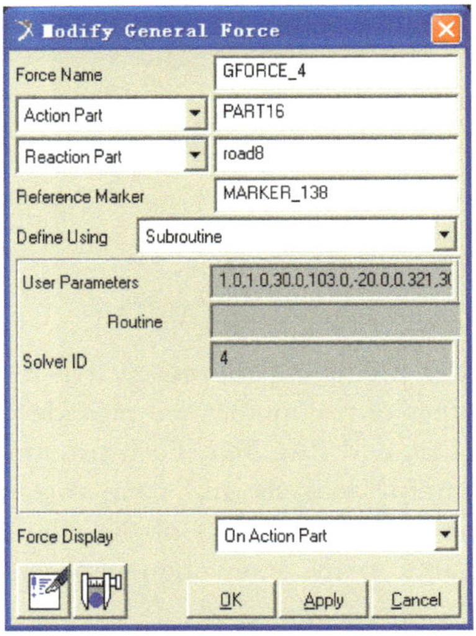

Fig. 13. GFORCE Subroutine

5.2 Matlab-based high-fidelity simulation platform for planetary WMRs (Ding, 2010a)

5.2.1 Implementation of simulation platform

A numerical simulation program based on a Matlab toolbox called SpaceDyn was developed (Yoshida 2000). The principle diagram is shown in Fig. 14. Given the DEM terrain, soil parameters, and rover model parameters, the program calculates the wheel-soil interaction area, predicts the external forces that act on the wheel, calculates the accelerations of the generalized coordinates on the basis of the dynamics model, and then integrates them to obtain their velocities and positions on the basis of kinematics equations.

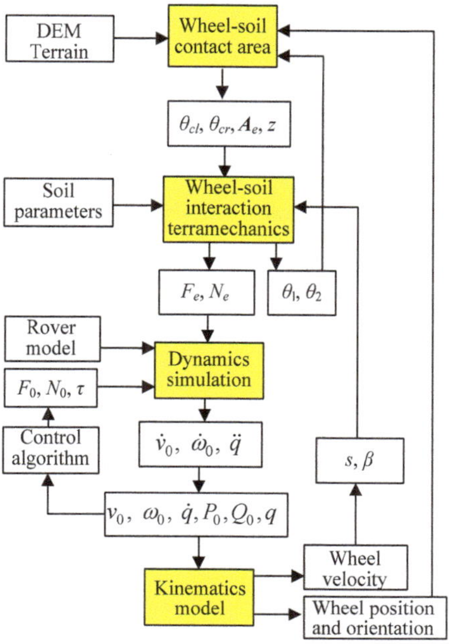

Fig. 14. Principle diagram of dynamics simulation

5.2.2 Experimental validation

El-Dorado II, a four-wheeled mobile robot developed at Space Robotics Laboratory of Tohoku University in Japan was used to validate the simulation. The robot uses four force/torque (F/T) sensors to measure the wheel-soil interaction terramechanics. A visual odometry system was developed to measure the position of the rover body and the slip ratio of the wheels. The entrance angles used for calculating the wheel sinkage were measured with an angle meter. Two groups of experiments were performed. In group 1, resistance forces were applied to the rover with counterweights from 0 to 60 N in intervals of 10 N, to generate different slip ratios. In group 2, the rover was controlled to climb slopes ranging from 0° to 15° in intervals of 3° (Fig. 15).

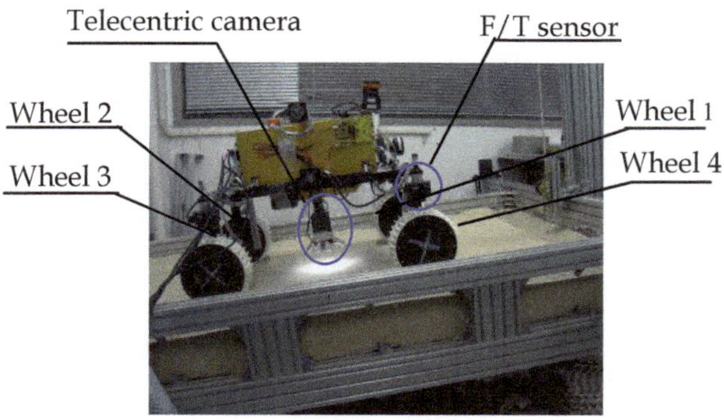

Fig. 15. Slope-climbing experiment using El-Dorado II robot

The parameters of Toyoura soft sand are identified from the experimental data: K_s = 1796 Kpa/m^N, c = 24.5 Pa, φ = 35.75°, and K = 10.45 mm. k_y is 19 mm. When the robot climbs a slope, the remaining parameters are n_0 = 0.66, n_1 = 0.72, c_{P1} = -0.379, c_{P2} = 0.616, c_{P3} = -0.448, and C_M = 0.214; on flat terrain, the parameters are n_0 = 0.63, n_1 = 0.72, c_{P1} = -0.276, c_{P2} = 0.633, c_{P3} = -0.304, and C_M = 0.354.

Comparisons of the simulation and experimental results are shown in Figs. 16 and 17. Not only can the motion of the robot be predicted with high fidelity, as indicated by the slip ratio, so too can the drawbar pull, moment of resistance, the normal force, and wheel sinkage.

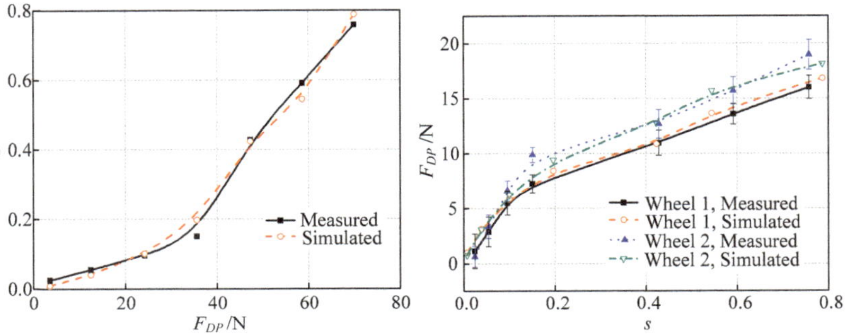

Fig. 16. Simulation and experimental results for robot moving on flat terrain

5.2.3 Simulation for deformable rough terrain

The robot was controlled to move from (0.5 m, 0.5 m) to (5 m, 5 m) on the randomly generated rough terrain shown in Fig. 17, with an initial yaw angle of 45°. While moving, the robot deviates from the scheduled path because of the inclination angle of the terrain. Figure 18 shows the slope angles that wheel number 4 traverses, the RPY (roll, pitch, and yaw) angles of the body and q_1 and q_2 joint angles (q_1 = -q_2), the slip ratios, and normal forces. The simulation platform was used to verify the slip-ratio-coordinated control (Ding, 2010b) and path-following strategy (Ding, 2009c), as shown in Fig. 19.

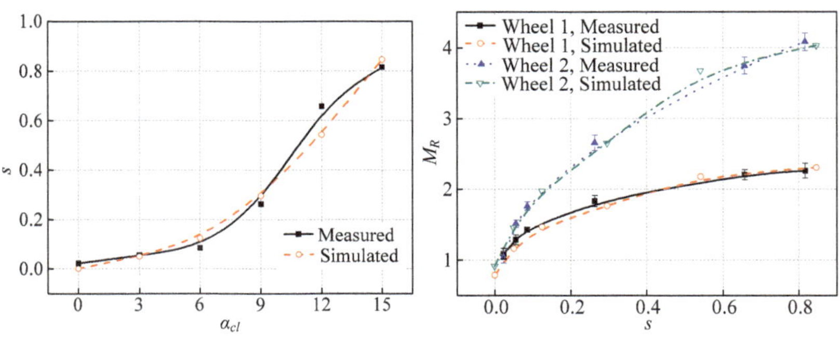

Fig. 17. Simulation and experimental results for robot climbing slope

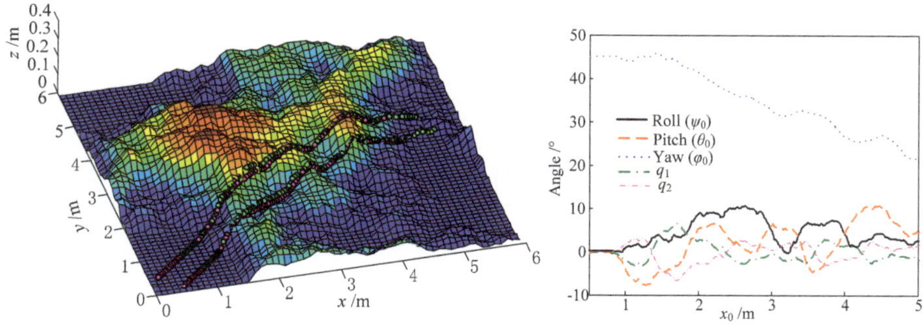

(a) Rough terrain and wheel trajectories (b) RPY and q1 and q2 angles

Fig. 18. Simulation results for El-Dorado II moving on deformable rough terrain

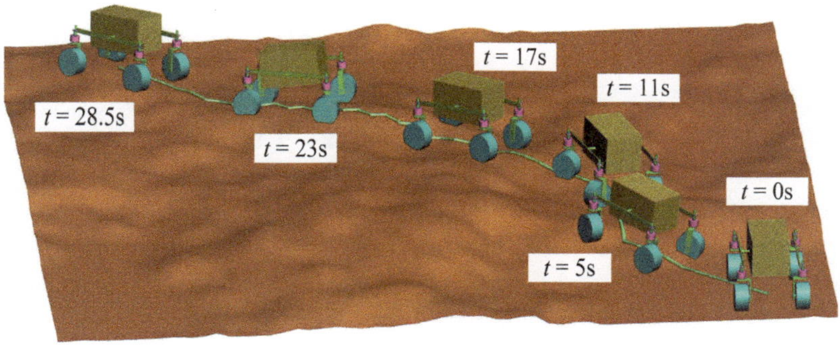

Fig. 19. Path-following simulation

5.3 RoSTDyn: Vortex-based high-fidelity and real-time simulation platform

5.3.1 Structure of RoSTDyn

RoSTDyn is composed of five modules as shown in Fig. 20 (Li et al., 2012). The module of the planetary rover model is used to create the simulation object-rover model, which is composed of a physical model and scenic model. The physical model is the real model that is used in the collision detection and dynamics calculation; the scenic model does not participate in any calculation, and it is driven according to the message transferred from the physical model, so it can be created vividly with 3D modeling software such as 3DS Max or Creator.

The controlling module focuses on realizing the interaction between the user and the simulation platform. This module is mainly used to control the movable joints of the rover and thus control, for example, the speed of the driving wheel, the turning degree of a wheel, and the posture of the solar panels and mast.

The module of the terrain model is used to simulate the terrain. It also includes a physical model and scene model. For the terrain, the physical model is a file of the node information including X, Y, and Z; the scenic model is generated by the 3D modeling software on the basis of node information.

The contact-area computing module is used to compute the parameters for the contact area between the wheel and terrain. These parameters are the precondition of the interaction force module, and this module will be introduced in detail in part III.

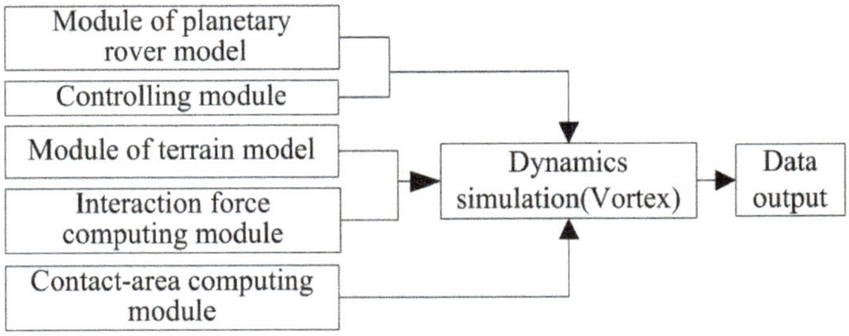

Fig. 20. Structure of RoSTDyn

The interaction force computing module is used to compute the interactive force between the wheel and terrain. These forces act on the rover model, and drive it to move and turn. The five modules are integrated in Vortex. Each module only needs to do its own work and transfer necessary information to other modules; in this way, the five modules work together in the simulation. Snapshots of rover simulation with RoSTDyn are shown in Fig. 21.

Fig. 21. Snapshots of rover simulation with RoSTDyn

5.3.2 Testing of real-time properties

To test the simulation speed, an experiment is designed. The rover's velocity is set to a constant value. At the moment that the simulation starts, a stopwatch is used to record the physical time, and the simulation time is exported by the program. At the same time, to test the calculation speed, another test is designed while the display of the scene is closed. Table 2 lists the time that the rover passes over three types of terrain. The display is opened for the first three and closed for the last three.

Terrain (display)	Simulation time t_s (s)	Physical time t_p (s)	t_s:t_p
Flat (Yes)	24.33	30.45	0.799:1
Slope (Yes)	34.50	42.91	0.804:1
Random (Yes)	43.50	53.76	0.809:1
Flat (No)	41.25	41.07	1:1
Slope (No)	38.50	38.43	1:1
Random (No)	45.00	44.82	1:1

Table 2. Time that the rover passes over three types of terrain

Here, because the rate is 80 Hz, the display wastes much time, and the time ratio is about 0.8; however, if the display is closed, the ratio reaches 1.0. Therefore, the calculation of

RoSTDyn is done in real time. To demonstrate this, the rate is multiplied by a factor of 10, and it is found that the time ratio decreases only to 0.8. This shows that it is still possible to increase the simulation speed of RoSTDyn.

6. Conclusions and future work

Virtual simulation can guarantee the successful exploration of planets by WMRs as it plays important roles in both the R&D phase and the exploration phase of the rovers. Customized simulation tools or tools with different simulation functions have been developed to support the R&D of planetary rovers, because general simulation tools do not meet the requirements of rover-based exploration missions well. The key technologies for developing high-fidelity comprehensive simulation systems include recursive dynamics, wheel–soil interaction terramechanics, and deformable rough terrain modeling. At SKLRS, different methods have been employed for rover simulation. ADAMS software is adopted along with the Contact model, Tire model, and self-developed GFORCE model during the design phase of rovers. Matlab is used to develop a high-fidelity simulation platform by embedding the terramechanics model and rough terrain model into support control strategy simulation. Vortex is used to develop RoSTDyn to realize real-time simulation and thus support the teleoperation of rovers.

In the future, successive teleoperation of planetary rovers using a 3D predictive display will be explored on the basis of the high-fidelity/real-time simulation platform. A faster-than-real-time simulation system will be developed to support the supervised control of rovers.

7. Acknowledgments

This work was supported by the National Natural Science Foundation of China (50975059/61005080), the Postdoctoral Foundation of China (20100480994), the Foundation of the Chinese State Key Laboratory of Robotics and System (grant No. SKLRS200801A02), the Key Natural Science Foundation of Heilongjiang Province in China (ZJG0709), the Postdoctoral Foundation of Heilongjiang Province, and the "111" Project (B07018).

8. References

Backes, P. G., Noms, J. S., & Powell, M. W. (2004). Multi-mission activity planning for Mars lander and rover missions. *IEEE Aerospace Conference Proceedings*, pp. 877-886, ISBN: 0-7803-8155-6, Big Sky, Montana, USA, March 2004

Bauer, R., Leung, W., & Barfoot, T. (2005). Development of a dynamic simulation tool for the exomars rover. *Proceedings of the 8th International Symposium on Artificial Intelligence, Robotics and Automation in Space*, Munich, Germany, September 2005

Chen, L., Zhang, Y., & Ren, W. (2005). *Dynamic Analysis of Mechanical System and ADAMS Application*, Tsinghua University Press, ISBN 7302100969, Beijing, China

Ding, L., Gao, H., Deng, Z., et al. (2008). Design of comprehensive high-fidelity/high-speed virtual simulation system for lunar rover. *Proceedings of 2008 IEEE Int. Conf. on Robotics, Automation and Mechatronics*, pp. 1118-1123, ISBN 978-1-4244-1675-2, Chengdu, China, September 2008

Ding, L., Yoshida, K., Nagatani K., et al. (2009a). Parameter identification for planetary soil based on a decoupled analytical wheel-soil interaction terramechanics model.

Proceedings of IEEE/RSJ International Conference on Intelligent Robots and Systems, pp. 4122-4127, ISBN 978-1-4244-3803-7, St. Louis, MO, USA, October 2009

Ding, L., Gao H., Deng Z., et al. (2009b). Slip ratio for lugged wheel of planetary rover in deformable soil: definition and estimation. *Proceedings of IEEE/RSJ International Conference on Intelligent Robots and Systems*, pp. 3343-3348, ISBN 978-1-4244-3803-7, St. Louis, MO, USA, October 2009

Ding, L. (2009c). *Wheel–soil Interaction Terramechanics for Lunar/planetary Exploration Rovers: Modeling and Application*. Doctoral Thesis of Harbin Institute of Technology (December, 2009), Harbin, China

Ding, L., Nagatani, K., Sato, K., et al. (2010a). Terramechanics-based high-fidelity dynamics simulation for wheeled mobile robot on deformable rough terrain. *Proceedings of IEEE International Conference on Robotics and Automation*, pp. 4922-4927, ISBN: 978-1-4244-5038-12010, Anchorage, Alaska, USA, May 2010

Ding, L., Gao, H., Deng, Z., & Liu, Z. (2010b). Slip-ratio-coordinated control of planetary exploration robots traversing over deformable rough terrain. *Proceedings of IEEE/RSJ International Conference on Intelligent Robots and Systems*, pp. 4958-4963, ISBN: 978-1-4244-6674-0, Taipei, Taiwan, China, October 2010

Gao, H., Deng, Z., Ding, L., & Wang M. (2011). Virtual simulation system with path following control for lunar rovers moving on rough terrain. *Chinese Journal of Mechanical Engineering*, Vol. 24, No. 5, (May 2011), pp. 1-10, ISSN 0577-6686

Ishigami, G. & Yoshida, K. (2005). Steering characteristics of an exploration rover on loose soil based on all-wheel dynamics model. *Proceedings of IEEE International Conference on Intelligent Robots and Systems*, pp. 2041-2046, ISBN 0-7803-8912-3, Roman, Italy, August 2005

Ishigami, G., Nagatani, K., & Yoshida, K. (2007). Path planning for planetary exploration rovers and its evaluation based on wheel slip dynamics. *Proceedings of IEEE International Conference on Robotics and Automation*, pp. 2361-2366, ISBN 1-4244-0601-3, Roman, Italy, April 2007

Jet Propulsion Laboratory. (a). http://marsrovers.jpl.nasa.gov/home/index.html

Jet Propulsion Laboratory. (b). http://marsprogram.jpl.nasa.gov/MPF/rover/sojourner.html

Jiao, Z. (2009). *Dynamics Modelling and (ADAMS) Simulation for Lunar Rover Based on Terramechanics*. Master Dissertation of Harbin Institute of Technology (July 2009), Harbin, China.

Kuroda, Y., Kawanishi, M., & Matsukuma, M. (2003). Tele-operating system for continuous operation of lunar rover. *Proceedings of the IEEE/RSJ International Conference on Intelligent Robots and Systems*, pp. 2565-2570, ISBN 0-7803-7860-1, Las Vegas, Nevada, USA, October 2003

Kunii, Y., Suhara, M., Kuroda, Y., & Kubota, T. (2001). Command data compensation for real-time tele-driving system on lunar rover: Micro-5. *Proceedings of the 2001 IEEE International Conference on Robotics and Automation*, pp. 1394-1399, ISBN 0-7803-6576-3, Seoul, Korea, May 2001

Lei, Z. (2004). *3D Predictive Display and Force Feedback Control of Tele-operation with Large Time Delay*. Dissertation of Master degree, Jilin University, Changchun, Jilin

Li, W., Gao, H., Ding, L., et al. RoSTDyn: A High-fidelity Real-time Simulation Platform for Planetary Rovers. 2012 IEEE International Conference on Robotics and Automation, submitted

Liang, B., Wang, W., & Wang, C. (2005). Primary concept on developing lunar rover. *Journal of Space International*, Vol. 2, (February, 2003), pp. 22-25, ISSN 009-2366

Luo, X., Ye, J., & Sun, Z. (2002). Research on the lunar rover simulation platform. *Journal of System Simulation*, Vol. 14, No. 9, (September 2002), pp. 1235-1238, ISSN 1004-731X

Maxwell, S., Cooper, B., Hartman, F., et al. (2005). The best of both worlds: integrating textual and visual command interfaces for mars rover operations. *Proceedings of IEEE International Conference on Systems, Man and Cybernetics*, pp. 1384-1388, ISBN 0-7803-9298-1, Waikoloa, HI, USA, October 2005

Michaud, S., Richter, L., Patel, N., et al. (2004). RCET: Rover Chassis Evaluation Tools. *Proceedings of the 8th ESA Workshop on Advanced Space Technologies for Robotics and Automation*, ESTEC, Noordwijk, Netherlands, November 2004

Neal, C. R. (2009). The Moon 35 years after Apollo: what's left to learn? *Chemie der Erde Geochemistry*, Vol. 69, No. 1, (February 2009), pp. 3-43, ISSN 0009-2819

Patel, N., Ellery, A., Allouis, E., et al. (2004). Rover Mobility Performance Evaluation Tool (RMPET): a systematic tool for rover chassis evaluation via application of Bekker theory. *Proceedings of the 8th ESA Workshop on Advanced Space Technologies for Robotics and Automation*, pp. 251-258, ESTEC, Noordwijk, Netherlands, November 2004

Schäfer, B. & Krenn, R. (2010). Multibody and contact dynamics: from satellite capturing to planetary rover terrainability. *Proceedings of the 1st ESA Workshop on Multibody Dynamics for Space Applications*, ESTEC, Noordwijk, Netherlands, February 2010

Schäfer, B., Gibbesch, A., Krenn, R., & Rebele, B. (2010). Planetary rover mobility simulation on soft and uneven terrain. *Journal of Vehicle System Dynamics*, Vol. 48, No. 1, (January, 2010), pp. 149-169, ISSN: 0042-3114

Tao J., Deng Z., Fang H., et al. (2006). Development of a wheeled robotic rover in rough terrains. *Proceedings of the 6th World Congress on Intelligent Control and Automation*, pp. 9272-9276, ISBN: 1-4244-0332-4, Dalian, China, June 2006

Van winnendael, M., Baglioni, P., Elfving, A., et al. (2008). The ExoMars Rover—Overview of phase B1 results. *Proceedings of 9th International Symposium on Artificial Intelligence, Robotics and Automation in Space*, Hollywood, Los Angles, California, USA, February 2008

Volpe, R. (2003). Rover functional autonomy development for the Mars Mobile Science Laboratory. *Proceedings of IEEE Aerospace Conference*, pp. 643–652, ISBN 0-7803-7651-X, Big Sky, Montana, USA, March 2003

Volpe, R. (2005). Rover technology development and mission infusion beyond MER. *Proceedings of IEEE Aerospace Conference*, pp. 971-981, ISBN 0-7803-8870-4, Big Sky, Montana, USA, March 2005

Wright, J., Hartman, F., Cooper, B., et al. (2006). Driving on Mars with RSVP: building safe and effective command sequences. *IEEE Robotics and Automation Magazine*, Vol. 13, No. 2, (March 2006), pp. 37-45, ISSN 070-9932

Yang, Y., Bao, J., Jin, Y., & Cheng, Y. (2008). A virtual simulation environment for lunar rover: framework and key technologies. *International Journal of Advanced Robotic Systems*, Vol. 5, No. 2, (June 2008), pp. 201-208, ISSN 1729-8806

Ye, P. & Xiao, F. (2006). Environment problem for lunar exploration engineering. *Spacecraft Environment Engineering*, Vol. 23, No. 1, (January 2006), pp. 1-11, ISSN 1673-1379

Yen, J., Jain, A., & Balaram, J. (1999). ROAMS: Rover Analysis Modeling and Simulation Software. *Proceedings of International Symposium on Artifcial Intelligence, Robotics and Automation in Space*, pp. 249-254, ESTEC, Noordwijk, the Netherlands, June 1999

Yoshida, K. (2000). The SpaceDyn: a MATLAB toolbox for space and mobile robots. *Journal of Robotics and Mechatronics*, Vol. 12, No. 4, (August, 2000), pp. 411–416, ISSN 0915-3934

Yoshida, K. & Hamano, H. (2002). Motion dynamics of a rover with slip-based traction model. *Proceedings of the IEEE International Conference on Robotics and Automation*, pp. 3155-3160, ISBN 0-7803-7272-7, Washington, D.C. USA, May 2002

18

Design and Prototyping of Autonomous Ball Wheel Mobile Robots

H. Ghariblu, A. Moharrami and B. Ghalamchi
Mech. Eng. dept., Mechatronics Lab.
Zanjan University
Iran

1. Introduction

Mobile robots are platforms that are able to move autonomously. Now a day, their use is increased in different areas like, autonomous vehicles, flexible manufacturing and service environments and lunar explorations. A large number of wheeled or tracked platform mechanisms have been studied and developed to prove their mobility and capability as autonomous robot vehicles (Pin, and Killough, 1994), (Kim, et al. , 2003), (Wada, et al. , 2000), (Jung, et al., 2000), (Mori, et al., 1999). For large and heavy outdoor robots, four-wheel car-like driving mechanisms or skid-steer platforms have traditionally been used. These vehicles are quite restricted in their motion (Jarvis, 1997), particularly when operating in tight environments.

In recent years, study of nonholonomic systems has been an area of active research. Nonholonomic systems are characterized by nonintegrable rate constraints resulting from rolling contact or momentum conservation. Nonholonomic behaviors are sometimes introduced on purpose in the design of mechanism, in order to obtain certain characteristics and performances such as those in. One advantage offered by nonholonomic systems is the possibility of controlling a higher number of configurations than the number of actuators actually employed in the system, which is sometimes useful in terms of reducing the system's weight and cost. The nonholonomic constraints cause complexities in trajectory planning and designing of control algorithms for feedback stability of the vehicle system. It is required that a suitable desired trajectory satisfying the above constraint be designed to control a nonholonomic mobile mechanism (Fierro & Lewis, 1997).

On the other hand, holonomic vehicles have been proposed with several advantages and disadvantages, so that there is introduced a control strategy to avoid a nonholonomic constraint of a wheel to implement a holonomic omnidirectional vehicle (Asada & Wada, 1998). Holonomic vehicles, also, have some problems in practical applications such as low payload capability, complicated mechanism and limited accuracy of motion (Ferriere & Raucent, 1998).

Several omnidirectional platforms have been known to be realized by developing a specialized wheel or mobile mechanism. From this point of view, such specialized mechanisms suitable for constructing an omnidirectional mobile robot are summarized as following:

1. Steered wheel mechanism (Chung, et al., 2010), (Wada, et al. , 2000)..

2. Universal wheel mechanism (Song & Byun, 2004).
3. Ball wheel mechanism (Ostrovskaya, et al., 2000), (West, & Asada, 1997), (Ghariblu, 2010), (Ghariblu, et al., 2010).
4. Orthogonal wheel mechanism (Pin, and Killough, 1994).
5. Crawler mechanism (Tadakuma. et.al., 2009).
6. Offset steered wheel mechanism(Udengaard & Iagnemma, 2007).

From the workspace view point, mobile robots must be able to reach at any position on its plane of motion, with any orientation. Thus, their frame must have the three Independent Coordinates of the general plane motion of the rigid body. This can be achieved by means of 2 DOF provided the robot frame kinematics is non-holonomic, but maneuvering is required. Maneuvering can be avoided whenever the mobile robot has 3 DOF. Mobility is enhanced by the use of omni-directional instead of conventional wheels. Omni-directional movement is the ability to travel in any path at any specified orientation. This type of drive system combines translation navigation along any desired path and desired orientation. This ability cause many researchers to study in the subject of omni-directional (holonomic) mobile robots and vehicles.

Generally, a special type of wheels called as omni-wheels are used to achieve Omni-directional movement. Omni-wheels are all based on the same general principle: while the wheel provides traction in the direction normal to the motor axis, it can roll passively in the direction of the motor axis. In order to achieve this, the wheel is built using smaller wheels attached along the outside edge of the wheel. Each wheel provides a torque in the direction normal to the motor axis and parallel to the floor.

The main drawback of such structures is vibrations induced into the complete robot system due to the successive shocks occurring when the contact with the ground shifts from one roller to the next. Other drawbacks of omni-wheels are limited load capacity and surmountable bump height that is limited by the diameter of the rollers, and not by the diameter of the wheels, as with conventional wheels. Therefore, these types of wheels are not appropriate to employ in outdoor applications. To overcome these difficulties, (West, & Asada, 1997), (Ostrovskaya, et al., 2000), and (Asada & Wada, 1998) introduced omni-directional ball wheel vehicles. All of these designs suffer from technical and functional issues include complex design, cost, nonholonomic structure or limited load capacity.

To overcome the above mentioned difficulties and based on our experiment in fabricating a ball wheel robot (BWR-1) (Ghariblu, 2010), the second generation of this robot (BWR-2) is developed (Ghariblu, et al., 2010). In the first generation, a new approach introduced that uses three spherical wheels as main wheels. This design was a good step to develop a new holonomic mobile robotic vehicle. But, after experimental tests, several difficulties were appeared. Most important problems in our former style robot were lack of suspension system, rigid structure of robot, non modular construction of driving system that leads to fixing all driving motor on a common rigid frame, and low effective space between main wheels chassis and floor. These problems are basis to weak operation of robot in passing through rough terrains.

Other drawback was the unwanted locking of ball wheels due to inappropriate structure of their supports to corrupt the overall robot motion on its desired path. To solve the above problems, second generation of a ball wheel mobile robot is designed and constructed.

To improve the robot motion in rough terrains several modification are made on our former robot. The first modification was the modular construction of driving system, which separately fabricated and assembled on this system. Also, proper suspension system consists of springs and dashpots are added to each driving system. Meanwhile, specific design of the robot solved the trouble of small space between the balls chassis and moving base.

To overcome bearing problem standard ball and socket bearings implemented to ease the balls motion.

The outline of the chapter is as following: Section 2, describes the mechanical and electrical construction of ball wheel robots consist of wheels driving mechanism, chasis and suspension system. Section 3 discusses on new elements of BWR-2 to overcome to drawbacks of BWR-1 robot. Sections 4 and 5 present kinematic and dynamic equations of both robots. Finally, in Section 6 with some simulation studies we show the ability of robot in traversing predefined desired paths.

2. The structure of ball wheel robot

This section introduces the mechanical structure of BWR-1 and troubles find out in that robot. Then we will discuss about modifications utilized in BWR-2 to solve these troubles.

Fig. 1 shows schematic views of BWR-1. The construction of the BWR-1 consisted of a triangular platform and three identical ball wheels that were fixed with bearings between two upper and lower fixing plates. Ball wheels driving were realized by the actuated omni-wheels mounted on a common base over the main base. The locomotion mechanism of this robot consists of three independently actuated ball wheels.

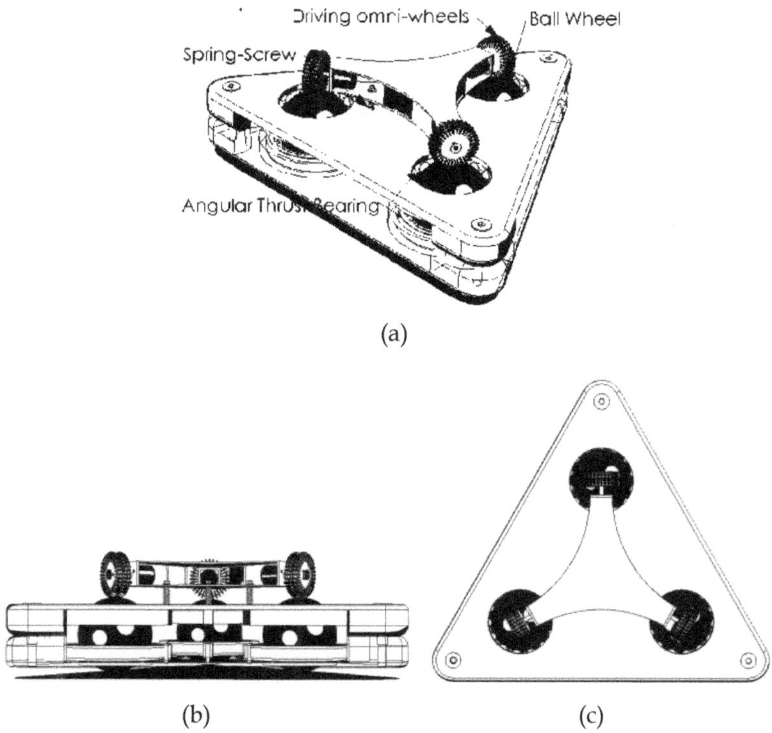

Fig. 1. First type Spherical wheel robot, a) Isometric View, b) Side View, c) Top View

After the design stage and constructing of first prototype of BWR-1, we determined some difficulties in application of this vehicle in real environment, as follows

a. In specific conditions owing to rigid structure of robot, in passing through uneven ground one of the robot wheels separated from the moving plane. This fact results unwanted deviation of robot motion from its desired path.
b. Driving wheels and their transmission system were rigidly assembled on a common frame. This fact directed road irregularities and acceleration and deceleration motions to the main body, with a possible degradation to the vehicle required tasks.
c. Unwanted locking of ball wheels due to inappropriate structure of their supports corrupt the overall robot motion.

Above mentioned troubles in BWR-1 degrades its function for performing requested motions such as rotation and lateral movements.

To solve the above problems and achieve the required characteristics for a high performance, its second generation named as BWR-2 is designed and constructed.

2.1 The architecture of BWR-2

Here, the design specification of new ball wheel omni-directional mobile robot BWR-2 is explained (Fig. 2). The objective of designing BWR-2 is to ensure that new robot meets design specifications and overcome weaknesses of BWR-1.

Performing some experiments on BWR-1 directed us to problems existed in robot structure that was mentioned in the previous section.

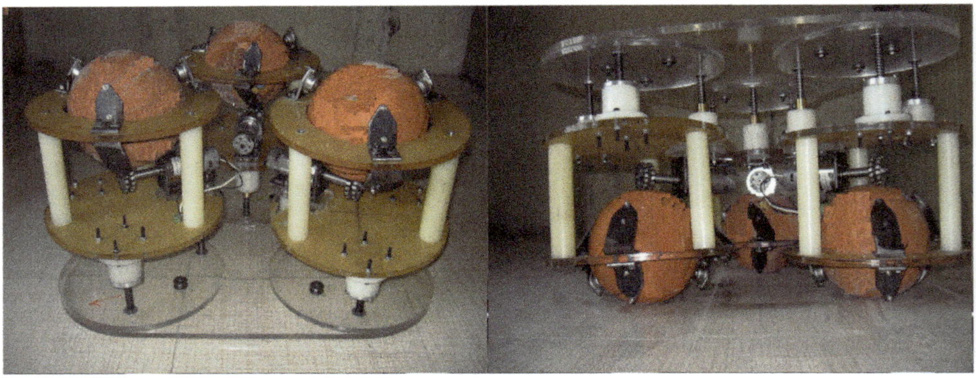

Fig. 2. Two different views of BWR-2

BWR-2 robot is an omni-directional mobile platform. Omni-directional design of this robot improves its characteristics for high maneuverability. Consequently, robot can simultaneously move along any desired path with rotation about its geometrical axis.

Same as BWR-1, the new robot BWR-2 is equipped with three balls as main wheels driving by two omni-wheels that transfer motor traction force to the balls. Moreover, three parallel suspension systems consist of springs and dashpots are added to BWR-2 structure. Maximum achievable speed for robot is 3.4 m/s.

To drive the robot, modular and mechanically independent driving system is employed (Fig. 3). Each driving system is driven by a DC motor with reduction ratios of 12:1, consists of a ball as main wheel, and two omni-wheels to transfer motor traction force to the ball, and three parallel suspension systems consists of springs and dashpots.

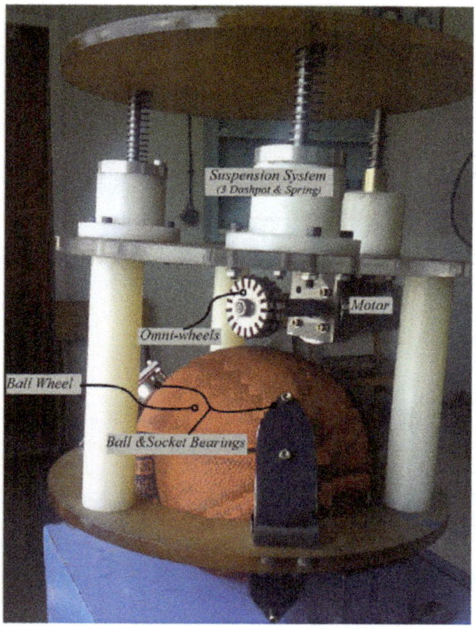

Fig. 3. BWR-2 modular driving system

A list of robot geometric and physical parameters is summarized in Table 1.

- **MAIN BASE**

Dimension: Equilateral Triangle with 750 mm sides
Material: polyamide
Weight: 8kg

- **Ball wheels**

Dimension: φ150 mm
Material: polyurethane resin
Weight: 0.25 kg,

- **Omni-wheels**

Dimension: φ 50 mm
Material: Aluminum, peripheral wheels covered with plastic o-ring
Weight: 0.1 kg

- **Geared DC motor**

120 rpm- 45 Watt, 12 volt

- **Total weight**: 20 kg, without batteries

Table 1. Geometric and physical characteristics

Fig. 4 shows the electrical architecture of the BWR. The BWR has an onboard notebook computer. The robot can be controlled remotely through the wireless connection or using a Microsoft game pad USB Sidewinder connected to another USB port of the notebook. Game pad buttons control the movement and rotation of the robot. Three Serial/USB converters are used to connect the robot base. Software drives the motor by the USB hub and the

Serial/USB converters. The robot is powered by two 24V, 7A-H lead-acid batteries. Motor drivers activate DC motors with two omni-wheels assembled on their shaft to transfer motor traction to the ball wheels.

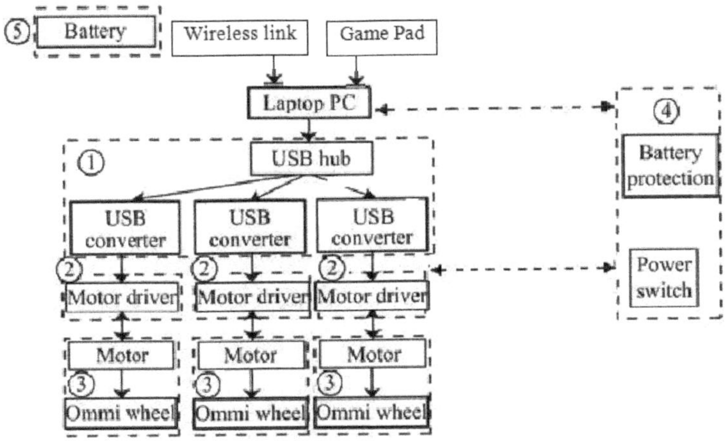

Fig. 4. Electrical architecture of BWR

3. Advantages of BWR-2 robot

Some significant modifications are performed to ensure better function of BWR-2 respecting BWR-1. As bellow

- Corrugated polyurethane resin spherical ball is utilized to generate relative high coefficient of friction regarding to driving omni-wheels and moving surface.
- Handmade bearings of ball wheels changed with standard ball and socket bearings. By this modification locking problem between bearings and ball wheels appropriately solved. Consequently, balls wheels are able to rotate freely according to vehicle desired motion. Each ball is held by six ball and socket bearings fixed on two sides of the frame disk on outside edge of ball wheel to make stable motion (Fig. 3).
- To make adequate driving force between driving omni-wheels and ball wheel doubled omni-wheels are used.
- Finally, differing from BWR-1 that was designed to navigate in indoor applications or flat surfaces, owing to employing new suspension for wheels BWR-2 will be able to navigate in outdoor applications. It is shown in Fig. 5 by adjusting suspension system according to obstacle existed in the robot path; it can easily traverse along uneven ground.

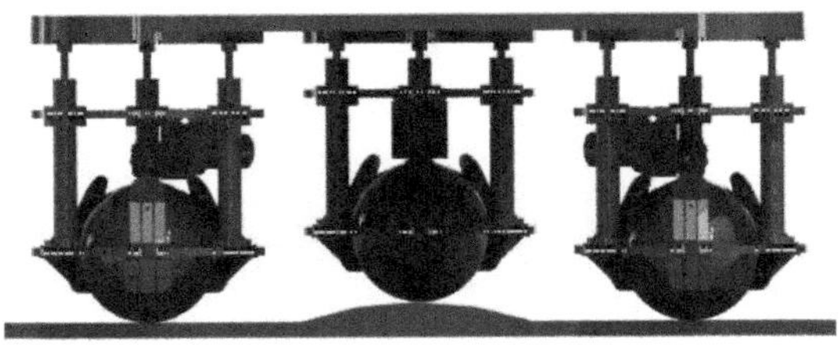

Fig. 5. Wheels adjustment with obstacle owing to proper action of suspension system.

4. Kinematics

Kinematics of both robots assuming horizontal plane movement are similar. The main focus lies in the connection between driving omni-wheels angular velocities and robot velocity. Fig. 6 illustrates a top view of the BWR-1 robot. The common radius of all omni-wheels is r_w and all ball wheels is r_s , and the distance from robot center to the contact points between wheels and floor is r_R. Four sets of coordinates are introduced: The mutually perpendicular unit vectors $\{e_x; e_y; e_z\}$ are fixed in a global non-moving reference frame, with e_x and e_y being parallel to the floor and e_z facing upwards. The unit vectors $\{e'_1; e'_2; e'_3\}$ are also mutually perpendicular. They are fixed on the robot, e'_1 denotes the robot forward direction and e'_3 is parallel to e_z. Note that e_z represents a translational dimension whereas e'_3 is used to express rotational quantities. The third set of coordinates consists of the unit vectors $\{e_{w1}; e_{w2}; e_{w3}\}$, each of them points in the respective wheel's axial direction. The unit vectors $\{e_1; e_2; e_3\}$, are pointing in the wheel peripheral directions. They are linearly dependent on e'_1 and e'_2 and compute as vector cross products:

$$e_i = e'_3 \times e_{wi} \quad ,i = 1,2,3 \tag{1}$$

The origin of the X´-Y´ coordinate system is located in the robot's geometrical center where the omni- wheel axes intersect. The robot orientation α is defined as the angle between ex and e'_1.

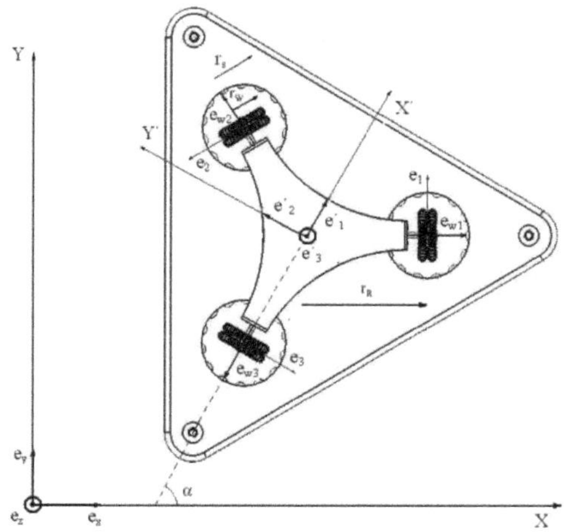

Fig. 6. Coordinate system definition

The robot velocity is expressed by $\dot{X}_R = [\dot{x}, \dot{y}, \dot{\alpha}]^T$ and $\dot{X}'_R = [\dot{x}', \dot{y}', \dot{\alpha}]^T$ in global and body coordinates, respectively. The angular velocity of the omni-wheels are expressed by $\dot{\phi} = [\dot{\phi}_1, \dot{\phi}_2, \dot{\phi}_3]^T$. Matrix T_R^W transforms body coordinates into omni-wheel coordinates, and T_G^R transforms global coordinates into body coordinates. It can be seen that body coordinates and wheels motion are expressed as

$$\begin{cases} -\dot{\varphi}_1 r_w = \dot{x}' \sin 60 + \dot{y}' \sin 30 + \dot{\alpha} r_R \\ -\dot{\varphi}_2 r_w = -\dot{x}' \sin 60 + \dot{y}' \sin 30 + \dot{\alpha} r_R \\ -\dot{\varphi}_3 r_w = 0 - \dot{y}' + \dot{\alpha} r_R \end{cases} \tag{2}$$

Matrix form of Eq. 2 is

$$\begin{bmatrix} \dot{\varphi}_1 \\ \dot{\varphi}_2 \\ \dot{\varphi}_3 \end{bmatrix} = T_R^W \begin{bmatrix} \dot{x}' \\ \dot{y}' \\ \dot{\alpha} \end{bmatrix}$$

Where

$$\begin{bmatrix} \dot{x}' \\ \dot{y}' \\ \dot{\alpha} \end{bmatrix} = T_G^R \begin{bmatrix} \dot{x} \\ \dot{y} \\ \dot{\alpha} \end{bmatrix} \tag{3}$$

Substituting Eq. 3 into Eq. 2, the relation between $\dot{\phi}$ and $\dot{X}_R$ is derived.

$$T_R^W = \frac{-1}{r_w} \begin{bmatrix} \frac{\sqrt{3}}{2} & \frac{1}{2} & r_R \\ -\frac{\sqrt{3}}{2} & \frac{1}{2} & r_R \\ 0 & -1 & r_R \end{bmatrix}, \quad T_G^R = \begin{bmatrix} \cos(\alpha) & \sin(\alpha) & 0 \\ -\sin(\alpha) & \cos(\alpha) & 0 \\ 0 & 0 & 1 \end{bmatrix} \tag{4}$$

5. Dynamics

In this section, to simplify the dynamic equations, it is assumed that robot motion lies in a horizontal surface without obstacles. Therefore, to derive the dynamic equations, the action of suspension system would be neglected. Fig. 7 shows the masses and inertias of the robot. The robot is modeled as a set of three different rigid bodies. The robot chassis is illustrated as a triangle. Its mass m_c contains the chassis, motors, battery, and all other parts rigidly attached to it, its moment of inertia with respect to the Z′ axis is $I_{z'}$, Its mass center is assumed to be located on the Z′ axis. The three omni-wheels are same, with masses of m_w. The mass center of the omni-wheels is located on their axis of rotation at the distance r_R from the robot's center. Also, moments of inertia about the axis passing through their center and parallel to Z′ axis and axis of rotation are $I_{wz'}$ and I_{wa}, respectively. Also, the three ball wheels are the same, with masses of m_s. their moments of inertia about principle axis of rotation is I_s. The robot total mass m_R consists of the chassis mass plus the wheels masses. Its center of mass is assumed to be located on the Z′ axis at the height h_M above the ground:

$$m_R = m_c + 3m_w + 3m_s,$$

$$I_R = I_{z'} + 3(I_{wz'} + I_s + (m_s + m_w)\, r_R^{\,2}) \tag{5}$$

Assuming horizontal motion for the robot, the gravitational potential energy may be omitted. The Euler-Lagrange equations of motion for the simplified robot model are

$$\frac{d}{dt}\left(\frac{\partial T}{\partial \dot{q}_i}\right)-\frac{\partial T}{\partial q_i}=Q_i,\quad i=1,2,3 \tag{6}$$

The robot's generalized coordinates is expressed with its omni-wheels angular motion $[\dot{q}_1,\dot{q}_2,\dot{q}_3]^T=[\dot{\phi}_1,\dot{\phi}_2,\dot{\phi}_3]^T$. Therefore, associated generalized force Q_i, are the motor torques applied to the actuators attached to the driving omni-wheels, $[Q_1,Q_2,Q_3]^T=[T_1,T_2,T_3]^T$. The robot's kinetic energy T is $\boldsymbol{T}=\boldsymbol{T}_{trans}+\boldsymbol{T}_{rot}$;

$$T_{trans}=\frac{1}{2}m_R(\dot{x}^2+\dot{y}^2),\ T_{rot}=\frac{1}{2}I_R\dot{\alpha}^2+3\sum_{i=1}^{3}(\mathrm{I}_{\mathrm{wa}}+((\frac{r_w}{r_s})^2+(\frac{r_w}{r_R})^2)I_s)\dot{\phi}_i^2 \tag{7}$$

Components of $\dot{x},\dot{y}$ and $\dot{\alpha}$ in the Eq. (7) are substituted from Eqs. (2) and (3) with respect to actuators angular velocities $\dot{\phi}_1,\dot{\phi}_2$ and $\dot{\phi}_3$. Computing the derivatives in the Euler-Lagrange equations and rearranging terms, the equations of motion can be expressed in its closed form as

$$\boldsymbol{M}(\phi)\ddot{\phi}+\boldsymbol{C}(\phi,\dot{\phi})+\boldsymbol{G}(\phi)+\boldsymbol{D}(\phi)=\boldsymbol{T} \tag{8}$$

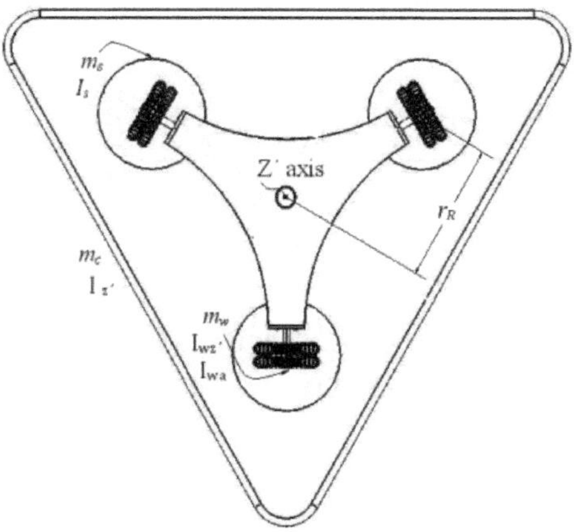

Fig. 7. Robot mass and inertia properties

6. Simulation study

Some numerical examples have been performed based on equations developed the previous section to simulate the behavior of the robot. Hence, a circular path is used with two scenarios of motion for the vehicle. Then, associated equivalent motor torques and wheels angular motions are computed, and overall characteristics of the robot are analyzed and evaluated.

The given trajectory is a circle with radius of 1.5 m . The motion of robot starts from the rest with constant acceleration and reaches to its maximum velocity V_{max} in the first 90°, then moves with constant velocity V_{max} in the next 180° of the path, and, finally decelerates constantly in the last 90° to reach at rest. The overall time of the motion is equal *t=6.0* s.

In the first scenario, as shown in the Fig. (8-a), during the motion X′ axis that shows forward direction, is directed tangent to the robot path. The corresponding actuators angular velocities and torques are shown in Fig's (8-b) and (8-c).

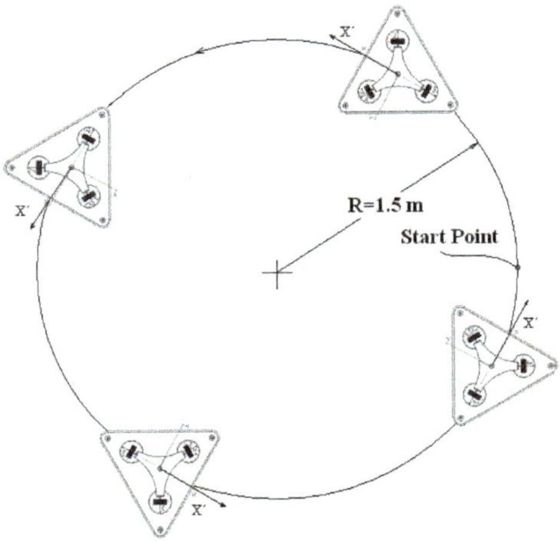

(a)

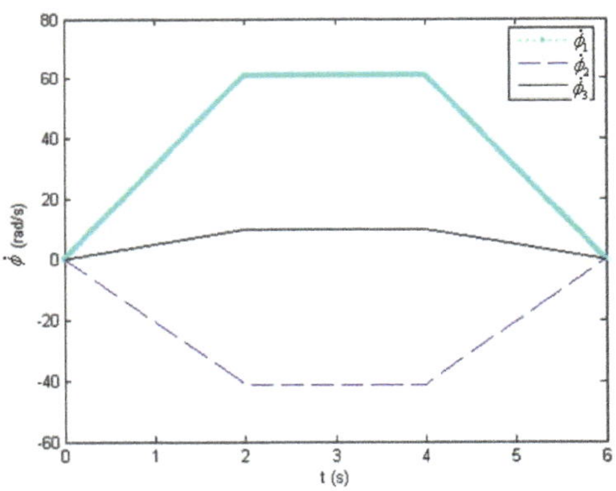

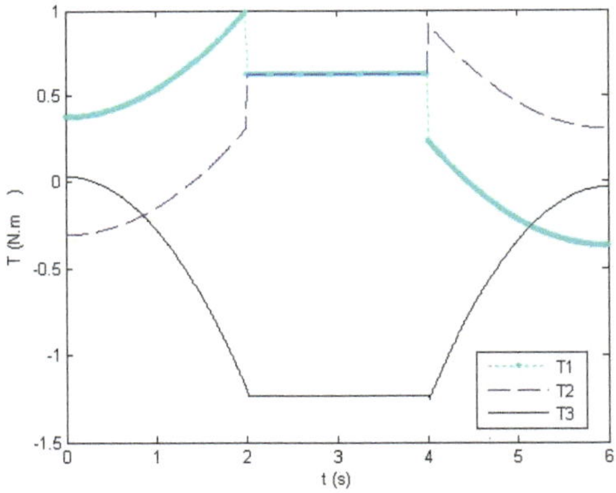

(b) (c)

Fig. 8. a) Robot moving with Orient tangent to the path, b and c) corresponding actuators angular velocities and torques

In the second scenario, as shown in the Fig. (9), direction of the robot during the motion is fixed. The corresponding actuators angular velocities and torques are shown in Fig's (9-a) and (9-b).

According to the simulation results, it can be seen that both ball wheeled platforms are able to move on a given path with different scenarios.

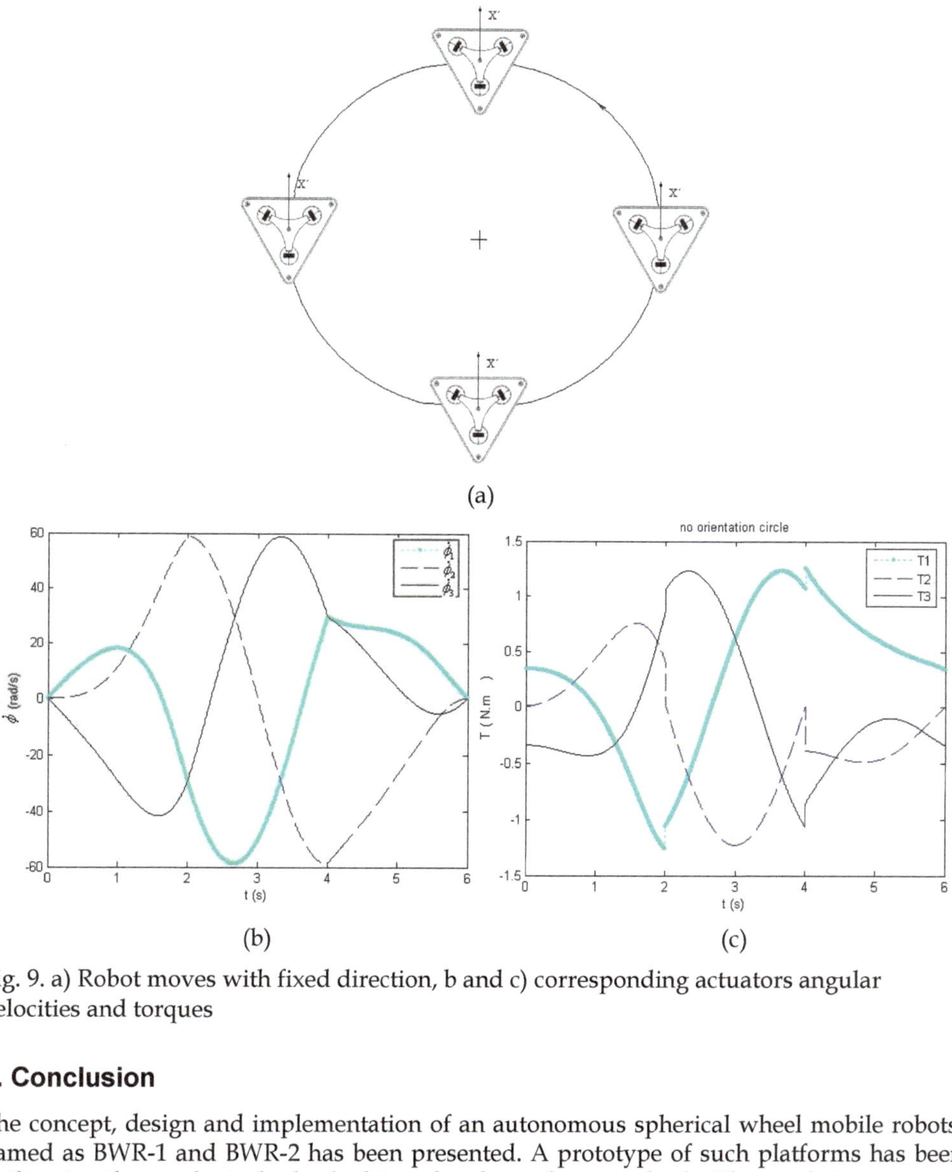

Fig. 9. a) Robot moves with fixed direction, b and c) corresponding actuators angular velocities and torques

7. Conclusion

The concept, design and implementation of an autonomous spherical wheel mobile robots, named as BWR-1 and BWR-2 has been presented. A prototype of such platforms has been built using three spherical wheels driven by classical omni-wheels. These robots are omni-directional, and so simultaneous longitudinal, transverse and rotational motions are possible. These robots have the same mobility as robots equipped with classical universal wheels that are suitable for in indoor applications. But, these robots owing to larger diameter of ball wheels are able to move in outdoor purposes, especially BWR-2 that provides a modular driving system equipped to suspension system. Future work will employ new features on the prototype for autonomous navigation such as optic encoders mounted on passive universal wheels and camera for vision.

8. References

Asada H., & Wada M., (1998), "The Superchair :A Holonomic Omnidirectional Wheelchair with a Variable Footprint Mechanism", *Total Home Automation and Health/Elderly Care Consortium* .

Chung W., Moon C., Jung C., & Jin J., (2010), "Design of the Dual Offset Active Caster Wheel for Holonomic Omni-directional Mobile Robots", *Int. J. of Advanced Robotic Systems*, Vol. 7, No. 4., pp. 101-106.

Ferriere L. & Raucent B., (1998), "ROLLMOBS, a new universal wheel concept", *Proceedings of the 1998 IEEE, International Conference on Robotics & Automation,* Leuven, Belgium, pp. 1877–1882.

Fierro R. & Lewis F. L., (1997), "Control of a Nonholonomic Mobile Robot: Backstepping Kinematics into Dynamics", *Journal of Robotic Systems,* Vol. 14(3), pp. 149–163.

Ghariblu H., "Design and Modeling of a Ball Wheel Omni-Directional Mobile Robot", *ICIET 2010*, ,March. 2010, Bali, Indonesia.

Ghariblu H., Moharrami A., Ghalamchi B., Bayat N., (2010), "Design and Modeling of a Ball Wheel Mobile Robot 2 (BWR-2)", *2010 1th International Conference on Mechanical Engıneerıng, Robotics and Aerospace*. Bucharest, Romania.

Jarvis R., (1997), "An autonomous heavy duty outdoor robotic tracked vehicle", *IROS '97*, Vol.1, pp. 352 - 359.

Jung M.-J., Kim H.-S., Kim S., & Kim J.-H., (2000), "Omni-Directional Mobile Base OK-II", *Proc. IEEE Int.Conf. on Robotics and Automation*, pp. 3449-3454.

Kim D. S., Kwon W. H., & Park H. S., (2003).' Geometric Kinematics and Applications of a Mobile Robot ", *Int. J. of Control, Automation, and Systems,* Vol. 1, No. 3, pp. 376-384.

Mori Y., Nakano E., Takahashi T., & Takayama K., (1999), "Mechanism and Running Modes of New Omni-Directional Vehicle ODV9", *JSME International Journal*, Series C, 42(1): 210-217.

Ostrovskaya, J. Angeles & Spiteri R., (2000), "Dynamics of a Mobile Robot with Three Ball-Wheels", *The Int. J. of Robotics Research,* Vol. 19; pp. 383-393.

Pin F. G., & Killough S. M., (1994), "A New Family of Omnidirectional and Holonomic Wheeled Platforms for Mobile Robots" *IEEE Transactions on Robotics and Automation*, Vol. 10, No. 4, pp. 480-489.

Song J. & Byun K., (2004), "Design and Control of a Four-Wheeled Omnidirectional Mobile Robot with Steerable Omnidirectional Wheels", *J. of Robotic Systems*, Vol. 21(4), pp. 193–208.

Tadakuma K.. et.al., (2009), "Basic Running Test of the Cylindrical Tracked Vehicle with Sideways Mobility", *The 2009 IEEE/RSJ Int. Conf. on Intelligent Robots and Systems*, pp.1679-1384.

Udengaard M.& Iagnemma K., (2007), "Kinematic Analysis and Control of an Omnidirectional Mobile Robot in Rough Terrain", . *IROS 2007*, pp. 795-800.

Wada M., Takagi A. & Mori S., (2000), "Caster Drive Mechanisms for Holonomic and Omni-directional Mobile Platforms with no Cover Constraints", *Proc. IEEE Int. Conf. on Robotics and Automation*, pp. 1531-1538.

West M., & Asada, H., (1997). "Design of ball wheel mechanisms for omnidirectional vehicles with full mobility and invariant kinematics". *ASME J. of Mechanical Design*, 119(2):153–161.

Permissions

The contributors of this book come from diverse backgrounds, making this book a truly international effort. This book will bring forth new frontiers with its revolutionizing research information and detailed analysis of the nascent developments around the world.

We would like to thank Prof. Dr. Zoran Gacovski, for lending his expertise to make the book truly unique. He has played a crucial role in the development of this book. Without his invaluable contribution this book wouldn't have been possible. He has made vital efforts to compile up to date information on the varied aspects of this subject to make this book a valuable addition to the collection of many professionals and students.

This book was conceptualized with the vision of imparting up-to-date information and advanced data in this field. To ensure the same, a matchless editorial board was set up. Every individual on the board went through rigorous rounds of assessment to prove their worth. After which they invested a large part of their time researching and compiling the most relevant data for our readers. Conferences and sessions were held from time to time between the editorial board and the contributing authors to present the data in the most comprehensible form. The editorial team has worked tirelessly to provide valuable and valid information to help people across the globe.

Every chapter published in this book has been scrutinized by our experts. Their significance has been extensively debated. The topics covered herein carry significant findings which will fuel the growth of the discipline. They may even be implemented as practical applications or may be referred to as a beginning point for another development. Chapters in this book were first published by InTech; hereby published with permission under the Creative Commons Attribution License or equivalent.

The editorial board has been involved in producing this book since its inception. They have spent rigorous hours researching and exploring the diverse topics which have resulted in the successful publishing of this book. They have passed on their knowledge of decades through this book. To expedite this challenging task, the publisher supported the team at every step. A small team of assistant editors was also appointed to further simplify the editing procedure and attain best results for the readers.

Our editorial team has been hand-picked from every corner of the world. Their multi-ethnicity adds dynamic inputs to the discussions which result in innovative outcomes. These outcomes are then further discussed with the researchers and contributors who give their valuable feedback and opinion regarding the same. The feedback is then collaborated with the researches and they are edited in a comprehensive manner to aid the understanding of the subject.

Apart from the editorial board, the designing team has also invested a significant amount of their time in understanding the subject and creating the most relevant covers. They scrutinized every image to scout for the most suitable representation of the subject and create an appropriate cover for the book.

The publishing team has been involved in this book since its early stages. They were actively engaged in every process, be it collecting the data, connecting with the contributors or procuring relevant information. The team has been an ardent support to the editorial, designing and production team. Their endless efforts to recruit the best for this project, has resulted in the accomplishment of this book. They are a veteran in the field of academics and their pool of knowledge is as vast as their experience in printing. Their expertise and guidance has proved useful at every step. Their uncompromising quality standards have made this book an exceptional effort. Their encouragement from time to time has been an inspiration for everyone.

The publisher and the editorial board hope that this book will prove to be a valuable piece of knowledge for researchers, students, practitioners and scholars across the globe.

List of Contributors

Georgios A. Demetriou
Frederick University, Cyprus

Cláudio Rodrigo Torres, Jair Minoro Abe, Germano Lambert-Torres and João Inácio da Silva Filho
Universidade Metodista de São Paulo, University of São Paulo and Paulista University - UNIP, Federal University of Itajubá, Universidade Santa Cecilia - UNISANTA, Brazil

Evangelos Georgiou, Jian Dai and Michael Luck
King's College London, United Kingdom

Lluís Pacheco, Ningsu Luo, Inès Ferrer, Xavier Cufí and Roger Arbusé
University of Girona, Spain

Chi Zhu and Masashi Oda
Department of Systems Life Engineering, Maebashi Institute of Technology, Kamisadori 460-1, Maebashi, Gunma, 371-0816, Japan

Haoyong Yu
Bioengineering Division, Faculty of Engineering, National University of Singapore, Singapore

Hideomi Watanabe
Department of Health Sciences, Faculty of Medicine, Gunma University, Japan

Yuling Yan
School of Engineering, Santa Clara University, CA, USA

Alonso A. Alonso, Ramón de la Rosa, Albano Carrera, Alfonso Bahillo, Ramón Durán and Patricia Fernández
University of Valladolid, Spain

Chika Hiroki and Wenwei Yu
Graduate School of Engineering, Chiba University, Japan

Yutaka Hiroi
Osaka Institute of Technology, Japan

Akinori Ito
Tohoku University, Japan

Shuro Nakajima
The Department of Advanced Robotics, Chiba Institute of Technology, Japan

Nicolae Alexandrescu, Despina Duminică, Constantin Udrea, Georgeta Ionaşcu and Lucian Bogatu
"POLITEHNICA" University of Bucharest, Romania

Tudor Cătălin Apostolescu
"TITU MAIORESCU" University of Bucharest, Romania

Alain Segundo Potts and José Jaime da Cruz
University of São Paulo, Brazil

Surachai Panich
Srinakharinwirot University, Thailand

Giuseppe Quaglia and Riccardo Oderio
Politecnico di Torino, Italy

Luca Bruzzone and Roberto Razzoli
University of Genova, Italy

E. Jauregi, I. Irigoien, E. Lazkano and B. Sierra
Department of Computer Sciences and Artificial Intelligence, University of Basque Country, Spain

C. Arenas
Department of Statistics, University of Barcelona, Spain

Jie-Tong Zou
Department of Aeronautical Engineering, National Formosa University, Taiwan, ROC

Wei Yu, Emmanuel Collins and Oscar Chuy
Florida State University, USA

Liang Ding, Haibo Gao, Zongquan Deng and Weihua Li
State Key Laboratory of Robotics and System, Harbin Institute of Technology, P. R. China

H. Ghariblu, A. Moharrami and B. Ghalamchi
Mech. Eng. dept., Mechatronics Lab., Zanjan University, Iran